THE CORRESPONDENCE OF ISAAC NEWTON

VOLUME VI

1713–1718

THE CORRESPONDENCE OF

ISAAC NEWTON

VOLUME VI
1713–1718

EDITED BY

A. RUPERT HALL
AND
LAURA TILLING

NULLIUS IN VERBA

PUBLISHED FOR THE ROYAL SOCIETY

CAMBRIDGE UNIVERSITY PRESS

CAMBRIDGE

LONDON · NEW YORK · MELBOURNE

Published by the Syndics of the Cambridge University Press
The Pitt Building, Trumpington Street, Cambridge CB2 1RP
Bentley House, 200 Euston Road, London NW1 2DB
32 East 57th Street, New York, NY 10022, USA
296 Beaconsfield Parade, Middle Park, Melbourne 3206, Australia

First published 1976

Printed in Great Britain
at the
University Printing House, Cambridge
(Euan Phillips, University Printer)

Library of Congress Cataloguing in Publication Data

Newton, Isaac, Sir, 1642–1727.
Correspondence.

Vol. 6 edited by A. R. Hall and L. Tilling.
Includes bibliographical references.
Contents–v. 1   1661–1675.–v. 2   1676–1687–v. 3   1688–1694. [etc.]
1. Scientists–Correspondence, reminiscences, etc.
I. Turnbull, Herbert Westren, 1885–1961, ed.
II. Hall, Alfred Rupert, 1920–   ed.
III. Tilling, Laura, ed.
QC16.N7A4   509'.2'4   [B]   59–65134

ISBN 0 521 08722 8 (v. 6)

CONTENTS

THE CORRESPONDENCE

LIST OF PLATES

PREFACE

The scope of this volume, which we expect to be the penultimate of this *Correspondence*, is explained in the following *Introduction*.

Many individuals have augmented whatever value our contributions to it may possess: besides those mentioned elsewhere, Dr Max Burckhardt (Basel University) and Dr E. A. Fellmann (Institut Platonæum, Basel), Mrs F. W. Harrison (Babson College), Mr P. S. Laurie (Royal Greenwich Observatory), Dr A. L. Murray (Scottish Record Office) and Mr P. E. Spargo (Johannesburg) gave us useful help and advice. We are (as invariably) grateful to the noble owner of a private collection who has graciously allowed us to reproduce documents. Would that all private owners would follow this generous example! Professors I. Bernard Cohen and R. S. Westfall kindly read proofs of volume v and made valuable comments, besides providing other aid and encouragement.

The following libraries (and their librarians) have courteously facilitated our work: The British Library and the Public Record Office, London; the Bodleian Library; the Basel University Library; the Babson College Library, Massachusetts, and the Burndy Library, Connecticut; in Cambridge the University Library and the libraries of King's and Trinity Colleges. Since the Keynes manuscripts at King's may be consulted on microfilm at the University Library we have added the appropriate references. Similarly we have employed a microfilm copy of the Leibniz manuscripts in the Niedersächsische Landesbibliothek at Hanover which is available in the University of London Library (Senate House); however, this does not include the manuscripts printed by Gerhardt a century ago. The Newton manuscripts in the Yahuda Collection of the Jewish National and University Library, Jerusalem, have also been transcribed from a microfilm in the University Library, Cambridge. To all the above and to the British Museum we are most grateful for permission to publish material in their collections.

To our friend Dr D. T. Whiteside we are hopelessly indebted for gifts of his good sense and scholarship. He has read both these two last volumes in typescript and proof, adding much and improving much in his own incomparable way. If there is much still to improve, it is not his fault.

Finally our thanks go to Mrs Frances Couch for her care in preparing the typescript of this volume.

A. RUPERT HALL
LAURA TILLING

SHORT TITLES AND ABBREVIATIONS

Baily, *Flamsteed*	Francis Baily, *An Account of the Revd. John Flamsteed . . . & Supplement to the Account.* London, 1835–7; reprinted, Dawsons, London, 1966.
Bodemann's catalogue	E. Bodemann, *Der Briefwechsel des G. W. Leibniz in der K. offent. Bibliothek zu Hannover.* Hannover, 1895; reprinted, Hildesheim, 1966.
Boss	Valentin Boss, *Newton and Russia.* Cambridge, Massachusetts, 1972.
Brewster, *Memoirs*	Sir David Brewster, *Memoirs of the Life, Writings and Discoveries of Sir Isaac Newton.* Edinburgh, 1855; reprinted, Johnson, New York and London, 1965.
C.S.P.	*Calendar of State Papers.*
Cal. Treas. Books	William A. Shaw (ed.), *Calendar of Treasury Books preserved in the Public Record Office.* London H.M.S.O. (There are two parts in this series for each calendar year, 1709–13. The second part is invariably referred to in this *Correspondence*.)
Cal. Treas. Papers	J. Redlington (ed.), *Calendar of Treasury Papers preserved in Her Majesty's Public Record Office.* London, Longman & Co., 1868–9.
Cohen, *Introduction*	I. Bernard Cohen, *Introduction to Newton's 'Principia'.* Cambridge, 1971.
Commercium Epistolicum	*Commercium Epistolicum D.* Johannis Collins, *et aliorum de Analysi promota: jussu Societatis Regiæ in lucem editum.* London, 1712.
Craig, *Newton*	Sir John Craig, *Newton at the Mint.* Cambridge, 1946.
Des Maizeaux	Pierre Des Maizeaux (ed.), *Recueil de Diverses Pièces sur la Philosophie, la Religion Naturelle, l'Histoire, les Mathematiques, etc., par Mrs Leibniz, Clarke, Newton et autres Auteurs célèbres.* The second edition, Amsterdam, 1740, is referred to unless otherwise stated.
Edleston, *Correspondence*	Joseph Edleston, *Correspondence of Sir Isaac Newton and Professor Cotes including Letters of other Eminent Men.* London, 1851, reprinted, Cass, 1969.
Flamsteed MSS.	The *Flamsteed* MSS. at the Royal Greenwich Observatory, Herstmonceux, have been renumbered since the publication of Newton's *Correspondence* began. To avoid confusion we have retained the old numbering.

Gerhardt, *Briefwechsel*	C. I. Gerhardt (ed.), *Der Briefwechsel von Gottfried Wilhelm Leibniz mit Mathematikern.* Berlin, 1899; reprinted, Georg Olms, Hildesheim, 1962.
Gerhardt, *Briefwechsel zwischen Leibniz und Wolf*	C. I. Gerhardt (ed.), *Briefwechsel zwischen Leibniz und Christian Wolff aus den Handschriften der Königlichen Bibliothek zu Hannover.* Halle, 1860; reprinted, Georg Olms, Hildesheim, 1963.
Gerhardt, *Leibniz: Mathematische Schriften*	C. I. Gerhardt (ed.), *G. W. Leibniz Mathematische Schriften*, 7 vols. Halle etc., 1849–63; reprinted, Georg Olms, Hildesheim, 1962.
Gerhardt, *Leibniz: Philosophische Schriften*	C. I. Gerhardt (ed.), *Die philosophischen Schriften von Gottfried Wilhelm Leibniz.* Vols. I–VII. Berlin, 1875–90.
Hall & Hall, *Oldenburg*	A. Rupert Hall and Marie Boas Hall, *The Correspondence of Henry Oldenburg.* University of Wisconsin Press, vol. I, 1965, to vol. IX, 1973: in progress.
Hall & Hall, *Unpublished Scientific Papers*	A. Rupert Hall and Marie Boas Hall, *Unpublished Scientific Papers of Isaac Newton.* Cambridge, 1962.
History of Parliament	[Romney Sedgwick], *History of Parliament: House of Commons, 1715–1754.* London, 1970.
Hofmann	J. E. Hofmann, *Die Entwicklungsgeschichte der Leibnizschen Mathematik während des Aufenthaltes in Paris (1672–76).* München, 1949. (Now available in an English translation by A. Prag under the author's name and the title *Leibniz in Paris 1672–1676.* Cambridge, 1974.)
Huygens, *Oeuvres Complètes*	*Oeuvres Complètes de Christiaan Huygens publiées par la Société Hollandaise des Sciences.* Vols. I–XXII. La Haye, 1888–1950.
The Leibniz–Clarke Correspondence	H. G. Alexander, *The Leibniz–Clarke Correspondence.* Manchester, 1956.
Mémoires de l'Académie Royale des Science	*Histoire de l'Académie Royale des Sciences, avec les Mémoires de Mathématique et de Physique.*
Mint Papers	Newton's private file of papers concerning Mint business, sold at Sotheby's in 1936, now bound in three volumes in the Public Record Office. (Mint/19, I–III.)
More	Louis Trenchard More, *Isaac Newton, a Biography.* New York and London, 1934.
P.R.O.	Manuscripts in the Public Record Office.
Raphson, *History of Fluxions*	Joseph Raphson, *The History of Fluxions Shewing in a compendious manner The first Rise of, and various Improvements made in that Incomparable Method.* London, 1715. (For the appendix of letters concerning the calculus controversy, added by Newton in 1718, see Letter 1170, note (2).)

Ravier	Emile Ravier, *Bibliographie des Oeuvres de Leibniz*. Paris 1937; reprinted, Georg Olms, Hildesheim, 1966.
R.G.O.	Manuscripts at the Royal Greenwich Observatory, Herstmonceux.
Recensio	'Recensio Libri Qui inscriptus est Commercium Epistolicum Collinii et aliorum, de Analysi Promota . . .' printed in the second edition of the *Commercium Epistolicum* (London, 1722), pp. 1–59 and also published separately in both English and a French translation in 1715. (See Letter 1162, note (2).)
Rigaud, *Correspondence*	S. P. and S. J. Rigaud, *Correspondence of Scientific Men of the Seventeenth Century . . . in the collections of the Earl of Macclesfield.* Oxford, 1841; reprinted, Georg Olms, Hildesheim, 1965.
Sharp Letters	A volume containing 'A Collection of Original Letters addressed to Mr. Abraham Sharp . . . by Mr. John Flamsteed' placed on permanent loan to the Royal Society by Mr. F. S. Edward Bardsley-Powell.
Shaw	William A. Shaw, *Select Tracts and Documents illustrative of English Monetary History, 1626–1730.* London, 1896.
Sotheby Catalogue	*Catalogue of Newton Papers Sold by Order of the Viscount Lymington . . . which will be sold by Auction by Messrs. Sotheby and Co.* [on 13 and 14 July 1936].
Taylor, *Contemplatio Philosophica*	Brook Taylor, *Contemplatio Philosophica: a posthumous work, of the late Brook Taylor, . . . To which is prefixed a Life of the Author by his Grandson, Sir William Young . . . with an Appendix containing sundry original Papers, Letters from the Count Raymond de Montmort, Lord Bolingbroke, Marcilly de Villette, Bernoulli &c.* Not published; printed London, 1793.
Turnbull, *Gregory*	H. W. Turnbull (ed.), *James Gregory Tercentenary Memorial Volume.* London, Bell, 1939.
U.L.C.	Manuscripts in the University Library, Cambridge (Portsmouth Collection).
Whiteside, *Mathematical Papers*	D. T. Whiteside (ed.), *The Mathematical Papers of Isaac Newton.* Cambridge, 1967 onwards: in progress.
Wollenschläger, *De Moivre*	Karl Wollenschläger, 'Der mathematische Briefwechsel zwischen Johann I Bernoulli und Abraham de Moivre', *Verhandlungen der Naturforschenden Gesellschaft in Basel*, Band XLIII, 1931–3, pp. 151–317.

INTRODUCTION

As Newton had by now entered his eighth decade, it can surprise no one that the correspondence of the present volume shows a marked decline in his activity and intellectual vigour. While the number of extant letters written by him on other than Mint business is relatively small, the majority of them is devoted to his controversy with Leibniz, Newton's dominant interest during the present period. The paucity of letters from Newton's pen may be partly accounted for by his residence in London—hence, in the ordinary way, he had little or no correspondence about Royal Society business—and partly by the activity of other men on Newton's behalf. The correspondence of Newton shades gradually into the correspondence of the Newtonians. Thus notably Keill, De Moivre, Chamberlayne, Brook Taylor, the Abbé Conti and Des Maizeaux interested themselves in the calculus dispute, all of them (except the first) having frequent opportunities for personal conversation with Newton. It is certain that letters written by Newton to Keill have been lost, and equally to be regretted are any letters that he may have exchanged with Samuel Clarke, whose relations with Newton (though obviously intimate) remain enigmatic. Since it is hardly likely that every one of any considerable number of letters written by Clarke to Newton would have been destroyed, it is to be presumed that the two men never corresponded extensively.

Nevertheless, Newton's habits remained unchanged and he was still tireless in preparing drafts. A great deal of paper was covered with abortive sketches leading to the anonymous printed article of 1715 which we here call the *Recensio*, drafts whose detailed consideration must await some future chronicler of the calculus dispute; and the more important matters of Mint business were treated by Newton in the same way. A conspicuous example of Newton's practice is Letter 1295 addressed to Des Maizeaux, where, since there is no definitive final version, it is very hard to determine what Newton's ultimate intentions were.

There is almost no material of interest bearing on Newton's private or family life in this period—there are no letters relating to Catherine Barton's marriage to John Conduitt (26 August 1717), though Newton retained copies of her will and documents related to her marriage,* nor to the gift of £500 which he apparently made to his relative Ralph Ayscough on 26 June 1716.† Perhaps we may take this opportunity, however, to amend and amplify

* *Sotheby Catalogue*, Lot 175. The location of these documents is at present unknown.
† *Ibid.*, Lot 204, also at present unlocated.

what was written about Newton's family in the Introduction to vol. v (pp. xliii–xliv). Katherine (Greenwood), wife of the ill-fated Col. Robert Barton, married as her second husband Robert Gardner by whom she had a daughter Joanna. She in turn became the wife of Cutts Barton whose grandfather was Robert Barton of Brigstock, parent of Col. Barton, by his first wife Elizabeth Pilkington. Thus the Cutts Barton family of Hampshire was doubly connected to the Newton clan.* A document relating to Newton's purchase of an estate settled upon Katherine's four children (Newton, Catherine, Robert and Joanna) is listed in the *Sotheby Catalogue*.

In this volume and the next we print a number of appeals to Newton from humble relatives, none of whom can now be identified since a complete table of Newton's connections has never been worked out. If nothing else they seem to indicate a pretty widespread benevolence.

It was remarked in the Introduction to the last volume that Newton, though a Whig, seems to have enjoyed, as Master of the Mint, a satisfactory relationship with the Tory Lord High Treasurer, even if he was to fail to satisfy the latter's wish for a resumption of the copper coinage. At the meeting at the Treasury on 8 May 1713 (vol. v, Number 996) Oxford had made his wish pretty clear, and had instructed the Mint to prepare a scheme for it. Unfortunately the next *dated* document referring to the project is of 22 January 1714 (Letter 1034). There is no paper on copper coinage in the government files and the drafts in Newton's Mint Papers (which are indeed addressed to Oxford) are without date and difficult to arrange in sequence. Possibly Newton was purposefully slow to hasten the business.

We have not found the evidence for Sir John Craig's statement (*Newton at the Mint*, pp. 96–7) that Oxford in the autumn of 1713 gave Newton definite instructions to strike some experimental coins, and that 'a few tons' were actually minted. There seems to be no documentary evidence for the existence of such trial coins before James Bertie's report on them on 4 March 1714 (p. 75). The many undated papers that appear to relate to this business do not permit a more precise interpretation; we have printed a selection as illustrating Newton's thoughts and actions at this time and during the following year up to the time of Oxford's fall from power, when the project collapsed. Newton insisted, as always, on an issue of high intrinsic value and aesthetic merit with as small an allowance as possible for the costs of machining and distribution to be added to the price of the bulk refined metal. His original notion, to buy refined copper ingots and draw them down at the Mint to the

* It may also interest collectors of old glass that the glass-engraver Frans Greenwood of Dordrecht (1680–1761), since he was born at Rotterdam of English parents, must surely have been connected thus remotely to Iasac Newton.

thickness from which coin blanks could be cut, was frustrated by the inadequacy of the Mint's machinery; hence he was forced to pay the manufacturers for drawing down the metal at their mills (of which there were, it seems, several in the neighbourhood of London). There was also a difficulty about the mixing of even a small quantity of tin with the pure copper, the alloy being cast much more easily. For all such reasons, and because of the lack of enthusiasm on Newton's part that may be assumed, no great quantity of copper coin was minted and none publicly issued, despite the building of new furnaces at the Mint and other steps. Furthermore the trial pieces that were struck (of which examples survive) did not prove wholly satisfactory (Letter 1066).

A strange instance in Newton's relations with Oxford of his persistence in arranging matters to his own satisfaction occurred in relation to the Trial of the Pyx in 1713 (7 August). It may be recalled that in 1710 Newton had been scandalized by the verdict that his gold coins made during the previous three years were slightly less than standard, a verdict which he attributed to the excessive purity of the 1707 gold test plate with which the coins were compared. Preparing for the next trial, and determined that his guineas should not again be measured against an over-refined standard, he drafted a 'Petition of the Merchants & Goldsmiths in behalf of themselves and other Importers of Bullion into her Majesty's Mint' (Mint Papers, I, 278). Opening with the finely Newtonian statement 'the Trial Piece of Gold made in the year 1707 being at the last Trial of the Pix found too fine by the Assay' (which was not the verdict of the jury of goldsmiths in 1710) the petitioners are made to say that there is danger in making the coins too fine also in order to conform to it; they therefore ask that a new trial-plate be made, or a correction applied to that of 1707, 'to the end that the coinage of gold may agree as well with the assay as with the Trial piece, the Assay being the rule by wch gold & silver is valued amongst Merchants & Importers of Bullion'. Whether Newton ever canvassed signatures for his draft and submitted it to Oxford is not known; perhaps he secured the side-stepping of the 1707 plate which actually occurred without flourishing his petition.

Curiously enough, the Tory Oxford seems to have been more ready than his Whig successors to appeal to Newton as the government's natural scientific adviser, just as he also supported Newton against Flamsteed. The most important consequence of Oxford's confidence in Newton was his involvement in the longitude business (Letter 1093). The documents printed here make it reasonably clear that Newton believed (correctly, in a sense) that the primary determination of longitude could only be effected by some astronomical method, such as one depending on an accurate theory of the Moon's motion;

other methods, such as those using mechanical clocks, could only be of second-ary value at best. 'Clockwork may be subservient to Astronomy,' he told one correspondent, 'but without Astronomy the longitude is not to be found. Exact instruments for keeping of time can be usefull only for keeping the Longitude while you have it' (Letter 1137). We may thus attribute to New-ton's authority, it seems, that distrust of chronometers from which William Harrison was to suffer later.

During almost all of Newton's government service under Anne he had been responsible to two Ministers only: Godolphin and Oxford. After the Duke of Shrewsbury, who replaced Oxford as Lord High Treasurer by Anne's dying choice, had resigned his staff in October 1714, the Whigs always placed the Treasury in commission and, until Robert Walpole became First Lord and in effect 'prime minister' in 1721, oscillations of power were frequent and usually marked by alterations in the Treasury commission. After the brief tenure of office as First Lord by Newton's old friend and patron Halifax had ended with his death in March 1715 (and little more than two years later Catherine Barton found a husband), Carlisle, Walpole, James Stanhope and Sunderland replaced each other rapidly as the various Whig factions struggled for ascend-ency. At least so far as the Mint was concerned, and disregarding a foolish and temporary interest in the elusive prospect of wealth to be won from Sir John Erskine's silver-mine (Letter 1200, etc.), the Treasury seems to have been content to leave the Mint to its routine, though large payments to the Moneyers caused by the heavy gold coinage of the years 1713–15 caused a temporary cash crisis in 1716 (Letter 1182). Even the business of Cornish tin was out of the Mint's hands—since there was now a Duke of Cornwall—though the old accounts, surpluses, and loans remained to be dealt with and occasioned Newton a little formal business. It was only after the more active Stanhope had become First Lord that the Treasury revived the issue of a copper coinage; in May 1717 Newton put his own thoughts about it on paper once more (Letter 1242) and in September the scheme, on Newton's lines and drafted by himself, was formally authorized (Number 1262). Tenders for the supply of rolled copper strip, from which the Mint would cut the moneyer's blanks, had resulted from an advertisement of the Treasury's intention which Newton had been ordered to insert into the *London Gazette* for 30 April 1717. A manu-facturer named Hines was chosen for the initial supply of 30 tons (equal to over three million halfpence) and, perhaps as an inevitable result, Newton was troubled by the complaints and fault-finding of almost all the others, not least the William Wood of Wolverhampton to be rendered so odious by Swift later. In the end Newton was compelled to allow more for the cost of manu-facture and machining then he had at first estimated, so that the coins (at 23*d* to the pound) were relatively light; of those issued, a good many proved defective, despite the Mint's efforts at standardization (Number 1262). The working out of Hines' contract took until early 1719, and then the issue was interrupted for a time.

Another product of Lord Stanhope's ministry was Newton's often-printed report on bimetallism (Letter 1267) which led to a slight devaluation of gold in terms of silver effected by proclamation on 21 December 1717. The affair has been fully discussed by Sir John Craig in his history of the Mint and in his special study of *Newton at the Mint*. When a state whose currency laws are in any case ineffectively enforced seeks to maintain a bimetallic currency, in which the circulation of the various coins of each metal rests essentially upon their bullion content, the relative values of (let us say) gold and silver fixed by law for the conduct of its Mint (which cannot conveniently fluctuate rapidly) will tend to get out of step with the commercial estimation of the relative value of gold and silver if there are *de facto* (not necessarily *de jure*) ways in which the metals are also treated as free commodities whose values are subjected to a variety of commercial pressures. This was the situation in England in the early eighteenth century. Gold and silver were freely used for the manufacture of plate, a marketable commodity. The precious metals were (whether lawfully or not) imported and exported as coin, bullion, and plate. Foreign gold coin freely entered the British isles and was widely used in the areas remote from London. Thus Britain was in a condition of essentially free metallic flow, not only with parts of Europe where any fixing of the relative values of gold and silver by law was beyond her control, but where these values were, in some cases, customarily different from those prevailing in Britain. Moreover, trade with India demanded an overall export of bullion, largely silver, a metal relatively more highly valued in the East than it was in Europe; while, at the same time, merchants in Europe employed actual movements of metal to settle the trading debts between European countries, which varied with the terms and the prosperity of trade.

Hence any fixed ratio between gold and silver in any one state was likely to be now lower than that prevailing in commerce, now higher. In practice this meant that if the commercial value (in terms of gold) of the bullion content of the English silver coins was higher than its nominal value there would be a tendency for the coins to be (unlawfully) melted down into bullion and, usually, exported. There was, as Newton pointed out, less tendency to melt down plate, presumably because of the relatively greater cost of the 'fashion' or workmanship of pieces of plate. This equally would mean that no one would wish to sell silver bullion to the Mint for conversion into coin; for both reasons the silver currency would tend to vanish, leaving only the gold coins in circulation and creating effectively a monometallic currency. This was what was happening in England after the Peace of Utrecht, and continued throughout the eighteenth century.

Newton understood the reasons for variations in the commercial valuation of gold and silver, and that the relative currency values fixed by law were not insulated from these variations by state prohibitions. He was more imaginative than some of his predecessors in realizing that the technical means of law enforcement of his day (which could not quell even such clumsy crimes as clipping and counterfeiting) were inadequate to protect the currency against commercial greed, although, of course, the pursuit of currency offenders remained one of Newton's official concerns, reflected in this correspondence. He seems to have understood that the margin of the profitability of currency offences was dependent on the efficiency of law enforcement—should the risk of conviction for the crime be great, a potential criminal would only melt down coin, for example, if the profit from his act was likely to be high. If the currency laws were well enforced, therefore, it did not matter much if the legal values of gold and silver were a little out of step from time to time with the commercial values, especially as the potential offender in any case incurred deterrent costs amounting to a few per cent of the actual value of silver (the vulnerable metal)—for example, the cost of furnaces and fuel, of transport, possibly of bribing officials, and so forth. This is the basis of Newton's argument for the small scale of the change in the margin he proposed in 1717.

Nevertheless, Newton's reduction in the value of the gold guinea by 6d was, as he admitted himself, too small to be effective. Twice as much—bringing the sovereign back to its old (and more modern) value—would perhaps have had some effect; the objection against making a devaluation of gold of this magnitude was, of course, that it effected a corresponding apparent reduction in value of every individual's stock of gold coins. As it was, the 6d reduction produced very sharp protests. In consequence the silver coinage almost vanished in England during the eighteenth century; only the worst, most worn and clipped coins, no longer worth melting, remained in circulation. The Government experimented with an issue of quarter-guineas to remedy the want of silver (Letter 1283); but these little coins were too small for the Moneyers and the public alike. Newton's clear analysis of the bimetallic currency problem was, presumably, vitiated for his own age by the weakness of law enforcement. The risk of detection in the crime of melting down the King's silver coin was so slight that a profit of 6d, 3d, or perhaps even less for every one pound's worth of coin melted made it well worth while. The analyst like Newton could hardly hope to hit, or retain, for the bimetallic currency so close a relation with the fluctuating commercial values of the precious metals.

Having surveyed Newton's considerable activity in Mint affairs and weighed the mass of paper concerned with Mint minutiæ which still remains, one may be inclined to ask: if Newton had not spent many hours in these latter years of his life drafting and redrafting administrative documents, would he have filled his paper with writing on other subjects? One may well feel that the answer to this question must be affirmative, for it is difficult (for example) to imagine Newton idling away his time in that gorgeous company who accompanied George I in his royal progress on the Thames in 1717 that was enlivened by frequent repetitions of Handel's *Water Music*. Would these unwritten sheets of Newton's manuscript have been of real significance to historians of science? We suggest, though with less confidence, that the answer must be no. They would have been devoted to the chronology of ancient kingdoms, or to the prophecies of Daniel and John, or to his dispute with Leibniz, rather than to natural philosophy. Newton preserved his mental determination and indeed acuity almost to the end of his life, his mind continued to work well in its familiar channels (as the third edition of the *Principia* demonstrates), but one could hardly expect genuine creativity from so aged a man, even a Newton. Of course he exaggerated when he wrote more than once of having abandoned mathematics forty years before (in 1676!), and no doubt his feeble response to mathematical challenges in 1716 in contrast to the vigour he had displayed so recently as the autumn of 1712 may be accounted for by lack of interest, as much as by advancing age. One may even argue that another Hauksbee or another Cotes might, even in the last dozen years of Newton's life, have provided a fresh creative stimulus. But it seems implausible.

We have inevitably devoted much space in this volume to the controversy between Newton and Leibniz, or rather between the British and the Continental mathematicians, for very few Britons were other than enthusiastic champions of Newton while the Continentals virtually all, though with varying degrees of warmth, entered the party of Leibniz. Its core, of course, was Germanic, the group of Leibniz's own friends and in some sense pupils as they had acquired the calculus from him. Since one may dismiss such a minor figure as Levesque de Pouilly as exceptional, the French may be considered on better terms with Leibniz and his friends than with the Newtonians. Yet neither Fontenelle nor Varignon joined the vendetta against Newton, and both strove to demonstrate public impartiality; indeed, Newton (and his niece) created pleasing impressions on all French travellers (Letter 1234). The Italians seem to have been somewhat more hostile to Leibniz, despite the establishment of a Leibnizian dynasty of mathematicians at Padua (Letter 1183); in particular the Abbé Conti, during his long visit to London, seems to have been won over to the English side (Letter 1190).

THE CALCULUS DISPUTE, 1713–18

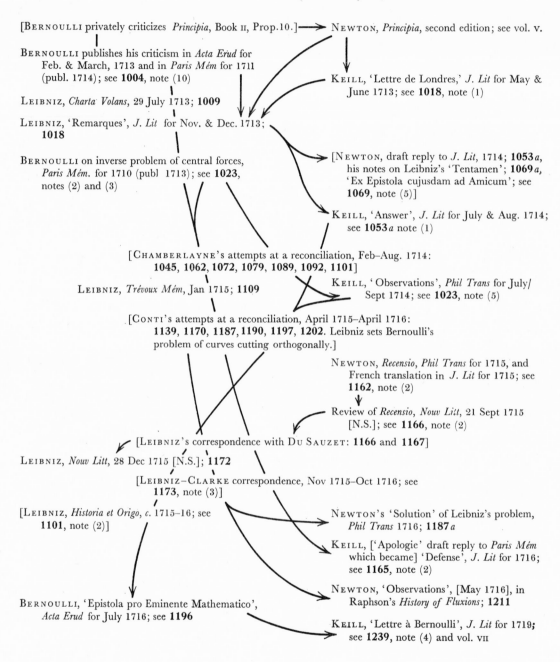

[BERNOULLI privately criticizes *Principia*, Book II, Prop.10.] ⟶ NEWTON, *Principia*, second edition; see vol. V.

BERNOULLI publishes his criticism in *Acta Erud* for
  Feb. & March, 1713 and in *Paris Mém* for 1711
  (publ. 1714); see **1004**, note (10)

KEILL, 'Lettre de Londres,' *J. Lit* for May &
  June 1713; see **1018**, note (1)

LEIBNIZ, *Charta Volans*, 29 July 1713; **1009**

LEIBNIZ, 'Remarques', *J. Lit* for Nov. & Dec. 1713;
  1018

[NEWTON, draft reply to *J. Lit*, 1714; **1053***a*,
  his notes on Leibniz's 'Tentamen'; **1069***a*,
  'Ex Epistola cujusdam ad Amicum'; see
  1069, note (5)]

BERNOULLI on inverse problem of central forces,
  Paris Mém. for 1710 (publ 1713); see **1023**,
  notes (2) and (3)

KEILL, 'Answer', *J. Lit* for July & Aug. 1714;
  see **1053***a* note (1)

[CHAMBERLAYNE's attempts at a reconciliation, Feb–Aug. 1714:
  1045, **1062**, **1072**, **1079**, **1089**, **1092**, **1101**]

LEIBNIZ, *Trévoux Mém*, Jan 1715; **1109**

KEILL, 'Observations', *Phil Trans* for July/
  Sept 1714; see **1023**, note (5)

[CONTI's attempts at a reconciliation, April 1715–April 1716:
  1139, **1170**, **1187**, **1190**, **1197**, **1202**. Leibniz sets Bernoulli's
  problem of curves cutting orthogonally.]

NEWTON, *Recensio*, *Phil Trans* for 1715, and
  French translation in *J. Lit* for 1715; see
  1162, note (2)

Review of *Recensio*, *Nouv Litt*, 21 Sept 1715
  [N.S.]; see **1166**, note (2)

[LEIBNIZ's correspondence with DU SAUZET: **1166** and **1167**]

LEIBNIZ, *Nouv Litt*, 28 Dec 1715 [N.S.]; **1172**

[LEIBNIZ–CLARKE correspondence, Nov 1715–Oct 1716; see
  1173, note (3)]

[LEIBNIZ, *Historia et Origo*, *c*. 1715–16; see
  1101, note (2)]

NEWTON's 'Solution' of Leibniz's problem,
  Phil Trans 1716; **1187***a*

KEILL, ['Apologie' draft reply to *Paris Mém*
  which became] 'Defense', *J. Lit* for 1716;
  see **1165**, note (2)

NEWTON, 'Observations', [May 1716], in
  Raphson's *History of Fluxions*; **1211**

BERNOULLI, 'Epistola pro Eminente Mathematico',
  Acta Erud for July 1716; see **1196**

KEILL, 'Lettre à Bernoulli', *J. Lit* for 1719;
  see **1239**, note (4) and vol. VII

If we had limited ourselves to publishing here the letters written by or to Newton on the calculus dispute we would have had little material and that little would have been quite devoid of context: plums, perhaps, but no pudding. We believe it is useful to print here, in addition, a number of other documents that have not been published before as well as many others well known to historians, so that although this volume by no means contains a complete documentation of the calculus dispute it does display the rôle of the principal participants on either side, and presents (we hope) a reasonably clear picture of the succession of printed papers in the journals. To this end we have devised a 'flow-chart' connecting the principal documents (public and private) with the names generally attached to them. It will be appreciated, of course, that varying periods elapsed between the publication of particular contributions to the polemic and their appreciation by the injured and insulted party of the other part.

One aspect of the intellectual controversies between Britain and the Continent we have omitted altogether, except for a few necessary allusions in the notes. We have not here republished the philosophical exchanges between Leibniz and Dr Samuel Clarke, partly because these are very well known indeed and have been excellently translated and edited, partly because there is no overt documentation of Newton's involvement in them. Samuel

---

The calculus dispute, 1713–18.

The chart on the left shows schematically the sequence of major papers and letters written in relation to the calculus dispute in the period 1713–18. Drafts of papers which were never published, and correspondence intended at least initially to be private, are referred to in square brackets. Where the document is printed in the present volume, the document number is given in boldface type; otherwise a reference is given to an editorial note concerning the document.

The chart is arranged chronologically as far as possible, but it must be noted that dates of publication may be considerably later than the date of writing; for example Keill's 'Lettre à Bernoulli', published in the *Journal Literaire de la Haye* for 1719, nevertheless appears on our chart because it was drafted by Keill as early as May 1717.

The dispute did not, of course, end in 1718; but subsequent material relating to September 1718 onwards, is the province of vol. VII of this correspondence, and so we deal with it there.

The short titles for papers are those we have used throughout vol. VI; the editorial note to which we refer in the chart will give the full title. The following abbreviations are used in the chart for the names of journals:

Acta Erud	*Acta Eruditorum*, Leipzig
J Lit	*Journal Literaire de la Haye*
Trévoux Mém	*Mémoires pour l'Histoire des Sciences et des Beaux Arts*, Trévoux
Nouv Litt	*Nouvelles Litteraires*, the Hague
Paris Mém	*Mémoires de l'Académie Royale des Sciences*, Paris
Phil Trans	*Philosophical Transactions*, London

Clarke nowhere figures even indirectly in Newton's correspondence, and it is not even certain that Newton played the behind-the-scenes rôle with Clarke that he certainly occupied with Keill. It is obvious that Clarke meant to set out, interpret and defend Newton's real thoughts and beliefs, and that his knowledge of Newton's mind may have derived from unrecorded conversations or even written communications as well as from those statements of Newton which the press had made accessible to everyone; to speculate upon this is not to admit that Newton was a ghost-writer to Clarke or ever saw his papers privately, as of course he did those prepared by Keill (which, equally, we have not reprinted *in extenso* here). In fact, just as we have no certain evidence that Samuel Clarke ever appealed to Newton for help or sanction (though he may have done so), equally Leibniz seems to have been reluctant to implicate Newton and force him to appear in a personal defence of his own philosophy of God and nature. Leibniz did nothing to compel Newton to take notice of *this* controversy or come to the rescue of this champion. Whether the situation would have changed, had Leibniz lived longer, it is impossible to say. But in this way the philosophical had a different character from the mathematical dispute or, rather, the former remained always in the state that the latter had been in before 1711, with the additional advantage of privacy until after Leibniz's death. Furthermore, although Leibniz challenged Newton's competence in philosophy and mathematics alike, it was only in mathematics that he charged him with treason against the republic of letters.

Since the succession of documents is set out in our chart it would be pointless to rehearse again every thrust and parry, but something may be said here about the changing nature of the dispute. By publishing his *Commercium Epistolicum* in 1712 Newton had hoped to bury Leibniz under the weight of testimony to the effect that he, Newton, had clearly possessed the concepts of the calculus in 1676 at latest. This gave him a comfortable eight years' priority over Leibniz's first *publication* of the calculus to make his own independence plain. Secondarily, the documentation was to show that Leibniz in 1676 had been far behind Newton in mathematics and had actually received instruction from him through their correspondence and indeed earlier. This made Newton's absolute priority in discovery plain. In Newton's mind, no one who clearly saw the force of the documents bearing on these two points could fail to do him justice. In 1711 Newton had seen the issue as a biographical and historical one (here, as in his Mint work, in his writings on chronology, and indeed in his alchemical lucubrations, Newton reveals an almost legalistic mentality with respect to the value of evidence). He continued to see it in this way. His own achievements in and before 1676 were set out in a clear, irrefutable record; what had Leibniz to set against them of his own?

Though often tempted to set out his own history from the ample though confused personal materials first studied only in the nineteenth century, Leibniz did not do so. The *Historia et origo calculi differentialis* which he began to prepare in 1714 remained fragmentary. Rather than engage Newton in a war of historical documents, Leibniz and his friends (the Bernoullis, Wolf, Hermann) broadened the dispute while Newton's partisans Keill, Taylor and Clarke followed the example of Cotes earlier in allowing them to do so. The original historical issue was confused by others unrelated, such as philosophical and mathematical competence. Was Newton not merely a plagiarist but also the general of a white army pledged to subvert the intellectual revolution of the seventeenth century?

The pre-eminent influence on Leibniz here seems to have been that of Johann [I] Bernoulli, lion by night and jackal by day. This great mathematician, the most powerful and dangerous of Newton's critics, convinced of Newton's turpitude yet made cowardly by his consciousness of Newton's cordiality towards himself, forged powerful new weapons for Leibniz's armoury, as the first letters in this volume reveal. It was he who put strength into Leibniz's *Charta Volans*, though persisting in direct denials that he had done so for many years afterwards. It was Bernoulli who asked, after reading the *Commercium Epistolicum*, precisely what Newton had achieved by 1676, or even 1687. He rightly observed that Newton's adoption of a pricked-letter symbolism came after Leibniz's first publication, and that fluxions had no effective rôle in the *Principia*. Though he might carelessly write of the Newtonian fluxion as though it were equivalent to the Leibnizian differential, he could pointedly ask: had Newton, like Leibniz, devised a new algorithm, a genuine calculus? Was he not rather an improver of the well-known method of infinitesimals, and of course of the method of series introduced by Mercator? Again, it was Bernoulli who argued the sterility of English mathematics against the fertility of the German; men like David Gregory, George Cheyne and Charles Hayes were (he held) either fools or clever apes. Newton had stolen the bare rudiments of the calculus, but neither he nor his idolaters had mastered its finest, highest attainments of recent years. This was a matter that could be put to the test by the old method of challenge problems. Finally, it was Bernoulli who insisted on Newton's mistakes in the *Principia* and, following the lead of his nephew Nikolaus, proclaimed the plausible but false accusation that Newton did not properly understand how to obtain the differential of a differential.

The fact that Leibniz never vindicated his original development as a mathematician in the years 1672 to 1676—a far more difficult task than the analogous task of Newton and one in which Leibniz himself was scarcely able to evaluate his own progress accurately—was certainly unfortunate. That he and his friends, and their English counterparts, moved from intellectual biography into a wilderness of irrelevancy was even more so. Viewed dispassionately, what had either the theory of gravitation, Newton's metaphysics, or the incompetence of the English mathematicians to do with the understanding of events occurring half a century earlier (Letter 1295)? The enlarged dispute was productive of extreme absurdity; of the claims that the intrusive 'm' in '[m]eam' or the 'ut' missing from a crucial passage in *De quadratura curvarum* were simple printer's errors; of supposed 'solutions' of mathematical problems that were not solutions at all; and worst of all, of a spreading disease of bitterness and falsehood between men.

Newton probably saw a copy of the anonymous *Charta Volans*, and so became aware of the new facets of the dispute, in the autumn of 1713; but only in the spring of the following year did he learn both of its republication in French (together with the added 'Remarques') and of Bernoulli's dynamical papers. These fresh events impelled him to prepare another long statement of his case (Letter 1053a), actually translated into French for publication abroad, which was suppressed when Keill took over Newton's defence though it provided a basis for Newton's *Recensio* of the following year. Although the assembly of relevant documents had presented him with no difficulty Newton, like Leibniz, seems to have found it almost impossible to write a satisfactory narrative account of the history and nature of his discoveries in mathematics, either in English or in Latin. Hence many of his autobiographical statements (concerning the influence of Fermat, for example) are misleading and hence also derives his failure to define his concepts and their historical development. In the text just mentioned, Newton has to face the problem Bernoulli had raised: what is a fluxion? To this question Newton possessed no rigorous answer. Whiteside has written of Newton's search for it in his (then unpublished) 'Geometria curvilinea' tract of *c.* 1680: 'The fundamental appeal here . . . to the concept of a limit is glossed over by the use of such conventional, none too precisely defined verbal forms as "first", "beginning", "last", "vanishing" and especially (instantaneous) "speed", which Newton hopes will be intuitively understood by his reader.'*

He was hardly more successful in 1714 (p. 82). Newton rightly insisted that fluxions are not quantities of the same kind with Leibnizian differences. 'Fluxions are velocities, & Differences are small parts of things generated by fluxion in moments of time: fluxions are always finite quantities & differences are infinitely little . . . [Fluxions] without the moment *o* either exprest or understood . . . never signify moments or differences, but are always finite

* Whiteside, *Mathematical Papers*, IV, p. 410.

quantities & signify velocities.' (That is, on writing a Leibnizian dt for o, Newton's moment of time, the fluxion $\dot{x}$ of a variable x is the finite speed dx/dt, and not the infinitely small 'differential' dx.) Later Newton continued, writing of himself: 'When he is demonstrating any Proposition he always writes down the moment o & takes it in the sense of ye vulgar for an indefinitely small part of time & performs the whole calculation in finite figures by the Geometry of the Ancients without any approximation, . . .' Hence, finally, he could maintain that his own method was geometrical, while the differential method was not. And, he added, in analysis the moment o could be omitted 'for making dispatch' and suitable approximations employed. However, in his own work, Newton never used this abbreviated method which, as he stated, was equivalent to the differential method.

What does this amount to? Newton seems to make one claim for his method in mathematical analysis, another for its use in synthetic demonstration. As regards the former, he admits that both the fluxional and the differential calculus are non-geometrical, non-rigorous, in this respect enjoying only the heuristic value of the 'method' of Archimedes; at this level the two methods are 'equipollent'. Newton, then, seems to be claiming much less for fluxions than the Continentals claimed for the differential calculus; moreover, there is something very *ad hoc* and ambiguous about his statement that 'wherever the pricked letters represent moments [rather than fluxions] and are without the letter o this letter is always understood'. That is, in analysis, for $\dot{x}$ read $\dot{x}o$. The superiority of fluxions, Newton asserts, lies in their capacity to yield rigorous synthetic geometrical demonstrations in the manner of Euclid and Apollonios (p. 82) such as he had employed in the *Principia*. But for these it is absolutely essential to introduce the moment o of time, and then (as Newton correctly writes): 'The rectangles under fluxions & the moment o being my marks for moments are to be compared with the marks dx & dy of Mr Leibnitz'. Where then was the difference and the advantage? Especially as the moment o ('taken in the sense of the vulgar') was as indefinite a concept as the differential. Newton could certainly claim that his method was older than that of Leibniz, and also that it was equivalent to the differential calculus, but it is by no means clear that he could argue for its equivalence and superiority at the same time. The fact that he preferred to utilize it as a geometrical method to obtain synthetic geometrical demonstrations was hardly relevant and appears an unnecessary brake on the development of the calculus.

Neither at this time nor later could Newton explain satisfactorily why none of his published mathematical writings contained a clear account of the manner of deriving second differentials, but he could claim with superficial truth that he had not explicitly employed them in *Principia*, Proposition 10, Book II.

Some of this argument passed into Keill's hands as he prepared his 'Answer' to Leibniz in the early summer of 1714 (Letters 1064 and 1077), an answer in which, amidst much else, Keill attempted to turn a *tu quoque* against him in the matter of second differentials (Letter 1069).

Independently, it seems, both Keill and Newton had found in Leibniz's 1689 paper 'Tentamen de motuum cœlestium causis' (directly inspired by a review of the 1687 *Principia*) errors which they attributed to his incorrect use of second differentials. (In fact, advised by Varignon, Leibniz had himself privately corrected the essay in 1706.) Newton placed his own critical notes on the 'Tentamen' in Keill's hands. Thus Keill answered irrelevancy with irrelevancy, though (it may be added) this malicious Newtonian riposte, if not accurate in every detail, came nearer to the heart of Leibniz's error than did the Bernoullian 'analysis' of Proposition 10, Book II, which was decidedly off the mark—as far off the mark as Bernoulli's carefully argued hypothesis that the letters in the *Commercium Epistolicum* had been falsified in order to bolster the English case, to which Leibniz fortunately paid no serious attention (Letter 1075). Whether because he really had not the papers with him, or for whatever reason, Leibniz was throughout reluctant to compete with Newton in documentary research and he must have known that Newton would in any case not be so idiotic as to falsify documents of which originals were in Leibniz's hands. For a time at least both Johann Bernoulli and Christian Wolf were so impressed by the weight and historicity of Newton's evidence that they feared Leibniz's case would go by default; 'most people may deduce from [your] silence that the English case is a good one' as Wolf put it (Letter 1140) in urging Leibniz, as did Bernoulli, to give his own version of the evolution of the calculus and so demonstrate Newton's plagiary. Leibniz, however, steadfastly refused to answer Keill's offensive papers (Letter 1136), instead amusing himself by pointing out Newton's philosophical absurdities to Bernoulli (Letter 1138). In effect, both Leibniz and Bernoulli now agreed that Newton's pre-eminence in the matter of series went back to the 1670s, and allowed him discovery of a 'præludium calculi differentialis' (Letter 1142); what they now firmly (but mistakenly) denied was that Newton had ever devised an algorithm, or formulated his procedures for tangents and quadratures into a systematic method: 'we judge rather that it [Newton's 'infinitesimal calculus'] was not well enough known to Newton himself before I published' (Leibniz in Letter 1136). That is to say, Newton was far enough advanced to profit enormously from Leibniz's rather obscure paper of 1684. And Leibniz's thoughts, again prompted by Johann Bernoulli, turned increasingly to the demonstration that Newton had never *to this day* understood the Leibnizian calculus thoroughly by the test of a challenge problem.

All this was, of course, highly unjust and inaccurate in its assessment of what Newton had accomplished from 1664 onwards.

If Bernoulli hoped to maintain friendly relations with England by concealing his partisanship towards Leibniz, he was sadly mistaken. After his three dynamical papers appearing in 1713–14, highly critical of Newton, had been read he was dropped by his older English correspondents, William Burnet and Abraham de Moivre, although—partly through the tact of Pierre Varignon, partly through his shameless willingness to perjure himself—Bernoulli long remained officially on terms of meaningless cordiality with Newton. Burnet and De Moivre, unlike Keill, seem not to have rejoiced, as it were, in the natural iniquity of all mathematicians outside the British pale, but rather to have attempted without betrayal of Newton's rights to lead Bernoulli to see the virtue of compromise; Chamberlayne too, dealing directly between Newton and Leibniz, demonstrated little liking for Keill's policy of demanding unconditional surrender and apology from the Continentals. Any moderating influence they may have exercised was to no effect, however, and the next attempt to create an understanding between Newton and Leibniz, if anything, made matters worse.

In 1715, at about the time of Newton's anonymous publication of the *Recensio* or self-styled *Account* of the *Commercium Epistolicum* in the *Philosophical Transactions* (which brought out no fresh aspect of the controversy) there arrived in London the Abbé Conti, a Venetian acquainted with Leibniz and other Continental mathematicians but not himself a man of any mathematical substance. Apart from satisfying himself that the documents printed in the *Commercium Epistolicum* had been accurately transcribed, Conti's rôle at best was that of a postman, at worst that of an *agent provocateur*; notably he introduced to London the challenge problem whose presentation Bernoulli and Leibniz had discussed for some time As soon as the former had sent Leibniz copies of their old correspondence relating to suitable problems (November 1715) Leibniz took occasion to address Conti on quite a different matter (Letter 1170), artfully introducing his challenge after a postscript intended for Newton's eye. This contained little that was new relating to the mathematical dispute, but launched into a lengthy criticism of Newton's philosophy, natural and divine; by some curious act of policy on Leibniz's part or malign chance the challenge intended 'to try the pulse of the English' (which he enjoined Conti to render anonymous) was so phrased that the English took as the whole only its particular exemplification: to find a curve cutting at right angles each member of a family of hyperbolas having the same centre and the same vertex. What he had meant to demand was a general way of determining the line cutting any determinate family of curves (Letter

1175). Although Leibniz tried to correct his inadvertence a few weeks later (Letters 1178 and 1179), the English mathematicians had already disposed of Leibniz's challenge as they understood it (Letter 1186)*—and Leibniz learnt of this from John Arnold, a former pupil of Bernoulli, who also retailed a conversation he had had with Newton about the matter. Newton himself also published a purported solution of Leibniz's problem in its general form, but this must be reckoned a complete failure (Letter 1187a).

In his reply to Conti (Letter 1187) Newton developed yet another facet of the dispute; he insisted that Leibniz had *once* accorded Newton priority in his private letters and ought to do so still before the public. Since Newton's letter naturally went to Leibniz, while Leibniz in turn pressed the justice of his case upon Conti for Newton's benefit, the two men were at last in direct communication, the only moment in the whole dispute when this was so, but to no profit. And soon Newton could accuse Leibniz of replacing his first easy problem by a second, harder one as the latter did in April 1716 (Letter 1202). Johann Bernoulli, again its author, generalized and elaborated a problem whose construction he had published—without proof—in 1697; no solutions of this new problem were offered until 1717, when Hermann gave one in the *Acta Eruditorum*, which was repeated by Brook Taylor, using fluxional notation, in the *Philosophical Transactions*. This new problem led on to the later challenges offered by Brook Taylor and Keill, which do not concern us here. As for Newton, he refused to respond to Leibniz's fresh review of his relationship with Newton over the years (addressed to Conti, Letter 1197) or other letters sent to London (Letter 1203) but he did again prepare—for his friends, as he explained to Des Maizeaux later—a long rebuttal of Leibniz's latest statements (Letter 1211) Only a few months later, as Conti informed Newton: 'M. Leibniz est mort; et la dispute est finie' (Letter 1231). Nevertheless the wrangle between England and the Continent continued, with Newton less and less overtly involved in it, it is true, yet far from unconscious of the later moves; indeed, soon after Leibniz's death he set about publishing the interchange of letters with Leibniz and Conti in an appendix to Raphson's *History of Fluxions*, and included the long paper of 'Observations' he had prepared in reply to Leibniz's last letter (see Number 1211). He still regarded Keill as his stoutest champion, and it was Keill (see Letter 1239) who realized the true situation with regard to the bizarre 'Epistola pro eminente mathematico' (Letter 1196); Johann Bernoulli had indeed supplied the matter for it, although the embellishments upon his own name and the egregious failure to convert a few first-person words to the third person were the work of another

* One who solved the easy problem was Henry Pemberton, editor of the third edition of the *Principia*.

hand—Wolf's, as we have known since 1800. From these later violent exchanges between Keill, Taylor and the Bernoullis, against whom Keill had launched his first attack in the *Philosophical Transactions* for 1714 (Letter 1153), the original historical issue of the discovery of the calculus has all but disappeared, to be replaced by disputes over the merits of Newton's contributions to the science of mechanics and particularly to rational fluid mechanics —a debate which has not entirely subsided at the present day. Keill charged Johann Bernoulli with misrepresentation and falsehood, for example in maintaining that his own solutions of certain problems were different from and independent of those already published by Newton in the *Principia*; Bernoulli rejected these accusations and in turn charged Newton and his champion with almost total mathematical incompetence. In qualifying its originality Keill's assessment of some of Bernoulli's 'new' mechanics was not unfair; but neither he nor his aged hero was—any more, at least—in mathematics and exact science an equal of their opponent.

A word should now be added concerning another enemy of much longer standing whose death (like those of Hooke and Leibniz) left Newton with the victory of survival. Flamsteed's letters to Abraham Sharp reveal no change in his feeling towards Newton, and show him as ready as always to collect every item of damaging gossip (Letter 1151). Early in 1714 he was again provoked by the Royal Society's letter to the Ordnance criticizing the state of the instruments at Greenwich—virtually all Flamsteed's property—and demanding their repair (Letter 1044). Nothing came of this initiative and Flamsteed went quietly on with his attempt to discredit the lunar equations of the second edition of the *Principia* (Baily, *Flamsteed*, pp. 302, 304–5 and 309–10). With the accession of George I a bright star rose in Flamsteed's heaven; as he wrote to Sharp: 'now all those that would do any injury to the observatory have ruined their own credit, and our friends are advanced' (*ibid*. p. 311). The death of Halifax removed another powerful patron of Newton who 'having been in with Lord Oxford, Bolingbroke and Dr Arbuthnot is not now looked on as he was formerly' (Letter 1151). In a short time Flamsteed was emboldened to demand the return of his manuscripts from Newton, and then the delivery into his hands of the remaining stock of *Historia Cœlestis*. He now had friends at Court, among them Thomas Newport, one of the Treasury Lords, and the Duke of Bolton, Lord Chamberlain. The Referees appointed in the previous reign were now compelled to release over three hundred copies of the book (Letter 1171) and to account for the late Prince's money disbursed in preparing it (Letter 1170a). Although Flamsteed objected strongly to the way in which the money had been expended, Newton's accounts seem to have been accepted. Pressed by the Treasury (Letter 1189), the Referees reported in March 1716 that some forty copies had gone to the Treasury (of which about half went to France), another ten or so had been distributed as private gifts, and the remainder of the edition of about 400 copies remained with Awnsham Churchill, the stationer (Letters 1189 and 1199). Three hundred of this stock (in sheets) were delivered to Flamsteed on 28 March, and by him 'some few days after . . . made a *Sacrifice . . . to Heavenly Truth*'.

Nearly six years after Flamsteed's death (on the last day of 1719) there appeared, through his wife's devotion, the three volumes of his *Historia Cælestis*. Perhaps stimulated by that event, Sir Robert Walpole sent one of the remaining copies of the Royal Society's 1712 edition to the Bodleian Library, Oxford, where it was duly endorsed as a survivor of the 'fire and wrath of Flamsteed'. Learning of this, Margaret Flamsteed vehemently addressed the Vice-Chancellor, demanding that the book be withdrawn from the library as an imposture (Baily, *Flamsteed*, pp. 363–4). This was not done. Thus, as with Leibniz, the regrettable animosities surrounding Newton pursued him throughout life, and long after his opponent's death.

THE CORRESPONDENCE

1004 J. BERNOULLI TO LEIBNIZ
27 MAY 1713

Extract from Gerhardt, *Leibniz: Mathematische Schriften*, III/2, pp. 910–12.[1]
For the answer see Letter 1005

. . . Attulit Agnatus meus Lutetia exemplar unum Commercii Epistolici Collinsii et aliorum de Analysi promota, quod Abbas Bignonius, qui plura Exemplaria Eruditis distribuenda, Londino missa acceperat, ipsi tradidit.[2] Legi illud, nec sine attentione sufficienti. Displicet imprimis modus procedendi parum urbanus; accusaris statim coram Tribunali, quod ut videtur ex ipsis Actoribus et Testibus consistit, tanquam plagii reus, postea productis documentis contra Te, fertur sententia; causa cadis, damnaris. Recensionem in Actis Lipsiensibus Januarii 1705 editam Libri Newtoni *De numero curvarum tertii generis deque quadratura figurarum*, Tuo stylo conscriptam pronuntiant; imo postea Tibi, tanquam Auctori, diserte imputant.[3] Hæc recensio inprimis conquerendi ansam præbuit, bilemque movit accusatoribus Tuis, utpote quam Newtoni inventis nimium derogare existimant. His itaque permotum fuisse Keillium, ut in Epistola in Philosophicis Transactionibus impressa,[4] Newtono quod suum est vindicaret ostenderetque fluxionum Arithmeticam inventam esse ante Calculum differentialem: imo hunc postea, mutatis tantum nomine et notationis modo, ab illa esse mutuatum et ita Newtono subreptum sub alia tantum facie in Actis Erud. a Te editum fuisse. Sed ut ego dicam, quod de hac re sentio, quantum quidem ex hac farragine Epistolarum constare potest, videtur Mercator primus Serierum inventor per continuam divisionem, Jac. Gregorius postea hanc materiam excolens incidit, ut apparet,

primus in Circuli Quadraturam Arithmeticam $1-\dfrac{1}{3}+\dfrac{1}{5}-\dfrac{1}{7}+$etc. quam Tu,

inscius sine dubio, quod jam ante Te fuerit detecta, tanquam Tuam in Actis edidisti, et revera Tua fuit æque ac Gregorii;[5] invenisti enim (licet posterior) æque ac Gregorius, nam invenisse est industriæ, sed primum invenisse est felicitatis, ut Wallisius alicubi dicit. Deinde[6] videtur Newtonus, occasionem nactus, Serierum opus multum promovisse per extractiones radicum, quas primus in usum adhibuit, et quidem in iis excolendis, ut verisimile est, ab initio omne suum studium unice posuit, nec, credo, tunc temporis vel somniavit adhuc de Calculo suo fluxionum et fluentium, vel de reductione ejus ad generales operationes Analyticas, ad instar Algorithmi vel Regularum Arithmeticarum et Algebraïcarum inservientes. Cujus meæ conjecturæ validissimum indicium est, quod de literis $\dot{x}$, $\ddot{x}$, $\dot{\ddot{x}}$; $\dot{y}$, $\ddot{y}$ etc. quas pro differentialibus dx, ddx, d^3x; dy, ddy etc. nunc adhibet, in omnibus istis Epistolis nec

I

volam nec vestigium invenias;[7] imo nequidem in Principiis Philos. Natural.
ubi Calculo suo fluxionum utendi tam frequentem habuisset occasionem, ejus
vel verbulo fit mentio, aut notam hujusmodi unicam cernere licet, sed omnia
fere per lineas figurarum, sine certa Analysi, ibi peraguntur, more, non ipsi
tantum, sed et Hugenio, imo jam antea dudum Torricellio, Robervallio,
Fermatio, Cavallerio, aliis usitato.[8] Prima vice hæ literæ punctatæ comparuerunt in tertio Volumine Operum Wallisii, multis annis postquam Calculus differentialis jam ubique locorum invaluisset. Alterum indicium, quo
conjicere licet, Calculum fluxionum non fuisse natum ante Calculum differentialem, hoc est, quod veram rationem fluxiones fluxionum capiendi, hoc
est differentiandi differentialia per gradus ulteriores Newtonus nondum
cognitam habuerit, quod patet ex ipsis Princ. Phil. Nat. pag. 263, ubi pro
differentiis vel incrementis primo, secundo, tertio, quarto etc. alicujus potestatis ex. gr. x^n, judicat ponendos esse secundum, tertium, quartum, quintum
etc. terminos ipsius $\overline{x+o}^n$ in Seriem expansæ per extractionem, nempe

$$x^n + \frac{n}{1} x^{n-1} o + \frac{n.n-1}{1.2} x^{n-2} o^2 + \frac{n.n-1.n-2}{1.2.3} x^{n-3} o^3 \text{ etc.}$$

(intelligit per o incrementum constans ipsius x, quod nunc notaret per $\dot{x}$).[9]
Vides autem, quod vera differentiandi methodus evincit, hanc regulam
Newtoni falsam esse. Nam, excepto primo et secundo termino, reliqui omnes
abludunt a differentialibus superioribus potestatis x^n, nam differentialis
secunda non est $\frac{n.n-1}{1.2} x^{n-2} o^2$, sed simpliciter $n.n-1.x^{n-2}o^2$, et differentialis
tertia non est $\frac{n.n-1.n-2}{1.2.3} x^{n-3} o^3$ sed tantum $n.n-1.n-2.x^{n-3}o^3$, et ita de
reliquis, et hoc ipsum est, quod in nupero meo schediasmate Actis Lipsiensibus
inserto jam notavi.[10] Sed ex eo tempore, quo hæc scripseram, animadverti
(quod Agnatus meus mihi ostendit) Newtonum in suo errore perseverasse
usque ad annum 1711, quo Libellus ejus, cui Titulus: *Analysis per quantitatum
series, fluxiones, ac differentias, cum Enumeratione Linearum tertii ordinis*, qui antea
Tractatui Optico erat adnexus, fuit recusus, utpote in quo (pag. 64) sicuti in
præcedente editione Tractatui Optico adnexa, asserit adhuc terminos secundum, tertium, quartum etc. hujus seriei

$$z^n + noz^{n-1} + \frac{nn-n}{2} oo z^{n-2} + \frac{n^3 - 3nn + 2n}{6} o^3 z^{n-3} \text{ etc.}$$

exprimere incrementa primum, secundum, tertium etc. ipsius z^n. Sed, quod
notandum, in exemplari quod mihi dono misit per Agnatum meum, ibi
calamo ascripsit altera vice voculam, *ut*: nam ubi habebantur hæc verba:

2

tertius (terminus) $\dfrac{nn-n}{2}\,ooz^{n-2}$ *erit ejus incrementum secundum etc. quartus*

$\dfrac{n^3-3nn+2n}{6}\,o^3z^{n-3}$ *erit ejus incrementum tertium* etc. interseruit *ut*, scribendo

nunc: *erit ut ejus* etc. adeo ut errorem suum non animadverterit, nisi brevi ante, et forte nonnisi post adventum Agnati mei in Angliam, ex quo alia quoque expiscati sunt Angli.[11] Hinc dubito, annon in Epistola Keillii ad Sloannium scripta et jussu Societatis Tecum communicata mense Majo 1711, duo paragraphi incipientes: *Sit incrementum* et *Præterea si differentia*, qui continent correctionem erroris Newtoniani, sint nunc demum intrusi in apographo impresso in Commercio Epistolico, quod non nisi post abitum Agnati mei ex Anglia in lucem prodiit; Tuum est videre, an hi paragraphi in originali Tecum communicato reperiantur.[12] Saltem constat, Newtono rectam Methodum differentiandi differentialia non innotuisse, longo tempore postquam nobis fuisset familiaris. Sed cogor abrumpere hac vice; rogo vero, ut quæ hic scribo, iis recte utaris, neque me committas cum Newtono ejusque popularibus; nollem enim immisceri hisce litibus, nedum apparere ingratus erga Newtonum, qui me multis benevolentiæ testimoniis cumulavit.[13] Alias plura; nunc vale et fave etc.

    Basileæ 7 *Junii* 1713 [N.S.]

Translation

. . . My nephew brought from Paris a single copy of the *Commercium Epistolicum Collinsii et aliorum de Analysi promota*, which the Abbé Bignon had handed to him, having received a number of copies sent from London for distribution to the learned.[2] I have read it, not without a fair amount of attention. This hardly civilized way of doing things displeases me particularly; you are at once accused before a tribunal consisting, as it seems, of the participants and witnesses themselves, as if charged with plagiary, then documents against you are produced, sentence is passed; you lose the case, you are condemned. They declare that the review of Newton's tracts *Enumeratio linearum tertii ordinis* and *De quadratura curvarum* published in the *Acta* [*Eruditorum*] of Leipzig for January 1705 is written in your style; moreover, they later on attribute it expressly to you, as its author.[3] This review provides a particular excuse for complaint and provokes the gall of your accusers, inasmuch as they judge it to detract excessively from Newton's inventions. And so Keill was stirred up by this to vindicate Newton's claim to his own discoveries (in a letter printed in the *Philosophical Transactions*[4]) and to show that the calculus of fluxions was invented before the differential calculus; indeed, that the latter was adapted from the former (with a simple change of name and method of notation), and so having been thus stolen from Newton was published by you under a disguise in the *Acta Eruditorum*. But to say what I think of this myself, so far as one can learn anything from this medley of letters, it seems that Mercator [was] the first inventor of series

3

by continued division, and James Gregory afterwards taking this subject further hit first, as it seems, upon that arithmetical quadrature $1 - \frac{1}{3} + \frac{1}{5} - \frac{1}{7} +$ etc. which you, doubtless in ignorance and because it had been discovered by you a long time before, published as though it were your own in the *Acta* [*Eruditorum*] and really it was as much yours as Gregory's;[5] for you did discover it (although later) as much as Gregory did, for to discover something is a matter of hard work whereas to discover something first is a blessing, as Wallis says somewhere. Then[6] it seems that Newton, seizing the opportunity, very much advanced the business of series by the extraction of roots, a method he first employed, and it is very likely that in the beginning of all his studies he devoted himself solely to developing them nor did he then, I believe, so much as dream of his calculus of fluxions and fluents, or of its reduction to the general operations of analysis in order to serve as an algorithm or in the manner of the arithmetical and algebraic rules. The strongest evidence for this conjecture of mine is that you can find in all those letters no trace or vestige of the symbols $\dot{x}, \ddot{x}, \dddot{x}; \dot{y}, \ddot{y}$ etc. which he now employs in place of the differentials $dx, d^2x, d^3x; dy, d^2y$ etc.[7] Indeed, you can find no least word or single mark of this kind even in the *Principia Philosophiæ Naturalis*, where he must have had so many occasions for using his calculus of fluxions, but almost everything is there done by the lines of figures without any definite analysis in the way not used by him only but by Huygens too, indeed by Torricelli, Roberval, Fermat and Cavalieri long before.[8] These pointed letters appeared for the first time in the third volume of Wallis's works, many years after the differential calculus had established itself everywhere. Another piece of evidence from which it may be conjectured that the calculus of fluxions was not born before the differential calculus is this, that the true way of deriving the fluxions of fluxions, that is of differentiating differentials to higher degrees, has not yet been understood by Newton, as is obvious from his own *Principia* page 263, where he supposes that for the first, second, third, fourth etc. differences or increments of some power, x^n for example, one should take the second, third, fourth, fifth etc. terms of $(x+o)^n$ expanded into a series by extraction, namely

$$x + \frac{n}{1} x^{n-1} o + \frac{n(n-1)}{1.2} x^{n-2} o^2 + \frac{n(n-1)(n-2)}{1.2.3} x^{n-3} o^3 \text{ etc.}$$

(he meant by o the constant increment of x, which he now denotes by $\dot{x}$).[9] You see, however, that the true method of differentiating proves this rule of Newton's to be false. For, with the exception of the first and second terms, all the rest differ from the higher differentials of x^n, for the second differential is not $\frac{n(n-1)}{1.2} . x^{n-2} o^2$ but simply $n(n-1) x^{n-2} o^2$ and the third differential is not $\frac{n(n-1)(n-2)}{1.2.3} x^{n-3} o^3$ but only $n(n-1)(n-2) x^{n-3} o^3$ and so with the rest, and this is the very same point that I made in my recent note inserted in the *Acta* [*Eruditorum*] of Leipzig.[10] But since the time when I wrote that, I have remarked (as my nephew showed me) that Newton persisted in his error up to the year 1711, when that little treatise called *Analysis per quantitatum series, fluxiones, ac*

4

differentias, cum enumeratione linearum tertii ordinis was reprinted, which had formerly been annexed to his treatise on optics, inasmuch as there (p. 64) just as in the preceding edition annexed to *Opticks,* he still asserts that the second, third, fourth etc. terms of this series

$$z^n + noz^{n-1} + \frac{n^2 - n}{2} o^2 z^{n-2} + \frac{n^3 - 3n^2 + 2n}{6} o^3 z^{n-3} \text{ etc.}$$

express the first, second, third etc. increments of z^n. But, what is noteworthy, in the copy that he sent me by my nephew he has on second thoughts written by hand the little word *ut,* for he has inserted where these words were: 'the third [term] $\frac{n^2 - n}{2} o^2 z^{n-2}$ will be its second increment etc. the fourth $\frac{n^3 - 3n^2 + 2n}{6} o^3 z^{n-3}$ will be its third increment etc.' the word *ut,* writing now: 'will be *as* its . . .' etc., so that he could not have noticed his mistake until a very little time before, and possibly not until after the arrival in England of my nephew, from whom the English have fished out other things.[11] Hence I wonder whether the two paragraphs in Keill's letter to Sloane communicated to you by order of the Society in May 1711 beginning: 'Let the increment . . .' and 'Moreover if the differences . . .', which contain a correction of Newton's error, are not now recently intruded into the printed copy in the *Commercium Epistolicum,* which only appeared after the departure of my nephew from England: you may observe whether these paragraphs are to be found in the original communicated to you.[12] At any rate it is clear that the true way of differentiating differentials was not known to Newton until long after it was familiar to us. But I am driven to break off for the present; I do indeed beg you to use what I now write properly and not to involve me with Newton and his people, for I am reluctant to be involved in these disputes or to appear ungrateful to Newton who has heaped many testimonies of his goodwill upon me.[13] More another time; for now, farewell, etc.

Basel, 7 *June* 1713 [N.S.]

NOTES

(1) Johann [I] Bernoulli's correspondence with Leibniz was first published in *Virorum Celeberr. God. Gul. Leibnitii et Johann. Bernoulli Commercium Philosophicum et Mathematicum* (Lausanne and Geneva, 1745); we have followed the texts as printed by Gerhardt (see Gerhardt, *Leibniz: Mathematische Schriften,* III/1, p. 132). The manuscripts of Bernoulli's letters to Leibniz are in the Niedersächsische Landesbibliothek, Hanover, and of Leibniz's letters to Bernoulli in Basel University Library. See *Der Briefwechsel von Johann Bernoulli,* ed. O. Speiss (Basel, 1955), I, p. 26.

(2) For the visit of Nikolaus Bernoulli to London see Letter 951*a,* note (2) (vol. v, pp. 349–50). Jean-Paul Bignon (1662–1743), an Oratorian, also Royal Librarian, was nominated either President or Vice-President of the Académie Royale des Sciences in most years from 1699 to 1734; he was a purely literary man, also a member of the Académies Françaises and des Beaux-Arts.

(3) This review was indeed by Leibniz, see Edleston, *Correspondence,* pp. lxxi–lxxii.

(4) See Letter 830, note (5) (vol. v, p. 116).

(5) Leibniz's series $\frac{\pi}{4} = 1 - \frac{1}{3} + \frac{1}{5} - \frac{1}{7}$ etc. discovered in 1673, was communicated to Huygens in the autumn of 1674 (see *Oeuvres Complètes*, VII, 393–5). For Gregory's earlier work see Turnbull, *Gregory*, especially his letters of 23 November 1670 and 15 February 1671, while for the prior discovery of this series by the fifteenth-century Hindu mathematician Nilakaṇṭha see Whiteside, *Mathematical Papers*, III, p. 34, note (5).

(6) The passage 'Videtur Newtonus . . . nobis fuisset familiaris' was printed, with some alterations, omissions and additions in the *Charta Volans* (see Number 1009 and the public acknowledgement of Bernoulli's authorship in Letter 1172).

(7) This remark is just in that Newton employed his standard dot notation only from December 1691 (see D. T. Whiteside in *Journal for the History of Astronomy*, **1** (1970), 118) although it was presaged in his mathematical notes as early as 1665. However, Bernoulli was mistaken in identifying fluxions with differentials, as he does here, for the fluxions $\dot{x}\left(= \frac{\dot{x}}{\dot{t}}, \right.$ $\left. t = 1 \right)$, $\ddot{x}\left(= \frac{\ddot{x}}{\dot{t}^2} \right)$ etc. are the Leibnizian derivatives $\frac{dx}{dt}, \frac{d^2x}{dt^2}$ etc. See, for Newton's own insistence upon this, Letter 1053*a*, note (10), p. 91.

(8) Again this judgement contains a partial truth; certainly the printed state of the *Principia* reflects accurately the mathematical method by which Newton derived it; see Whiteside, *loc. cit.*, p. 119. But Bernoulli fails to credit Newton with his sophisticated and original use of the geometrical limit-increment of a variable line segment, where this becomes equivalent to the use not only of the first but of higher-order derivatives, so falsifying Bernoulli's claim that Newton (before 1712) had no notion of such derivatives.

(9) Bernoulli returns once more to Proposition 10 of Book II in the first edition, and Newton's series $e - \frac{ao}{e} - \frac{n^2 o^2}{2e^3} - \frac{an^2 o^3}{2e^5} \ldots$ still following his nephew Nikolaus' suggestion that the later terms in this series had been wrongly interpreted by Newton as successive differentials (see Letter 951, vol. V, note (3) and Letter 951*a*, note (2)). It was a pure accident that the expressions employed by Newton differed by factors of 2 and 6 in the denominators from successive differentials; moreover *o* is not to be indentified with $\dot{x}$.

(10) Bernoulli first signalized Newton's mistake in the scholium (pp. 93–5) to a long paper 'De motu corporum gravium, pendulorum et projectilium' in the *Acta Eruditorum* (February 1713, pp. 77–95; March 1713, pp. 115–32); he also dealt with it in the *Mémoires de l'Académie Royale des Sciences* for 1711 (Paris, 1714), pp. 47–56.

(11) Bernoulli refers to the concluding scholium of *De quadratura curvarum* as printed by William Jones in 1711. As he says, Newton continues (after remarking generally that the fluxions 'sunt ut Termini Serierum infinitarum convergentium') 'tertius $\frac{nn - n}{2} ooz^{n-2}$ erit ejus Incrementum secundum, seu Differentia secunda, cui nascenti proportionalis est ejus Fluxio secunda'; there is no *ut*, and the word *proportionalis* renders it redundant so far as *fluxio* is concerned. Newton says always that the fluxion is proportional to the term of the series, which is correct if not explicit. See also Letter 1063. Newton's own 1706 copy of *De quadratura curvarum* was amended in the same way as that which he sent to Bernoulli.

(12) Bernoulli refers to Keill's letter to Sloane for Leibniz of May 1711 (Letter 843*a*, vol. V). The passages he mentions do of course appear in the letter as sent to Leibniz.

(13) The request is somewhat pusillanimous, after Bernoulli had not hesitated to provide Leibniz with ammunition. Leibniz ignored it; see Letter 1172.

6

1005   LEIBNIZ TO J. BERNOULLI
17 JUNE 1713

Extract from Gerhardt, *Leibniz: Mathematische Schriften*, III/2, pp. 913–14.
Reply to Letter 1004; for the answer see Letter 1008

Cum heri demum Tuas literas Hanovera[1] acceperim, iisque statim respondere velim, cætera differo.

Anglicanum libellum[2] mihi oppositum nondum vidi; merentur illæ insulsæ rationes, quas afferri ex Tuis conjicio, sale satyrico defricari. Poterant Newtonum suum in compossessione inventi Calculi conservare, sed apparet non magis, eum cognovisse Calculum nostrum, quam Apollonius cognovit calculum Vietæ et Cartesii speciosum. Fluxiones cognovit, non Calculum fluxionum, quem demum, ut recte judicas, nostro jam edito conflavit. Itaque plus justo ei attribui ipse, et humanitatis meæ hoc pretium fero.[3]

Scribit mihi Dominus Agnatus Tuus, in recensione Lipsiensi, de qua queruntur, valde attolli Tschirnhausii inventa. Ego non bene memini, sed si ita est, facile judicas, illam Newtoniani libri recensionem non esse a me; ego enim tam magnifica mihi de Dno. Tschirnhausio, ut scis, non promittebam.[4]

Ubi domum reversus fuero, Keillii dissertationem manuscriptam[5] mihi missam inspiciam, quam ego, ut par est, sprevi. Volebat, credo, ille ut serram cum ipso reciprocarem. Respondi simpliciter ad amicum, hominum novum et parum versatum in Historia inventorum anteriorum, somnia sibi fingere de modo, quo me mea cognovisse putat. Jussu autem Societatis has nugas ad me missas non poteram suspicari, etsi miserit Sloanius.[6]

Nunc primum audio Gregorio quoque attribui meum inventum magnitudinis Circuli: Hugenius et quoque alii Parisiis testes sunt inventionis meæ et modi, quo inveni. Hugenii hac de re literas adhuc inter schedas meas esse puto. Quin ipse Newtonus inventum meum Oldenburgio tunc communicatum in literis suis laudavit, modumque inveniendi meum singularem esse fassus est.[7] Ignorabat ergo tunc inventionem Gregorii.[8]

Satis apparet, Newtonum id egisse suis blanditiis, ut benevolentiam tuam captaret; conscium sibi quam non recto stent talo quæ molitus est. Ego tametsi nolim ut in mei gratiam Tibi negotium facessas, expecto tamen ab æquitate Tua et candore, ut profitearis apud amicos quamprimum, et publice data occasione, Calculum Newtoni nostro posteriorem Tibi videri . . .

Velim nosse quid judicent Parisini. Varignonium pro me stare, vix dubito; sed nonnulli alii invidia ducti libenter fortasse accipient occasionem carpendi sibi datam, sed hæc, credo, a parum intelligentibus fient.

Jam a multis annis hæc Anglis nonnullis, etiam insignibus Viris, vanitas est inolita, ut captarent occasionem res Germanorum involandi et pro suis venditandi. Boylius Glauberianum inventum Nitri regenerati sibi tribuit;[9] idem totum inventum Anthliæ Pneumaticæ a Gerikio habuit et tantum minuta quædam in ejus structura mutavit, et tamen Angli, et eorum exemplo alii, verum inventorem ignorantes, Machinam Boylianam appellarunt. Ita nunc libenter Nicolaum Mercatorem Holsatum gloria primæ inventionis Serierum privare vellent, et mihi indignati sunt optimi Viri et mihi amici decus vindicanti. Sic Hugenius Heuratii inventum contra Wallisii oppositiones vindicavit Neilio cuidam id tribuentis.[10]

Dicis a Dno. Agnato Tuo non tantum Newtonum habuisse videri quandam erroris sui correctionem, sed et alia nonnulla ab eo expiscatos esse Anglos; hæc qualia sint rogo ut indices. Inservient enim nobis ad cautelam. Spero eis nondum innotuisse meum modum Tangentes inveniendi curvarum quarundam non vulgari modo compositarum, ex quo deinde præclaras consequentias duxisti. Nondum enim, quod sciam, publicatas puto, spero etiam cum Italo illi, cui cum Dno. Hermanno lis fuit, non fuisse communicatum . . .[11]

Translation

. . . Since I at last received yesterday your letter from Hanover,[1] and wished to reply to it at once, I am setting other matters aside.

I have not yet seen the little English book[2] directed against me; those idiotic arguments which (as I gather from your letter) they have brought forward deserve to be lashed by a satirical wit. They would maintain Newton in the possession of his own invented calculus and yet it appears that he no more knew our calculus than Apollonius knew the algebraic calculus of Viète and Descartes. He knew fluxions, but not the calculus of fluxions which (as you rightly judge) he put together at a later stage after our own was already published. Thus I have myself done him more than justice, and this is the price I pay for my kindness.[3]

Your nephew writes to me that in the Leipzig review of which they complain the inventions of Tschirnhaus are greatly praised. I do not clearly recollect, but if it is so you may readily judge that that review of Newton's book was not by me; for I could not promise, as you know, to think so well of Mr Tschirnhaus.[4]

I will examine the manuscript essay[5] by Keill, which was sent to me, after I have returned home; this I treated with contempt, as was only right. It was intended, I believe, that I should bandy words with him. I merely replied to a friend, that a 'new man' little acquainted with the history of past discoveries might dream to himself of the way he thinks I have learned what I know. I could not suspect that these trifles had been sent to me by order of the Society, however, even though Sloane sent them.[6]

I now hear for the first time that my discovery of the magnitude of the circle is to be attributed to Gregory also; Huygens and others at Paris too are witnesses to my discovery and the way in which I made it. Huygens' letters on this business are still among my papers, I think. Indeed Newton himself praised my discovery in his letter at the time when it was imparted to Oldenburg, and it is admitted that my own way of discovering it was a strange one.[7] Therefore he did not then know of Gregory's discovery.[8]

It is obvious enough that Newton has been so forceful with his blandishments in order to capture your goodwill; knowing within himself that he has built upon an insecure foundation. As for myself, although I have no wish that you should make trouble for yourself for my sake, I expect from your honesty and sense of justice that you will as soon as possible make it evident to our friends that in your opinion Newton's calculus was posterior to ours, and say this publicly when opportunity serves . . .

I would like to know what the Parisians think. I can scarcely doubt that Varignon is on my side but several others, moved by envy, will perhaps gladly seize the opportunity given them of carping, but these people I think will be found only among the undiscerning.

For many years now the English have been so swollen with vanity, even the distinguished men among them, that they have taken the opportunity of snatching German things and claiming them as their own. Boyle attributed to himself Glauber's discovery of the redintegration of nitre; the same person had the whole invention of the pneumatic pump from von Guericke and only modified some details of its structure,[9] and yet the English and others following their example who are ignorant of the true inventor call it 'Boyle's machine'. So now they mean to deprive Nicolaus Mercator of Holstein of the glory of the first discovery of series, and are displeased with me for vindicating the honour of a very good man, [who is] my friend. Thus Huygens vindicated Heuraet's discovery against the opposition of Wallis, who attributed it to a certain Neile.[10]

You say that Newton not only seems to have had from your nephew a certain correction of an error of his, but that the English have fished several other things out of him too; I beg you to indicate what these are. For they serve [to advise] us to caution. I hope they are not yet aware of my method of finding the tangents of certain curves not compounded in the common way, from which I have deduced some remarkable consequences. For these are not yet published, so far as I know, and I hope too that they were not communicated to that Italian who had a dispute with Mr Hermann . . .[11]

NOTES

(1) Leibniz being at Vienna, Bernoulli had presumably written to Hanover.

(2) The *Commercium Epistolicum*; see vol. v, pp. xxv–xxvii.

(3) It is obvious that this letter is written in a passion of resentment, and this paragraph though emphatic as to sense is not very clear in style.

(4) This is a prevarication; Leibniz just manages to avoid a downright lie about his authorship of the review.

(5) That is, Letter 843a, vol. v.

(6) See Letter 884, vol. v, written (as was proper) to Sloane.

(7) Leibniz first wrote of the series to Oldenburg on 6 October 1674, without stating it;

it was a very simple series, he said, whose sum exactly yields the circumference of a circle. In reply Oldenburg reminded him of Gregory's and Newton's work on curvilinear quadratures, and advised him that Gregory, as he thought, had proved the impossibility of obtaining an *exact* value for π. Leibniz imparted the series itself to Oldenburg on 17 August 1676 (vol. II, p. 60). Newton replied to Leibniz in his *Epistola Posterior* of 24 October 1676, commencing: 'Leibniz's method of arriving at convergent series is indeed extremely elegant and is sufficient evidence of the author's ingenuity if he had written nothing else. But the matters scattered elsewhere through his letter are most worthy of his reputation, and cause us to have the highest hopes of it.'

(8) The statement is unworthy of Leibniz. To the question of Gregory's prior knowledge of the series Newton's ignorance or otherwise of what Gregory had done is irrelevant.

(9) This is a very bitter paragraph. Boyle's essay on the redintegration of nitre had appeared in *Certain Physiological Essays* (1661) the various parts of *New Experiments Physico-Mechanicall touching the Spring of the Air* in 1660, 1669 and 1680.

(10) William (1637–70), son of Sir Paul Neile, rectified the semi-cubic parabola in 1657; this was the first rectification of an algebraic curve, but the nearly-simultaneous investigation by Hendrik van Heuraet (whom Huygens in *Horologium oscillatorium* improperly called Johannes), published in the second Latin edition by Frans van Schooten of Descartes' *Geometrie* (Amsterdam, 1659), was much more significant. For the controversy between John Wallis and Huygens over this rectification see Hall & Hall, *Oldenburg*, IX and X *passim* and Hofmann, § 8.

(11) Jakob Hermann (1678–1733), of Basel, was a pupil of the Bernoullis, and commended himself to Leibniz by taking his part against the Dutch philosopher and mathematician Bernard Nieuwentyt (1700). At this time he was Professor of Mathematics at Padua, whence he was very soon to move to Frankfurt-an-der-Oder and later to St Petersburg. He kept up a considerable correspondence with both Leibniz and Bernoulli on current mathematical developments. The letters he exchanged with Leibniz, in which there is frequent mention of Newton's work, are published in Gerhardt, *Leibniz: Mathematische Schriften*, IV, pp. 255–413; his correspondence with Bernoulli remains unpublished; see Wollenschläger, *De Moivre*, pp. 155–65.

The Italian whom Leibniz mentions is probably Ercole Corazzi (1669–1726), a Benedictine monk, who in 1719 was to obtain the chair of algebra at Bologna University, and later also taught the theory of fortifications. In 1720 he became Professor of Mathematics at Turin. His dispute with Hermann was over the appointment to the Chair of Mathematics at Padua in 1707; both were candidates for the Chair and Hermann succeeded. Later in 1713, after Hermann's return to Germany, Corazzi again tried to obtain the Chair but this time was ousted by Nikolaus [I] Bernoulli, appointed in 1716 (see Letter 1240).

1006 OXFORD TO NEWTON
3 JULY 1713
From the minute in the Public Record Office[1]

Let the Master & Worker of Her Mats. Mint take Care that her Mats pleasure Signifyed in the above written Order of Councill be duely Complyed with so far as appertains unto him [2]

Whitehall Trea[su]ry Chambers 3d July 1713   OXFORD

NOTES

(1) T/54, 22, p. 87.

(2) The Peace for which the Tory Party had so long sought was restored to Europe in April 1713, at Utrecht. An Order of Council was made at Kensington Palace on 24 June approving an issue of medals 'to perpetuate the Memory of the happy Conclusion of the Peace', and the Queen approved a design for it. The responsibility for preparing such medals rested on the Master of the Mint, under the direction of the Treasurer. Accordingly, a Royal Warrant was issued to Oxford on 14 July to enable him to instruct Newton to strike 812 medals in gold; the issue actually began two days later (*Cal. Treas. Books*, XXVII (Part II), 1713, pp. 285 and 360; Mint Papers, III, fo. 321). The issue was completed by 24 September. The weight of gold used by the Mint amounted to 591 ounces while the whole cost of the operation amounted to over £2754. Peers, the Court and the Diplomatic Corps received 250 of the medals, the remainder being conveyed to the Speaker for distribution to members of the House of Commons. The Peace Medal is shown on Plate I (facing p. 40).

1007 JONES TO COTES
11 JULY 1713
Extract from the original in Trinity College Library, Cambridge[1]

July 11th 1713

Dear Sr

'Tis impossible to represent to you, with what pleasure I receiv'd your inestimable Present of the Principia, and am much concern'd to find my self so deeply charg'd with Obligations to you; and such, I fear, as all my future endeavors will never be able to requite. This Edition is indeed exceeding beautifull, and interspers'd with great variety of admirable discoverys, so very natural to its great Author; but is much more so, from the additional advantage of your excellent Preface prefix'd; which I wish might be got publish'd in some of the foreign Journals; and since a better account of this Book cannot be given, I suppose it will not be difficult to get it done.

Now this great Task being well over, I hope you'l think of publishing your own Papers, & not let such valuable pieces lye by: . . .

Sr I am

your most obedient

humble Servt

W: JONES

NOTE

(1) R.16.38, no. 318, printed in Edleston, *Correspondence*, p. 225. Cotes had last written (Letter 995, vol. v) to indicate that the *Principia* would be finished about the beginning of June.

1008 J. BERNOULLI TO LEIBNIZ
18 JULY 1713

Extract from Gerhardt, *Leibniz: Mathematische Schriften*, III/2, pp. 916–17.
Reply to Letter 1005; for the answer see Letter 1010

Qualia si tentaminis loco Anglis proponerentur, foret meo judicio brevissima via ad os obturandum illis, si nimirum infirmitatem suam et calculi sui, cujus tantopere jactant antiquitatem, insufficientiam proderent, atque sic inviti palmam cedere cogerentur, quamobrem problemata quædam excogitare deberes hac sola methodo solubilia, ut inde descerent esse quædam saltem, quæ nos suo calculo fluxionum non debemus: sicuti Cheynaeus [1] quondam inepte jactavit, nihil nempe intra hos 20 vel 30 annos prodiisse in lucem quæ non sint iteratæ repetitiones vel ad summum levia tantum corollaria eorum, quæ Newtonus jam pridem invenerit, quasi nobis nihil amplius relictum fuisset, vel nullius esset pretii, quod subinde a nobis publicatum extat, et cujus in Newtonianis ne vestigium quidem videre est: qualia sunt quæ de Catenariis, Velariis, Isochronis Paracentricis, Brachystochronis, de novis proprietatibus Cycloidis, de ejus segmentis innumeris quadrabilibus, de Calculo exponentialium seu percurrentium eosque differentiandi modo, de Coëvolutarum dimensione, de Motu tractorio, de reptorio, de Curvarum reductione ad circulares, de earum transformatione, et de innumeris aliis, quæ Angli pro parte tentarunt, sed omni suo calculo fluxionum adjuti irresoluta reliquerunt, quod vel ex solo problemate Catenariæ et Curvarum transformandarum patet, cui pertinaciter et longo tempore insudantes, aliud nihil quam turpes paralogismos produxerunt. [2] Reliqua vero nostra si spernere affectant, qui fit quod ea omnia corradere et in suam linguam convertere non dedignentur, exemplo illius Angli nobilis Des-Hayes, [3] cujus extat liber in folio, anglice conscriptus (nescio an videris) continens Marchionis Hospitalii *Analysin* [4] et reliqua nostra hinc inde ex Actis Lipsiensibus aliisque Diariis compilata, et omnia verbotenus translata. Præterea nihil fere, vel parum admodum, de Newtonianis suisve aliis, editor admiscuit, licet prodigus admodum sit in laudes Newtoni, nostri vero parcus laudator, nec nisi in præfatione, et quidem ita, ut qui nostra non cognoverit, non facile judicaturus sit, illa quæ tractat inventa ad nos magis pertinere, quam ad alios obscuri nominis, quorum integram phalangem nobiscum memorat, et tantam multitudinem obtrudit, ut, quemadmodum persuadere conatur, plagii suspicionem a se moveat. Sed contrarium facit manifeste, dum veros Auctores et Inventores sub nube reliquorum, a quibus nihil mutuatus est, abscondit, et ita quod unicuique debet, subdole dissimulat. Num quid iniquius censes? Sed hæc est horum hominum indoles, ut quod a peregrinis proficiscitur, eo quidem utantur et tacite laudent, sed publice vel spernant vel supprimant vel ad suos deferant.

Cæterum non denegavi Tibi gloriam inventionis quadraturæ circuli per seriem $1-\frac{1}{3}+\frac{1}{5}-\frac{1}{7}$ etc.[5] Dixi tantum ex literis Gregorii (si non sint suppositæ) patere, Gregorium fuisse primum hujus seriei inventorem: sed ipse videbis, quid hac de re sit credendum, ubi ad manus habueris Libellum Epistolarum. Potest esse, ut Gregorius invenerit prius, quod Tu postea etiam Tuo marte invenisti, et inscius quod ante Te jam fuerit ab alio inventum. Interim, si per literas probare potes, ipsum Newtonum Te primum credidisse Inventorem et inventum laudasse, hoc in rem Tuam apprime faciet, simulat enim Newtonus, in aliquibus literis in Libello isto publicatis, se non adeo magnifacere hanc seriem ob segnem admodum advergentiam, ut et sibi dudum innotuisse, quod primo inventa fuerit a Gregorio.

Translation

[*Bernoulli reassures Leibniz, telling him that Varignon is offended by the English procedures, and that Nikolaus has let slip nothing of concern to Leibniz. He then refers generally to some recent advances of his own.*] If such things were proposed to the English by way of a trial, it would be in my opinion the quickest way of stopping their mouths, particularly if they should reveal their extraordinary feebleness and the inadequacy of their calculus of whose antiquity they boast so greatly, and if they were reluctantly forced to concede defeat, for which reason you ought to work out some problems resolvable only by this [new] method [of yours], so that thence they might at least learn that there are some things for which we are not indebted to their calculus of fluxions; as formerly Cheyne [1] was foolishly boasting that nothing had been published during the last 20 or 30 years which is not a repetition or at most a trivial corollary of what Newton had discovered long ago, as though there were nothing more left to us or nothing were of value in what has been published by us in that time, in things of which no trace at all is to be found among the Newtonians; examples are the catenaries, stretched surfaces, paracentric isochrones, brachystochrones, new properties of the cycloid and innumerable quadratures of its segments, the calculus of exponentials or transcendents and the way of differentiating them, the quantification of co-evolutes, drawing and creeping motion, the reduction of curves to circular ones, their transformation, and numberless other things which the English did their best to tackle but for all the aid of their fluxional calculus leave unresolved, as is obvious from the single problem of the catenary and of the curves to be transformed over which they sweated long and hard to produce nothing but disgraceful errors. [2] If they affect to despise the rest of us why is it that they do not scruple to scrape together everything of that kind and translate it into their own language, as for example the noble Englishman 'Des-Hayes' [3] whose folio we have (I do not know whether you have seen it), written in English, containing the *Analysis* of the Marquis de l'Hospital [4] and other things of ours gathered here and there from the *Acta* [*Eruditorum*] of Leipzig and other journals, and all of it literally translated. Moreover, the editor has intermingled almost nothing or very little of the Newtonians, or others of his [own people] although he is full of praises for Newton and sparing in praise of us, except in the preface, and even then in such a way that anyone not acquainted with us would not easily judge that the matters he treats belong rather to us than to others of little repute of whom he records a whole host along with ourselves, introducing so great a multitude [of names], it seems, as though seeking to persuade [his readers] that no suspicion of plagiary attaches to himself. But the contrary is obviously true, while he conceals the true authors and discoverers beneath a cloud of others from whom he has taken nothing, and so cunningly disguises the person to whom he owes everything. Now can you think of anything worse? But this is the nature of these people, that they employ and by implication applaud the achievements of foreigners, but in public they despise or suppress them, or claim them for themselves.

For the rest, I have not denied you the glory of the discovery of the series $1 - \frac{1}{3} + \frac{1}{5} - \frac{1}{7}$ etc. for the quadrature of the circle. [5] I have only said that it is clear from Gregory's letter (if this has not been faked), that Gregory was the first to discover this series; but you yourself will see what is to believed in this context when you have the *Commercium Epistolicum* in your hands. It may be that Gregory first discovered what you afterwards also found out by your own efforts, [you yourself] being ignorant of what another had discovered before you. Meanwhile, if you can prove from the correspondence that Newton himself believed you to be the first to discover it and praised your discovery this serves your purpose exactly, for Newton pretends in some letters published in that book that he did not much admire this series because of its gradual convergence, as also because it was long known to him because it was first discovered by Gregory.

NOTES

(1) George Cheyne (1671–1743) studied medicine at Edinburgh with Pitcairne, and in 1702 set up practice in London. He became an active member of the Royal Society, and an enthusiastic exponent of British Newtonianism. His interests extended to mathematics, and in 1703 he published *Fluxionum methodus inversa* which was severely criticized by David Gregory and others, including De Moivre, who wrote to Bernoulli of his objections in a letter of 6 July 1708 (see Wollenschläger, *De Moivre*, p. 256). Gregory claimed that it was the publication of Cheyne's work that provoked Newton into printing his *De quadratura curvarum* with the 1704 edition of his *Opticks*. In later life Cheyne returned to the study of medicine.

(2) The problem of defining the catenary ('Invenire, quam curvam referat funis laxus & inter duo puncta fixa libere suspensus') was proposed by Jakob Bernoulli as a public challenge in the *Acta Eruditorum* for May 1690; see C. Truesdell, *The Rational Mechanics of Flexible or Elastic Bodies*, 1638–78 (in *Leonhardi Euleri Opera Omnia*, Series 2, **11** (2) (Zurich, 1960), 1–435). Newton made no attempt to respond to the challenge; see Whiteside, *Mathematical Papers*, v, pp. 520–3. The name 'catenary' for this curve was first employed by Huygens (see *Oeuvres Complètes*, IX, p. 537); it was Huygens also who first, in 1646(!), solved the related problem of the weightless cord loaded with equal weights in such a way that the verticals through

the weights are equidistant. In this case the angles of the chord lie on a parabola; this theorem was first published by Ignace-Gaston Pardies in 1673, who derived it independently. Independent correct solutions of Bernoulli's problem were published by Johann Bernoulli, Leibniz and Huygens in the *Acta Eruditorum* for June 1691. David Gregory printed an elaborate purported demonstration of their solutions in *Phil. Trans.* **19**, no. 231 (August 1697), 637–52 (and *Acta Eruditorum* for July 1698, pp. 305–21). Unfortunately Gregory's paper was founded on a mistake as Leibniz pointed out in the same journal for February 1699 (pp. 87–91) ascribing it to the incompetence of fluxions. Gregory's attempted rejoinder was 'pitiful' (Truesdell, *op. cit.*, p. 85). Newton took up the problem of the catenary near the end of his life: see Whiteside, *loc. cit.*

(3) Charles Hayes (1678–1760), of whom little is known, published in 1704 *A treatise on fluxions, or, an Introduction to Mathematical Philosophy*. This is the first English book to employ the Leibnizian notation and is 'a remarkable work for its time, which ought to be better known; it is much more up-to-date than Newton's *De quadratura curvarum* published in the same year.' (D. T. Whiteside, personal communication). However, it is heavily derivative from articles in the *Acta Eruditorum* and the Marquis de l'Hospital's *Analyse* (see the next note) with frequent citations from 'The excellent Geometer M. *Jo. Bernoulli*' etc. There is a long list of the mathematicians who have contributed to the new calculus in the Preface.

(4) Guillaume François Antoine de l'Hospital (1661–1704) was a pupil of Johann Bernoulli; his *Analyse des infiniment petits* (Paris, 1696), the first textbook of the calculus, was based on Johann Bernoulli's lectures of 1691 without public acknowledgement, but the style and exposition were L'Hospital's own and he was capable of improving on his master as in the treatment of diacaustics (see Whiteside, *Mathematical Papers*, III, p. 491, note). His *Analyse* was rightly esteemed.

(5) See Letter 1004, note (5), p. 6 and Letter 1075 (for the discussion between Bernoulli and Leibniz). The series was included, as Gregory's, among those mentioned in a letter sent by Collins to Oldenburg on 10 April 1675 (Royal Society MS. 81, no. 23, fos. 1–9) and transmitted, in Latin, by Oldenburg to Leibniz on 15 April (cf. vol. I, p. 341, where this date mistakenly appears as 12 April).

1009   THE *CHARTA VOLANS*
18 JULY 1713
From the printed paper in King's College, Cambridge[1]

29. *Julii* 1713 [N.S.]

L....us nunc Viennæ [2] Austriæ agens ob distantiam locorum nondum vidit libellum in Anglia nuper editum, quo N....o primam inventionem Calculi differentialis vindicare quidam conantur. Ne tamen commentum mora invalescat, quam primum retundi debere visum est. Equidem negare non poterunt novam hanc Analyticam Artem primum a L....o fuisse editam (cum diu satis pressisset) & publice cum amicis excultam; & post complures demum annos a N....o aliis notis & nominibus, quendam quem vocat Calculum Fluxionum, Differentiali similem, fuisse productum; qui tamen

15

tunc nihil contra L....um movere ausus est. Nec apparet quibus argumentis nunc velint L....um hæc a N....o didicisse, qui nihil tale unquam cuiquam quod constet communicavit, antequam ederet. L....us tamen ex suo candore alios æstimans, libenter fidem habuit Viro talia ex proprio ingenio sibi fluxisse dictitanti; atque ideo scripsit N....um aliquid calculo differentiali simile habuisse videri. Sed cum postremo intelligeret, facilitatem suam contra se verti, & quosdam in Anglia præpostero gentis studio eousque progressos, ut non N....um in communionem inventi vocare, sed se excludere non sine vituperii nota vellent, & N....um ipsum (quod vix credibile erat) illaudabili laudis amore contra conscientiæ dictamen tandem figmento favere; re attentius considerata, quam alias præoccupato in N....i favorem animo examinaturus non fuerat, ex hoc ipso processu a candore alieno suspicari coepit, Calculum Fluxionum ad imitationem Calculi Differentialis formatum fuisse. Sed cum ipse per occupationes diversas rem nunc discutere non satis posset, ad judicium primarii Mathematici,[3] & harum rerum peritissimi, & a partium studio alieni recurrendum sibi putavit. Is vero omnibus excussis ita pronuntiavit literis 7 Junii 1713 [N.S.] datis:[4]

Videtur N....us occasionem nactus serierum opus multum promovisse per Extractiones Radicum, quas primus in usum adhibuit, & quidem in iis excolendis ut verisimile est ab initio omne suum studium posuit,[5] nec credo tunc temporis vel somniavit adhuc de Calculo suo fluxionum & fluentium, vel de reductione ejus ad generales operationes Analyticas ad instar Algorithmi vel Regularum Arithmeticarum aut Algebraicarum.[5] Ejusque meæ coniecturæ [primum][6] validissimum indicium est, quod de literis x vel y[5] punctatis, uno, duobus, tribus, &c. punctis superpositis, quas pro dx, ddx, d³x; dy, ddy, &c. nunc adhibet, in omnibus istis Epistolis [Commercii Epistolici Collinsiani, unde argumenta ducere volunt][6] nec volam, nec vestigium invenias. Imo ne quidem in principiis Naturæ Mathematicis N....i, ubi calculo suo fluxionum utendi tam frequentem habuisset occasionem, eius vel verbulo fit mentio, aut notam hujusmodi unicam cernere licet, sed omnia fere per lineas figurarum sine certa Analysi ibi peraguntur more non ipsi tantum, sed & Hugenio, imo jam antea [in nonnullis][6] dudum Torricellio, Robervallio, Cavallerio, aliis, usitato. Prima vice hæ literæ punctatæ comparuerunt in tertio Volumine Operum Wallisii, multis annis postquam Calculus differentialis iam ubique locorum invaluisset. Alterum indicium, quo coniicere licet Calculum fluxionum non fuisse natum ante Calculum differentialem, hoc est, quod veram rationem fluxiones fluxionum capiendi, hoc est differentiandi differentialia,[5] N....us nondum cognitam habuerit,[7] quod patet ex ipsis Principiis Phil. Math. ubi non tantum incrementum constans ipsius x, quod nunc notaret per x punctatum uno puncto, designat per o [more vulgari, qui calculi differentialis commoda destruit] sed etiam regulam circa gradus ulteriores falsam dedit [quemadmodum ab eminente quodam Mathematico dudum notatum est][8]......Saltem apparet, N....o rectam Methodum differentiandi differentialia non innotuisse longo tempore, postquam aliis fuisset familiaris &c. Hæc ille.

Ex his intelligitur N....um, cum non contentus laude promotæ *synthetice* vel linealiter *per infinite parva*, vel (ut olim minus recte vocabant,) *indivisibilia Geometriæ*; etiam inventi *Analytici* seu *calculi differentialis* a L...o in Numeris primum reperti, & (excogitata *Analysi infinitesimalium*) ad Geometriam translati, decus alteri debitum affectavit, adulatoribus rerum anteriorum imperitis nimis obsecutum fuisse, & pro gloria, cujus partem immeritam aliena humanitate obtinuerat, dum totam appetit, notam animi parum æqui sincerique meruisse: de quo etiam Hookium circa Hypothesin planetariam, & Flamstedium circa usum observationum, questos ajunt.[9]

Certe aut miram ejus oblivionem esse oportet, aut magnam contra conscientiæ testimonium iniquitatem, si accusationem (ut ex indulgentia colligas) probat, qua quidam ejus asseclæ etiam seriem, quæ arcus circularis magnitudinem ex tangente exhibet, a Gregorio hausisse L....um volunt.[10] Tale quiddam Gregorium habuisse, ipsi Angli & Scoti, Wallisius, Hookius, Newtonus, & junior Gregogorius [*sic*], prioris credo ex fratre nepos, ultra triginta sex annos ignorarunt, & L....i esse inventum agnoverunt. Modum, quo L....us ad seriei Nicolai Mercatoris (primi talium inventoris) imitationem invenit seriem suam, ipse statim Hugenio B. Lutetiæ agenti communicavit, qui & per Epistolam laudavit.[11] Eundem sibi communicatum laudavit ipse mox N....us, fassusque est in litteris hanc novam esse Methodum pro Seriebus, ab aliis quod sciret nondum usurpatam.[12] Methodum deinde generalem series inveniendi, pro curvarum etiam transcendentium ordinatis in Actis Lipsiensibus editam, non per Extractiones dedit, quibus N....us usus est, sed ex ipso fundamento profundiore Calculi differentialis L....us deduxit. Per hunc enim calculum etiam res serierum ad majorem perfectionem deducta est. Ut taceam *Calculi exponentialis*, qui transcendentis perfectissimus est gradus, quem L....us primus exercuit, Johannes vero Bernoullius proprio marte etiam assecutus est, nullam N....o aut ejus discipulis notitiam fuisse: & horum aliquos, cum etiam ad Calculum differentialem accedere vellent, lapsus subinde admisisse, quibus eum parum sibi intellectum fuisse prodiderunt, quemadmodum ex junioris Gregorii circa Catenariam paralogismo patet. Cæterum dubium non est, multos in Anglia præclaros viros hanc N....-ianorum Asseclarum vanitatem & iniquitatem improbaturos esse; nec vitium paucorum genti imputari debet.

Translation

29 *July* 1713 [N.S.]

As Leibniz is now living at Vienna [2] in Austria, he has not yet, because of the distance between the places, seen the little volume lately published in England in which certain people endeavour to claim the first discovery of the differential calculus for Newton. It seems advisable, lest comments should be weakened by delay, to make a retort as soon as possible. They have not in fact been able to deny that this new analytical art was first published by Leibniz (since it was printed a pretty long time ago) and developed by him and his friends before the public, and that then after many years there was produced by Newton something that he calls the calculus of fluxions similar to the differential calculus but with other notations and terminology; yet Newton then did not dare to urge anything against Leibniz. Nor does it appear by what arguments they now suppose Leibniz to have learned it from Newton who (as it appears) communicated nothing at all of this to anyone, before he went into print. Leibniz on the other hand, judging others according to his own honest nature, readily believed the man [Newton] when he declared that such things had come to him from his own ingenuity, and so he wrote that it appeared that Newton possessed something similar to the differential calculus. But when he learned later that his own simplicity had been turned against himself and that certain persons in England with an unnatural xenophobia had gone so far that they meant not merely to embrace Newton among the discoverers but to exclude himself from their number (not without some abusive remarks) and in the end, though this was scarcely credible, by a fiction to give their favour to Newton against the dictates of conscience, because of an unworthy desire to please, [Leibniz] considered the question more carefully, which otherwise he would not have examined because he was prejudiced in Newton's favour, and began to suspect from that very procedure [of the English] which was so remote from fair-dealing that the calculus of fluxions had been developed in imitation of the differential calculus. But as his business prevented him from looking sufficiently deeply into the matter at the time he decided to resort to the judgement of a leading mathematician [3] most skilled in these matters and free from bias. After considering everything the latter declared himself as follows in a letter dated 7 June 1713 [N.S.] [4]:

'It seems that . . . [*for the translation see Letter 1004*] . . . as is obvious from his *Philosophiæ Principia Mathematica* where he not only denotes by o the regular increment of x (which he now denotes by $\dot{x}$) [following the usual custom which destroys the convenience of the differential calculus] but even states the rule concerning the ultimate degrees falsely [as a certain eminent mathematician has recently observed] [8] . . . At any rate it is clear that the true way of differentiating differentials was not known to Newton until long after it was familiar to others.' Thus [the writer].

From these words it will be gathered that when Newton took to himself the honour due to another of the analytical discovery or differential calculus first discovered by Leibniz in numbers and then transferred (after having contrived the analysis of infinitesimals) to Geometry, because Newton was not content with the fame of advancing [geometry] synthetically or directly by infinitely small quantities (or as they were formerly but less correctly called, the indivisibles of geometry), he was too much influenced by flatterers ignorant of the earlier course of events and by a desire for renown; having undeservedly obtained a partial share in this, through the kindness of a stranger, he longed to have deserved the whole—a sign of a mind neither fair nor honest. Of this Hooke too has complained, in relation to the hypothesis of the planets, and Flamsteed because of the use of [his] observations.[9]

Surely either his forgetfulness must seem miraculous or his iniquity against the testimony of his conscience enormous, if he approves the accusation (as you may suppose from his complacency) by which some of his partisans seek to claim that Leibniz also derived from Gregory the series by which the magnitude of an arc of circle is obtained from the tangent.[10] For that Gregory had such a thing the English and Scots themselves—Wallis, Hooke, Newton, and the younger Gregory (I believe the former's brother's son)—were ignorant of for more than thirty-six years, and they have acknowledged that Leibniz is its discoverer. Leibniz himself at once communicated the way in which, imitating the series of Nicolaus Mercator (the first to discover such things), he had found his series, to Huygens who was then resident at Paris, who wrote a letter praising it.[11] Soon Newton himself praised the series when it was made known to him, and he admitted in a letter that (so far as he knew) this new method for series was not yet employed by others.[12] Then the general method for finding series even for the ordinates of transcendental curves was published by Leibniz in the *Acta* [*Eruditorum*] of Leipzig, but not employing the extraction [of roots] as Newton has done, for Leibniz deduced it from the deep foundations of the differential Calculus. The business of series has been taken to an even greater perfection by means of this calculus. I shall say nothing of the exponential calculus which is the most perfect degree of the transcendent, which Leibniz employed for the first time and which Johann Bernoulli has also pursued on his own initiative, [and of which] nothing was known to Newton or his disciples; and some of these when they have tried to attain to the differential calculus also have at once made mistakes by which they revealed that they understood little of it, as may be seen in the younger Gregory's error concerning the catenary. Moreover, there is no doubt that in England many distinguished persons deplore this vanity and injustice among Newton's disciples; and the bad conduct of the few should not be imputed to the whole nation.

NOTES

(1) Keynes Ec 7.2.27; an example of the complete paper containing two printings of the text of the fly-sheet ('Charta Volans') is in U.L.C. Add. 3968(34). The paper was so imposed that two texts were obtained by dividing it through the middle. It was published anonymously but was clearly the work of Leibniz and was acknowledged as such in the German translations (see below). Leibniz had not yet seen the *Commercium Epistolicum*, but Johann Bernoulli had, and on 27 May 1713 sent Leibniz a brief account of its nature (see Letter 1004) and included his own comments, which Leibniz prints here in slightly modified form. Bernoulli had asked Leibniz not to involve him personally in the controversy, a request which Leibniz at first

complied with in the *Charta Volans*, by printing Bernoulli's letter anonymously, but later ignored (see Letters 1172 and 1203 and below). Leibniz had also received details of the *Commercium Epistolicum* from Christian Wolf (see Letter 1107, note (1)) with a short extract from it, in a letter dated 1 July 1713 N.S. (see Gerhardt, *Briefwechsel zwischen Leibniz und Wolf*, pp. 149–52); in subsequent letters between the two men the controversy over the calculus was a constant subject of discussion, and Wolf was later to act as an intermediary between Keill and Leibniz.

Further, it was Wolf who undertook to print and circulate the *Charta Volans* for Leibniz (see Gerhardt, *ibid.*, pp. 157–8 and Letter 1075) and after Leibniz had read Keill's 'Lettre de Londres' published in the *Journal Literaire de la Haye* for May and June 1713 he wrote Wolf a further defence, in French (see Letter 1018), which Wolf transmitted to the *Journal Literaire* together with a French version of the *Charta Volans* (see Ravier, p. 151, no. 313). It was this now double apology for Leibniz in the *Journal Literaire* which first came to Keill's attention (see Letters 1039 and 1039*a*) though Newton, seemingly, had already received copies of the Latin fly-sheet from Germany but had decided to ignore it (see Letter 1053). When he found the same claims openly made in the *Journal Literaire* he felt that a reply must be made.

The *Charta Volans* was also published, in Latin, in the *Deutsche Acta Eruditorum oder Geschichte der Gelehrten*, **19** (Leipzig, 1713), pp. 591–4, where it was preceded (pp. 586–91) by a German translation of the Royal Society report which had appeared in the *Journal Literaire* for May and June 1713, pp. 210–13 (see above). Leibniz's *Remarques* (Letter 1018) also appeared in German in the same volume of the *Deutsche Acta Eruditorum*, pp. 915–18.

The French translation of the excerpt from Bernoulli's letter in the *Charta Volans* appeared for a second time in the *Nouvelles Literaires* for 28 December 1715 [N.S.], where it was explicitly attributed to Johann Bernoulli (see Letter 1172).

(2) Although Leibniz remained all his life, from 1676, in the service of the Duke (from 1692, Elector) of Brunswick, he had spent much time in Berlin during the years 1700 to 1710, where the Electress of Brandenburg was of the Brunswick house and Leibniz became President of the Academy. And from 1712 to 1714 he lived in Vienna where he was granted the title of Freiherr and the office of Privy Councillor.

(3) Johann Bernoulli; see note (1) above. It has been traditional to accuse Bernoulli of duplicity in thus taking Leibniz's part, after superficially cordial relations with Newton in the autumn of 1712 when Nikolaus was in London. Yet Bernoulli's long friendly correspondence with Leibniz and his use of the calculus must have been known to Newton and his friends, not least because some of them (De Moivre and Burnet, for example) were in direct correspondence with him—and Bernoulli made no secret of his feelings about the controversy. It was *because* Newton knew of Bernoulli's allegiance to Leibniz that the business of Proposition 10, Book II in the *Principia* worried him so much. That Bernoulli should have written such a letter to Leibniz, in the then context of the dispute, can have suprised no one who was at all familiar with mathematical affairs.

(4) For notes on this passage, and its translation, see Letter 1004, note (6) etc., p. 6; here we point out only differences between the original letter written by Bernoulli, and the version printed in the flying paper.

(5) C.f. Letter 1004; the omission is trivial.

(6) The square brackets are in the original; the words enclosed, absent in Bernoulli's original letter, were presumably interpolated by Leibniz.

(7) The wording of the remainder of the passage differs considerably from Bernoulli's version. It is effectively a précis of the mathematical section of Bernoulli's letter.

(8) The eminent mathematician is, of course, Bernoulli again; see Letter 1004, note (10), p. 6. Leibniz thus effectively makes Bernoulli refer to himself in these terms of praise.

(9) Presumably this back-hander implies some private knowledge on Leibniz's part. For Hooke and the *Principia* see A. Rupert Hall, in *Isis*, **42** (1951), 219 and vol. II, pp. 431–47. The ill-feeling between Flamsteed and Newton over the latter's use of the former's observations became acute in 1698; see vol. IV, especially Letters 599, 600 and 601.

(10) Leibniz had written to Oldenburg about his rational quadrature of the circle $\left(\frac{\pi}{4} = 1 - \frac{1}{3} + \frac{1}{5} - \frac{1}{7} + \frac{1}{9}\ldots\text{etc.}\right)$ on 5 July and 6 October 1674 (for extracts of his letters see vol. I), making the claim: 'hoc in Circulo efficere hactenus potuit nemo'. An extract of the latter letter was printed in the *Commercium Epistolicum*, with the assertions that Newton's work on series was well known to his friends through Collins, and that Gregory too had made his results widely known (see his letters to Collins of 23 November 1670, 19 December 1670, and 15 February 1671; see vol. I, pp. 45–55). Both Newton and Gregory were indeed ahead of Leibniz in the use of series, though Leibniz knew nothing of their work.

(11) Huygens' delighted comments on the rational quadrature ('fort belle et fort heureuse . . . une voye nouvelle qui semble donner quelque esperance de parvenir a sa veritable solution') of 28 October 1674, printed in *Oeuvres Complètes*, VII, pp. 393–5. Later Huygens became very dubious of Leibniz's mathematical aims and methods. (The 'B.' in this sentence is presumably a misprint for 'P.', standing for 'Parisiorum'.)

(12) Oldenburg replied somewhat coolly to Leibniz's claims on 8 December 1674 (Letter 130, vol. I); Newton wrote first to Leibniz (via Oldenburg) on 13 June 1676 (Letter 165, vol. II). Leibniz's circle series (note (10) above) was communicated by him to Oldenburg on 17 August (Letter 172, vol. II) and discussed by Newton in the long *Epistola Posterior* of 24 October 1676 (Letter 188, vol. II), which opens with a polite statement of admiration for Leibniz's discoveries; however, there is no admission in either of the two first letters that Leibniz's method of series was previously unknown. On the contrary, Newton gave an explanation of his own work in this branch of mathematics and alluded to that of Gregory; he showed how Leibniz's series was obtainable from a general analysis of quadratures by series, and discussed the advantage of computing the sum of one series rather than another.

1010 LEIBNIZ TO J. BERNOULLI
8 AUGUST 1713

Extract from Gerhardt, *Leibniz: Mathematische Schriften*, III/2, p. 919.
Reply to Letter 1008; for the answer see Letter 1014

Certissimum est, omnes in Anglia ad novissima usque tempora ignorasse, Jacobum Gregorium Tetragonismum meum etiam habuisse; certe ipse ejus ex Fratre Nepos, David Gregorius, hoc ignoravit, et inventum in Libro suo *De Quadraturis* mihi adscripsit.[1] Ubi ad meas schedas reversus fuero, inspiciam veteres literas, quæ adhuc extant, in quibus, ni fallor, manifesta erunt vestigia, Newtonum Analysin nostram parum exploratam habuisse tunc, cum aliquod ei mecum per Oldenburgium commercium erat.[2] Ego, neglecto Keilio, aliisque ejusmodi Newtoni adulatoribus, hominibus obscuris, cogar Newtono ipsi exprobrare animum parum sincerum et acta testimonio conscientiæ contraria: nam ipse ignorare non potest, Analysin istam infinitesimalem ab ipso ad me proficisci non potuisse, et tamen nugatoribus imperitis talia asserentibus favet et indulget. Ita dum nimia, cum alterius injuria, affectat, cogit me ei jam negare, quæ, ex meo candore alios æstimans, nimis liberaliter concesseram, credideramque asserenti, in Calculum nostrum de suo eum incidisse, quod verum non esse re diligentius excussa satis apparet.

Intelligo in Anglis esse Viros Eruditos, qui processum Viris gravibus et bonis indignum non probant. Et quod conjicis, non ab omni veri specie abest, eos, qui parum Domui Hanoveranæ favent, etiam me lacerare voluisse; nam amicus Anglus ad me scribit, videri aliquibus non tam ut Mathematicos et Societatis Regiæ socios in socium, sed ut Toryos in Whigium quosdam egisse.[3] Sed ego exigua, credo, scheda efficiam, ut pœniteat eos nugarum. Utar inter alia argumentis Tuis, sed a Te nominando abstinebo.[4] Quod superest vale et fave etc.

Translation

It is most certain that until very recently everyone in England was ignorant of the fact that James Gregory also possessed my tetragonismus; it is certain that his own brother's son, David Gregory, did not know it and in his book *De quadraturis* he attributed it to me.[1] When I am back with my papers again I will examine the old letters that still exist in which, if I am not wrong, there is unmistakeable evidence that Newton had but little explored our analysis at that time, when we had some correspondence through Oldenburg.[2] Paying no attention to Keill and others of Newton's toadies of the same sort, obscure men, I am compelled to blame Newton himself for insincerity and acts contrary to the dictates of conscience, for he himself cannot but know that that infinitesimal analysis could not have been furnished by him to me, and yet he caresses and indulges the ignorant triflers who assert such things. Thus so long as he protests too much, with complaints against someone else, he forces me to deny him that which (judging others in my generous fashion) I would with too great liberality have conceded: I would have believed his claim that he had himself stumbled upon our calculus, which, when the matter is more diligently sifted out, appears clearly to be untrue.

I understand that in England there are learned men who do not approve this procedure which is an affront to grave and good men. And your guess just about hits the nail on the head, that is, that those who have little love for the House of Hanover have also meant to wound me; for an English friend writes to me that it seems to some that certain persons have acted not as mathematicians and Fellows of the Royal Society against a Fellow, but as Tories against a Whig.[3] But I shall, I believe, get out such a little paper as will make them smart for their nonsense. Among other things I use your arguments, but refrain from giving your name.[4] For the rest, farewell and flourish. etc.

NOTES

(1) By his tetragonismus, Leibniz means the quadrature of the quadrant of a circle of unit radius by the series $1 - \frac{1}{3} + \frac{1}{5} - \frac{1}{7} \ldots$; by David Gregory's book *De quadraturis* he means the *Exercitatio Geometrica de Dimensione Figurarum*. See Letter 1008, note (5), p. 15.

(2) See Letter 165, vol. II, the *Epistola Prior* and Letter 188, vol. II, the *Epistola Posterior*. In these letters to Oldenburg for Leibniz, Newton made no direct use of fluxional methods, except in indecipherable anagram form; see also Whiteside, *Mathematical Papers*, IV, pp. 666–74. Hence Leibniz's statement, in terms of the information available to him, is justified.

(3) If Leibniz was attributing Jacobitism to Newton and his friends, he could hardly have been further from the mark. Newton had been an opponent of James II, and his whole political life had been in the Whig interest. His great patron, Halifax, was a Whig, a Hanoverian, almost 'Prime Minister' after the Succession. Although his official business with the Tories was conducted on perfectly cordial terms, there is no evidence at all to suppose that Oxford or Bolingbroke won the slightest personal allegiance in politics from Newton. Moreover, as Leibniz (if he had thought seriously) must have known the majority of English Tories were Hanoverians, if reluctant Hanoverians, and Bolingbroke's Jacobite Wing had little solid strength even in his own party.

(4) Leibniz refers here to the *Charta Volans* (Number 1009) and his anonymous publication of Bernoulli's letter there.

IOII T. HARLEY TO THE MINT
15 AUGUST 1713

From the original in the Public Record Office.[1]
For the answer see Letters 1013 and 1035

Gentlemen

My Lord Treasurer is pleased to direct You to consider the Peticon which I send you inclosed of Charles Tunnah and Wm Dale proposing the Coyning of Farthings and half pence of a new mettle that toucheth like Gold[2] & to make your report to his Lordp thereupon when you are prepared to lay your thoughts before his Lordp about Coyning Farthings and half pence in the Method discoursed of when you last attended here. I send You also by his Lordps Order a Memo[ria]l & Certificate in behalfe of Catesby Odam who prays the Office of Assay Master for Your Consideracon[3] I am &c. 15th August 1713.

T. HARLEY

NOTES

(1) T/27, 21, p. 15.

(2) The memorial from Tunnah and Dale follows the reply from the Mint at T/1, 172, no. 25A. They claim that they will coin their new alloy (presumably a brass) at the rate of

2s. 8d. a pound although it sells at 10s. a pound for making sword hilts etc., and wished to coin one hundred tons per annum for ten years. The metal could not be counterfeited and was readily malleable.

(3) See Letter 991 etc., vol. v, and Letter 1038.

1012 THE ROYAL SOCIETY TO THE OFFICERS OF THE ORDNANCE
24 AUGUST 1713
From the copy in the Letter book of the Royal Society of London[1]

Gentlemen

Her Majesty having last winter was two Years constituted the Royal Society Visiters of the Royal Observatory at Greenwich, was pleased a few days since, in pursuance to that her Order voluntarily with her own mouth to give fresh Commands to our President that he and the rest of the Gentlemen of the Society should take care of Mr. Flamsteeds Observatory.[2] Whereupon the society sent thither some of their members to view the same and they have reported the state thereof in respect of the Instruments and what repaires they need, and that some of them are not the Queens nor capable of being made sufficiently fit for use.[3] The Observatory being supported and repaired by your Office, we the President and Council of the society, take the liberty to lay these things before you, being ready on our part to do what in us lies for the putting her Majesty's Commands in execution in the best manner, and this tending to make the Observatory more usefull and Creditable, we pray the favour of your Answer and remain Gentlemen

Your humble servants &c [4]

NOTES

(1) xv fo. 26. The date is given in the heading to the letter. It was drafted by Newton himself (U.L.C., Add. 4006, fo. 33)—the draft shows insignificant variants and is headed 'Crane Court in Fleetstreet 24 Aug. 1713'. The letter was printed in Baily, *Flamsteed*, p. 304.

(2) This, Newton's second version of the sentence, appears on the verso of the draft. See Letter 814, vol. v.

(3) John Machin informed Flamsteed by a letter dated 30 July 1713 that the Visitors intended to inspect his house and observatory on the Saturday following, 1 August (Baily, *Flamsteed*, p. 303). Flamsteed reported on these proceedings to Abraham Sharp, see Letters 1020 and 1044.

(4) In their reply of 4 September 1713 the Ordnance Officers expressed their failure to understand the terms of this letter, disclaiming any responsibility for the instruments. Compare Letter 1044, note (2), p. 70.

1013 THE MINT TO OXFORD
26 AUGUST 1713
From the original in the Public Record Office.[1]
Reply to Letter 1011

To the most Honorable the Earle of Oxford and Earle Mortimer
Lord High Treasurer of Great Britain

May it please your Lordship

In obedience to your Lordship's order we have perused the Memorial and
Certificate in favour of Mr Catesby Oadham against Mr. Brattell [2] communicated to us the 15th instant, and having fully considered the same we find
no reason to alter our opinion concerning either of them certified in our
former Reports to your Lordship,

Which is most humbly submitted to your Lordships great Wisdom

Mint Office the 26th. August 1713

C. PEYTON

IS. NEWTON

P. PHELIPPS

NOTES

(1) T/1, 163 no. 50. There are several longer holograph drafts by Newton preparatory to
this brief reply in the Mint Papers, I, fos. 90–100.

(2) Charles Brattle (see vol. v) was acting Assay-master in succession to his dead brother;
Catesby Oadham the defeated non-Mint candidate for the office. Compare Letter 1038.

1014 J. BERNOULLI TO LEIBNIZ
29 AUGUST 1713
Extract from Gerhardt, *Leibniz: Mathematische Schriften*, III/2, pp. 921–2.
Reply to Letter 1010; for the answer see Letter 1017

Pro voto meo facies, si in scheda, quam paras contra libellum Anglicanum,
argumentis a me suggestis ita utaris, ut Angli a me profecta non sentiant;
nisi forte quædam invenias, quæ publice extant in Annotationibus meis in
Actis Lipsiensibus nuper editis, de quibus aliquis Anglus, Discipulus antehac
meus, nunc Parisiis agens, sequentia mihi scribit:[1] 'La semaine passée',
inquit, 'il arriva un de mes amis d'Angleterre, qui m'informa qu'on n'avait
pas encore achevé d'imprimer la Nouvelle Edition des Principes de Monsieur
Newton; pour moy, je croy que le Chevalier tarde à dessein, pour voir, s'il peut,
les remarques que vous avez faites sur la première Edition de ce livre: ce qui

me confirme dans cette opinion, est que j'ay rencontré l'autre jour un Ecossois, qui se mêle un peu de Mathematiques, qui m'a informé qu'une faute, que vous aviez trouvée dans ce livre, faisait beaucoup de bruit en Angleterre; mais que les amis de Mr. Newton la faisaient passer pour une petite faute de calcul etc.' Qualem vero intelligat errorem, quem detexerim (detexi enim plures) non facile conjiciam, nisi sit ille, de quo Agnatus meus eum præmonuit, cum nuper in Anglia versaretur, antequam schediasma meum fuisset impressum et cujus erroris correcturam postea, singulari scheda, suo libro nondum edito inseruit Newtonus. Habet autem error iste, qui respicit determinationem resistentiæ medii pro data curva describenda a projectili, originem suam ex eo ipso, quod paralogizat Newtonus (ut ego ostendi) in differentiandis potestatibus ad ulteriores gradus, seu in capiendis fluxionum fluxionibus omnium ordinum, ope potestatis indefinitæ in seriem expansæ.

Utrum facile patiatur Varignonius, ut aliquid in Gallia fiat, de quo queri possis, in hac præsertim lite ab Anglis Tibi intentata, et quid ei hanc in rem perscripserim, intelliges ex sequentibus, quæ responsi loco nuperrime mihi reposuit. [2] 'Je suis,' inquit 'comme vous, fort mecontent de la mauvaise querelle que Monsieur Keill vient de susciter à Monsieur Leibnitz. Il me parait, comme à vous, que le *Commercium Epistolicum* prouve seulement que Monsieur Newton, au temps des lettres qui y sont rapportées, avoit connaissance des infiniment petits, mais il n'y parait pas qu'il en eût le calcul tel que Monsieur de Leibnitz l'a publié en 1684, et que Monsieur Newton l'a donné 3 ans après dans les pages 251, 252, 253 de ses Princip. Mathem. où il reconnait que ce Calcul luy avoit été communiqué 10 ans auparavant par Monsieur Leibnitz auquel temps il dit qu'il l'avait aussy, ainsi que la phrase renversée le prouve, sans dire à quel point il l'avoit. Avant vous, Monsieur de Leibnitz et feu Monsieur vôtre Frère, je ne sçay point, qu'on eût passé les premières differences, employées dans les pages précedentes de Monsieur Newton, qui n'en a fait mention que longtemps depuis dans son Traité *De Quadraturis*. Je suis, dis-je, très faché de voir Monsieur de Leibnitz forcé de se distraire de ses occupations si utiles au public, pour se defendre, d'un mauvais procès dont le public n'a que faire etc.'

Translation

You will suit my wishes if you will so employ the arguments suggested by myself in the paper you are preparing against the English that the English will not detect their origin with myself; unless perchance you come across those which have been published in my remarks lately printed in the *Acta* [*Eruditorum*] of Leipzig, concerning which a certain Englishman,[1] formerly a pupil of mine but now living at Paris, wrote to me as follows: [*the passage will be found on* pp. 25–6]. I cannot easily guess what error he means, that I had detected (for I detected many) unless it be that of which my nephew warned him when he recently stayed in England before that paper of mine was printed; a correction of this error has since been made by Newton on a single sheet inserted into his as yet unpublished book. However, that error (which related to the determination of the resistance of the medium for a given curve to be described by a projectile) had its origin in the fact that Newton had made a mistake (as I showed) in differentiating powers to higher degrees, or in taking the fluxions of fluxions of all orders by means of the indefinite expansion of the power into an infinite series.

Whether Varignon will readily permit anything to be done in France of which you might complain, particularly in this dispute with you which has been created by the English, and also what I had written to him as to this business, you may easily understand from what follows, which he very recently returned to me by way of a reply:[2] [*the remainder of the extract will be found on* p. 26].

NOTES

(1) Probably John Arnold, who matriculated in medicine at Leyden on 6 August 1708 N.S., aged 20. He was also inscribed at Basel University on 17 October 1711, and finally proceeded M.D. et Phil, Padua, 17 January 1715, spending at least two years there. He was from Exeter and returned to practise there, being admitted Extra L.R.C.P. 13 December 1720. He was in correspondence with Johann Bernoulli from 1713 to 1719, and also with Leibniz; see Letter 1181. It is reasonable to suppose that he spent some time in Paris in 1713.

(2) Varignon had written directly to Leibniz in a very similar vein on 9 August 1713 N.S.; (see Gerhardt, *Leibniz: Mathematische Schriften*, IV, p. 195):

'Je suis tres faché du mauvais procès que M. Keill vient de vous susciter en Angleterre: on est ici d'autant plus surpris que M. Newton lui-même, dans les Princ. Math. vous reconnoist aussi pour l'Inventeur du calcul en question, et que depuis pres de 30 ans vous jouissez paisiblement de cette gloire que vous vous êtes jusqu'ici reciproquement accordés avec une civilité qui edifioit les honnêtes gens: gloire aussi grande pour chacun de vous deux que s'il étoit le seul inventeur de ce calcul. C'est ce qui fait qu'on ne cesse point ici de vous en rendre honneur comme à M. Newton.'

1015 NEWTON TO JOHN GRIGSBY
1 SEPTEMBER 1713
From the original in the Babson College[1]

Pray pay to Dr. Francis Fauquier the three per cent Dividend due on my two thousand five hundred pounds south sea stock at Midsummer last past, & his Receipt shall be your discharge from

1st Sept. 1713

Your humble Servant

ISAAC NEWTON

To Mr. John Grigsby
Accomptant General For
the South Sea Comp.

(1) MS. 425. The South Sea Company had been formed in 1711 under Oxford's auspices to develop trade with the Spanish South American colonies; investors were guaranteed a return of 6 per cent. The first *Asiento* ship was in fact to be despatched only in 1717; meanwhile the Company engaged in government finance. Newton's investments in it are discussed in R. de Villamil, *Newton: the Man* (London, n.d.) pp. 19 ff. but de Villamil was not aware of this early participation by Newton. Fauquier seems to have been Newton's regular agent in this and other business—he was soon to take over active management of the 'tin affair'—see Newton's letter to him of 27 July 1720 (de Villamil, *ibid.*, pp. 19–20; More, p. 652).

1016 H. H. STRINGER TO NEWTON
6 OCTOBER 1713
From the original in the University Library, Cambridge[1]

Sr Isaac Newton
Sir,

You are hereby Desired and Summoned to Meet the rest of your Brethren, Members of the *Society of the City* of London, *of and for the Mines, the Mineral, and the Battery Works*, at the *Mineral-Office-General* in *Black-fryers*, at 2 of the Clock in the afternoon on Thursday next being the 8th Day of this instant Month of October *Anno Dom.* 1713. Per Cur' H. H. Stringer [2]

Dated at the *Mineral-Office-General*
　　the 6th Day of October

Serjeant.

(1) Add. 3968(41), fo. 136. This is a printed notice adorned by an engraving of the seal of the joint Society, with the dates and signature added in ink. Both the Society of the Mines Royal and the Society of the Mineral and Battery Works were incorporated in 1568 (not the following year, as shown on the seal). The former was privileged to search and work for precious metals as well as copper in certain counties of England; the object of the latter was to make iron and brass wire for industrial purposes from English copper and calamine (zinc ore). By the eighteenth century both Societies were moribund; a Minerals Act of 1693 had effectively destroyed the privileges of the Mines Royal. The two Societies amalgamated in 1710, when some manufacture of brass was still continuing.

Moses Stringer, author of *English and Welsh Mines and Minerals discovered in some proposals . . . for employing the poor to gain the hidden treasures of this Kingdom* (London, 1699), and of *Opera mineralia explicata* (London, 1713), a history of the Mineral and Battery Works and the Mines Royal, was the moving spirit in the renewed enterprise to which (it is said) Newton restored some of the account-books of the old companies which he had bought (see William Rees *Industry before the Industrial Revolution*, II (Cardiff, 1968), pp. 658–66). Moses Stringer was a strange figure who quite barefacedly proclaimed himself Governor of Tobago, Trinidad etc. and also Professor of Chemistry and Experimental Philosophy in the University of Oxford.

The only action taken at the meeting to which Newton was summoned was to offer a reward to any person giving information of any Mines Royal (silver or gold) being worked in England and Wales.

(2) 'Per curam', that is 'by the hand of . . .' H. H. Stringer was presumably related to Moses.

1017 LEIBNIZ TO J. BERNOULLI
14 OCTOBER 1713

Extracts from Gerhardt, *Leibniz: Mathematische Schriften*, III/2, pp. 922–3.
Reply to Letter 1014; for the answer see Letter 1026

Vellem doctrinam Serierum, in qua potissimum versatus fuit Newtonus, promovisset longius, inprimis circa modum agnoscendi, utrum advergant, qui transcendentibus æque ac ordinariis quadret. Ita enim multa in transcendentibus agnosci possent, quæ alias non facile paterent. Ideo Dominum Agnatum Tuum hortatus sum, ut huic argumento diligentius incumbat, repetiique monitum in adjecta Epistola, vel ideo quod ille de mente mea ex priore præcipitantius judicavit, cum explicationem verborum meorum convenientiorem, credo, mererer.[1]

Universalissima (id est ordinariis æque ac transcendentibus quibuscunque communis) hæc regula est, ut omnis valor per Seriem sit advergens, cum partes Seriei in infinitum decrescentes sunt alternis affirmativæ et negativæ.[2] Et videtur methodus excogitari posse, quamvis Seriem advergentem transformandi in talem, quanquam et alias vias video, sed quæ amplius excoli mererentur . . .

Gratias ago, quod Dni. Varignonii sententiam sane æquitati consentaneam mecum communicasti. Is qui litem ab Anglis mihi motam in Diario Parisino attigit, merito irrisit judices in propria causa . . .[3]

Hugenius etiam ad marginem Exemplaris[4] sui quosdam in Newtono errores notaverat, ut mihi narravit, qui in Batavis Exemplar vidit, cum Bibliotheca Viri distraheretur.

Translation

I could wish that Newton had developed the theory of series further, as this was the area in which he was most skilful, especially as regards the manner of discovering whether they converge when they correspond with transcendental as well as ordinary [functions]. For in this way much may be learnt about transcendentals which is not easily disclosed otherwise. I have exhorted your nephew on that account to work more diligently at this problem, and I have repeated the advice in an additional letter lest he should for that reason leap to too hasty an opinion about my intention in the first one since I deserved, I think, a more convenient explanation of my words.[1]

This rule is quite universal, that is, it is common to ordinary and any kind of transcendental [series]: the total value over a series is convergent when the terms of the series decreasing to infinity are alternately positive and negative.[2] And it seems that a method may be worked out of transforming any convergent series into such a one although I also see other ways which deserve to be developed further . . .

I thank you for your imparting to me Mr Varignon's opinion which is surely only fair. He who has touched on the quarrel against me begun by the English in the Paris *Journal* [*des Sçavants*] [3] has properly ridiculed them as judges in their own cause . . .

Huygens too noted certain errors of Newton's in the margin of his copy,[4] as someone reported to me who saw that copy in Holland when [Huygens'] library was dispersed.

NOTES

(1) The extant correspondence between Nikolaus Bernoulli and Leibniz, which deals mainly with the subject of the convergence of series, may be found in Gerhardt, *Leibniz: Mathematische Schriften*, III/2, pp. 979–94.

(2) Leibniz gave a proof of this theorem (that a series with terms which alternate in sign and decrease in absolute value monotonically towards zero, converges) in his letter to Johann Bernoulli of 30 December, 1713. See Gerhardt, *Leibniz: Mathematische Schriften*, III/2, pp. 926–7.

(3) Leibniz presumably meant the following passage (quoted in its entirety) from the *Journal des Sçavans*, no. 21 (22 May 1713 [N.S.]), p. 335, under the heading *Nouvelles de Littérature: De Londres*: 'La Societé Royale de Londres a prononcé sur le different qui estoit entre M. Leibnitz & M. Newton, au sujet de quelques découvertes en matiere de Calcul differentiel. La Societé Royale en attribue tout l'honneur à ce dernier, & elle vient de rendre sa décision publique avec les Pieces qui y ont servi de fondement. Tout cela fait un petit in 4o. intitulé *Commercium Epistolicum* [&c]. Comme il n'y a pas apparence que M. Leibnitz s'en tienne à cette décision, le Public recevra sans doute de sa part de nouveaux eclaircissemens sur ce sujet.'

As there is no other reference to the dispute in the three years 1712–14, it was evidently not regarded as interesting to the readers of the *Journal des Sçavans*.

(4) See Cohen, *Introduction*, pp. 186–7, and Letter 1070, note (7), p. 125.

1018 LEIBNIZ TO THE EDITORS OF THE
JOURNAL LITERAIRE DE LA HAYE
[*c.* OCTOBER 1713]

From the *Journal*, November and December 1713, pp. 445–8

Remarques sur le different entre M. de Leibnitz, & *M.* Newton

La Lettre inserée dans le premier Tome du *Journal literaire*, p. 206 & qui ren[f]erme un recit de ce different, contient plusieurs choses, qui font voir,

30

que l'Auteur de cette Lettere a été mal informé.[1] Il n'y a point eu autrefois de dispute sur ce sujet entre ces deux Messieurs. M. *Newton* n'avoit jamais donné à connoître qu'il prétendit ravir à M. *de Leibnitz*, la gloire d'avoir inventé le *Calcul des differences*; & ce n'est que par ceux qui ont vû le *Commercium Litterarium* imprimé à *Londres* il n'y a pas long-tems, que M. *de Leibnitz*,[2] & qui a sçû que M. *Newton* prenoit part, à ce que quelques personnes mal informées avoient avancé sur ce sujet. M. *de Leibnitz* qui est à Vienne, n'a pas encore vû lui-même cet Ecrit.

Ce sçavant Mathematicien n'a jamais communiqué ses raisons à la *Société Royale d'Angleterre*, croyant l'affaire trop évidente, pour que cela fût nécessaire: Il avoit seulement écrit qu'il ne doutoit point que la *Société* & M. *Newton* même, ne désaprouvassent ce procédé. Ainsi la *Société* n'a pas pû examiner les raisons de part & d'autre, pour prononcer là-dessus.

Voici maintenant un raport véritable de ce qui s'est passé. Il y a environ quarante ans qu'il y eut un Commerce de Lettres entre M[essieu]rs *de Leibnitz*, *Oldenbourg*, *Newton*, *Collins* & autres. Quelques-unes de ces Lettres ont été publiées dans le troisiéme Volume des Oeuvres Mathématiques de M. *Wallis*. On y voit que M. *Newton* faisoit mistére d'une chose qu'il disoit avoir découverte, & que depuis il a voulu faire passer pour le Calcul des différences. M *de Leibnitz* au contraire lui communiqua franchement les fondemens de ce Calcul, comme ces mêmes Lettres, publiées par M. *Wallis*, en font foi; quoi qu'il se soit trouvé, que M. *Newton* ne l'ait pas bien compris, sur tout par raport aux différences des différences. On a encore trouvé depuis, d'autres Lettres de M. *Collins* & de ses Amis, & on les a publiées maintenant à *Londres*, avec des Additions, dans lesquelles on prétend, sur de simples conjectures & sur de fausses suppositions, que le Calcul des différences est dû à M. *Newton*, & que M. *de Leibnitz* l'a apris de lui: quoique le contraire se voye clairement, & en termes exprès, dans leurs Lettres, publiées par M. *Wallis*.

L'Auteur de ces Additions a jugé avec témérité sur des choses dont il n'étoit pas bien instruit, & il a fort mal rencontré, quand il a voulu deviner, comment M. *de Leibnitz* étoit parvenu à son invention. Il s'est trouvé de plus, que M. *Newton* n'a pas encore connu le véritable Calcul des différences en 1687. lors qu'il a publié son Livre intitulé, *Philosophiæ Naturalis principia Mathematica*: car outre qu'il n'en a rien fait paroître, quoi qu'il eût de belles occasions de la faire valoir, il a de plus fait des fautes, qui ne pouvoient pas être compatibles avec la connoissance de ce Calcul; ce qu'un illustre Mathématicien fort impartial a remarqué le premier.[3] M. *de Leibnitz*, avoit déja publié son Calcul, quelques années auparavant, en 1684. & M. *Newton* n'a jamais communiqué rien d'approchant, à qui que ce soit que l'on sache, ni en public, ni en particulier, que longtems après la publication de ses *Principes*,

31

c'est à dire, lors que M. *Wallis* en 1693. publia ses *Oeuvres Mathématiques*, & lors que l'invention de M. *de Leibnitz* étoit déja célébre, & pratiquée publiquement avec beaucoup de succès & d'applaudissement, sur tout par MM. *Bernoulli*, Freres. Quand on considere ce qui a été publié par M. *Wallis*, on voit d'abord que l'Invention de M. *de Leibnitz* y paroît sous d'autres noms & d'autres caracteres, mais bien moins convenables. Cependant M. *Newton* ni alors ni long-tems après, n'a pas troublé M. *de Leibnitz* dans la possession de l'honneur de sa découverte: Il n'en a parlé, qu'apres la mort de MM. *Huygens* & *Wallis*, qui etoient bien instruits & auroient pû être juges impartiaux de cette affaire. M. *Leibnitz* avoit cru jusques à present sur la parole de M. *Newton*, que ce dernier pouvoit avoir trouvé quelque chose de semblable au *Calcul differentiel*, mais on voit maintenant le contraire. On a publié là-dessus le jugement impartial d'un Illustre Mathematicien: ce jugement est fondé sur le long silence & sur les fautes de M. *Newton*.[4]

NOTES

(1) In the May and June issue of the *Journal Literaire de la Haye*, pp. 206–14, there had been published a long, anonymous *Lettre de Londres* written by Keill much on the lines of Letter 843*a*, vol. v. Keill here traced the origin of the dispute between Leibniz and Newton to Fatio's *Dissertatio* of 1699, explaining his own part in it and the letters received from Leibniz, leading to the appointment of the *Commercium Epistolicum* committee. His letter is followed by a French translation of the 'Committee's' report and by an extract from Newton's letter to Collins of 10 December 1672 (Letter 98, vol. i) also rendered into French. The communication is moderate and factual in tone. Leibniz's reply in the paper here printed was sent to the *Journal Literaire* by his friend Wolf (see Ravier, p. 151, no. 313); it was published anonymously. For the German publication see Number 1009, note (1), p. 20. For Newton's reaction to these 'Remarques' see Letters 1045, note (3), p. 72, and 1053*a*.

(2) The text is clearly at fault here, presumably it should read somewhat as follows: 'and it is only from those who have seen the *Commercium Epistolicum* printed at London not long ago, that Mr. Leibniz [learned of Newton's claim to the invention, and that he] learned that Mr Newton had a hand . . .'

(3) Johann Bernoulli.

(4) Here follows, in French, the text of the *Charta Volans* (see Number 1009). It was Wolf also who had this French version prepared for the *Journal Literaire de la Haye*.

1019 NEWTON TO JOHN THORPE
19 OCTOBER 1713
From the original in Trinity College Library, Cambridge[1]

These are to order You to Summon a Council of the Royal Society to meet at there House in Crane-Court, Fleetstreet, on Thursday next the 22d. of October (1713.) at Twelve of the Clock—And for so doing this shall be sufficient Authority

Octob. 19. 1713.

Is. NEWTON P.R.S.

To Jo: Thorpe, M.D.
   These

NOTE

(1) R.16.38A⁵. Only the date, signature and address are in Newton's hand. John Thorpe (1682–1750), who took his M.D. at Oxford in 1710, was elected F.R.S. on 30 November 1705. He acted as Clerk to the Royal Society (in succession to Humfrey Wanley) from 1706 and in addition (after the death of Henry Hunt in June 1713) as Housekeeper and Keeper of the Repository and Library. He resigned all these offices in November 1713 when he took up the practice of medicine at Rochester. He published two notes in the *Philosophical Transactions* but was mainly active in studying the antiquities of Kent.

1020 FLAMSTEED TO SHARP
31 OCTOBER 1713
Extract from the original in the Library of the Royal Society[1]

... S[ir]. I. N[ewton] still continues his designes upon me under pretence of takeing Care of ye Observatory & hinder[s] me all he can but I thank God for it hitherto without success[.] lately he was for makeing me New Instruments which I want not[.] by the way he has given me me [*sic*] occasions to prove that all ye Instrumts in ye house are my own. & I have good evidence that Sr J Moore gave me the sextant & clocks & yt they are at my disposall but I hope I shall not long be troubled with him[.] I think the New Princip. worse then the Old save in ye Moon & there he is fuller but not so positive & seemes to refer much to be determined by observations, the book is realley worth about 7 or 8*sh* it cost 4*s.* 4*d* a peice printing & paper [2] Dr Bently puts ye price 18*sh.* & so much mine cost me. I am told he sent S.I.N. half a dozen & made him pay 18*sh.* a peice for them phaps this was Contrivance possibly it is not true.

NOTES

(1) Sharp Letters, fo. 86; the letter is fully printed in Baily, *Flamsteed*, p. 305. For Newton's plans for new instruments, see Letter 1044.

(2) On the price of the *Principia*, see Number 1002, vol. v, and Appendix.

1021 ———— TO THE BISHOP OF WORCESTER [1]
7 NOVEMBER 1713
From the copy in the British Museum[2]

Novbr. 7. 1713.

My Lord,

I had the honour to receive, & the pleasure to read the Papers your Lo[rdshi]p directed to the Dean of Norwich:[3] and before I sent ym forward I communicated ym to Sr Isaac Newton according to your Lo[rdshi]ps order by Mr Archdeacon. When Sr Isaac brought them back, he told me yt he found many excellent Observations in them about the ancient year, & at the same time acquainted me that he had formerly discoursed with your Lordship about that year of 360 daies,[4] & represented to you yt it was the Kalendar of the ancient Lunisolar year composed of the nearest round number of Lunar Monthes in a year, & daies in a Lunar Month: that the Ancients corrected this Kalendar monthly by the new Moons, & yearly by the returns of the 4 seasons, dropping a day or two, when they found the Kalendar Month of 30 daies too long, for the return of the moon, & adding a month or two [5] at the end of the year when they found the year of 12 Lunar Monthes too short for the return of the seasons & fruits of the Earth: that Moses in describing the flood uses the Kalendar Monthes not corrected by the course of the Moon the Cloudy rainy weather not suffering her then to appear to Noah: that when Herodotus [6] or any other Author reckons 30 daies to ye month, & 360 daies to the year, he understands the Kalendar Month & year without correcting them by the course of the Sun & moon: that when Herodotus reckons by years of 12 & 13 Monthes alternately for 70 years together, he understands the Dieteris [7] of the Ancients continued 70 years without correcting it by the Luminaries: and that when we read of a week, or a month or a year, consisting of any other daies than the naturall, we are to reckon 7 daies, or 30 daies, or 360 daies according to the Kalendar; because where the daies are not natural ones, the Kalendar cannot be corrected by the courses of the Sun & moon; & if the daies be taken mystically for the years of any Nation, we are to take these years in the Vulgar sense for 7, or 30, or 360 practical years of that Nation, such as they commonly use in their Civil Affairs.

Sr Isaac said further that he met with nothing in your Lo[rdshi]p's papers, wch in his opinion made against what he then represented to your Lo[rdshi]p, that Suidas [8] (in Σάρος) tells us that the monthes of the Chaldees were Lunar, their ordinary years composed of 12 lunar Monthes, & their Sarus composed of 18 such years & six Monthes, wch Monthes he takes to be intercalary: (the end of all Cycles of years being to know when to intercale the Monthes of the Lunisolar year for keeping the year to the seasons;) & yt Censorinus [9] mentions a Chaldean Cycle of 12 years, & that the Jews in returning from Captivity called their own Monthes by the names of the Chaldeans & that the feast Saccæa [10] of the Babylonians was celebrated upon the 16th day of a Lunar Month, & kept to the same season of the year, & that in all Antiquity he meets with no other sorts of years than the Lunisolar the Solar & the Lunar, & their Calendars & Cycles, [11] & that the proof of any other sort of year must be upon him yt affirms it.

This, My Lord, is the Substance of what I had from Sr Isaac, wch I thought my self obliged to report to your Lordship in discharge of that Commission you was pleased to honour me with.

NOTES

(1) William Lloyd (1727–1717), one of the famous 'seven bishops' unsuccessfully prosecuted by James II, successively occupied the sees of St Asaph, Lichfield and Coventry, and Worcester (from 1700). He was a voluminous writer, among his unpublished books being *A System of Chronology*.

(2) Add. 6489, fos. 67–8. We print this letter here because the main body of it is based on a draft in Newton's hand (fo. 71), from which it differs only in orthographical detail. Edleston suggests (*Correspondence*, p. lxxv, note (163)) that the letter may have been written by Charles Trimnell (1663–1723), the Bishop of Norwich, who earlier acted as intermediary between the Bishop of Worcester and the Dean of Norwich, to whom a copy of the letter was apparently sent. Newtons' draft (printed in Edleston, *Correspondence*, pp. 314–15) is in fact a summary of a much longer holograph manuscript (fos. 69–70), which was later printed in the *Gentleman's Magazine* for January 1755, pp. 3–5, under the heading 'A Letter from Sir Isaac Newton to a Person of Distinction . . .'. (Compare Newton's *The Chronology of Ancient Kingdoms Amended* (London, 1728), pp. 71ff. dealing with the same topic.) Earlier holograph drafts both of the longer manuscript and of its summary are to be found in the Jewish National and University Library, Jerusalem (Yahuda Collection, Newton MSS. 24, packet 2).

(3) Humphrey Prideaux (1648–1724), Dean of Norwich since 1702, had some reputation as an oriental scholar.

(4) Newton's draft (fo. 71) begins 'Sr Isaac Newton represents that he did formerly discuss wth your Lordp about the ancient year of 360 days . . .' and then continues, with minor orthographical variations, as the version printed here.

(5) 'or two' is an accidental insertion; it does not appear in Newton's draft.

(6) Herodotus of Halicarnassus (d. *c.* 425 B.C.), celebrated Greek historian.

(7) A period of two years.

35

(8) Suidas was a Byzantine lexicographer of the tenth century. He was the first writer to use the word 'saros' (Sumerian *šár*) in a sense other than '3600 years', thereby causing much confusion (see Otto Neugebauer, *The Exact Sciences in Antiquity* (Providence, 1957), pp. 141–3).

(9) Censorinus was a Roman astrologer, who wrote in 238 *De die natali*, part of which deals with the divisions of time.

(10) The *Sacæa* resembled the Roman *Saturnalia* in that it was a festival at which masters and slaves changed rôles.

(11) Here Newton's draft ends.

1022 THORPE TO NEWTON
9 NOVEMBER 1713
From the original in the University Library, Cambridge[1]

Hon[oure]d Sir,

I take leave to put You in mind of bringing with You to the Royal Society Your Key of the Iron Chest which contains the Common Seal, the Lease being ingross'd and ready for the Seal to be affixt to it.[2]

> I am
> Your most Obliged
> and Obedient Servt.
> Jo: Thorpe[3]

Crane-Court
Nov. 9th. 1713.

To
Sir Isaac Newton

NOTES

(1) Add. 3968(41), fo. 9. Newton used this paper for a *Commercium Epistolicum* draft.

(2) At a meeting of the Council of the Royal Society on 9 November 1713 'A Lease of the Royal Society's Sellar under the Repository to Mr. Samual Clements Cheesemonger for 7 years commencing from [*blank*] at the Rent of five pounds per annum was read sealed and delivered, and the duplicate lockt up in the Iron Chest.' (see the Royal Society *Council Minute Book*, II, p. 214). The letting of the cellar was an attempt to recoup some of the cost of the new repository, on which Richard Waller had spent over-lavishly, to the financial embarrassment of the Society. Later the Council had cause to regret its action for 'the Repository was very much annoyed with the ill scent of the Goods loged in that Cellar.' (*ibid.*, II, p. 248; 14 April 1719).

(3) See Letter 1019.

1023   KEILL TO NEWTON
9 NOVEMBER 1713
From the original in the University Library, Cambridge[1]

Oxford November 9th
1713

Honoured Sr

Since I left London I have considered Mr Bernoulli's solution of the Inverse Probleme about Centripetal forces,[2] and I am amazed at his impudence and could not forbear makeing the following remarks. The demonstration of the 40th prop.[3] you have made plain and easy, and yet he sayes it is intricate, and therefore he puts down one of his own wch is much harder than yours,

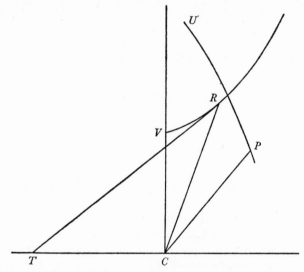

he afterwards gives a formula for the Element of the angle at the centre wch seemed to be more intricate than yours, but I find it to be only your's disguised, so that his general solution is exactly taken from yours, and he has done nothing but what you had done better before.[4] In his application of it to the particulare case he has with a great deal of Labour showed that the curve described must be a Conick Section when the thing may be demonstrated in a few lines, for to shew you this I have sent you the inclosed remarks,[5] to wch I intend to add the solution of the Probleme where the force is reciprocally as the Cubes of the distances, and from thence I can deduce the 3d corolary of the 41st proposition [6] and show that it gives universally the formation of all the Curves that can be described with a force reciprocal to the cubes of the distance, where the velocity is either greater or less than that wch is acquired by a bodies comeing from an infinite distance, but I observe that where the velocity is greater there are tuo cases one of wch is constructed by the sector of a circle or an Ellipse and the other by the sector of an Hyperbola, the figure of wch is left out in the Principia where the angle VCP is still proportional to the hyperbolick sector, but the distance CP is to be measured by the line CT taken upon the conjugate Axis.[7] If the velocity be increased so far as that it bear the same proportion to the velocity acquired by falling from an infinite distance as the Radial CP is to the perpendicular from the center to the tangent of the curve at P the curve becomes the Hyperbolick Spiral. If the velocity is greater than in that proportion, the Curve is constructed by the circulare sector so that the Hyperbolick Spiral is the mean between the tuo sort of curves, and may be imagined to be formed by either hyperbolick or circulare sector by takeing the semi Axis VC infinitely small, I shall be much obliged to you if you will send me your thoughts on this matter and informe me if I am right. I am Sr with all respect

your most obliged Humble Servant

JOHN KEILL

For
Sr Isaack Newton
    These.

NOTES

(1) Add. 3985, no. 2. Keill had been appointed Savilian Professor of Astronomy at Oxford in the previous year.

(2) Printed in the *Mémoires de l'Académie Royale des Sciences* for 1710 (first published at Paris in 1713; reprinted in 1732), pp. 521–33. Johann Bernoulli sent to the Academy (where they were read on 13 December 1710) (i). *Extrait d'une Lettre de M. Herman à M. Bernoulli* concerning the latter's solutions of the inverse problem of central forces; (ii). *Extrait de la Reponse de M. Bernoulli à M. Herman*, dated 7 October 1710 and explaining his solutions in detail. This is followed by a communication from Varignon showing that analogous solutions followed by integration from his own demonstrations of the direct law (printed earlier). Both Bernoulli and Varignon, of course, employed the calculus of Leibniz.

(3) Bernoulli in his *Réponse* first criticizes Hermann's solution of the inverse problem of central forces as being factitious; he could not have found it without knowing in advance that conic sections were the sought-for curves. Also, he writes, other curves might satisfy Hermann's differential equation without his having considered them. Presenting his own solution, which he claims to be general for any force-law, Bernoulli commences with a Lemma: 'That if two bodies . . . begin to descend from the same point A with equal speeds, and with equal forces [acting upon them] towards the same point O; the one directly, following the straight line AO, and the other obliquely, following the trajectory ABC which it will describe; I say that at all equal distances of one or the other [body] from the centre of force O, as at B and E (imagining BE the arc of a circle described about O) these two bodies will always have equal

speeds.' This lemma, he admits, is demonstrated by Newton (*Principia* 1687, Book I, Prop. 40, pp. 125–7), but he calls Newton's demonstration 'trop embarrassée: la voici plus simplement'.

Bernoulli then argues that the ratio of the force components acting along the trajectories at B and E is $Ee:Bb$. If we assume the speeds are the same at B and E, then the ratio of the times taken to traverse the distance increments Bb and Ee will be $Bb:Ee$; whence the ratio of the speed increments will be $Ee.Bb:Bb.Ee$, or unity, so that the speeds at b and e will also be equal. Since the initial condition is that the speeds should be equal at A, the speeds will always remain equal.

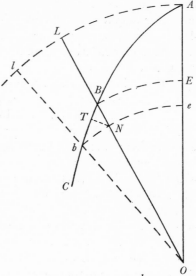

Thus far Bernoulli's approach differs little from Newton's; Newton's is seemingly longer because of the introduction of the normal TN, and the resolution of the force along the curved trajectory into components BT and TN. But later he uses the geometrical result $BT.Bb = (BN)^2 = (Ee)^2$ which is implicit in Bernoulli's first assumption that the ratio of the forces is $Ee:Bb$.

Now writing $OA = a$, $OB = x$, $NB = dx$, $AL = z$ and $Ll = dz$, so that $Nb = \dfrac{x.dz}{a}$, Bernoulli combines the result $v = \sqrt{(ab - \int \phi . dx)}$ (where a factor of 2 is intentionally ommitted, for simplicity, in front of $\phi(x)$, the central force, and where ab is a constant) with the Kepler area law $v = \dfrac{Bb}{(x^2/a).dz}$ to give

$$\left(dx^2 + \frac{x^2}{a^2}.dz^2\right)^{\frac{1}{2}} = Bb = \frac{x^2.dz}{a}(ab - \int \phi . dx)^{\frac{1}{2}},$$

and so

$$\frac{a^2 c . dx}{(abx^4 - x^4 \int \phi \, dx - a^2 c^2 x^2)^{\frac{1}{2}}} = dz,$$

where c is a constant introduced to render the equations homogeneous. If the integration can now be carried out, the trajectory ABC will be defined by a polar equation in x and z.

Bernoulli now completes the solution in the case of the inverse square law, putting $\phi = (a^2 g)/x^2$ (where g is a constant), and using a number of substitutions to simplify the integration. The result, expressed analytically, is clearly always a conic.

(4) Bernoulli did not claim (so far) that he had in any way *differed* from Newton; his mathematical approach is different, however. On p. 524 he simply claims that he has achieved the result 'more conveniently than Mr. Newton on p. 127 ff. of his *Principia*' (that is, Prop. 41, Book I).

(5) These are not now with the letter. Keill later published 'Observations' on the subject ('Observationes de inverso problemate virium centripetarum') in *Phil. Trans.* **29**, no. 340 (July–September 1714), 91–111 strongly critical of Bernoulli's *Réponse*, and repeatedly insisting that Newton's failure to publish certain demonstrations in no way reflected a private inability to perform them (see also Letter 1155, note (2)). Bernoulli's chief claim (or 'impudence') had been (p. 532) that Newton, in the first edition of the *Principia*, had followed Bernoulli's demonstration in showing that if the action of a central force causes a body to move in a conic section, then the force obeys the inverse-square law, by the mere supposition (in his first

corollary, p. 55) that the converse was true—that is, that in every case the inverse-square law would generate motion in a conic section and in no other curve. Bernoulli maintained that Newton's supposition would be justified only by a proof that whenever motion occurred in a curve other than a conic, the inverse-square law did not apply. To make his 'if and only if' point clearer, he instanced the curvilinear motions appropriate to an inverse-cube law of force, that is, the logarithmic and hyperbolic spirals. It would be true to say that all bodies moving in a logarithmic spiral obey the inverse-cube law, but the converse—that all bodies subject to the inverse-cube law move in logarithmic spirals—would be false. See also D. T. Whiteside in the *Journal of the History of Astronomy*, **7** (1970), 125–6.

(6) Proposition 41 gives a general method for determining the curves described by bodies when the centripetal force obeys any law; its third corollary constructs, without proving, the particular case of the inverse-cube law afforded by the general secant and hyperbolic secant spirals (these spiral orbits were investigated by Cotes in 'Logometria'; see Whiteside, *op. cit.* and *Mathematical Papers*, VI, pp. 352–3, notes (213) and (214)).

(7) Keill's fluxional treatment of the inverse-cube law case appeared on pp. 98–111 of his article mentioned in note (5) above, and included the two solutions described here. The construction is similar to those in the *Principia*, Book I, Prop. 41, Coroll. 3; thus P is the locus of the gravitating particle describing PV, and is such that $CP = TC$ where T is the intersection of the tangent to the elliptical or hyperbolic arc VR with the axis TC. The angle VCP maintains a constant proportionality with the angle VCR (the 'hyperbolick sector').

1023 A THE ABBÉ BIGNON TO NEWTON
19 NOVEMBER 1713
For the answer see Letter 1023 B

This letter, thanking Newton for 'un exemplaire de vostre dernier ouvrage', that is, a copy of the second edition of the *Principia* sent to him in Paris, was sold at Sotheby's in 1936 (*Sotheby Catalogue*, Lot 125). We have not succeeded in tracing it. The same Lot included a draft of Newton's reply. For the Abbé Bignon, see Letter 1004, note (2), p. 5.

1023 B NEWTON TO BIGNON
[late 1713]
From the holograph draft in the Jewish National and University Library, Jerusalem.[1]
Reply to Letter 1023 A

Illustrissimo Domino
Dno Abbati de Bignon
Is Newton salutem

Quod munusculum meum vobis non ingratum fuerit quam maxime gaudeo. Et vestro judicio apprime tribuendum esse censuero si viris doctis posthac non displicuerint quæ in Libro meo tractantur. Quippe quæ nimia

PLATE I. Commemorative Medals by John Croker for which Newton was responsible: see Letters 1006 and 1103–6. They were minted in gold, silver and copper (see the printed list on Plate V, facing p. 233). Specimens of the gold medals were presented to the Officers of State, Members of the Diplomatic Corps, and Members of Parliament. *Top left to bottom right*: Peace Medal, 1713, obverse and reverse; Coronation Medal, 1714, reverse and obverse. (Reproduced by courtesy of the Trustees of the British Museum.)

brevitate redduntur subobscura, et ab Hypothesibus Philosophicis vulgo receptis impugnanantur [*sic*], et defectu explicationum hypotheticarum minus grata sunt, & ob rerum difficultatem a paucissimis legi possunt. At auspicijs vestris scientias florere in Gallia in omnium ora est, et ijsdem auspicijs Mathesis ad Philosophiam spectans, uti spero, florebit

Vale

Translation

I rejoice exceedingly that my little gift was not unwelcome to you. And I shall think it particularly attributable to your judgement if, in the future, the learned do not take a dislike to those matters which are discussed in my book. For they can be read over but by very few indeed because of the difficulty of the topics [treated]; they have been made rather obscure by excessive brevity, and run counter to the philosophical hypotheses commonly received, and [also] are rendered the more unwelcome by the lack of hypothetical explanations. But every one says that under your leadership the sciences flourish in France, and as I hope, [the study of] mathematics as related to philosophy will flourish likewise. Farewell.

NOTE

(1) Yahuda Collection, Newton MSS. 25. This draft reply to Letter 1023A (having religious notes elsewhere on the same paper) is probably not that mentioned in *Sotheby Catalogue*, lot 125.

1024 PIERRE VARIGNON TO NEWTON
24 NOVEMBER 1713
From the original in King's College Library, Cambridge[1]

Equiti Nobilissimo,
Viroque Doctissimo D. D. Newtono,
Regiæ Societatis Anglicanæ Præsidi Dignissimo
S.P.D.
Petrus Varignonius.

Paucis abhinc diebus, Vir Nobilissime ac Doctissime, D. Anissonius [2] reddidit a te mihi alteram aurei tui Libri *De Philos. Natur. Princ. Math.* editionem, quam impertire mihi benevole Dignatus es, quaque nihil mihi erat exoptatius, adeo ut jam plusquam ab anno duos ex amicis meis Angliam petentes, & a sex circiter mensibus Clariss. D. Moivræum enixe rogaverim ut eam mihi emerent statim ac ipsa prodiret in publicum, nulla sane ratione præsumens eam ignoto mihi a te fore donandam. Hoc ardens in me desiderium excitaverat prima exquisitissimi hujus tui libri editio; quam incredibili cum voluptate legi, miratus semper summam Ingenij tui vim & aciem qua Claustra Naturæ reserasti, penetralia ejus altiora pervasisti, & detectas inde leges ejus occultiores rationibus e Sublimiori Geometria petitis Solertissime Demonstrasti, quas nemo videat non acutissimus. Isthæc etiam editio prima magis eo placuit quod ex ejus Lectione multa mihi nova Nascerentur, quibus additis ad egregii ejusdem tui libri margines, eos fere omnes & integros notis hisce qualibuscunque meis maculavi. Hinc longe plura mihi oriunda fore spero ex ejusdem editione nova, in qua raptim evoluta multas jam accessiones & elucidationes perspexi, quas ubi totum librum accurate legam, debita cum attentione meditabor. Hanc, inquam, novam editionem cursim duntaxat pervolvi, ne diutius impediretur mea grati erga te animi testificatio. Gratias igitur ago tibi quam plurimas, non solum pro dato mihi tuo libro quem pecunia quantacunque emissem, sed maxime pro honore quo hoc a te Donum me afficit. Vale, vir Nobilissime, Doctissime, ac humanissime; Celeberrimique Nominis tui cultori perpetuo mihi observatissimo favere perge.

Dabam Parisiis V. D. Decemb. An. M.DCC.XIII [N.S.]

Translation

Pierre Varignon presents a grand salute
to the most noble
and learned Mr Newton
Worthy President of the English Royal Society

A few days ago, most noble and learned Sir, M. Anisson [2] brought me from you the second edition of your splendid book *On the Mathematical Principles of Natural Philosophy*, which you have kindly given to me, and than which there was nothing I would have wished for more, in that more than a year ago now I earnestly asked two of my friends who were travelling to England, and about six months ago I entreated de Moivre, to have them buy it for me immediately it was published, having no reason at all to presume that you would give it to an unknown person. This burning eagerness had been aroused in me by the first edition of this exquisite book of yours; which I had read with exceedingly great pleasure, wondering always at the supreme strength and sharpness of your genius, with which you unlock the door of nature, penetrate into her inmost recesses, and most skilfully demonstrate by reasoning derived from a sublime geometry, laws which no one who was not most acute could perceive, which you have discovered there.

This very same first edition pleased me the more because the reading of it provoked many new ideas in my mind; by adding these to the margins of your extraordinary book, I have stained all of them completely with notes of this sort, for what they are worth. Thus I hope that far more will spring up within me from this same new edition, in which by hastily running through it I have already observed many additions and clarifications, which I will reflect upon with due attention when I have read the whole book carefully. As I say, I have only perused this new edition cursorily, lest the expression of my gratitude towards you should be prevented for some time. Accordingly I return you most hearty thanks not only for the gift of your book for which I would have paid untold gold but most of all for the honour your gift does me. Farewell, most noble kind and learned Sir, and continue to show kindness to me as a constant and most zealous admirer of your famous name.

Paris, the fifth day of December 1713 [N.S.]

NOTES

(1) Keynes MS. 142(A); microfilm 1011.26. Pierre Varignon (1654–1723) was Professor of Mathematics at the Collège Mazarin and Professor of Philosophy at the Collège de France. He was elected to the first Académie Royale des Sciences in 1688, and nominated by Louis XIV to the refounded Académie in 1699. He was an exponent of the Leibnizian calculus (compare Letter 1005) who wrote many memoirs on mechanics, without ever realizing the extent to which not only his results but his procedures also had been anticipated by the geometrical calculus of Newton.

(2) Jean Anisson was Director of the Imprimerie Royale at the Louvre from 1690 to 1707, when his brother-in-law, Claude Rigaud, succeeded him. Anisson remained closely connected with the printing world. In 1713 he visited London and was thus able to carry back with him copies of the *Principia* for distribution in France.

1025 CHAMBERLAYNE TO NEWTON
25 NOVEMBER 1713
From the original in the Mint Papers[1]

Petty France Westmr
25 Novemb. 1713

Honored Sr

I beg the Favor of you to mark the inclos'd List for me between this & Munday next,[2] just as you intend to do your own both for New Councelors & New Officers all but one, whom I desire to choose Freely, & whom I would make Perpetual Dictator[3] of the Society, if that depended only on the vote of

Your most faithful Humble Serv

JOHN CHAMBERLAYNE

NOTES

(1) II, fo. 334r. This side of the folio has been used by Newton for 'priority' drafts mentioning Wallis and Mercator. The other side (fos. 334v–335r) has been devoted to drafts related to Letters 1035 and 1044.

For the writer of this letter see Letter 799, note (1) (vol. v, pp. 59–60).

(2) 30 November, St Andrew's Day—traditionally the date on which the Council and Officers of the Royal Society were chosen for the year following.

(3) Newton was re-elected at the meeting on 30 November (see the Journal book of the Royal Society of London, xi, p. 389).

1026 J. BERNOULLI TO LEIBNIZ
25 NOVEMBER 1713
Extracts from Gerhardt, *Leibniz: Mathematische Schriften*, iii/2, pp. 924–5.[1]
Reply to Letter 1017

Tecum sentio fore ut aliquando pœniteat Dn. Newtonum, quod tam faciles aures præbuerit adulatoribus; consultum interim erit, ut quam responsionem moliris contra Commercium Epistolicum, mature absolvas et in publicum edas, ne ob retardationem gloriandi causam habeant. Inprimis Keylius et Cheynæus mereri videntur, ut probe defricentur, omnium quippe Idolo-latrorum Newtonianorum acerrimi et in Exteros iniquissimi. Sed vereor, ne non vacare possis huic negotio, antequam ad Lares Tuos redieris.

Bene mones de Doctrina serierum, multa per eas agnosci posse in Trans-cendentibus, quæ alias non facile paterent; Agnatus meus qui ob negotia quædam extra urbem agit, super hoc argumento ad Te scribet;[2] interim nescio, an demonstrari possit quod asseris, omnem valorem per seriem esse advergentem, per consequens finitum, cum partes seriei continuo decrescentes sunt alternatim affirmativæ et negativæ.[3]

A longo jam tempore nihil literarum accepi a Moivræo et Burneto, quan-quam uterque responsionem mihi debeat,[4] sed aliunde audivi Novam Edi-tionem Princip. Phil. Newtoni jam a 4 aut 5 mensibus publice prostare; eam vero nondum vidi, etsi aliquis Anglus ejus exemplar ex Gallia mittendum quam primum acciperet, mihi promiserit.[5] Quem notavi errorem Newtoni circa determinationem medii, cujus resistentia data curva describatur, correxit in Nova Editione, antequam publicaretur, per interpolationem alicujus schedæ idque ex monitu Agnati mei,[6] reliquos errores suos potuit etiam corrigere ex scripto meo, quod publicavi in Actis Lips. sub initium hujus anni et quod ille vidit aut sane videre potuit, antequam lucem aspexisset nova ipsius Editio;[7] sed sicuti priorem correctionem ita inseruit, quasi a semet ipso haberet, nulla facta mentione mei vel ejus, a quo fuerit monitus, ita haud dubie, quod ad cæteros, eadem sinceritate egit.

Errores illi, quos dicis, Hugenium ad marginem exemplaris sui in Newtono notasse, sunt forte non alii, quam qui jam dudum publice extant ad calcem Historiæ Cycloidis a quodam Groningio Wismariensi [8] editæ; forte etiam ab ipso Groningio Tibi hoc narratum fuit, nam et mihi narravit, cum ex Batavis veniens Groninga transiret et in transitu Gradum Doct. Juris capesseret; sed nihil eorum omnium, quæ ego notavi in Newtono, ab Hugenio notatum fuit. Et certe haud valde magni momenti sunt notæ Hugenianæ, possetque Newtonus gloriari, si cætera omnia in opere suo recte se haberent.

Translation

Like you, I think that Mr Newton will some time smart for so easily lending his ear to flatterers; meanwhile it will be wise for you to concentrate on your reply to the *Commercium Epistolicum*, finish it in good time and lay it before the public, lest they should have reason to rejoice in the delay. Especially it seems that Keill and Cheyne deserve to be publicly satirized because they have been the most eager of Newton's toadies and the most unjust to foreigners. But I fear that you will have no leisure for this affair before you see your own fireside.

Your advice about the theory of series is good, much can be learnt from them as to transcendentals which is not readily disclosed otherwise; my nephew (who is busy away from the city on some business) will write to you about this question;[2] meanwhile I do not know whether what you affirm can be demonstrated, that the whole value over a series is convergent, and consequently finite, when the continually decreasing terms of the series are alternatively positive and negative.[3]

I have for a long time now received no letter from de Moivre or Burnet although both owe me an answer,[4] but I have heard from another quarter that the new edition of Newton's *Principia Philosophiæ* has been on sale for four or five months now; I have not yet actually seen it even though a certain Englishman has promised me a copy of it coming from France, as soon as he has received it.[5] That error of Newton's that I had remarked (concerning the determination of a medium by whose resistance a given curve may be described) he has corrected in the new edition, before publication, by the interpolation of a sheet,[6] and at that it was after advice from my nephew; he could also correct the rest of his mistakes from my paper published in the *Acta* [*Eruditorum*] of Leipzig about the beginning of this year, which he saw (or surely could have seen) before his new edition was released;[7] but, just as he has inserted that first correction in a manner as though he had made it himself without any mention of myself or of the person who advised him of it, so I have no doubt he will behave in an equally candid way with regard to the remainder.

Those errors which, as you say, Huygens recorded in the margin of his copy of Newton are perhaps none other than those long ago made public in an appendix to the *Historia cycloidis* published by a certain Groening of Wismar;[8] perhaps also this was related to you by the same Groening, for he told me it too, when he passed through Gröningen from Holland and collected an LL.D. in passing; but nothing of all those things which I had remarked in Newton was noted by Huygens. And certainly Huygens' notes are not of great importance and Newton might congratulate himself if all the rest of his work were correct.

NOTES

(1) For Leibniz's answer see Gerhardt, *Leibniz: Mathematische Schriften*, III/2, pp. 926–7. There Leibniz tells Bernoulli that he has used Bernoulli's Letter 1004 in the *Charta Volans* (Number 1009), but without disclosing its authorship.

(2) This letter does not appear to be extant.

(3) See Letter 1017, note (1), p. 30.

(4) William Burnet was to write to Bernoulli on 3 December 1713 (see Letter 1028) and De Moivre on 28 June 1714 (see Letter 1088).

(5) Bernoulli might mean Arnold (see Letter 1014, note (1)). Both De Moivre (see Letter 1046, note (1)) and William Burnet (see Letter 1055) had promised Bernoulli copies of the two books; he eventually obtained them from Burnet.

(6) See Letters 951 and 951*a*, vol. V; the interpolated page was the correction to Book II, Proposition 10.

(7) Newton did not in fact see the *Acta Eruditorum* for February and March 1713 until April 1714; see Letters 1055 and 1064.

(8) See Letter 1070, note (7), p. 125.

1027 CHAMBERLAYNE TO NEWTON
[*c.* END OF NOVEMBER 1713]
From the original in King's College Library, Cambridge[1]

Friday Morning

Honored Sr

I beg Pardon for coming so early & for my abrupt going away from hence, but being obliged to meet the B[isho]p of Norwich & Chester & some other Gentlemen just now at the ABp's Library hard by, I am forced to go without introducing the Honest Dutchman, whom I have made come out of the City this morning to shew you K Ch & Qu. Mary's Effigies curiously don with a Hammer as he says;[2] he has desired me to get them shewn to her Majesty, in order to make a Present of them to her, & I have undertaken to do it, in case you judge them worthy; which is what I beg to know from you, & remain Honor'd Sr

Your most Humble servant

JOHN CHAMBERLAYNE

I hope you don't forget Mess Leveret & Brattle, they are the most learned Men in N. Engl: as that Colony is the best provided [?] of all the Queens Plantations: nor Mr Berkeley neither.[3]

NOTES

(1) Keynes MS. 141E, microfilm 1011.25; presumably written a little before 3 December 1713 (see note (3) below).

(2) We have not been able to identify the Dutch artist.

(3) On 3 December 1713 Newton read at the meeting of the Royal Society a letter (not traced) from Chamberlayne proposing the election of the two New Englanders; their election in fact occurred on 11 March 1713/14. William Brattle (1662–1717) was a tutor at Harvard College in Cambridge, Massachusetts, and Colonel John Leverett (1662–1724) was President of the College. 'Mr Berkeley' may be George Berkeley (1684–1753), the philosopher and prelate, but he was never F.R.S.

1028 WILLIAM BURNET TO J. BERNOULLI
3 DECEMBER 1713
Extracts from the original in the University Library, Basel.[1]
For the answer see Letter 1040

a Salisbury ce 3 *dec*: 1713

... J'ay toujours etè icy a la compagne depuis le recu de votre lettre; et pour cette raison je n'ay pu trouvèr moyen de vous envoyer le livre de Mr Neuton.[2] mais asteur par hazard il s'est presentè une occasion que je n'ay garde de perdre. deux Messrs de Geneve, proposons en Theologie, nommès de la Rive, et Rilliet, ont ecrit a mon pere de Londres qu'ils devoient aller bientot a Paris, c'est par cet occasion que je leur recommande cette lettre aussi bien que la 2de edition de la phil: de Mr Newton, que je vous pris d'accepter de ma part; ...

Comme cette lettre vous viendra par la poste et peutetre quelques semaines devant le livre, je vous copie une petite preface qui nomme, a peu pres tous ce qu'il y a dé nouveau dans cette edition ...[3] [*A copy of the* Authoris Præfatio *from the second edition of the* Principia *follows*.]

NOTES

(1) LIa 654 no. 18*. The writer of the letter, who lived from 1688 to 1729, was the second son of Gilbert Burnet, Bishop of Salisbury; he was educated at Trinity College, Cambridge and at Leyden. His correspondence with Johann Bernoulli arose through his connections with the mathematician John Craige (cf. vol. v, p. 118, note (3)); in 1708 Bishop Burnet had awarded Craige a living at Durnford in the diocese of Salisbury, and William lived in Craige's house in 1708–9. The Bernoulli–Burnet correspondence began in September, 1708, instigated by Burnet, and the early letters centre largely on a discussion of Bernoulli's objections to Craige's work on the calculus. In 1709, on a visit to the continent, Burnet was able to meet Bernoulli.

In a letter of 5 March 1712 N.S. Bernoulli asked Burnet if he could send him a copy of the second edition of the *Principia* as soon as it appeared in print; Burnet promised to do so. In letters of 13 August 1712 N.S. and 8 December 1712 N.S. (to which the present letter is the reply) Bernoulli reiterated his anxiety to see the new edition, and in particular to see in what manner Newton had incorporated the corrections suggested to him by Nikolaus Bernoulli.

(2) The second edition of the *Principia*. The failure of Newton himself to supply a complimentary copy aroused considerable chagrin in Bernoulli. A list of intended recipients of free

copies (U.L.C. Add. 3965(12), fo. 358, printed in Cohen, *Introduction*, p. 247), including Bernoulli's name, shows that Newton had originally meant one for him; but as Burnet later explains (see Letter 1055) Newton, because of Bentley's control of the edition, was in the peculiar position of having very few copies available for distribution. De Moivre later (Letter 1088) repeated this same excuse to Bernoulli.

(3) Bernoulli was of course particularly anxious to see the *Authoris Præfatio*—first to see what major changes Newton had made to the old edition, and second to see if he himself was acknowledged as originator of any of these changes. Bernoulli eventually received the copy of the *Principia* which Burnet sent in the spring of 1714 (see Letter 1070, note (3)).

1029 COTES TO NEWTON
22 DECEMBER 1713
From the original in the University Library, Cambridge[1]

Cambridge December. 22d. 1713

Sr

I lately received from You by Mr Crownfeild a paper of *Errata, Corrigenda & Addenda* to be printed & bound up with Your Principia.[2] I take leave to send You some Observations upon them.

By comparing Yr Catalogue with my Table of *Corrigenda*, I find You have omitted that of Pag. 3 lin. 14.[3] I think it convenient to make that alteration, that You may not seem to assert what is false.

You have also omitted that of Pag. 47. lin. penult. which I think is requisite to determine Your meaning.[4] Whilst that Sheet was printing I remember I did not understand what it was that You there asserted, & not having then time to examin the thing to the bottom, I was forc'd to let it go. Soon after I consider'd it & found in what sense Your words could be true & accordingly made the alteration. Since Yr Book has been published I have been ask'd the meaning of that place by one, who told me, he knew not what sense to put upon Yr words; I referr'd him to the Table of Corrigenda & then I perceiv'd he understood You.

Your addition in pag. 47 lin. 4, should I think be omitted.[5] For if that addition be made, the 8 preceding lines are to no purpose & ought to be omitted. Tis very evident by lin. antepenult. pag. 46 that PV is equal to $\frac{2DCq}{PC}$.

In Pag. 109 You direct to put H in the Figure instead of O. You mean instead of the lower O, which bisects the transverse diameter of the Hyperbola. If this be done, then the figure will not agree with the 2d line of this Page, nor indeed with the whole Demonstration as it relates to the Hyperbola.

In Pag. 148 lin. 7 I think the Alteration should not be made. There are three different *distantiæ* & three different *termini* & one common angular motion.

In Pag. 151. You change *prima* the Feminine into *primum* the Neutre. Tis my Opinion that this alteration is not necessary. I understand the printed text thus. *prima duarum medie proportionalium quantitatum.* If it were adviseable to make an alteration, I should rather chuse the Masculine & put it; *primus duorum medie proportionalium terminorum inter &c.*

Pag. 191. lin. 7 I think wants no correction. I cannot understand by what reasoning You make one; You will be pleas'd to reconsider it. If Your correction be true, it will be very necessary to explain it more fully.

Pag. 463 in the beginning of the 2d Table, I suppose You intended to put ♍ 12.25.50, not ♏ 12.25.50, as it is in Your written copy.[6]

You order the three last lines of page 460 & the two first lines of page 461 to be struck out, & in their room You place what follows [Eodem die ad horam quintam matutinam Ballasoræ —Londini, erat in ♎ 28 gr, 11′ cum Latitudine australi 1 gr. 16′ circiter.] I suppose You intended to make this addition at the end of the Paragraph which begins with (*Nov. 21. Ponthæus*), but would not have the 5 first lines of the following Paragraph struck out.[7]

I observe You have put down about 20 Errata besides those in my Table. I am glad to find they are not of any moment, such I mean as can give the Reader any trouble. I had my self observ'd several of them, but I confess to You I was asham'd to put them in the Table, lest I should appear to be too diligent in trifles. Such Errata the Reader expects to meet with, & they cannot well be avoided. After You have now Your self examined the Book & found these 20, I beleive You will not be surpris'd if I tell You I can send you 20 more, as considerable, which I have casually observ'd, & which seem to have escap'd You. And I am far from thinking these forty are all that may be found out, notwithstanding that I think the Edition to be very correct. I am sure it is much more so than the former, which was carefully enough printed; for besides Your own corrections & those I acquainted You with whilst the Book was printing, I may venture to say I made some Hundreds with which I never acquainted You.

I am Sir

Your very Humble Servt

ROGER COTES

For
Sr Isaac Newton
  at His House
in St Martin's Street
near Leicester Feilds
    London

NOTES

(1) Add. 3983, fo. 37, printed (from Cotes' copy in Trinity College Library, Cambridge) in Edleston, *Correspondence*, pp. 166–8. The brackets are in the original.

(2) Newton's paper is in Trinity College, Cambridge, MS. R.13.38, fos. 285–7; it is printed in Edleston, *Correspondence*, pp. 160–5. See also Cohen, *Introduction*, pp. 248–9. As Cotes remarks, the actual *errors* in the second edition here detected by Newton are almost all trivial; the list is largely composed of Newton's additional words, phrases and sentences, or other afterthoughts for which Cotes certainly was not to blame.

The longest addenda are to Book III. To page 459 Newton wished to add several lines on the comet of 1680; to page 483 the following passage: 'A necessitate Metaphysica, quæ utique eadem est semper et ubique, nulla oritur rerum variatio. Omnis illa quæ in mundo conspicitur pro locis ac temporibus diversitas a voluntate sola Entis necessario existentis oriri potuit. Dicitur autem Deus per Allegoriam videre, audire, loqui, ridere, amare, odio habere, cupere, dare, accipere, gaudere, irasci, pugnare, fabricare, condere, construere, & intelligentes (vitam infundendo)* generare. Nam sermo omnis de Deo a rebus humanis per similitudinem aliquam desumi solet. Et hæc de Deo; de quo utique ex phænomenis disserere ad Philosophiam experimentalem pertinet.'   *Job. 38.7 / Luc. 3.38

('No variety of things could arise from a metaphysical Necessity, which must of course be everywhere and at all times the same. All that diversity in terms of time and place which is observed in the world, could arise only from the will of a necessarily existing Being. Now metaphorically speaking God is said to see, hear, speak, smile, love, detest, desire, accept, rejoice, become angry, fight, make, establish, build, and create intelligences (by infusing them with life). For all discourse concerning God must be borrowed from the human realm by means of similitude. And thus far concerning God, to discourse of whom from the phenomena is surely within the province of experimental philosophy.')

Newton added this famous passage to his interleaved copy of the second edition and it was printed (in a modified form) in the third edition. (See Koyré & Cohen, *Principia*, II, pp. 763–4.)

(3) Speaking of the gravitational deflection downwards of a projectile Newton wrote that it is the less as the gravity of the body *in proportion to its quantity of matter* is less, and also as its velocity is greater. The italicized words were added by Cotes in the *Corrigenda*, and remain in the third edition.

(4) As corrected by Cotes, Newton wrote: 'the forces towards any centre whatever placed on the abscissa, *from any two places of the figure at which are terminated the ordinates erected from corresponding points of the abscissae*, are increased or diminished in proportion to the distance from the centre.' Cotes introduced the italicized qualifying words; in the third edition they were reduced to *along each ordinate*.

(5) Newton had proposed to say: '. . . and this circle will be of the same curvature as the conic section at P, and *so (from the nature of conic sections) the chord of this circle*, PV, will be equal to . . .' The italicized words were not employed by Newton in the third edition when the sentence was slightly recast.

(6) Newton proposed to add a cometary observation of 9 November 1680, but wrote the symbol for Scorpio in mistake for the symbol for Virgo (it is easy to see from the table that Virgo must precede Libra, since Scorpio follows).

(7) The references to the observations of Ballasora were ultimately made in the third edition in the manner here proposed by Cotes.

1030 NEWTON TO OXFORD
[c. DECEMBER 1713]
From the holograph draft in the Mint Papers[1]

To the most Honble. the Earl of Oxford & Earl Mortimer
Lord High Treasurer of great Britain

May it please your Lordp

In obedience to your Lordps Order of Reference upon the Petition of John Pery & others for supplying the Mint with either Blancks or Plates of fine copper to be coined into half pence & farthings,[2] We humbly represent that copper money is at present very little wanted, but if it shall be thought fit to put the coinage of such money into a standing method, We are humbly of opinion

That the whole coinage including the making of the blancks be done in the Mint it being unsafe to have coining tools & coinage abroad.

That it be done of the cheapest fine copper wch will hammer when red hot & is worth about 11d or 12d per pound weight. In finer & dearer copper we may be easily deceived, there being no certain test of the higher degrees of fineness; & the great price will tempt fals coiners to counterfeit the money.

That it be done out of copper either hammered into plates at the copper mills, or cast into barrs at the Mint with an addition of two or three ounces of Tinn to an hundred weight of copper in fusion to make the metal run close. Both ways may be tried, but The last way is most conformable to the coinage of gold & silver, & is cheapest by two pence in the pound weight, & seems therefore to be preferred. For there will be least got by counterfeiting that money whose workmanship is cheapest.

That this money have a fair & bold impression & such an edging as may be fittest to prevent counterfeiting it by casting & with the stamp [as] shall be directed by the Queen & council[3]

That an Importer be appointed to buy & import the copper by weight, & receive back the new money by weight & tale & put the same away. And that the Master & Worker for the time being, be charged & discharged by his Note as in the coinage of gold & silver & have power to retain the Coinage whenever upon the Assay it proves not good.

That all the charges of Copper; coining tools, coinage, wages & incidents be paid out of the profits of the coinage & that either the Importer or the Master be accountant, & that there be no perpetual salaries to increase the extrinsic value of the moneys, but all services be paid for by the pound weight.

51

That a coinage of about twenty or thirty tunns once in three or four years, or of fifty Tunns once in six or eight years is sufficient for supplying the daily loss & wast of the moneys already coined, & may prove too much if the counterfeiting of the money encreases. And that a coinage of twenty or thirty or at the most fifty Tunns seems to be abundantly sufficient at present.

That a coinage of such money may be performed from time to time by one & the same standing Commission & that it be left in the power of the Ld H. Treasurer to appoint by a particular Warrant the quantity of copper money to be coined at any time which quantity should never be so great at once as to endanger any clamour.

And that when the coinage of such money shall be resolved upon by her Ma[jes]ty, the Petitioners & others who have copper works be treated with, & his copper chosen wch is best coloured & most malleable & cheapest of those sorts of copper which will hammer when red hot.

All which, & whether a coinage shall be set on foot till there be a greater want of such money, is most humbly submitted to your Lordps great wisdome.

NOTES

(1) II, fo. 306; there are other drafts at fos. 431 and 438.

(2) On 12 February 1713 John Pery and others, proprietors of the Temple Mills copper and brass works near Great Marlow in Bucks., offered to deliver fine copper in blanks (not exceeding 21 to the pound weight, and selling for $17\frac{1}{2}d$. the pound) or in plates for 15 pence the pound (see *Cal. Treas. Books*, XXVII (Part II), 1713, pp. 118–19). Newton's (or the Mint's) reply to this offer does not seem to survive among the official records, hence we cannot date this (undated) draft.

(3) Three interlineated words here are hard to read, but the meaning is clear from other, deleted lines; the dies for the copper coins should remain the same, except as directed by the Queen in Council.

1031 NEWTON TO OXFORD
[JANUARY 1714]
From the holograph draft in the Mint Papers[1]

To the most Honble the Earl of Oxford & Earl
Mortimer Ld H. Trea[sure]r of great Britain.

May it please your Lordp

The last Autumn I laid before your Lordp a Memorial wherein (to the best of my memory) I represented that if Copper worth $11\frac{1}{2}d$ per Lwt could be wrought into money by casting,[2] the Copper & Coynage without edging the money would amount unto $17\frac{1}{2}$ per Lwt, & allowing $2\frac{1}{2}d$ more for the charge of putting off, buying & setting up coyning tools, repairing the buildings, building furnaces, paying Clerks & a Comptroller, & other Incidents; the whole would amount unto 20d per lwt. And after so much was coyned as was sufficient to supply the present want of copper money (wch I recconed might be about 80 or 100 Tunns) the surplus above all charges, if there were any, might be paid into the Exchequer. And in the next coynage after a copper Mint was set up the money might be made heavier. But by further experience it appeared afterwards that such Copper could not be wrought by casting but must be wrought into barrs at the battering Mills.[3]

I now beg leave to represent to your Lordp that in my humble opinion the best method of coyning such bars into money, is to receive the same by weight & assay, giving Bills to the Importers & taking back the bills upon delivering back the same weight of Copper in scissel & money together, the Importer paying for the coynage of the money by the pound weight a certain Seigniorage to be accounted for by the Master & Worker. That the Importers will expect about 17d per pound weight above the Seigniorage for their copper & workmanship & for putting off the Copper money. And that the Moneyers insist upon $2\frac{1}{2}d$ per lwt for their work, & it will cost an halfpenny per lwt to the Graver & Smith & a Clerk, in all 3d per lwt, to be paid out of the Seigniorage besides the allowance to the Master or his Deputy & such other Officers as shall be appointed, & besides the repairs of the buildings & charge of the coining Tools.

All which &c

NOTES

(1) II, fo. 308. Like the previous draft, this is undated and hard to place. One might reasonably suppose, but for the first sentence, that it was written after Letter 1034. Since the reference to autumn as the time of the last memorial excludes that possibility, one must suppose it to have been written in early January.

(2) See the undated 'Considerations about the Coynage of Copper Moneys' (Number 1003, vol. v) and another undated holograph memorandum addressed to Oxford which we have arbitrarily printed here as Letter 1030. It will be observed that the numerical estimates in these various drafts are highly discrepant.

(3) Hammer mills; alternatively, the copper would be rolled from the ingot. Casting into bars—still less into coin—did not work well on a large scale because of the tendency of the metal to become porous unless much tin was added.

1032 OXFORD TO NEWTON
5 JANUARY 1714
From a copy in the Public Record Office[1]

After my hearty Commendations, These are to signify to you her Majties Pleasure That you prepare or cause to be made and prepared with all the Speed that maybe the number of fifty Medals of Gold according to the Draft approved by her Majty. the 24th Day of June last [2] each Medal to contain in Weight about 14dwt. & 15gr. of fine Gold and to be in value (with three shillings a piece for workmanship and Wast) about three pounds ten shillings which said Medals you are to deliver to the Cofferer of her Majties Household [3] whose acquittance for the same shall be your sufficient Discharge herein, And for so doing this shall be your Warrant.

Whitehall Trea[su]ry Chambers 5th. Janry 1713

To my very Loving Freind,   OXFORD
Sr. Isaac Newton Knt. Master
& Worker of her Majties Mint

NOTES

(1) Mint/1, 8, p. 99; there is another in T/54, 22, p. 139. Masham's receipt—immediately following the order—is dated 16 January. It records the weight of the fifty medals as 36 oz 13 dwt 17 gr and their total cost as £170 less a halfpenny.

(2) See Letter 1006, note (2), p. 11. The medal commemorated the Peace of Utrecht.

(3) Samuel Masham (1679?–1758), whose wife was the redoubtable Abigail Hill; he had been appointed Cofferer to the Household of Queen Anne in 1711 and was one of the twelve peers created in 1712.

1033 BENTLEY TO NEWTON
6 JANUARY 1714
From the original in Trinity College Library, Cambridge[1]

Sir,

Your borrowing ye [Manuscript] for me at ye hazard of your own Bond was a favour extraordinary: [2] but upon Mr Crownfield's [3] report concerning ye Order made by ye Society, I had drawn a Bond here before ye Receipt of Your Letter; [4] and have sent it by ye Bearer our Cambridge Carrier, who will deliver it you, and bring me ye [Manuscript] safe. I acquainted Mr Cotes with ye Contents of your Letter. I am, with my Hearty Wishes to you of many happy New Years,

Your very Obliged humble servt

RI: BENTLEY

Jan 6th 17$\frac{13}{14}$

For Sr Isaac Newton
at his House in Martin street
near Leicester fields
   London

NOTES

(1) R.4.47, fo. 25.

(2) At a meeting of the Royal Society on 10 December 1713 it was 'Ordered that a Manuscript in folio Being Suetonius about the year 1480 he delivered to Dr. Richd. Bentley he complying with the order of Council in such Cases provided by giving Bond for the Return thereof.' (Journal book of the Royal Society of London, XI, p. 394).

Clearly the bond was not immediately forthcoming, for at a meeting two weeks later it was 'Ordered that a Manuscript of Suetonius be lent to Dr. Bentley for which the President was pleased to offer himself for security for the Return of it.' (*ibid.*, XI, p. 400).

This action on Newton's part occasioned the present letter. Caius Suetonius Tranquillus was a historian of the first century A.D., whose major work was *C.S.T. de vita XII Caesarum Liber primus* . . . First published in 1470, the book went into over thirty editions by 1500. Presumably the MS. was part of the Arundel Library; these books were later sold by the Society.

(3) Presumably Cornelius Crownfield, Inspector of the Press for Cambridge University; see Letter 777, note (2) (vol. V, p. 30).

(4) Newton's letter to Bentley is lost.

1034 THE MINT TO OXFORD
22 JANUARY [1714]
From a copy in the Public Record Office[1]

To the Most Honble: Robert Earle of Oxford &c.

In Obedience to Your Lordships Verbal Order of Reference [2] concerning the best manner of Importing Copper into the Mint to be Coined into Copper money of a certain Standard and whether such an Importation may be made free [3] We humbly represent to Your Lordship that if Copper be mixed with any other base Mettal or Semimetal it will not endure the Hammer when red hott butt will fly in peices. So soon as it is refined by the Copper Workers to

55

that degree as to be pretty well purged from all other base Mettals it begins to endure the Hammer when red hott without flying in peices butt nott without cracking, And for making Vessells and other Utensills of Copper there is no need to refine it higher.

They that Work Copper from the Oar when they have brought it to such a degree as they call fine Copper, sell it to those who have Mills for manufacturing it and expose it to sale in their Warehouses in London and such Copper is worth from £95 to £100 per Tun and very little of it is worth above £100 per Tun. This is commonly called fine Copper But that of £95 per Tun will scarce hammer without cracking and then it is worth about $11\frac{1}{2}$ [d] per pound weight or about £107 per Tun or thereabouts.

Refiners of Copper by refining a small part of any Mass can make an Estimate of the Charge of Refining the whole Mass and how much fine Copper it will produce But the best Way of making such an Assay is not yet Agreed upon, when it shall be agreed upon & brought into Common Use it may be then considered whether the Master and Worker shall be allowed to buy course Copper by such an Assay and putt it out to Refine wth: publick Money and what he shall be allowed in his Acc[oun]t for the refineing thereof according to the several degrees of the Coursness, and what for the waste by evaporation. In the mean time the Master or any other person may buy course Copper, put it out to be refined and lend it to the Mint to be Coined.

The Malleability of the Copper depends not only on the fineness of the metal or freedom from other Metals but also upon the manner of refining it for if it be refined with Seacole it will nott be Malleable and fitt for working though it be fine. It may be refined with Sea coale till it begins to be fine and then it must be wrought with charr'd coale till it be fully fine and the charr'd Coale of Wood is better than Sea coal charred. Also the melting diminishes the Malleability, especially if it be melted with too much heat And for these Reasons the Assay by the Hammer is the best and Surest for the Mint.

If it shall be thought fitt that Copper Money be made of such Copper as in hammering when red hott will crack but not fly in peices, it may be bought at the Copper warehouses in London; if of such Copper as will hammer red hott without cracking it must be had of those who refine Copper.

In the Reign of King Charles the Second a pound weight of Swedish Copper was cutt into 20d. The Copper and making the Blanks cost 18d the Stamping 1d and a penny remained for other charges; This Copper was malleable so as to hammer red hott without cracking.

If it be thought fit that the Money now to be Coined be of like fineness so as to endure the same teste the Copper will cost $11\frac{1}{2}$[d] per pound weight as above and Coynage about 6d or $6\frac{1}{4}$d without edging or 7d. with edging And if a

56

PLATE II. Experimental copper coins struck in 1713–14. The dies were designed by John Croker; special pieces were minted from them (even in precious metal) into George I's reign. No copper was ever issued in Anne's reign, but it is believed that some hundreds of coins were struck. *Top to bottom*: (1) Double-headed halfpenny, showing some splitting and cracking of the metal (see Letter 1034); (2) Halfpenny, the reverse symbolizing the Union of England and Scotland (see Letter 723, vol. v); (3) Farthing, the reverse symbolizing the Treaty of Utrecht, dated 1713. The obverse shows the same head of the Queen as the double-headed halfpenny. (Reproduced by courtesy of the Trustees of the British Museum.)

pound weight not edged be cutt into 20*d* or a pound weight edged be cutt into 21*d* there will be an Excess of $2\frac{1}{2}$[*d*] per pound weight for purchasing Mills and presses and Cutters and flatters and setting up a Copper Mint and paying Clerks and incident Charges of Assaying, weighing, telling, porterage, baggs paper and packthread, putting off &c. But if it be thought that the Copper be only so fine as to endure the Hammer when red hott without flying in peices tho not without cracking a pound weight may be cutt into 19*d* not edged or 20*d* edged.

The Mills and Presses and other Engines for Setting up a Copper Mint will cost Six or Seven hundred pounds and three farthings per pound weight in Coining an hundred Tuns will pay that charge. And when that charge is paid the weight of the money may be a little augmented.

If the Blanks be so thick or the Impression rise so high as to straine the Dyes or Cutters and make them more apt to breake then in the Coinage of Gold and Silver or the Casting into Barrs prove so difficult as to make above one half of the Barrs become Scissile,[4] the Charge of Coinage must be proportionably augmented for which reasons the Charges of Coinage cannott be positively sett without Experience in Coining some Tuns of Copper Money.

The Weight of all the Copper received and the Weight and Tale of all the Copper money Coined may be Entered into Books and in the Accts. of the Master and Worker and the Surplus above all Charges may be paid into the Exchequer.

All which is most humbly submitted to Your Lordships great Wisdome

Janry: 22*d*: 1713

NOTES

(1) Mint/1, 7, pp. 60–1. There is a draft (not in Newton's hand) in the Mint Papers, II, fo. 413. Compare Letter 1031.

(2) The question of a copper coinage had been in suspense since the autumn of 1713, and was to remain so until the autumn of 1714 (see Letter 1113) and beyond. Newton was willing enough to discuss the technicalities in general (as in this letter) but highly reluctant to make a bargain with any particular copper manufacturer which would compel him to accept so much in such a condition at a stated price. He constantly emphasized the slight need for more coppers (in his own view) and the great good sense of doing nothing. No copper coins were minted for currency during Queen Anne's reign, but a number were struck for trial purposes (see Plate II).

(3) Viz: whether any merchant, rather than specific contractors, might offer copper for purchase by the Mint.

(4) Scissel, metal scrap left after cutting or punching shapes from sheet or strip.

1035 NEWTON AND PEYTON TO OXFORD
23 JANUARY 1714
From the original in the Public Record Office.[1]
Reply to Letter 1011

To the most Honble the Earl of Oxford
& Earl Mortimer, Lord High Treasurer
of Great Britain.

May it please your Lordp

In obedience to your Lordps Order of Reference upon the annexed Memorial of Mr Charles Tunnah & Mr William Dale for coyning in ten years a thousand Tunns of halfpence & farthings of an artificial metal wch toucheth like ordinary gold: & for cutting a pound weight Averdupois into 32 pence: We humbly represent

That the selling blancht copper or making it for sale is forbidden by law upon pain of death because of its fitness to be used in counterfeiting the silver moneys: & for the same reason it may be of dangerous consequence to encourage the making of an artificial metal which toucheth like gold, & is used in making sword hilts & other wares in imitation of gold. The halfpence made of this metal & melted down with a little fine Gold may make a composition very dangerous for counterfeiting the Gold moneys.

That in the last coynage of copper moneys an hundred Tunns per annum at the end of six years occasioned great complaints in Parliament so as to cause the coinage to be stopt for a year. And after another hundred Tunns were coined, the nation was overstockt for four or five years. And therefore six hundred Tunns may be deemed sufficient for the use of all England, whereof there seem to be about 500 Tunns already current.[2]

That the secret of making this metall being known only to the Petitioners it has no known intrinsic value or market price: whereas halfpence & fart[h]ings (like other money) should be made of a metal whose price among Merchants is known, & should be coyned as neare as can be to that price including the charge of coynage.

And that the people are not nice & curious in taking good copper money but may be imposed upon by money made of princes metal instead of the metal here proposed: and that the cutting a pound weight into 32 pence may be a great temptation to counterfeit such money.

All wch reasons incline us to preferr a coynage of good copper according

58

to the intrinsic value of the metall. But we most humbly submit our opinion to your Lordships great Wisdome

CRAV: PEYTON

IS. NEWTON

Mint Office
23 Jan. 171¾

NOTES

(1) T/1, 172, no. 25, written by Newton, signed by him and Peyton. There is a holograph draft by Newton in the Mint Papers, II, fo. 435. This report was read on 27 January and endorsed 'My Lord agrees with the report'. It was first printed in Shaw, pp. 186–7.

(2) At 44800*d* to the ton of copper, six hundred tons would have provided some ten pennyworth of small change for every man, woman and child in England and Wales—perhaps a sufficient supply.

1036 BERNARD LE BOVIER DE FONTENELLE TO NEWTON
24 JANUARY 1714
From the printed version in Brewster, *Memoirs*[1]

Monsieur,

Je suis chargé par l'Académie Royale des Sciences d'avoir l'honneur de vous remercier de la nouvelle Edition que vous lui avés envoyée de vos Principes des Mathematiques de la Philosophie Naturelle. Il y a déja plusieurs années que cet excellent ouvrage est admiré dans toute l'Europe savante, et principalement en France, où l'on sait bien connoistre le merite étranger. Mais presentement, Monsieur, que vous avés une place dans notre Academie, nous prétendons, en quelque façon que vous n'êtes plus étranger pour nous, et nos Savants qui ont quelque droit de vous appeller leur Confrere prennent une part plus particuliere a votre gloire. On peut sans temerité vous prédire qu'elle sera immortelle par les deux Livres que vous avés publiés, ou il brille de toutes parts un si heureux genie de découvertes, et ou ceux-même qui savent le plus trouvent tant a apprendre.

l'Academie vous prie, Monsieur, de lui faire quelque fois part de vos nouvelles productions, ainsi que font Ms. Leibnits, Bernoulli, et les autres Savants étrangers qu'elle a adoptés. Il n'est pas surprenant qu'elle cherche a se faire honneur de ce qu'elle vous possede. Je suis,

Monsieur,

Votre très humble et très obéissant serviteur,

FONTENELLE

Sec. Perp. de l'Ac. Roy. des Sc.

à Paris, ce 4 *Fév.* 1714 [N.S.]

NOTE

(1) II, p. 518; we have been unable to find the original of this letter but the version we print here is taken from the copy of Brewster's *Memoirs* in the University Library, Cambridge, which H. R. Luard collated with the original. For Bernard Le Bovier de Fontenelle see Letter 1084, note (1), p. 146.

1037 LOWNDES TO THE MINT
27 JANUARY 1714
From the copy in the Public Record Office.[1]
For the reply see Letter 1051

Gentlemen

My Lord Trea[sure]r directs You with all speed possible to send his Lordp an Acco[un]t of the whole Quantity of Tynn, which Her Ma[jes]ty has paid for, that now remains unsold As well in Your Custody in the Tower as in the Hands of the Agents for Tyn in Cornwall and Devon And at Hamburgh Holland or elsewhere, belonging to Her Ma[jes]ty wth as near an Estimate as You can of the Value thereof or what the same when sold may produce into the Excheqr.[2] 27. Janry. 1713

WM LOWNDES

NOTES

(1) T/27, 21, no. 122.
(2) Compare Letters 867 and 870, vol. v.

1038 THE MINT TO OXFORD
3 FEBRUARY 1714
From the copy in the Public Record Office[1]

To the Most Honourble the Lord High
Treasurer of Great Britain

May it please your Lordsp.

By the Indenture of the Mint the Warden and Master & Worker are directed to Account yearly, and by a late act of Parliament they are ordered to exhibit copyes of such accounts to the Comm[issione]rs appointed as other Accomptants are.

Her Majties. Assay Master having been dead above twelve months,[2] and his place having been supplied ever since and long before his death by Charles Brattle his Brother for enabling the Warden & Master to complete their Accounts for the year ending at Chr[ist]mas last, we humbly pray that your Lordsp. will be pleased to appoint such allowance as your Lordsp. shall think fit to the said Mr. Brattel & his Clerk for their service for the said year. The Salary of the Queens Assaymaster upon the Indenture of the Mint being £200 per annum & that of his Clerk £20 per annum

All which &c

CRAV. PEYTON

3. *Feb.* 1713

IS. NEWTON

ED. PHELIPS

NOTES

(1) Mint/1, 8, p. 98; the original (a clerical copy) differing only in spelling is in T/1, 172, no. 35.

(2) See Letter 966, vol. v and Letter 1013; Letter 1041 is the response to the present request.

1039 KEILL TO NEWTON
8 FEBRUARY 1714
From the original in the University Library, Cambridge.[1]
For the answer see Letter 1053

Oxon February 8th 171$\frac{3}{4}$

Honoured Sr

I lately received the Inclosed [2] from Holland. It seems the Gentleman I intrusted my letter with keept it a long time by him, before he delivered it, It seems Mr Leibnits has already made some answer to the Commercium wch I doubt not but you have seen, [3] I would gladly have your opinion what you think is needfull further to be done in answer to Mr Leibnits, I cannot conveniently (having some bussiness that detains me here) come to London at present, [4] but I wish I could see the journal[s], where the answers are, if you could send them me, you may order your servant to deliver them either to the Oxford Coach or wagon directed for me and I will take care to return them to you. and I will observe your orders as far as I can. You see we may have what we please printed in these French Journals and I am of opinion that Mr Leibnits should be used a litle smartly and all his Plagiary and Blunders showed at large. [5] However Sr I expect your directions and I am with all respect

your

most obliged

and obedient Humble Servant

JOHN KEILL

For
Sr Isaack Newton
at his house in
St Martins street near
Licester feilds
    Westminster

NOTES

(1) Add. 3985, no. 3.
(2) Letter 1039*a*.
(3) Keill—or rather Johnson in the enclosed letter—alludes to Leibniz's 'Remarques' in the *Journal Literaire de la Haye* (see Letter 1018) which Newton had presumably not yet seen, though he was aware of the *Charta Volans* (Number 1009) (see Letter 1045, note (3)).

(4) Keill had been Savilian Professor of Astronomy at Oxford since May 1712.

(5) After detailed discussion with Newton in subsequent correspondence, Keill eventually had a reply published in the *Journal Literaire de la Haye*, for July and August 1714, pp. 319–58 (see Letter 1053*a*, note (1)); a few copies of the article also appeared as a separately printed pamphlet.

1039*a* JOHNSON TO KEILL
29 JANUARY 1714
From the original in the University Library, Cambridge.[1]
Enclosure with the preceding letter

Sir

I had the honour of yours by Dr Cull last month, & have communicate[d] it to the gentlemen that write the Journal Literaire.[2] They will always be glad to receive any thing curious in any part of Mathematicks or natural Philosophy from so good good [*sic*] a hand as yours & bid me, with their respects assure you, Sir, that they'l always endeavour to make the best use of every thing you'l be pleased to communicate to them & that they are per- fectly well-disposed to doe justice to the ingenious & learned of Great Brittain.

You'l see by the Journal of November & December what is said in behalf of Mr Leibnitz by some of his friends, by which you'l easily judge what may be further necessary for clearing up that matter.[3] Whatever you send new upon it will be very welcome, as well as the Demonstrations of the two propositions mentioned in your Letter. When you are so kind as to send me any thing, let it be given in (sealed & directed for me) to M. John Darbys in Bartholomew Close; I shall take care henceforth to send you always one of the Journals the same way.

This Journal takes very much all over Europe, & I have letters from learned men of all quarters, testifying their approbation of it & promising their assistance to enrich it. I hope Sir what you contribute to it shall be none of the least. I shall always be glade to see the learned of Great Britain make the most considerable figure in it, & more particularly those of our Country, for there is nothing I desire more than to contribute to advancing the honour & advantage of every ingenious & worthy Scots man.

I am with very much respect

Sir

Hague 9 *febry*. 1714   Your most obedient &

most humble Servant

T. JOHNSON

P.S. Let me beg of you, as often you favour me with any letters, to add what new books are coming out in your University

For
The Ingenious & learned
Doctor Keill, Professor
of Astronomy in Oxford
    Great Brittain

NOTES

(1) Add. 3985, no. 4. The writer, evidently a Scot like Keill, was a bookseller at The Hague and publisher of the *Journal Literaire de la Haye*.

(2) According to J. N. S. Allamand (*Oeuvres de 'sGravesande*, (Amsterdam, 1774)) they were, besides Willem 'sGravesande himself, Marchand, van Essen, Sallengre, Alexandre and St. Hyacinthe. This volume of the *Journal Literaire de la Haye* first appeared in May 1713.

(3) See Letter 1039, notes (3) and (5), pp. 62–3.

1040 J. BERNOULLI TO BURNET
8 FEBRUARY 1714
Extract from the original in Basel University Library.[1]
Reply to Letter 1028; for the answer see Letter 1055

... Mr. Rilliet qui m'a envoyé Vôtre lettre de Paris en l'accompagnant de la sienne, me mande qu'il m'enverra par la premiere occasion le livre de Mr. Newton dont Vous avez la bonté de me faire présent ... Vous trouverez mes remarques sur le livre de Mr. Newton de la premiere Edition dans les Actes de Leipsic du mois de Février et de Mars de l'année passée, j'avois esperé qu'elles paroitroient dans le méme temps que la nouvelle edition de Mr Newton, mais celle cy n'a eté rendue publique, comme Vous dites Vous memes que 4 ou 5 mois aprez, ensorte que Mr. Newton a pû avoir vû mes remarques avant la publication de son livre;[2] en effet il y a inseré une correction à ce qu'on m'a dit, qu'il a emprunté de moi, quoiqu'il soit vrai que mon Neveu la Luy a deja communiquée dés le temps qu'il etoit en Angleterre;[3] au reste j'ay parlé de Mr. Newton dans ces remarques avec beaucoup de respect et avec toute la consideration que l'on doit avoir pour son merite: cependant je ne sçai s'il ne l'a pas pris en mauvaise part puisque depuis qu'il m'a fait reçevoir dans la Societé Royale [4] et qu'il m'a promis par Mr. de Moyvre de m'envoyer le *commercium Epistolicum* [5] touchant l'affaire entre Lui et Mr. Leibnits comme aussi un exemplaire de la nouvelle Edition de son Livre, non seulement il ne m'a rien encore envoyé de tout cela, quoyqu'il en ait envoyé a Mr. Varignon et à d'autres Mathematiciens à Paris, mais je ne reçoi pas meme des lettres de Mr. de Moivre qui faisoit l'office d'internonce entre Lui et moi, vû qu'on me doit une reponse depuis prés d'un an . . .[6]

NOTES

(1) LIa 654 no. 12. Bernoulli enclosed this letter in one to Rilliet (Basel University MS. LIa 654, no. 12), thanking him for forwarding Burnet's letter of 3 December, and saying how anxious he was to receive the second edition of the *Principia*.

(2) *Acta Eruditorum* for February 1713, pp. 77–97 and March 1713, pp. 115–32. Newton did not in fact see the volume until April 1714.

(3) See Letters 951, 951*a*, 953 and 961, vol. v.

(4) Bernoulli was elected a Fellow on 1 December 1712, as de Moivre had informed him in a letter of 17 December 1712 (see Wollenschläger, *De Moivre*, p. 277).

(5) See Wollenschläger, *De Moivre*, pp. 273 and 280.

(6) De Moivre's last letter to Bernoulli (see Wollenschläger, *De Moivre*, p. 277) was dated 17 December 1712. He finally wrote again on 28 June 1714 (see *ibid.*, p. 289; we give extracts in Letter 1088).

1041 OXFORD TO THE MINT
9 FEBRUARY 1714
From the copy in the Public Record Office.[1]
Reply to Letter 1038

Anne R.

Whereas Our high Treasurer of Great Britain hath laid before us your Memorial of the third Instant wherein you represent That Charles Brattle Gent. Brother to the late Assay master of Our Mint hath ever since the death of his said Brother performed the duty of that Imploymt & doth still continue to execute that Trust, Wherefore you have proposed that he may be paid for that service after the rate of two hundred pounds per annum for himself and Twenty pounds per annum for a Clerk being the Salarys allowed by the Indenture of our Mint to the Assay master and his Clerk, To which we being gratiously pleased to condescend Our Will & pleasure is, And we do hereby direct Authorize and Command you or such of you to whom it may concerne to pay or cause to be paid unto the said Charles Brattle or his Assigns so much as the said Allowances after the rate of Two hundred pounds per annum for himself and Twenty pounds per annum for his Clerk hath amounted to from the decease of his said Brother untill Chrmas last, and also to pay the like allowances for the future for so long time as he shall continue to Execute the said Office of Assay master. And this shall be as well to you for making the said payments as to the Auditors of our Imprests and all others concerned in

65

allowing thereof a sufficient Warrant. Given at Our Court at Windsor Castle the 9th day of February 1713 In the twelfth year of our Reign

By her Majties Commd

OXFORD

To our trusty and Wellbeloved the warden
Master & Worker and Comp[trolle]r of our Mint

NOTE

(1) Mint/1, 8, p. 98, a clerical copy.

1042 THE DUKE D'AUMONT
TO THE ROYAL SOCIETY
14 FEBRUARY 1714
Extracts from the original in the Library of the Royal Society.[1]
For the answer see Letter 1082

A Versailles le 25: *fevrier* 1714

Je ne puis Messieurs perdre de Veüe l'honneur que Vous m'avez fait de me recevoir dans vostre Societé: apres les bontez qu'il a plu a la Reyne d'avoir pour moy Rien ne ma paru plus precieux que ces marques Singulieres de Vostre Estime, je les ay regardées comme la plus flateuse de toutes les distinctions . . . en rendant compte au Roy mon Maistre de tout ce que j'avois trouvé en Angleterre de celebre et de respectable je n'ay pû oublier une compagnée composée de tant de sujets excellens . . . J'ay trouvé tous nos sçavants prevenus sur l'Elevation de Vostre genie sur l'estendue de vos connoissances et la precision de vos idées, je n'ay eu a loüer que vostre modestie et vostre bon foy qualitez toujours inseparables des plus grand[es] Lumieres. Tout ce qui me reste a desirer est que Vous vouliez bien me faire part de vos decouvertes et me mettre par la non seulement a portée de demontrer les Sentiments qui m'attachent a la gloire de l'Academie, mais aussy d'estre utile a ceux de mon pays que le goust des Sçiençes et le desir de la perfection ont jettés dans les memes recherches. Je suis plus parfaitement que personne du monde Messieurs Vostre tres humble et tres Obeissant Serviteur

LE DUC DAUMONT

NOTE

(1) Early Letters, A 55. Louis-Marie d'Aumont-Rochebaron (1667–1723) was third duke, and Gouverneur du Boulonnais from 1704–23. He was a frivolous and thriftless man, of little

use as Governor; there is no evidence of any deep interest in scientific matters. He came to London, with a magnificent entourage, in December 1712. He was elected Fellow of the Royal Society on 21 May 1713, and he himself was present on the occasion, the usual business of the day being deferred and experiments made for his entertainment. (See Journal book of the Royal Society of London, XI, pp. 356–7 and 364).

The Duke left England in December 1713 under something of a cloud, for the house of Lord Powis, where he had been staying, was unaccountably burnt to the ground, and some suspected him of arson. (See A. Hamy, *Essai sur les ducs d'Aumont, Gouverneurs du Boulonnais, 1622–1789* (Boulogne-sur-Mer, 1906 and 1907), pp. 154–70).

A translation by Halley of the Duke's letter of thanks (see Early Letters, A 56) was read to the Society at a meeting on 13 May, and on 27 May Newton's reply was read and approved.

1043 J. BERNOULLI TO LEIBNIZ
17 FEBRUARY 1714
Extract from Gerhardt, *Leibniz: Mathematische Schriften*, III/2, p. 929

Optime facies, si quæ suppeditavi, ea a me profecta dissimules in Apologia, quam paras contra Commercium Epistolicum ab Anglis editum, in quo quidem repetuntur quædam ex literis, quas Wallisius edidit. Nescio autem, annon quædam, quæ in Tui defensionem facerent, sint omissa, neque hactenus vacavit omnia inter se conferre. Ante annum et ultra promisit Dn. Moivræus, nomine Newtoni, hunc mihi missurum hoc Commercium una cum exemplari novæ Editionis suorum Princip. Philos.[1] Jam vero dudum exemplaria distribuenda misit in Galliam; ad me vero nihil hactenus, nec libri, nec literarum amplius pervenit. Unde animum subiit suspicio, an forte Newtonus mihi succenseat, quod animadversiones quasdam meas superioris Anni mensibus Februario et Martio in Actis Lipsiensibus publicaverim in veterem Editionem Princip. Phil.[2] Keilius Newtoni simia videtur; quidquid ab eo vidi ex Newtono compilatum est. Cheynæus,[3] quondam magnus Newtoni Idololatra, valedixit, ut audio, rebus Mathematicis; victurus posthac inter nescio cujus sectæ Visionarios, Fatii[4] exemplum secutus, nisi omnino Wistoni,[5] qui Arianismum resuscitare conatur.

An Newtonus non erraverit in Enumeratione Linearum tertii ordinis,[6] omittendo forte quasdam, quasdam alias bis sumendo, pro diversis quæ eædem sunt habendo, ut accidit Craigio et aliis, qui problema de transformatione Curvarum solvere volentes, eandem cum data exhibuerunt ad diversum tantum axem;[7] an, inquam, Newtono non simile quid contigerit, asseverare non ausim: ejus enim Tractatum hac de materia ut examinarem, nondum a me impetrare potui, quia non libenter hisce tricis, utpote haud valde utilibus me immisceo. Frater meus aliquando hoc vadum tentavit; quo vero successu non memini, fortasse in Scriptis ejus aliquid invenire est.

67

Translation

You do best if you conceal the origin with me of what I have furnished, in the answer which you are preparing against the *Commercium Epistolicum* published by the English, where they do indeed repeat some things from the letters which Wallis published. I do not know, however, whether or not the ones which help in your defence are omitted, nor have I until now had leisure to make collations on every point. More than a year ago Mr de Moivre promised to send me, in Newton's name, this *Commercium* together with a copy of the new edition of his *Principia Philosophiæ*.[1] A little while ago he sent copies for distribution in France. But nothing has come to me so far; neither books, nor any more letters. Hence a suspicion creeps into my mind, that perhaps Newton is angry with me, because of some observations which I published last year in the *Acta* [*Eruditorum*] of Leipzig for February and March on the old edition of the *Principia Philosophiæ*.[2] Keill seems to be Newton's ape; whatever I see of his is compiled from Newton. Cheyne,[3] once a great idolater of Newton, has given up mathematics, so I hear; he is to live for the future as a member of some visionary sect or other, after the example of Fatio,[4] not to say Whiston himself,[5] who strives to revive Arianism.

Whether Newton has not erred in his *Enumeratio Linearum tertii ordinis*,[6] omitting some by chance, counting others twice, treating as different those which are to be treated as the same, as happened to Craige and the rest, who, wishing to solve problems of curve transformation, only referred the same [curve] as that given to a different axis [7]— whether, I say, something of the same sort happened to Newton is a statement I would not dare to assert; for I have not yet been able to bring myself to examine his treatise on this subject because I do not concern myself willingly with these trifles, since they are by no means of great use. My brother at one time braved this sea, but I do not remember his success; perhaps something may be found in his writings.

NOTES

(1) See Letter 1046, note (1), p. 74, and Letter 1026, note (4), p. 46. Bernoulli had not heard from De Moivre since December 1712.

(2) See Letters 951 and 951*a*, vol. v.

(3) See Letter 1008, note (1), p. 14. Cheyne was, in fact, a neo-Platonist and an exponent of natural religion; see Hélène Metzger, *Attraction Universelle et Religion Naturelle chez quelques commentateurs anglais de Newton* (Paris, 1938), pp. 140–51.

(4) See vol. II, p. 477, note (1); vol. III, *passim*; Fatio certainly became a religious fanatic and passed his last years (he died only in 1753) in religious dementia. His sin of befriending the wretched Camisards of the Cévennes, for which he was pilloried in London, expelled from Rotterdam, and imprisoned at the Hague, was if theologically unorthodox at least suggestive of a warm heart.

(5) For which reason Whiston had been expelled from Cambridge; see Letter 879, note (2) (vol. v, p. 202).

(6) On this see Letter 914, vol. v and its note (19), p. 281; in fact, Newton's enumeration of the cubics (in the *Enumeratio* published as an appendix to *Opticks* in 1704) did not duplicate any species, though it omitted six minor ones.

(7) Johann Bernoulli proposed the problem of finding 'innumerable curves equal in length to a given geometrical curve' in *Phil. Trans.* **24**, no. 289 (1704), 1527. A solution by Craige, sent by De Moivre shortly afterwards (Wollenschläger, *De Moivre*, p. 179, and Letter 1008, note (1)), was rejected by Bernoulli as erroneous, as was De Moivre's own solution sent in 1705 (Wollenschläger, *De Moivre*, pp. 210 and 222–3)—both solutions merely changing the coordinate axes without altering the curve. De Moivre had admitted that his 1705 solution had satisfied Newton; hence the failure of Craige and De Moivre became, in Bernoulli's eyes, the failure of Newton.

1044 THE ROYAL SOCIETY TO THE OFFICERS OF THE ORDNANCE
18 FEBRUARY 1714
From the holograph draft in the University Library, Cambridge[1]

Gentlemen

Her Ma[jes]ty having authorised our President or Vice-President & such others as the Council of the Royal Society shall nominate to be Visiters of the Royal Observatory at Greenwich, & We the said Visitors understanding that a Letter dated 12th Decemb. 1710 [2] was then by her Ma[jes]tys Order sent to your Board by [the Right Honourable the Lord Viscount Bolingbroke] [3] signifying her Mats pleasure that you do receive & take notice of such Representations as the said Visiters should make to your Board concerning Her Mats Instruments in the said Observatory, & that you should order them to be repaired erected or changed as there shall be occasion, or purchased for the Observatory if any be there wch do not belong to her Ma[jes]ty; & her Ma[jes]ty having sent to ye R. Society fresh Orders to take care of the sd Observatory; We the Visiters aforesaid take the liberty to represent to you

That [4] in the great Room of the Observatory up one pair of stairs there are two Clocks wch Sr Jonas Moor [5] the elder caused to be made for the Observatory as we understand by the inscriptions upon them but wch are claimed by Mr Flamsteed as given him by Sr Jonas Moor the younger. [6] If they be not the Queens we desire that they may be purchased as being necessary for the Observatory.

That in the same room there is a brass Quadrant of four foot Radius belonging to Mr Flamsteed. And such an Instrument is necessary there for observing the altitudes of the starrs. This instrument is not well divided, but if it can be divided anew we desire that it may be purchased for ye Observatory; or else that a new one be made of the same size.

That in the same room there is wanting also a Telescope of about [eight] [3] feet Radius furnished with a good Micrometer [and another good Telescope of about sixteen feet Radius] [3].

That in the Garden there being a [house] [7] with a sextant, a wall Quadrant & a clock therein: the [house] [7] should be removed six or eight yards further from the brow of the Hill that the grownd may not sink under it, & the western wall should be thick & firm with a broad foundation that it may not warp, because the wall Quadrant is to fixed upon it. The Sextant is grown rusty & should be cleaned; & there should be a new wall Quadrant made the old one being much worn by long usage & belonging [as we hear] [3] to Mr Flamsteed. The Clock is also Mr Flamsteeds & a better clock would be more usefull.

If you please to give order to an able workman to repair these Instruments & make new ones where they are wanting, & to another workman to take care of removing the house in the Garden some of us will go with them to Greenwich & shew them what is wanting to be done, [and give the best advice they can for doing everything after the best manner] [3]   We are

Gentlemen

Your most humble Servants

NOTES

(1) Add. 4006, no. 31, written by Newton as President of the Royal Society; it went to Sloane [probably] who wrote on the back 'the Rt. Honble. the Lord Visct. Princ. Secretary of State' (see note (3)) and (twice) 'shed' (see note (7) below). There is another draft in the Mint Papers, II, fo. 334v. A copy of the final version (slightly differing from the drafts) was entered in the Royal Society *Council Minute Book* for 18 February 1713/14 (II (copy), pp. 274–6). The drafts are not dated, and we give the letter the date of the Council meeting at which it was approved. The letter is printed (from the Royal Society copy) in Baily, *Flamsteed*, pp. 307–8.

(2) See Letter 814, vol. v, the Royal Warrant for the appointment of Visitors to the Observatory. This was sent to the Ordnance with a covering letter signed by Henry St John (then Secretary of State) of the same date, particularly insisting that the Officers of the Ordnance should repair the instruments, purchase those not already Royal property, and 'have regard to any Complaints the said Visitors may make to you of the misbehaviour of her Ma[jes]tys Astronomer.' Since that time the Royal Society had vigilantly pursued Flamsteed; notably, on 16 October 1711 Flamsteed had been summoned to report to the Council on the state of his instruments at Greenwich, and on 12 June 1712 the instrument-maker John Rowley had been commissioned to report to the Visitors on 'the Number and Condition of the Instruments at the Observatory at Greenwich'. Further, on 30 July 1713 when Newton informed the Society that the Queen had expressed 'Her Desire that hee and the rest of the Gentlemen of the Society would take care of Mr. Flamsteed's Observatory at Greenwich' a committee, including Halley and Rowley, was appointed to visit Greenwich and report on the instruments. The next step was the first approach to the Ordnance (Letter 1012).

(3) The draft leaves a blank. The words in square brackets are supplied from the copy in the Royal Society *Council Minute Book*.

(4) The final version omitted the mention of Flamsteed in this and the following paragraph; it reads thus

That in the Great Room of the Observatory up one pair of Stairs, there are two Clocks with Inscriptions upon them signifying that Sir Jonas [Moore] caused them to be made, those two Clocks or two others as good are requisite for that Room.

That in the same Room there is a brass Quadrant of four feet Radius for Observing the Altitudes of the Sun and Stars. It is not well divided either this or a new one of the same size is requisite for that Room.

(5) Sir Jonas Moore (1617–79), friend and patron of Flamsteed, had invited him to London in 1674 with the intention of setting up a small observatory at Chelsea College. Instead, plans were made for the Royal Observatory at Greenwich, and Moore provided Flamsteed with a 7 ft sextant and two clocks by Tompion (still extant). It is not clear whether these were a personal gift to Flamsteed, or merely for his use while at Greenwich.

(6) Unless this was the first Sir Jonas Moore's son (of whom nothing is known) it must have been his grandson (?1691–1741), also a military engineer.

(7) The word is deleted in the draft. The final version reads *shed*.

1045 CHAMBERLAYNE TO LEIBNIZ
27 FEBRUARY 1714

Extract from the original in the Niedersächsische
Landesbibliothek, Hanover.[1]
For the answer see Letter 1062

Westminster

I send this to you by Mr. Hasberg [2] an Agent for the D. of Wolfenbuttel and who has the Honor to Correspond with you; 'twas from him that I have been inform'd of the Differences Fatal to Learning between two of the greatest Philosophers & Mathematicians of Europe, [3] I need not say I mean Sr. Isaac Newton and Mr. Leibnitz, one the Glory of Germany the other of Great Britain, and both of them Men that Honor me with their Friendship which I shall always Cultivate to the best of my Power, tho' I can never deserve it; now altho I ought to say with the Poet: *Non nostrum est Tantas componere Lites*, [4] yet as it would be very Glorious to me, as well as Advantageous to the common Wealth of Learning, if I could bring such an Affair to a happy end, I humbly offer my Poor Mediation, and shall esteem myself exceeding Happy if by this or any other Instance I can convince You with how great passion & Truth I am

Most Honored Sr.

Your Obliged, Faithfull

humble Servant

JOHN CHAMBERLAYNE

NOTES

(1) Leibniz letters and manuscripts; see Bodemann's catalogue, no. 149, fo. 20. The correspondence between John Chamberlayne and Leibniz, begun in January 1710, was initially concerned with political matters only. With his last known letter to Leibniz, of August 1713, Chamberlayne had sent a copy of the latest edition of his annual directory, *The Present State of Great Britain*. No letter from Leibniz to Chamberlayne between the summer of 1713 and the present time is recorded.

(2) Or Hasperg; see Letter 1061.

(3) In a letter to Varignon of January 1721 Newton wrote: 'In autumn 1713 I received from Mr Chamberlain, (who then kept up a correspondence with Mr Leibnitz) a flying paper in Latin dated 29 July 1713 [N.S.] . . .' (see also Brewster, *Memoirs*, II, p. 53). If we rely upon Newton's recollection, it may be supposed that Leibniz sent some copies of the *Charta Volans* (see Number 1009) to Hasperg in London, and that one or more of these came to Newton himself *via* Chamberlayne. Then Chamberlayne himself must have been 'informed of the differences fatal to learning' some three or four months before writing the present letter. However, if we interpret the present letter as indicating that Chamberlayne's knowledge of the calculus dispute was fresh, and that he wrote at once to Leibniz, Newton's recollection must have been faulty in part, but not necessarily as to the date of his own first learning of the *Charta Volans*.

Newton's first allusion to the *Charta Volans* is in Letter 1053 of 2 April. At that time Keill was only aware of its republication in the *Journal Literaire de la Haye*; but Newton knew of the original printing, and seems to hint that he had known of it for some time.

(4) 'It is not our rôle to resolve such great disputes.'

1046 J. BERNOULLI TO ABRAHAM DE MOIVRE
9 MARCH 1714
Extracts from the copy in the University Library, Basel.[1]
For the answer see Letter 1088

à Bale le 20. *Mars*
1714

Monsieur

Depuis que Vous m'avez notifié ma reception dans Votre Societé Royale,[2] je n'ay plus reçu de Vos nouvelles, quoi qu'il y ait plus d'un an, que Vous me devez une réponse sur une lettre fort ample dattée du 18 fevrier 1713 [N.S.],[3] que je Vous ai envoyée sous couvert adressé a un certain Mr. Arnold[4] Frere d'un autre qui faisoit dans ce temps là ses etudes de mathematiques sous moi, et qui se trouve présentement a Paris; je l'ai prié il y a quelques temps de s'informer auprés de son Frere si cette Lettre là Vous a eté rendue, craignant qu'elle ne se soit perdue, ce que ne me chagrineroit pas peu: Mon Neveu se plaint aussi de Votre silence disant qu'il Vous a ecrit plus d'une fois et en dernier lieu par la voye de Paris, la quelle seroit la plus sûre et la plus courte[;]

si Vous vouliez nous honorer de la continuation de Votre correspondance, en adressant les lettres *à Mr. Varignon de l'Academie Royale des sciences et Professeur des mathematiques au College des 4.Nations, a Paris*; Nous ne pouvons pas penetrer si Votre silence provient de quelque mécontentement sans que nous en sçachions le sujet: Estce peutetre qu'épousant l'interest de Mr Newton Vous nous soupçonnéz d'avoir pris part contre Lui à la querelle qu'on a suscitée a Mr. Leibnits; assûrement Vous nous feriez grand tort, car nous estimons également ces deux grands Hommes, et la bienveuillance de l'un et de l'autre nous est si chere et si pretieuse, que nous ne voudrions pas la perdre auprez de l'un en nous attachant uniquement à l'autre, ainsi nous nous tiendrons dans la neutralité; nous souhaiterions seulement que la bonne intelligence pût etre retablie entre eux, et que chacun relachât un peu de ses pretensions; puisqu'il ne s'agit ici que de la gloire d'etre le premier inventeur, la quelle ce me semble peut fort bien etre partagée, en sorte que chacun en ait assez: pour moi je suis de cette nature que je laisserois entrer en partage avec moi quiconque voudroit me disputer l'honneur de la premiere invention des decouvertes que je croi avoir faites le premier, car j'aime mieux ceder le mien propre que de me tourmenter en querellant, d'autant plus que le lecteur intelligent et desinteressé sçaura toujours ce qu'il en doit penser, et le jugement qu'il doit porter de la vanité et de l'opiniatreté avec la quelle on tache de s'arroger les decouvertes d'autrui: il y a bien encore à inventer, que de pays encore in-connus dans le monde des mathematiques! que ne vont-ils chercher de nou-veaux thresors, ceux qui nous veulent ravit les nôtres, ils en trouveront assez pour s'en passer des vieux: Mais pour revenir à la cause de Votre silence, je ne sçai si je ne le doit pas attribuer à quelque mesentendu, d'autant plus que je n'ai point encore reçû les deux livres que Vous m'avez promis de m'envoyer dès qu'ils sortiroient de la presse sçavoir le *Commercium Epistolicum de Collins* et les *Princip. philosoph.* de Mr. Newton, les quels ont eté publiés depuis long-temps; on ne sçauroit être plus impatient que moi de voir ce dernier livre, parceque j'apprens qu'il y a beaucoup d'additions dans cette nouvelle Edition qui ne sont pas dans l'ancienne; j'apprens aussi que l'Auteur y a mis la correc-tion d'une erreur, suivant ma remarque que mon Neveu lui communiqua lorsqu'il étoit en Angleterre: Je ne doute pas que du depuis Vous n'ayez vû mon memoire inseré dans les Actes de Leipsic des mois de fevrier et de Mars de l'année passée, qui contient plusieurs choses sur le mouvement des corps pesants et quelques autres remarques sur les Princip. de Phil. de Mr. Newton, mais Vous aurez aussi vû que je parle de ce grand homme avec toute la moder-ation possible et avec un air respectueux et accompagné des eloges que l'on doit à son rare merite; en sorte que je n'ai pas lieu de croire, que cette piece l'aura offensé, ni qu'elle Vous aura donné sujet de garder le silence: au contraire, j'espere d'apprendre que Vous y aurez trouvé au moins quelque chose qui aura merité Votre approbation, ce qui me fera tout le plaisir que l'on peut attendre de son travail . . .

NOTES

(1) LIa 664, no. 8; printed in Wollenschläger, *De Moivre*, pp. 286–9. Wollenschläger prints the whole of the extant correspondence between De Moivre and Bernoulli, which began in the spring of 1704 as a result of arguments with Cheyne and Craige. Abraham de Moivre (1667–1754) had published his first paper on the method of fluxions in the *Philosophical Transactions* for 1695; and other mathematical papers followed. George Cheyne (see Letter 1008, note (1)) in 1703 published his *Fluxionum methodus inversa sive quantitatum fluentium leges generaliores*, a general exposition of the method of fluxions which drew heavily on the work of both De Moivre and Bernoulli, but which contained numerous errors. De Moivre pointed these out in his *Animadversiones in D. G. Cheynæi tractatum de fluxionum methodo inversa* (1704). But meanwhile Cheyne had added 'Addenda et adnotanda' to his work, but without acknowledging private help he had obtained from Bernoulli. He replied to De Moivre's *Animadversiones* with another pamphlet, which De Moivre declined to answer. Later Cheyne gave up his mathematical interests.

As a result of this minor clash with Cheyne, De Moivre initiated in the spring of 1704, a correspondence with Johann Bernoulli, sending him a present of the *Animadversiones*, and assuring Bernoulli that he realized that Cheyne had also misrepresented him. Letters immediately following deal largely with the work of Cheyne and Craige, but as the correspondence progresses we find De Moivre supplying Bernoulli with a lucid account of the English mathematical scene, and conveying to him the compliments of many of his learned acquaintance.

In a letter of 18 October 1712, De Moivre reports on the visit of Bernoulli's nephew Nikolaus to London, and discusses the problem of Proposition 10 (see Letters 951 and 951*a*, vol. v). He mentions also the intended publication of the *Commercium Epistolicum* and promises to send Bernoulli a copy of this, and also of the new edition of the *Principia*, as soon as these should appear in print. Three more letters pass between De Moivre and Bernoulli, carrying compliments to and from Newton, before the present letter, which is the first to carry intimations of real controversy between Bernoulli and Newton. (See Wollenschläger, *De Moivre*, pp. 165–8.)

(2) Johann Bernoulli had been elected F.R.S. on 1 December 1712; De Moivre informed him of this in his letter of 17 December 1712. See *ibid.*, p. 277.

(3) Printed in *ibid.*, pp. 280–6.

(4) See Letter 1014, note (1), p. 27.

1047 LOWNDES TO THE MINT
12 MARCH 1714
From the copy in the Public Record Office[1]

Gent[lemen]

My Ld Trea[su]rer having recd. the Enclosed Letter from Wm Bertie containing his Observations on what passed when you attended here last upon the Affair of the Coinage of Copper Halfpence and Farthings [2] His

L[ordshi]p Commands me to send the same to You, that You may be prepared at Your next attendance here to make Answer to what Concerns Your P[ro]posals in relation to the said Coinage 12 March 171$\frac{3}{4}$ [3]

WM LOWNDES

NOTES

(1) T/27, 21, p. 151. James [not William] Bertie was a manufacturer or merchant of copper; his letter (dated 4 March 1713/14) is in the Mint Papers, II, fos. 385–6. In it he claims that his offer to supply copper coinage was the most favourable, and he would use a good die to strike it. He criticizes Newton's specimen copper coins—this is the first evidence here that any had been struck—on the grounds that they are not made of fine malleable copper, but have been given a high artificial gloss which will soon wear off. Bertie is ready to take the specimen of copper provided by the Warden, Craven Peyton, as his standard. Finally, he does not approve of the Queen's arms on the reverse of Newton's coins.

Bertie's letter was referred by the Treasury to the Mint on 10 March.

(2) James Bertie (who had previously submitted a proposal for the coinage of 1000 tons of copper) was summoned together with the Mint Officers to the Treasury on Friday 26 February. There Bertie's proposal was read, followed by an unspecified Mint report. The minute continues: 'My Lord Treasurer opens [to them] what passed at a former attendance of the Officers of the Mint and some owners of copper works [see Letter 996, vol. v]; wherein his Lordship was of opinion that the Mint should be a free coinage for all copper that should be imported thither for making farthings as near as might be to intrinsic value, allowing for the charge of coinage, and orders the Officers of the Mint to prepare the draft of a warrant to authorise the making an experiment, at her Majesty's charge, of reducing copper to a proper standard for coining the said farthings and halfpence.' (Cal. Treas. Books, XXVII (Part II), 1713–14, p. 18 from T/27, 20, p. 181).

(3) We have found no reply to this letter.

1048 NEWTON TO OXFORD
12 MARCH 1714
From the holograph draft in the Mint Papers[1]

To the Most Honble the Earl of Oxford & Earl Mortimer
Lord High Treasurer of great Britain.

May it please your Lordp

I humbly beg leave to lay before your Lordp that the Master & Worker of her Majts Mint is not obliged to receive all the Gold & Silver brought into her Mats Mint to be coyned. If any gold be brought in which is not tough, he returns it back to be toughned at the Importers charge tho it be standard. If any gold or silver be not eavenly mixed he returns it back to be remelted at the Importers charge. If it be not neare to standard he returns it back to be refined at the Importers charge. And to judg whether it be fit to be received or returned back is left to his discretion. And by parity of reason he should not be obliged to receive all sorts of copper to be coyned. If it be not fine or not tough & malleable or ill coloured or otherwise faulty, he should be at liberty to return it back to be made fit for the Mint at the Importers charge. Otherwise it will be difficult & scarce practicable to coyn the money of good malleable copper without allowing for the charge hazzards & loss attending such an undertaking.

There is an assay of copper by refining a small parcel & then recconing what will be the wast charges & trouble in refining a Tunn of such copper. And such an assay is usefull in buying coarse copper to be refined but is of little or no use in buying fine copper, nor proper for the Mint. The price of fine copper depends upon the malleability, & two parcels of copper equally fine, may differ very much in their malleability & by consequence in their price. Copper refined to that degree & in that manner as to be malleable without cracking when red hot is the fittest material for manufacturing into all sorts of copper vessels & by consequence for money. The Swedish copper money is of this standard. And such copper is usually valued at about $11\frac{1}{2}d$ per lwt. And if it be made still more soft & malleable the wire-drawer may value it at two or three shillings per lwt because of its fitness for his use. Tis the ductility that makes it usefull & the usefulness that sets a price upon it & the trial by hammering & bending hot & cold that determins the ductility. This is the assay by which Refiners of Copper know when their Copper is fully fine & ready to vitrify & by consequence the proper assay for receiving fine copper into the Mint. For it determins the fineness & the malleability at once, no coarse copper being malleable.

By the estimates of Workmen the charges for repairing & fitting up the houses in the Irish Mint for a coynage of copper, will amount to about £146. And the putting up a furnace in the melting house with all things answerable for making an experiment by casting will cost about £32 more. And a small parcel of copper for making an experiment may cost about £20 more. If your Lordp please to impress £200 to me for this service upon accompt it may be repaid out of the copper coynage.

All wch is most humbly submitted to your Lordps great Wisdom

Is. NEWTON

Mint Office
12 *Mar.* 1713

NOTE

(1) II, fo. 305r. An earlier draft is at fo. 312r. This bears the same ambiguous date, which we assume Newton wrote in due form.

1049 NEWTON TO HUMFREY DITTON
16 MARCH 1714
From the original in the British Museum[1]

Sr

If you please to call on me on friday morning next about ten of ye clock you will find me at home. I am

Your most humble Servant

Is. NEWTON

16 *Mar.* 171¾
For Mr Ditton

NOTE

(1) MS. Stowe 748, fo. 103. Presumably the recipient of this note was Humfrey Ditton (1675–1715) who had attracted Newton's attention by publishing a *General Treatise on the Laws of Nature and Motion . . . being a part of the great Mr. Newton's Principles* (1705) followed by *An Institution of Fluxions . . . according to Sir Is. Newton* (1706). Newton obtained for him a special post as teacher of mathematics at the Christ's Hospital Mathematical school; Ditton also taught mathematics in London and assisted both the elder and the younger Francis Hauksbee in their courses of experimental philosophy.

Another friend was William Whiston, with whom Ditton devised in 1713 (see their pamphlet, *A New Method for the Discovery of the Longitude at Sea* of the next year) a way of determining longitude at sea, depending on the firing of signal cannon from hulks moored in shallow waters; rather as with a lighthouse, the sound would give seamen warning of their position. A few years earlier the wreck of Sir Cloudesley Shovell's fleet on the Scillies (1707)—to which the inventors refer—with much loss of life and treasure (some now recovered) had emphasized defects in navigation; Ditton and Whiston's proposal aroused much interest, leading ultimately to a parliamentary enquiry in which Newton was involved (see Letter 1093a).

Presumably Ditton wished to consult Newton about the longitude proposal.

1050 LOWNDES TO THE MINT
19 MARCH 1714
From the copy in the Public Record Office[1]

Gent[lemen]

I am commanded by my Lord Treasurer to transmit to You the enclosed Proposal of James Maculla of Dublin Pewterer for disposing of Three Thousand Pounds worth of Block Tyn yearly in the Kingdome of Ireland upon the Terms therein menconed My Lord is pleased to direct You to consider the said Proposal and let him have Your Opinion what is fit to be done therein I am &c. 19th March 171¾

WM LOWNDES

(1) T/27, 21, p. 161. The proposal is naturally not with this minute of its reference to the Mint, nor is it to be found among the Mint Papers.

1051 THE MINT TO OXFORD
22 MARCH 1714
From the original in the Public Record Office.[1]
Reply to Letter 1037

To the most Honble. the Earl of Oxford & Earl Mortimer
Lord High Treasurer of Great Britain

May it please your Lordp.

In obedience to your Lordps. Order signified to us by Mr Lownds his Letter of the 27th of January last, namely that we should send your Lordp. an Account of the whole quantity of Tyn wch her Majty hath paid for, that now remains unsold as well in our Custody in the Tower as in the hands of the Agents for Tyn in Cornwal & Devon at Hamburgh Holland or elsewhere with as near an estimate as we can of ye value thereof, or what the same may produce into the Exchequer. We humbly represent that upon the 19th Instant there were in the Tower 1919 Tunns 6 cwt 2 qrs 9 lb wch at 3£ 16s per cwt (the selling price at present,) amounts to 145869£ 0s 1d. In the hands of the Agents in Cornwall were 835 tuns 3 cwt 3 qrs 11 lb. wch at 3£ 15s per cwt (the selling price there) amounts unto £62639 8s 7d. In Devonshire there were 34 Tuns. 18 cwt. 1 qr. 23 lbs. which after the same rate amounts unto £2619 4s 2d. In the hands of Mr Beranger at Amsterdam there were 1540 Tuns wch at 4£ per cwt the present selling price amounts unto £123200. And in the hands of Sr John Lambert & Captain Gibbon at Hamburgh there were 960 Tunns, which if they can be sold at the same rate will amount unto £76800. The Total quantity of the Tyn is 5289 Tunns 8 cwt 3 qrs 15 lb. And the total price thereof according to the rates above mentioned, is £411127 12s 10d.

All wch is most humbly submitted to your Lordps. great Wisdome

C: PEYTON

IS. NEWTON

E. PHELIPPS

Mint Office
22 *March* 17¹³⁄₁₄.

(1) T/1, 174, no. 16. The letter is in a clerical hand and signed by the officers. There is a draft in the Mint Papers, III, fo. 554, wholly written by Newton. The various contracts mentioned in the letter are discussed in vol. v.

1052 LOWNDES TO THE MINT
25 MARCH 1714
From the copy in the Public Record Office[1]

Gentlemen

I am Commanded by My Lord Trea[sure]r to transmitt to you the inclosed Copy's of the Representations from Charles Pembruge Landwaiter at Bristoll and from the Pewterers there touching frauds Committed to the prejudice of her Mats. Revenue arising from Tinn together with Copys of Affidavits relating thereunto his Lordp is pleased to Direct you to Examine into the said Frauds and to consider of proper Method for the preventing thereof and that you give his Lordp an Acco[un]t: of what you shall do therein I am. 25th March 1714

WM LOWNDES

NOTE

(1) T/27, 21, p. 165. Again, the original papers have vanished and we have not found any reply from the Mint. Perhaps this was not handled by Newton.

1053 NEWTON TO KEILL
2 APRIL 1714
From the original in Trinity College Library, Cambridge.[1]
Reply to Letter 1039; for the answer see Letter 1063

Sr

Your Letter of Feb. 8th I delayed to answer till the Journal Literaire for November & December should come out. It is just come from Holland & I desired Mr Darby to send you a copy wch I doubt he has not done because he sent one to me this morning wch I reccon to be for you & I designe to send it to you the first opportunity by the Carrier. Mr Leibnitz in August last, by one of his correspondents published a paper in Germany conteining the judgment of a nameless Mathematician in opposition to the judgment of the Committee of the Royal Society, with many reflexions annexed.[2] This paper hath been sent to Mr Johnson [3] with remarks prefixed to it. [4] And the whole is printed in the journal Literaire pag. 445. And now it is made so publick I think it requires an Answer. It is very reflecting upon the Committee of the Royal Society, & endeavours to derogate from the credit of some of the Letters published in the Commercium Epistolicum as if they were spurious. If you please when you have it, to consider of what Answer you think proper, I will within a Post or two send you my thoughts upon the Subject, that you may compare them wth your own sentiments & then draw up such an Answer as you think proper. You need not set your name to it. You may write either in English or in Latine & leave it to Mr Johnson to get it translated into F[r]ench. Mr Darby will convey your Answer to the Hague. I am

Your most humble Servant

Is. NEWTON

London. 2 Apr.
   1714

For Dr John Keill, Professor of
   Astronomy, at his house in
            Oxford

NOTES

(1) R.16.38, fo. 428v, printed in Edleston, *Correspondence*, pp. 169–70.
(2) The *Charta Volans*, Number 1009. See Letter 1045, note (3), p. 72.
(3) See Letter 1039*a*.
(4) See Letter 1018.

1053*a* NEWTON TO [JOHNSON]
[1714]
From drafts in the University Library, Cambridge[1]

I have seen in your journal ⟨Literaire⟩ the translation of a Latin piece dated 29 July 1713 [N.S.] & published in Germany, & the Remarks upon it. These pieces are full of assertions without proof & wthout the name of the author & so are of no ⟨credit or⟩ authority. [ce pendant pour ne les laisser pas tout a fait sans reponce, Vous etes prié, Monsieur, de vouloir bien imprimer dans votre journal les reflexions suivantes.] [2]

   The author of the Latine piece represents that Mr Leibnitz had not then seen the Commercium Epistolicum, & this he could not know without keeping a correspondence with Mr Leibnitz. [3] But a copy of this book was sent to Mr Leibnitz by the Resident of the Elector of Hannover above a year ago &

several other copies were then sent to Lip[s]ic one of wch was for him ⟨and he knew how to write to a friend who had a copy⟩.

That Author tells us further that Mr Leibnitz not being at leasure to examin this affair himself had referred it to the judgment of a Mathematician of the first rank very skilfull in these things & very free from partiality. So then this paper was writ by the correspondents of Mr Leibnitz, & Mr Leibnitz himself desiring the judgmt of the great Mathematician & sending it to his correspondent to be published was the first mover & the credit of the Mathematician for candor & ability depends upon the credit of Mr Leibnitz. And for these reasons this paper must be looked upon as the best defence that he & his correspondents were able to make: especially if this paper be writ in the stile of Mr Leibnitz himself as some think. By his Letters against Mr Keill it appears that he is too much concerned to neglect this matter, & his appealing from a numerous Committee of the Royall Society to a nameless Mathematician of his own chusing is no better then appealing to himself. For he has wrote to the Society that it would be injustice to question his candour, that is, to deny him to be both witness & judge in his own cause.

Now this great Mathematician [dans la reponce à Mr Leibnitz datée du 7e juin 1713] conjectures that Mr Newton spent his first years in cultivating the method of series without thinking of the calculus of fluxions or reducing it to general rules. That is he will not allow the Analysis communicated by Dr Barrow to Mr Collins in the year 1669 to be a genuine piece. And he brings too [read two] arguments for his conjecture.

First, saith he, in all the Letters published in the Commercium Epistolicum & in all the Principia Philosophiæ, the letters with pricks wch Mr Newton now uses are not to be met with. But in all those Letters (except the Analysis [ou cela etoit necessaire]) & in all the Principia there was no occasion to make use of the fluxional calculus. And Dr Keill hath given a further Answer to this argument long since in his Letter dated 24 May 1711. *Observo ipsum Newtonum*, saith he, *sæpius mutasse nomen et notationem calculi.*(4) *In tractatu de Analysi Æquationum per series infinitas, incrementum abscissæ per literam o designat, et in principiis Philosophiæ, Fluentem quantitatem Genitam vocat ejusque incrementum Momentum appellat: illam literis majoribus A vel B, hoc minusculis a et b designat.* ⟨I may add that in one & the same book, The Book of Quadratures he sometimes uses prickt letters sometimes not.(5) For in the Introduction to that book where he describes his Method of fluxions & illustrates it with examples he makes no use of prickt letters.⟩ Mr Leibnitz confines his Method to the symbols dx & dy, so that if you take away his symbols you take away ⟨the characteristick of⟩ his method. Mr Newton doth not confine his method in such a manner. As he uses any symbols for fluents so he uses any others for

81

fluxions, & whether he uses letters wth pricks or other symbols for fluxions his method is still the same. In his Letter of 24 Octob. 1676 [6] he represents that he had a method of extracting Fluents out of equations involving their fluxions. Will Mr Leibnitz say that he had no such method unless he then used letters with pricks? If so, his letters with pricks must be allowed as old at least as the year 1676, & by consequence older then the differential Notes of Mr Leibnitz.

But its further to be observed that Fluxions & Differences are not quantities of the same kind. Fluxions are velocities, & Differences are small parts of things generated by fluxion in moments of time: fluxions are always finite quantities & differences are infinitely little. Mr Newton uses sometimes prickt letters sometimes other symbols for fluxions, Mr Leibnitz uses no symbols for fluxions to this day. The symbols of fluxions used by Mr Newton whether wth pricks or without, are therefore the oldest in the kind. [7] These he multip[l]ies by the moment o to make them infinitely little [8] & puts the rectangles for moments or differences, [9] & without the moment o either exprest or understood they never signify moments or differences, but are always finite quantities & signify velocities. The fluxion of time or of any exponent of time he usually represents by an unit, & the moment thereof by the letter o. [10] ⟨wch is equipollent to ye rectangle $1 \times o$. In his Analysis above mentioned he represents fluents by the areas of curves, fluxions by their ordinates & moments by the rectangles under the ordinates & the moment of the common Abscissa. And these rectangles he uses instead of the Indivisibles of Cavallerius, & thereby makes his method Geometrical. For in Geometry there are no Indivisibles. When he is demonstrating any Proposition he always writes down the moment o & takes it in the sense of ye vulgar for an indefinitely small part of time & performs the whole calculation in finite figures by the Geometry of the Ancients without any approximation, & so soon as the calculation is ended & the equation reduced he supposes the moment o to decrease in infinitum & vanish. Examples of this you have in the end of his Analysis & in the first Proposition of his Book of Quadratures. But when he is not demonstrating but only investigating a Proposition, he supposes the moment o to be infinitely small & usually for making dispatch neglects to write it down & proceeds in the calculation by any approximations wch he thinks will create no error in the conclusion. But this last way (to wch the Differential method is equipollent) is not Geometrical. For Geometry admits not of approximations nor of lines & figures infinitely little.⟩

The second reason of the great Mathemat[ic]ian for his conjecture is that Mr Newton understood not the differences of differences till after the writing of his Principia. For there, saith he, the constant increase of the letter x he represents not by a prickt letter as at present but by the letter o after the vulgar manner, wch destroys the advantages of the differential method. Here our great Mathematician commits two mistakes; one by supposing that Mr Newton represents differences by prickt letters, another by supposing that the method used in the ⟨Scholium of the⟩[11] tenth Proposition of the second Book of the Principia is Mr Newtons method of fluxions: Tis only a branch of his method of converging series. In his Letter dated 10th Decemb. 1672, where he speaks of a method whereof the method of Tangents there described is a branch or Corollary, he represents that this method (wch is the method of fluxions) extended to Questions about the curvature of curves; & thence it is manifest that he then understood the second fluxions or differences of differences.[12] [pour mettre cecy dans son jour, il est bon de scavoir que Mr Newton dans L'introduction a son traitté de la quadrature des Courbes a demontré, que si une quantité indeterminée quelconque $\frac{x}{1}$ croit ou coule uniformement, ou d'une maniere proportionnée au temps, et que sa fluxion, c'est-a-dire la Vitesse avec laquelle elle croit, soit representée par l'unité et que son moment, ou l'accroissement engendré dans un moment de temps soit appellé o, et que la dignite $\overline{x+o}|^n$ dont l'indice est n soit resolu en suite convergente, la suite sera

$$x^n + nox^{n-1} + n \times \frac{n-1}{2} oox^{n-2} + n \times \frac{n-1}{2} \times \frac{n-2}{3} o^3 x^{n-3} + \&\text{c.}$$

Il a demontré aussy dans la méme introduction que la fluxion d'une fluente quelconque x^n est nx^{n-1} d'ou il suit que la fluxion de nx^{n-1} est $n \times \overline{n-1} \times x^{n-2}$, et que la fluxion de cette fluxion est $n \times \overline{n-1} \times \overline{n-2} \times x^{n-3}$, et ce sont la les premieres secondes et troisiemes fluxions du premier terme de la fluente x^n selon Mr Newton, car il dit au commencement de son traitté que la fluxion de la premiere fluxion d'une quantité quelconque en est la seconde fluxion, et que la fluxion de la seconde fluxion en est la troisieme fluxion. Et dans la seconde section du dernier *Scholium* du dit traitté il dit que ces fluxions peuvent etre representées par les ordonnées, *BD, BE, BF, BG, BH*, des courbes qui sont là mentionnées, chacune desquelles est la fluxion de la precedente; Ce qu'il dit dans les deux sections suivantes du méme *Scholium* au sujet des secondes et troisiemes fluxions est touta fait conforme a cecy et ne scauroit etre vray dans aucun autre sens. Mais pour rendre la premiere section du *Scholium* comforme au reste du livre, le mot *ut* qui a eté omis d'une fois par accident ou peutetre par la faute de l'imprimeur,[13] doit etre retabli dans la 11e et dans la 13e ligne du *Scholium*, alors les termes de la suite cy dessus mentionnée, commencant au second terme, doivent etre multipliez par la progression 1[,] 1×2, $1 \times 2 \times 3$, $1 \times 2 \times 3 \times 4$ &c, pour produire les premiers, seconds, troisiemes et autres moments du premier terme de la dite suite, selon les notions et la methode de Mr Newton. On doit entendre la méme chose de la suite dont il est fait mention dans la dixieme proposition du second livre des principes. Mais les seconds et troisiémes moments ne sont pas considerez dans cet endroit, aussy n'ont ils rien affaire avec la methode qu'on y propose de resoudre les Problemes par des suites convergentes. La methode des fluxions est differente de celle des suites convergentes, et les deux methodes ensembles ne sont selon Mr Newton qu'une methode generale.]

⟨The elements of his method of fluxions he described in the second Lemma of the second book of his Principles & subjoyned this Scholium. *In literis quæ mihi cum Geometra peritissimo G.G. Leibnitio annis abhinc decem intercedebant, cum significarem me compotem esse methodi determinandi maximas et minimas, ducendi Tangentes et similia peragendi, quæ in terminis surdis æque ac in rationalibus procederet, et literis transpositis hanc sententiam involventibus,* [Data æquatione quotcumque Fluentes quantitates involvente Fluxiones invenire & vice versa] [14] *eandem celarem; rescripsit Vir Clarissimus se quoque in ejusmodi methodum incidisse & methodum suam communicavit a mea vix abludentem præterquam in verborum et notarum formulis. Utriusque fundamentum continetur in hoc Lemmate.* [15] The Letter here referred unto is that of 24 Octob. 1676 printed by Dr Wallis. [6] In this Letter Mr Newton distinguishes between the Method of infinite series & that of fluxions & represents that he had writ a treatise of both these methods five years before, [16] & that the method of fluxions readily gives the method of Tangents of Slusius & sticks not at equations affected with radicals. And in a Letter to Mr Collins dated 10 Decem 1672 Mr Newton writing of the method whereof the Method of tangents of Slusius is a branch or Corollary & which sticks not at surds, & by consequence was the Method of fluxions, represented it a very general method reaching to ye abstruser sorts of Problemes & among others to the determining of the Curvature of Curves, a Probleme wch requires the consideration of the second fluxions. And therefore he had then extended the method to the second fluxions or fluxions of fluxions [des l'année 1672 ou méme auparavant]. [17] And it is further observable that Mr Newton in the 2d Book of his Principles makes frequent mention of the increase of the velocities wherewith lines are described. The lines are the fluents, the velocities their fluxions & the increase of the velocities the fluxions of their fluxions or second fluxions. And particularly in demonstrating the 14th Proposition of the second Book of his Principles, he has these words. *Est igitur differentia momentorum, id est momentum differentiæ arearum* &c [18] Where differentia arearum is the first difference & momentum differentiæ is the second difference of the areas. So then Mr. Newton, when he wrote his Principles of Philosophy & a great many years before, had extended his method to the consideration of the second fluxions of quanti[it]ies. And indeed, to say that he then understood not second fluxions is all one as to say that he then understood not how to consider motion as a quantity increasing & decreasing.[19]

And whereas the great Mathematician represents that Mr Newton uses the Letter o in the vulgar manner wch destroys the advantages of the Differential method: he uses it, and has used it ever since the writing of his Analysis, in such a manner as makes his method more beautiful more Geometrical & more advantageous then the Differential, & (by joyning the methods of series & fluxions together) much more universal.

The Differential method is nothing else then the method of Tangents published by Mr Gregory in ye year 1668 & by Dr Barrow in the year 1670, disguised by changing Dr Barrows symbols a & e into dy & dx, improved by the instructions wch Mr Leibnitz received by the Letters of Mr Newton, & taken from them all by pretending that Mr Leibnitz found it long before he did. For in his Letter dated 21 July 1677 [N.S.] [20] he pretended to have found it jam a multo tempore, & yet he had not found it the year before. For in his Letter dated 27 Aug. 1676 [N.S.] [21] he wrote that there were many Problems wch could not be reduced to Equations or Quadratures such as were those of the inverse method of Tangents & many others. This method without the use of the Letter o is not demonstration, without the method of series is not universal, nor has any advantages wch are not to be found in the method of Fluxions, nor has Mr Leibnitz added any thing to it of his own besides a new name & a new notation. And thus much in answer to the great Mathematician.⟩

[On peut remarquer de plus que la premiere proposition du traitté de la quadrature des Courbes communiqué au Dr Wallis au mois D'aoust 1692,[22] s'etend aux secondes et troisiemes fluxions et aux suivantes, comme il paroit clairement par les exemples dont on se sert pour l'expliquer. Cette proposition etoit connue à Mr Newton dans le temps qu'il ecrivoit ses principes et dix ans auparavant dans le temps qu'il ecrivit sa lettre datée du 24e Octobre 1676, et cinq ans avant ce temps la, lorsqu'il ecrivoit un traitté sur la methode des suites à laquelle etoit annexée une autre methode fondée sur cette proposition, dont il est fait mention dans le dite lettre. L'inverse de cette proposition, scavoir la methode d'extraire les fluentes, des equations qui enveloppent leurs fluxions s'etend aussy aux secondes et troisiemes fluxions et aux suivantes, comme il paroit par les papiers que Mr Newton envoya au Docteur Wallis en Aoust et Septembre 1692, et par ce que le docteur avoit imprimé sur cette matierre dans le second Volume de ses ouvrages pag. 396; cecy etoit aussy connu à Mr Newton dans le temps qu'il ecrivit la lettre surdite du 24e Octobre 1676 c'est a dire avant que Mr Leibnitz scut rien de plus de la methode differentielle, que ce qu'on en trouve dans la methode des Tangentes du Dr Barrow.]

The letter *o* was used by Mr Newton in the manner above mentioned in his Analysis communicated by Mr Barrow to Mr Collins in the year 1669, & in his Book of Quadratures & is still used by him in the very same manner. And as it is the oldest notation for moments or differences so it is the best, the method being thereby more convenient more elegant & more suitable to Geometry then by the differential notation & as universal, & does justice to the memory of Mr. Fermat who first brought in the use of this letter *o*. [23]

To signify the summ of the ordinates or area of a Curve Mr Leibnitz prefixes the letter ∫ to ye Ordinate & Mr Newton in his Analysis communicated to Mr Collins in the year 1669, inclosed the Ordinate in a square. Mr Newtons notation of this kind is also much the oldest.

Dr. Barrow published his method of Tangents in the year 1670, & that very candid Gentleman the Marquess de L'Hospital, in the Preface of his Analysis, [24] represents that Dr Barrow stopt at fractions & surds, & where Dr Barrow left off Mr Leibnitz began. His method of Tangents is the same with Dr Barrows except that he has changed his letters *a* & *e* into ye symbols *dx* & *dy*, & (being admonished by Mr Newtons Letters of 10 Decem. 1672, 13 June 1676 & 24 Octob. 1676) taught how to avoyd fractions & surds.

And thus much in answer to the great Mathematician.

As to what the Author of the Latin paper saith of Mr Hook & Mr Flamsteed: Mr Hook indeed claimed one of Mr Newton's Propositions but could never produce a Demonstration thereof, [quoy qu'il en fut sollicité plusieurs fois], Mr Leibniz claimed it also but the Demonstration by wch he claimed it is erroneous. [quoyque jusqu'a present on ait eu l'honneteté de ne luy en rien dire] Mr Leibnitz claimed also an Invention from Mr Tschurnhause, & who is in ye right may be a question but Mr Newton always acknowledged the use of Mr Hooks Observations. [25]

This Author in the next place complains of the Committee of the R. Society for representing that Mr Leibnitz had from Mr James Gregory the series for finding the Arc of a circle by the Tangent given, that is, he represents that the Letters of Mr Gregory, Mr Collins, Mr Oldenburg & Mr Leibnits examined & approved [26] by a numerous Committee of the Royal Society were fourged. The Letter of Mr Gregory dated 15 Feb. 167$\frac{0}{1}$ is still extant in his own handwriting & conteins this series with several others then sent to Mr Collins. [27] That of Mr Oldenburg dated 15 Apr. 1675 is extent in the Letterbook of the Royal Society left by Mr Oldenburg in their Archives & conteins this series with several others then sent from Mr Collins by Mr Oldenburg to Mr Leibnitz at Paris. [28] The answer of Mr Leibnitz dated from Paris May 20th. 1675 [N.S.] was found in the same Letter-book; [29] & the original Letter in the hand-writing of Mr Leibnitz was also found in the Archives of the R.

Society & conteins his acknowledgment of the Receipt of Mr Oldenburghs Letter above mentioned. It begins thus. *Literas tuas multa fruge Algebraica refertas accepi pro quibus tibi et doctissimo Collinio gratias ago. Cum nunc præter ordinarias curas Mechanicis imprimis negotijs distrahar, non potui examinare series quas misistis ac cum meis comparare.*[30] By these words its plain that Mr Leibnitz at this time knew none of the series then sent him to be his own, tho before the end of the year he communicated to his friends at Paris one of those series then sent him as his own, vizt that of Gregory then dead, & by vertue of that communication has ever since claimed it as his own.[31] The collection of the papers of Mr Gregory made by Mr Collins after the death of that Gentleman, is still extant in the hand-writing of Mr Collins, & at the request of Mr Leibnitz was sent to Paris in June 1676,[32] & conteins a copy of the aforesaid Letter of Mr Gregory. But upon the death of that Gentleman Mr Leibnitz pretended in his Letter dated 28 Decem. 1675 [N.S.] [33] that he had communicated it at Paris above two years before & that it was the series whereof he had wrote before to Mr Oldenburg, that is, in his Letter[s] of 15 July [N.S.] [34] & 24 Octob. 1674 [N.S.].[35] And under this pretence he sent it back to Mr Oldenburg as his own in his Letter dated 27 Aug. 1676 [N.S.]. And yet the Series wch he wrote of in his said two Letters dated 15 July & 24 Octob 1674 was not this Series for finding the arc by the tangent but a Theoreme or Method for finding the Arc by the sine. This Theoreme or Series Mr Collins had received from Mr Newton in July 1669 [36] & communicated it soon after to his friends very freely. Mr Leibnitz was in London in the years 1671, 1672 & the beginning of 1673 [37] & having met with this series either in London or soon after in France pretended in his said Letters of 15 July & 24 Octob 1674 to have found it himself, & yet in his Letter dated 12 May 1676 [N.S.] [38] desired Mr Oldenburg to procure from Mr Collins the Demonstration thereof, that is, the method of finding it. And when he had received the method with some of Mr Newtons series, he pretended to have found three of those series before, tho he did not yet understand the method of finding them. For in his Letter of 27 Aug. 1676 [N.S.],[39] he wrote back for a further explication of the method. Mr Newton therefore in his Letter of 24 Aug. 1676 [40] explained it further & added another method of the same kind, & Mr Leibnitz in his Letter dated 21 June 1677 [N.S.] [41] still desired a further explication, but so soon as he understood it, he wrote in his Letter dated 12 July 1677 [42] that he found by his old papers, that he had used one of those methods before. And by the same power of invention, when he had newly found the Differential method (wch he might do by the help of Gregories & Barrows methods of Tangents & Newtons Letters) he wrote back: *Clarissimi Slusij methodum Tangentium nondum esse absolutam Celeberrimo Newtono assentior. Et jam a multo tempore rem Tangentium generalius tractavi, scilicet per differentias Ordinatarum.* [43] And yet its very certain that he had but newly found it. For in his Letter dated 27 Aug. 1676 [N.S.], he wrote. *Quod dicere videmini plerasque difficultates (exceptis Problematibus Diophantæis) ad series infinitas reduci, id mihi non videtur. Sunt enim multa usque adeo mira et implexa ut neque ab æquationibus pendeant neque ex quadraturis. Qualia sunt ex multis alijs Problemata methodi tangentium inversæ: Quæ etiam Cartesius in potestate non esse fassus est.* [44] These words are a Demonstration that he did not then understand the Differential method. He was then composing & polishing his Quadrature of the Circle vulgari more, & left of that way of writing as soon as he found the Differential method. I pass by his claiming the Inventions of Mouton & Paschal & a considerable part of Mr Newton's Principia Philosophiæ.

Our Author tells us further that Mr Leibnitz published in the Acta Eruditorum a general method for finding the Ordinates of transcendent Curves not by extraction of roots but deduced from a profounder foundation of the differential calculus, [45] by wch the business of series was brought to a greater degree of perfection. But Mr Newton many years before (vizt in his Letter to Mr Oldenburg dated 24 Octob. 1676, [)] communicated the very same method in this sentence, *Altera* [methodus consistit] *tantum in assumptione seriei pro quantitate qualibet incognita ex qua cætera commode derivari possint, & in collatione terminorum homologorum æquationis resultantis ad eruendos terminos assumptæ seriei.* [46] Mr Leibnitz therefore has no title to any part of the method of converging series.

Our Author tells us further that Mr Leibnitz was the first who used the exponential calculus while Mr Newton knew nothing thereof. Certainly Mr Newton was one of the first who introduced into Analysis fractions radicals & negative quantities for the indices or exponents of dignities, & thereby very much enlarged Analysis & laid the foundation of making it universal. In his Letter dated 24 Octob. 1676 he mentioned such Exponents of Dignities & thereupon Mr Leibnitz in his Answer dated 21 June 1677 [N.S.] proposed indeterminate exponents of Dignities. [47] And this seems to have been the Original of the Exponential Calculus. But such a calculus has hitherto been of no use. [48]

Our Author tells us also that the English & Scotch, Wallis, Hook, Newton & Gregory junior, [ont ignoré pendant 36 ans que Gregory fut auteur de la suite] acknowledged 36 years ago the series for finding the arc of a circle by the Tangent to be the invention of Mr Leibnitz. That is, he complains of Mr Oldenburg for not letting the English & Scotch know that he had sent this series with several others to Mr Leibnitz in April 1675. Tis sufficient that the Letters between Mr Oldenburg & Mr. Leibnitz were left by Mr Oldenburg in the Letter-book of the R. Society & that the Original Letter of Mr Leibnitz is still extant in his own hand-writing.[49]

And thus much concerning the printed paper.

In the Remarks its represented that Mr Leibnitz never communicated his reasons to ye Royal Society of England & so the Society hath not examined the reasons on both sides for giving judgment. ⟨And upon this pretence the Author of the Remarks gives a judgment contrary to that of the Committee of the R. Society.⟩ But the truth is, Mr Leibnitz ⟨absolutely⟩ refused to give any reasons at all, representing it injustice to expect that he should defend his candor, detracting from the candor of Mr Keil & pressing the R. Society to give judgment without hearing reasons: & the Committee of the R. Society grounded their Report not upon plausible & slippery reasons but upon the matter of fact conteined in the Letters & Papers found in the Adversaria of the R. Society & in those of Mr Collins & in the Acta Eruditorum; & published those Letters & Papers in the Commercium Epistolicum that all the world might see the grownd & justice of their Report.[50] And those Records are sufficiently plain to any man that considers them impartially.

⟨But the Author of the Remarks has laid aside the Records of the first seven years wch make for Mr Newton & begins his Report wth the years 1676 & 1677, & thereby confesses that he has no way to defend Mr Leibnitz but by laying aside the Records wch make against him.

In the Remarks its said further that Mr Newton did not speak of this matter till after the death of Mr Hugens & Dr Wallis who were well informed & able to judge thereof: Which implyes that Mr Newton began this dispute. Whereas Mr Leibnitz began it nine years ago by giving an abusive reflecting account of Mr Newtons book De Quadratura Figurarum, & Dr Keil retorted the charge upon Mr Leibnitz before Mr Newton knew what Mr Leibnitz had done. As for Mr Hugens he never was well informed about this matter nor doth it appear that he gave any judgment about it. And as for Dr Wallis, he gave his judgment against Mr Leibnitz 19 years ago in his Preface to the first Volume of his Mathematical works published A.C. 1695. For there he saith that Mr Newton in his Letters of June 13 & Octob. 24 1676 *Methodum hanc* [de Fluxionibus] [51] *Leibnitio exponit, tum ante decem annos nedum plures ab ipso excogitatam,* explained to Mr Leibnits the method of Fluxions invented by him ten years before or above, that is, in the year 1666 or 1665. And in a Letter dated from Oxford Apr. 20 1695 [52] & extant in the Archives of the R. Society Dr Wallis represented that he had intimation from Holland that Mr Newton's Papers relating to the Method of Fluxions should be printed because his notions of fluxions passed there wth great applause under the name of the Differential method. And tho Mr Newton has in this matter neglected his reputation abroad, yet in the second book of his Principles written 28 years ago, he mentioned the method of Fluxions as known to him in the year 1676, & Mr Leibnitz has hitherto allowed it without going about to make it appear that the Differential method was known to him before the year 1677.

But [53] because Mr Leibnitz & his correspondents or some of them have composed & published in Germany the Latin Paper without a name where by they defame Mr Newton accuse the committee of ye R. Society of partiality, affirm & deny things without proof, & endeavour to set aside Records & bring things to a wrangle: I intend to give you hereafter a fuller account of these matters out of the Records themselves.⟩ [54]

NOTES

(1) There are in U.L.C. Add. 3968(35) a number of Newton's holograph drafts, in English, of a reply to Leibniz's attack in the *Charta Volans* (see Number 1009) and the accompanying 'Remarques' as published in the *Journal Literaire de la Haye* (see Letter 1018). There is also (U.L.C. Add. 3985(20), fos. 1–5) a French version (not corresponding word-for-word with any one of the holograph drafts) in the hand of Abraham de Moivre, who would doubtless have translated it from English. Draft A (fos. 494–5), which we mainly follow in what we print here, actually has some French equivalents interlineated in another hand which may also be De Moivre's. However, considerable portions of the English corresponding to the French text are to be found only in draft B (fos. 496–7). Drafts C to G (fos. 498–504) seem to be preliminary states. There is a further preliminary draft in Mint Papers, II, fo. 340v.

In our version above we print in the first place draft A, adding within angle brackets, ⟨ ⟩, those passages from draft B that are additional to A, and within square brackets, [ ], still further passages from the French text. We have not attempted to record minor verbal variants of little significance, but we have recorded in our notes some interesting variations, and the significant passages in English missing from the French text are also given.

The paper was never published in the *Journal Literaire de la Haye*, and, indeed, was probably never sent to Johnson, its Editor. Although not perhaps strictly a 'letter', we print it here for two reasons. First, it indicates clearly Newton's own involvement in the calculus controversy, an involvement which is not explicitly revealed by the extant correspondence up to this point. Second, it is possibly the clearest, and most complete, exposition of the state of the controversy at this time.

Newton's paper was superseded by Keill's 'Answer' ('Réponse de M. *Keill*, M.D. . . . aux Auteurs des Remarques sur le Differént entre M. de *Leibnitz* & M. *Newton*', *Journal Literaire de la Haye* for July and August 1714, pp. 319–58; a few copies were also issued separately as pamphlets). We do not reprint this 'Answer', which covers much the same ground as Newton's paper, but Keill's language is far more emotional and provocative and his account is less factual than Newton's drafts are in making precise references to letters. As in his earlier reply to Leibniz (Letter 843a, vol. v) he is more concerned to give elementary mathematical explanations. Keill includes a criticism of Leibniz's 'Tentamen' (see *Journal Literaire de la Haye* for July and August, 1714, pp. 350–2, and Number 1086a), whereas Newton in draft makes only incidental mention of that issue.

Inevitably, much of the material Newton calls in evidence against Leibniz had already been discussed in Keill's letter to Leibniz of 24 May 1711 (Letter 843*a*, vol. v) and in the *Commercium Epistolicum* itself. However, in the present document Newton was also concerned, as a result of the difficulties over Prop. 10, Book II, of the *Principia* (see vol. v, Introduction p. xxxiv and Letters 951 and 951*a*), to defend himself against the charge that he did not properly understand second differences. He also took pains to emphasize the *difference* between his fluxional notation and Leibniz's use of differentials. He includes a lengthy discussion of Leibniz's $\pi/4$ series for the quadrature of the circle, and the possibility that Leibniz learnt it from Gregory

Much of the matter in the paper was to be reiterated yet again in the *Recensio* (published in *Phil. Trans.* **29**, no. 342 (February 1715), 173–225, as 'An Account of the Book entitled Commercium Epistolicum, . . .' and later prefaced to the second edition (1722), of the *Commercium Epistolicum*; see Letter 1162, note (2)). Leibniz was perhaps justified in writing scathingly to Bernoulli of the *Recensio*: 'Crambem commercii recoxerunt' (see Letter 1173).

(2) Perhaps this French phrase was added by de Moivre as it is not found in any draft.

(3) In the French version, the phrase, '& this he . . . Leibniz' is effectively transferred to the next paragraph, where the fact that both authors are correspondents of Leibniz is more fully elaborated.

(4) Letter 843*a*, vol. v, p. 141: 'I observe that Newton himself changed the name and notation of his calculus pretty often. In the treatise on the analysis of equations by infinite series he denoted the increment of the abscissa by the letter o; and in the *Principia Philosophiæ* he speaks of the fluent quantity as *generated* and calls its increment a *moment* denoting the former by upper-case letters (A,B) and the latter by lower-case (a,b).'

(5) In fact Newton devised his standard fluxional dot-notation for *De quadratura curvarum* in December 1691 (D. T. Whiteside in *Journal for the History of Astronomy*, **1** (1960), 118–19). See also note (10) below and Letter 1211.

(6) The oft-mentioned *Epistola Posterior* (vol. II, pp. 110–29).

(7) In the French text the last five words are replaced by 'anterieurs aux marques de Mr Leibnitz.'

(8) The last five words are omitted in French.

(9) In draft A 'or differences' has been deleted and corrected, in another hand, to read 'of ye fluents' as in both draft B and the French version.

(10) The whole of the next passage (found in draft B only) 'wch is equipollent . . . infinitely little' is omitted in the French version. Newton probably first used o notation as early as September 1664 (see Whiteside, *Mathematical Papers*, I, p. 557, note (21)). By October, 1665, he had overcome the problem of infinitesimals (at least in his own mind) by the introduction of instantaneous speeds (see *ibid.* pp. 369ff). However, there was still considerable variation in the notation he used for fluxional analysis, both in his published and unpublished works. As Whiteside points out, however, (*ibid.*, p. 147) his frequent implication that 'pricked' letters were used in his early mathematical researches was somewhat misleading (perhaps intentionally) as this notation had varying significations until the introduction of his standard dot notation in 1691 (see above, note (5)). Leibniz's accusation that Newton's notation was clumsy was understandably based on the *published* account in Book II, Lemma 2 of the *Principia*, where Newton used upper-case letters for quantities, and lower-case letters for their moments, a notation which would be difficult to extend to moments of moments and higher orders. See also Whiteside, *Mathematical Papers*, I, p. 344, note (4), and IV, Introduction to Part 3, especially p. 409. Many of the details are also given in D. T. Whiteside, *The Mathematical Works of Isaac Newton* (Johnson, New York, 1964), I, pp. viii–xix.

(11) The words 'Scholium of the' added in draft B are simply a mistake, and are omitted in French.

(12) The two sentences: 'Tis only a branch . . . differences of differences' in draft A are replaced in the French version by the more elaborate discussion which follows next. For Newton's letter to Collins of 10 December 1672 see vol. I, pp. 247–52. All the correspondence mentioned here and below had been printed in the *Commercium Epistolicum*.

(13) The issue is whether the successive differentials of x^n are equal to the successive terms of the binomial expansion of $(x+o)^n$ with the powers of o removed, or whether they are only proportional to ('ut') these successive terms. As the text goes on to say, the successive differentials are in fact the latter multiplied by 1, 2, 6, 24 . . . etc. No edition of *De quadratura curvarum*—not even in the 'Jesuits' edition of Newton's writings (Lausanne and Geneva, 1744, p. 240)—contains the crucial little word 'ut'. See Letter 1004, note (11), p. 6, and Letter 843*a*, vol. V.

(14) The brackets here are Newton's own.

(15) See *Principia*, 1687, pp. 253–4:

In the exchange of letters that took place between the very skilled geometer G. G. Leibniz and myself ten years ago, when I mentioned that I was master of a method of determining maxima and minima, of drawing tangents and performing similar operations, which was valid for surd quantities as well as rational ones, and concealed the same in transposed letters containing this sentence: 'Given any equation whatever containing flowing quantities [fluents], to find the fluxions, and vice-versa', that famous man replied to me that he also had fallen upon a method of this nature, and he imparted to me a method of his that hardly differed from mine save in words and notation. The foundation of both is contained in this Lemma.

(16) The French version omits 'that he had writ . . . years before' and instead describes the uses of the method (as in the Latin just quoted).

(17) From here through the next two paragraphs the English of draft B is omitted from the French text, which has been (roughly) following draft B. The French continues directly with 'On peut remarquer de plus . . .' (p. 85).

(18) See *Principia* 1687, p. 281, ll.17–18: 'Therefore the difference of the moments, that is the moment of the difference of the areas, is . . .' The expression occurs, of course, in a geometrical rather than an algebraic argument; nevertheless it is Newton's point that the concepts of the calculus of flowing quantities are implicit in the geometrical methods of the *Principia* even in the absence of an explicit fluxional notation or algorithm (see Whiteside, *loc. cit.* note (5) above, pp. 118–20).

(19) Motion here means velocity, the first derivative of distance, acceleration being the second derivative.

(20) Leibniz to Oldenburg, 22 July 1677 N.S.; see vol. II, pp. 231–2.

(21) Leibniz to Oldenburg, 27 August 1676 N.S., vol. II, pp. 57–64.

(22) For Newton's letter to John Wallis of 27 August 1692 and related documents see vol. III, pp. 219–28.

(23) The phrase '& does justice . . . this letter o' is omitted in the French, and rightly so, since Newton's attribution of this use of the letter o to Fermat is a mistake, based on his reading of Van Schooten long before.

(24) 'in the Preface to his Analysis' is not in the French. In the Preface to his *Analyse* (Letter 1008, note (4)) the Marquis de l'Hospital (or rather, here, Fontenelle, compare Letter

1084, note (1)) described the evolution of the method of tangents from Descartes to Barrow, noting that the latter invented 'une espece de calcul propre à cette Methode' though unable to cope with fractions and roots. 'Au défaut de ce calcul', he goes on, 'est survenu celui du célèbre, M. *Leibnitz*; & ce savant Géometre a commencé ou M. *Barrow*, & les autres avoient fini. Son calcul l'a mené dans les pays jusqu'ici inconnus; & il y a fait les découvertes qui font l'étonnement des plus habiles Mathématiciens de l'Europe . . . L'étendue de calcul est immense . . .' It is evident, while giving Barrow his due as a giant on whose shoulders Leibniz stood, l'Hospital is far from qualifying the differential calculus as a mere adaptation of Barrow's method of tangents.

(25) The reference to Tschirnhaus is omitted in the French; in the English manuscript another hand has (rightly) corrected 'Hooks' to 'Flamsteeds'. For Leibniz's just contention against Tschirnhaus see Hofmann, p. 101–18, 121–2 etc.

(26) '& approved' is not in the French.

(27) Turnbull, *Gregory*, pp. 168–72; the letter is still preserved in the Royal Society's *Commercium Epistolicum* manuscript guardbook (no. 6).

(28) Gerhardt, *Leibniz: Mathematische Schriften*, I, pp. 60–9; the version printed there is dated 12 April 1675.

(29) *Ibid.*, I, pp. 69–71.

(30) 'I have received your letter replete with algebraical fruits, for which I thank you and the very learned Collins. Since I am now particularly sidetracked by some questions of mechanics besides my ordinary responsibilities, I could not examine the series that you sent and compare them with my own.'

(31) Leibniz had in fact sent the $\pi/4$ series to Huygens in Paris before 7 November 1674, hence before Oldenburg's letter to him was written (see Huygens, *Oeuvres Complètes*, VII, pp. 393–4; Gerhardt, *Leibniz: Mathematische Schriften*, I, pp. 88–132; Hofmann, pp. 46–8).

(32) 'June' is missing in French.

(33) Vol. I, pp. 396–403.

(34) Vol. I, pp. 313–14.

(35) Gerhardt, *Leibniz: Mathematische Schriften*, I, pp. 53–6. Gerhardt dates the letter 26 October 1674 [N.S.].

(36) Newton first met Collins on 27 November 1669, and his first letter to him is dated January 1669/70 (vol. I, pp. 16–20). The *De analysi* draft had been sent to Collins by Isaac Barrow on 31 July 1669 (*ibid.*, p. 14).

(37) Whatever this means (did Newton think that Leibniz had made three visits to London, or spent nearly three years on a visit there?) it is wrong. Leibniz's first visit to London lasted from January to March 1672/3; he was elected F.R.S. on 22 January (see Hall & Hall, *Oldenburg*, IX, p. 384ff.). The French here reads more weakly 'qui peut donc douter qu'il n'eust receu cette suite . . .'

(38) Gerhardt, *Leibniz: Mathematische Schriften*, I, p. 88.

(39) See vol. II, pp. 57–64.

(40) *Sic* both in English and French, but a mistake for 24 *October* 1676, the *Epistola Posterior*.

(41) See vol. II, pp. 212–19.

(42) Dated 22 July 1677 N.S., *ibid.*, pp. 231–2.

(43) 'I readily agree with the celebrated Newton that the famous Sluse's method of tangents is not yet perfected. And for a long time now I have dealt with tangents in a more

general way, that is, by [employing] the differences of the ordinates.' (Leibniz to Oldenburg, 21 June 1677 N.S.; vol. II, p. 213).

(44) 'What you [English] seem to say, that most difficulties (Diophantine problems excepted) may be reduced to infinite series, does not appear [obvious] to me. For there are many [problems] so amazing and so involved that they do not depend either upon equations or upon quadratures. Among many others, such are the problems of the inverse method of tangents, which even Descartes admitted exceeded his powers'. (vol. II, p. 64). The editorial comment must be that Descartes said nothing of the sort, nor had he reason for such a confession: see Whiteside, *Mathematical Papers*, III, p. 84, note (109).

(45) In his paper 'Nova Methodus pro maximis et minimis . . .' *Acta Eruditorum* for October 1684, pp. 467–73.

(46) Newton uses the word 'communicated' in a strange sense, for this sentence too was expressed in 'transposed letters' (that is, effectively, in an insoluble cipher) in the *Epistola Posterior*; see vol. II, p. 129 and p. 159, note (72). The 'clear' version was first published by Wallis in 1699. 'The other method consists only of assuming a series for any unknown quantity whatever, from which the remainder can conveniently be derived, and in collecting the homologous terms of the resulting equation in order to elicit the terms of the assumed series.'

(47) See vol. II, pp. 212–31.

(48) The preceding two paragraphs are condensed into two sentences of the French text.

(49) The whole paragraph is slightly re-written in the French without important change of sense.

(50) This sentence is again much condensed in the French.

(51) Newton's brackets.

(52) See Letter 498, vol. IV, where the letter is dated (correctly) 10 April 1695.

(53) This whole paragraph has been deleted in the French version.

(54) Draft E consists of an alternative conclusion to the letter, continuing as follows after the quotation of the 'transposed letters'—'*Methodum hanc . . . excogitatam*' (p. 89 above):

But since Mr Leibnitz began these disputes, & detracts from the candor of those who oppose him & in opposition to them represents it unjust to question his candour, making himself a witness in his own cause contrary to the laws of all nations, & appeals from the Report of a large Committee of the R. Society, to the judgment of a nameless Mathematician chosen by himself, wch is the same thing as to make himself a Judge as well as a witness in his own cause; & since his correspondents endeavour to set aside the consideration of the ancient Letters & Papers & bring matters to a wrangle: I desire you to print (in French) the Letter of Mr James Gregory dated 15 Feb. 1671 to the words *secundum vulgaris Algebræ præcepta*, a copy of wch letter was sent to Paris in June 1676 to be communicated to Mr Leibnitz. I desire you to print also the two Letters of Mr

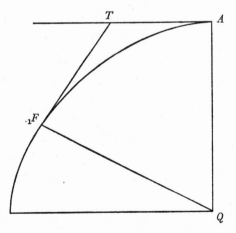

Leibnitz dated 15 July & 26 Octob 1674 concerning a Theoreme or Method for finding the sector or Arc of a Circle whose sine is given; & Mr Oldenburgs Letter of 15 Apr 1675

wherein he sent several series to Mr Leibnitz amongst wch was the series of Gregory; & the Answer of Mr Leibnitz dated 20 May 1675 wherein he acknowledged the receipt of those series; & the latter part of his Letter dated 28 Decem. 1675 beginning with the words, *Habebis & a me instrumentum* &c. All wch five Letters were left entered in the Letter books of the R. Society by Mr Oldenburg. Then print the Letter of Mr Leibnitz dated 12 May 1676 [N.S.] wch is still extant in his own handwriting & that part of his Letter of 27 August. 1676 [N.S.] wch begins with these words. sit QA_1F Sector duabus rectis &c, & ends with these maximeque afficiens mentem. And then leave it to the Reader to make his judgment upon those Letters concerning the pretence of Mr Leibnitz to the series of Mr Newton for finding the Arc by the sine & to that of Mr Gregory for finding the Arc by the Tangent & to some other series sent to him by Mr Newton. After wch the Reader will be better able to make a judgment of his pretence to the original invention of the method of moments or differences.

A closely related draft is also found in Mint Papers, ii, fo. 353v.

1054 CHARLES PARRY TO NEWTON
8 APRIL 1714
From the original in the Mint Papers[1]

Aprill, 8. 1714

Sr

Having some business which will unavoidably keep me in the Country for three weeks, & fearing some progress may be made towards coining of Farthings in that time; I give you the trouble of this to inform you, that since I was in London I have been considering of the lowest rate they can possibly be hammer'd at, which is $21\frac{1}{2}d$ to ye pound, & which rate I compute as follows viz

For ye Fillets	0. 1. $4\frac{1}{2}$
& To ye Tower . . .	4
To a Comptroler . . .	$\frac{1}{4}$
& for vending & dis-	
tributing	$\frac{3}{4}$
	1. $9\frac{1}{2}$

This last article will certainly be a very expensive one, for let us manage what we can, we must have a house to lodge them in, as well as a Teller or two to pay them to such as come to demand them. As for ye coinage being free I am of opinion you will find abundance of inconveniencies therein, (too long to set forth herein) whereas on ye contrary I hope you will be of opinion that a contract, so limited & restrain'd as we propose it may, will be subject to none. And I am sure in respect to the publick it will be ye same thing when ye price is once fix'd [.] I cannot be able to think of such a contract [by] myself but have desir'd Mr Essington (who has a mill as well as myself, & who is willing to join wth me,) [2] to wait on you, in case any thing should offer before my return. I am

Sr yo most humble sert

CHA. PARRY

NOTES

(1) II, fo. 325. The writer was proprietor of the Tower Mills near Mitcham, south of London, where copper plate for braziers was made. He claimed to have supplied copper for the last coinage (compare Letter 972, vol. v).

(2) This copper-manufacturer is also mentioned by Newton in Mint Papers, II, fo. 349.

1055 BURNET TO J. BERNOULLI
8 APRIL 1714

Extract from the original in the University Library, Basel.[1]
Reply to Letter 1040; for the answer see Letter 1070

...J'ay demandèr a Mr. Newton s'il avoit vu les actes de leipsick [2] ou [sont] vos remarques mais il[s] ne s'en trouvent pas encore en ce pais; les libraires ne les font venir qu'en annees, entieres; et il m'a priè de vous fais ses complimens, et de vous assurer que quand [il] verra vos remarques, il sera [possible] d'apprendre ses fautes, s'il en a faites[.] il m'a dit que Mr votre neveu lui en a indiquer une, sans pourtant avoir donnè la correction, et il l'a corrigè dans sa nouvelle edition luimeme. il m'a demandèr d'ecrire en hollande pour ces deux mois des actes de Leipsick, ce que j'ay fait, pour qu'il les eut au plustot.[3] Pour son livre, comme il ne l'imprime pas luimeme, mais qu'il a donné toute l'edition au docteur Bentley, il n'en a fait aucuns presens, qu'a Mr Anisson,[4] qui lui en a demander une copie pour le faire reimprimer a Paris, la dessus il n'a pu se despenser d'en envoyer 3 autre a l'Academie Francoise, qui vouloient entreprendre l'edition... [5]

NOTES

(1) LIa 654 no. 19*.

(2) *Acta Eruditorum* for February 1713, pp. 77–97 and March 1713, pp. 115–32, containing Johann Bernoulli's 'De motu corporum, gravium, pendulorum et projectilium' with its criticism of the *Principia*; by 20 April Newton had seen the volume for 1713 (see Letter 1064).

(3) For the response, see Letter 1083.

(4) For Jean Anisson see Letter 1024, note (2), p. 43.

(5) Burnet means the Académie Royale des Sciences. There was no such French reprint, which was rendered unnecessary by that of Amsterdam, Sumptibus Societatis, 1714, already in press in July/August 1714 according to the *Journal Literaire de la Haye* for those months.

96

1056 [WILLIAM POPPLE] TO NEWTON
8 APRIL 1714
From the copy in the Public Record Office.[1]
For the answer see Letter 1058

To Sr Isaac Newton

Sr.

There having been a Proposal made to the Lords Commissioners for Trade and Plantations, for Coining 1500 Tuns of Copper into halfpence and Farthings; Their Lordships have commanded me to send you [the] inclosed Estimate, and thereupon to desire your Opinion, and any Observations you may think proper thereupon

NOTE

(1) C.O.5, 913, p. 477. The writer was secretary to the Committee on Trade and Plantations, presumably succeeding his father (d. 1708, a friend of John Locke) in this office and being himself father of the dramatist of the same name (see *Dictionary of National Biography*).

On 2 April 1714 (see *C.S.P. Colonial: America and West Indies, 1712–14*, p. 330) the 'officers petitioners for settling a colony in N. America to be called Nova Anna' wrote seeking a licence to coin 1500 tons of copper farthings and halfpence for distribution in England and the Plantations. They hoped thus to make a profit in support of their venture computed as follows (from the estimate sent by Popple to Newton in P.R.O. C.O.5, 866, no. 16):

One ton of copper, at 20*d.* to the pound, yields £205.6.8

Deduct cost of copper	£93.6.8.	
exchanging coin, loss	11.0.0	
cost of casting	35.0.0	
		139.6.8
Net profit per ton		£ 66.0.0

Hence the gross total profit would be £99000, to be reduced by a further £2000 allowance for clerk's wages and other overheads.

As Newton pointed out in Number 1058*a*, the arithmetic was not quite correct.

1057 NEWTON TO OXFORD
12 APRIL 1714
From the holograph draft in the Mint Papers [1]

To the most Honble the Earl of Oxford & Earl Mortimer
Lord High Treasurer of great Britain

May it please your Lordp

Since I attended your Lordp last, [2] I caused a new Furnace to be built in order to a further trial of what may be done by casting of copper into Barrs & coyning copper money out of those barrs. But in the mean time upon assaying the half pence of wch I shewed your Lordp a specimen, I found the copper coarser then it was by the assay before casting. Whereupon I ordered Mr Bagley the founder to supply me wth such barrs as would fully endure the assay: but he has not yet produced any tho it be about three weeks since I gave him the order. Whence I suspect that in ye specimen of half pence wch I shewed your Lordp he put in some Tynn wthout my knowledge tho I stood by to see him cast the copper & he pretends another cause.

Whether the Fillets be made by hammering or casting it will be requisite to repair the Mill-rooms & other rooms in the Irish Mint for cutting flatting scouring & nealing the money, & the repairs by the Workmen's estimate will come to about 145£. If your Lordp please to let me have a Warrant for repairing them, It may be done in six weeks time.

We do not receive gold & silver into the mint to be coyned untill they be made fit to be received. If your Lordp shall think fit that the copper be made into fillets by hammering, & received into the Mint by weight & assay & the money delivered back by weight & assay the coynage being paid for by the Importer: a coynage may be set on foot in this manner so soon as the charge of coynage, the number of pieces in a pound weight & the Reverse of the money shall be setled. And the money made by this method will be of the same fineness wth that of Sweden. [3]

All wch is most humbly submitted to your Lordps great Wisdome

Is. NEWTON

Mint Office
12 *Apr.* 1714.

NOTES

(1) II, fo. 311; there are drafts at fos. 309, 338.

(2) Compare Letter 1047. There seems to be no record of the Mint Officers visiting the Treasury between 26 February and the date of this letter.

(3) Compare this paragraph with Letter 1048, which perhaps therefore was never sent.

1058 NEWTON TO POPPLE
13 APRIL 1714
From the original in the Public Record Office.[1]
Reply to Letter 1056

Mint Office 13 Apr. 1714

Sr

I have considered the annexed Estimate made upon a Proposal offered to the Lords Commissioners for Trade & Plantations for coyning 1500 Tonns of Copper into halfpence & farthings in five years and according to their Lordps Order signified to me by your Letter of 8th Instant, have sent you the inclosed considerations upon them with my opinion therein. I beg the favour that you will please to acquaint their Lordps that my Lord High Treasurer hath the coynage of copper under consideration.

I am

Sr

Your most humble Servant

Is. NEWTON

Wm Popple Esq

NOTE

(1) C.O.5, 866, no. 17. There is a draft in the Mint Papers, II, fo. 323r. The enclosure follows.

1058a OBSERVATIONS ON COPPER COINAGE
From the original in the Public Record Office [1]

Observations upon the Estimate of the neat Profit of coyning 1500 Tonns of Copper into half pence and farthings

Obs. 1   In the last coynage of copper money an hundred Tonns per annum in six years made a great complaint in Parliament whereby the coynage was stopt all the seventh year by reason of too great a quantity of copper money, & after the coynage of another hundred Tonns the nation was fully stockt during the next five or six years. Therefore six or seven hundred Tonns is abundantly sufficient to stock the nation of England & a coynage of 1500 Tonns in five

99

years time is not practicable by reason of the clamours it would make amongst the people. At present there wants not above 80 or 100 Tonns in all.

Obs. 2   If a pound weight of copper be cut into 20d, a Tonn in coyn will amount only to 186£ 13s. 4d. It must be cutt into 22d that a Tonn may make 205£ 6s. 8d. But its better to coyn the money nearer to the intrinsic value.

Obs. 3   Casting drawing cutting flatting scouring nealing blanching drying & coyning cannot be done for 35£ per Tonn. And 11£ per Tonn for changing the copper money is something too much. In the last coynage of copper money, 5d per £wt was allowed by the Patentees for casting drawing cutting flatting scouring nealing blanching drying & coyning, including the work of the Graver & Smith. There was also 40s per Tonn allowed to a Comptroller. And if 7£ 6s 8d per Tonn be allowed for putting off, the whole charge including the price of ye copper at 10d per £wt leaves a profit of 6d per £wt. And this profit in coyning 1500 Tonns amounts unto 85000£, out of wch something may be abated for housrent, Clerks, coyning tools, & incidents.

Obs. 4   He that assays sizes & coyns the copper money should not be impowered to make any profit by coyning it too light or too coarse, & therefore should have nothing to do with buying or providing the copper or distributing it by tale to the people, but should only receive it by weight & assay & deliver it by weight & assay after coynage, & have it in his power to refuse bad copper.

NOTE

(1) C.O.5, 866, no. 17(i) printed in *C.S.P. Colonial: America and West Indies, 1712–14*, p. 330; there are drafts in the Mint Papers, II, fos. 322, 353 etc. The paper may be compared with that printed as Number 1003, vol. v.

1059   SIR JOHN NEWTON TO NEWTON
13 APRIL 1714
From the original in the University Library, Cambridge[1]

Sr

Pray let me have the satisfaction of seeing you here on thursday morning next about 10, or 11 a-clock, to see me sign some papers of moment, in wch you will oblige.

Your Most Humble

Servant

J: NEWTON

Ap: 13*th*. 1714
For Sr Isaac Newton

NOTE

(1) Add. 3968(41) fo. 113. From its peremptory tone it is clear that the writer of this note was Sir John Newton, third baronet and Isaac's distant relation, whom Isaac regarded as head of the family (see vol. II, p. 335; vol. IV, p. 265, p. 461, note (2), and pp. 488–9). Sir John presumably still had a house in Soho Square. In the Indexes of previous volumes he has been confused with John Newton and Samuel Newton, both mediocre mathematicians.

The paper has been used for rough notes of a rejoinder to Leibniz's *Remarques* (Letter 1018).

1060   LOWNDES TO THE MINT
16 APRIL 1714
From the copy in the Public Record Office.[1]
For the answer see Letter 1065

Gentlemen

My Lord Treasurer Commands me to send you the Enclosed Memorial of Henry Smithson relating to several Prosecutions against false Coiners together with an Account of his Charges in carrying on the same;[2] His Lordp is pleased to direct you to consider the matters contained therein and let him have a true State thereof together with your Opinion what is fit to be done thereupon

I am

Gentlemen

Your most humble Servt.

WM LOWNDES

Treasury Chambers
16 *April* 1714

The Petitioner having since the writing of this Letter presented to my Lord another the same is here also enclosed for your Consideration & to hear him thereupon if you think fit [3]

W.L.

Officers of the Mint

NOTES

(1) Mint/1, 7, p. 64; there is another copy in T/27, 21, p. 170. Smithson's memorials are recorded at pp. 65–6.
(2) Smithson claimed (on 25 March 1714) that he had been employed in apprehending and prosecuting coiners for fourteen years; that several arrested coiners had been released because prosecution was not pressed against them and others carried on their trade with impunity; that when he attempted to see the Warden of the Mint with intent to press pro-

secutions he found only his deputy Richard Barrow 'who told him . . . that he . . . had not received a shilling so would not disburse a shilling to carry on any such prosecutions but they must drop.' Smithson then went to Newton who offered to lend the Warden money to finance prosecutions. Smithson adds his account for expenses in pursuit of his duties amounting to £95.17s.6d.

(3) Smithson's second memorial (dated 16 April 1714) repeats the allegation that the Warden is lethargic in his pursuit of coiners, and claims that 'great numbers of false coiners' have gathered on the Scottish Border where their activities may endanger the Kingdom. One John Thompson, a reliable witness, rode 260 miles to London to give information of these coiners, but the Warden rejected him 'with the poor reward of a Crown'.

1061 —— HASPERG TO LEIBNIZ
16 APRIL 1714
Extract from the original in the Niedersächsische Landesbibliothek, Hanover [1]

. . . Un Marchand libraire me dit l'autre jour qu'il avoit rencontré par hazard Mon. Newton, et qu'il lui avoit demandé quelques exemplaires du livre nommé, Collinsii Commercium Epistolicum, mais qu'il avoit repondu qu'il ne vouloit faire public, ni que tout le monde le sçache, se flattant, que vous de[s]avoueriré et recanterer comme il dit, [2] que vous etier le Premier qui aie inventé ce, dont il est question dans le dit livre. Je lui dis donc que Mons[ieu]r Neuton se flatteroit en vain et je l'ay dit à Mons[ieu]r le Docteur Woodward [3] qui se mocque aussi de cette expression du Marchand Libraire, et elle ne lui semble moins ridicule qu'à moy, pourtant je ne me puis pas imaginer que Mons[ieu]r Newton se puisse figurer une telle ideé. Mons[ieu]r Woodward tachera d'avoir un exemplaire du dit livre, je puis dire avec verité que celuici est fort de vos Amis, aussi bien que Mons[ieu]r Flamstadd [4] chez qui je dinai l'autre jour, ou il fit mention honorablement de Votre Excellence, et vous fait ses compliments. il fera publier une livre touchant l'Histoire cœleste, il m'a montré les figures qui sont deja desinees. il me dit que Collins est mort, depuis 25 ou 30 ans je croy, et qui n'aye fait point mention de ce dont il est question entre V.E. et Monsr. Newton. il m'a montré dans le livre Philosophiæ Natural. Princip. Mathemat. Newtonii in 2a Edit. pag. 423 et pag 425 [5] que cette demonstration la etoit fausse, et que ce qu'il debite pag. 402. in fine [6] etoit encor incertain et fort douteux. . .

NOTES

(1) Bodemann's catalogue, no. 374. When he wrote this letter Hasperg, the secretary of the Société des Arts et Sciences of Wolfenbüttel, was staying in London. He had a considerable correspondence with Leibniz. In a later letter, dated 22 May 1714 N.S., he mentioned that he had obtained a copy of the *Commercium Epistolicum* via Woodward and Waller, and would

send it to Leibniz. In a further letter of 9 December 1714 N.S. he wrote that he had also purchased a copy of the second edition of the *Principia* for Leibniz.

(2) The phrase is difficult to read, and the French unorthodox; but the meaning is clear.

(3) Woodward was no friend to Newton; see Letter 785, vol. v, and notes, and Letter 1174.

(4) Flamsteed, of course.

(5) The end of the Scholium to Proposition 35, Book III, on the motions of the Moon.
Here Newton proposes his Horrocksian hypothesis of epicycles.

(6) Proposition 29, Book III, on the magnitude of the Moon's variation.

1062 LEIBNIZ TO CHAMBERLAYNE
17 APRIL 1714
Extract from the printed version in Des Maizeaux.[1]
Reply to Letter 1045; for the answer see Letters 1089 and 1092

Vienne ce 28. *Avril* 1714 [N.S.]

Monsieur,

Je vous suis obligé tant de la communication de la Lettre de l'insigne Mr.
Wotton, qui m'est plus favorable que je ne pouvois espérer, & que je vous prie
de remercier de ses bons sentimens; que de votre offre obligeante de moyenner
une bonne intelligence entre Mr. *Newton* & moi. Un nommé Mr. *Keill* inséra
quelque chose contre moi dans une de vos *Transactions Philosophiques*; j'en fus
fort surpris, & j'en demandai réparation par une Lettre à Mr. *Sloane* Secrétaire
de la Societé.[2] Mr. *Sloane* m'envoya un Discours de Mr. Keill,[3] ou il jus-
tifioit son droit d'une maniére qui attaquoit même ma bonne foi. Je pris cela
pour une animosité particuliére de ce personnage, sans avoir le moindre
soupçon que la Societé, & même Mr. *Newton* y avoit part: & ne trouvant pas à
propos d'entrer en dispute avec un homme mal instruit des affaires anté-
rieures; & supposant d'ailleurs que Mr. *Newton* lui-même, mieux informé de
ce qui s'étoit passé, me feroit rendre justice, je continuai seulement à demander
la satisfaction qui m'étoit due.

Mais je ne sai par quelle chicane & quelle supercherie, quelques-uns firent
en sorte qu'on prît la chose comme si je plaidois devant la Societé, & me
soumettois à sa jurisdiction, à quoi je n'avois jamais pensé; & selon la justice,
on devoit me faire savoir que la Societé vouloit examiner le fond de l'affaire,
& l'on devoit me donner lieu de déclarer si j'y voulois proposer mes raisons, &
si je ne tenois aucun des Juges pour suspect. Ainsi, on n'y a prononcé qu'*una
parte audita*,[4] d'une maniére dont la nullité est visible. Aussi ne crois-je pas que
le Jugement qu'on a porté puisse être pris pour un Arrêt de la Societé.

Cependant Mr. *Newton* l'a fait publier dans le Monde par un Livre imprimé
exprès pour me décréditer, & envoyé en Allemagne en France, & en Italie,

comme au nom de la Societé. Ce Jugement prétendu, & cet affront fait sans sujet à un des plus anciens Membres de la Societé même, & qui ne lui a point fait deshonneur, ne trouvera guère d'Approbateurs dans le Monde; & dans la Societé même, j'éspére que tous les Membres n'en conviendront pas. Des habiles François, Italiens, & autres desaprouvent hautement ce procédé, & s'en étonnent: & on a là-dessus des Lettres en main; les preuves produites contre moi leur paroissent bien minces.

Pour moi, j'en avois toujours usé le plus honnêtement du monde envers Mr. *Newton*, & quoiqu'il se trouve maintenant qu'il y a grand lieu de douter s'il a su mon Invention avant que de l'avoir eue de moi; j'avois parlé comme si de son chef il avoit eu quelque chose de semblable à ma Methode. Mais abusé par quelques flateurs mal avisez, il s'est laissé porter à m'attaquer d'une maniére très-sensible. Jugez maintenant, Monsieur, de quel côté doit venir principalement ce qui est nécessaire pour faire cesser cette contestation.

Je n'ai pas encore vu le Livre publié contre moi, étant à Vienne, qui est l'extrémité de l'Allemagne, où de tels Livres sont portez bien tard. Et je n'ai point daigné le faire venir tout exprès par la poste. Ainsi je n'ai pas encore pu faire une Apologie telle que l'affaire demande. Mais d'autres ont déja eu soin de ma reputation. J'abhorre les disputes desobligeantes entre les Gens de Lettres, & je les ai toujours évitées; mais à présent on a pris toutes les mesures possibles pour m'y engager. Si le mal pouvoit être redressé, Monsieur, par votre entremise à laquelle vous vous offrez si obligeamment, j'en serois bien aise; & je vous en ai déja beaucoup d'obligation par avance.

...

NOTES

(1) II, pp. 120–5. A draft of the letter in the Niedersächsische Landesbibliothek, Hanover (see Bodemann's catalogue, no. 149, fos. 121–2) is almost word-for-word identical, but is dated 21 April. Newton, to whom Chamberlayne forwarded this letter, prepared an English translation which we also print. Later he sent the letter to Des Maizeaux, for inclusion in the *Recueil* (British Museum, MSS. Birch 4284, fos. 212–13).

(2) Letter 822, vol. v.

(3) Letters 843 and 843*a*, vol. v.

(4) 'after hearing one party only'.

1062a NEWTON'S TRANSLATION
[MAY 1714]
From the holograph manuscript in the University Library, Cambridge[1]

Sr

I am obliged to you as well for the communication of the Letter of the excellent Mr Wotton (who is more favourable to me then I could hope, & I pray return my thanks for his good opinion) as for your obliging offer to mediate a good understanding between Mr Newton & me. It was not I that interrupted it. One Mr Keil inserted something against me in one of your Philosophical Transactions. I was much surprised at it & demanded reparation by a Letter to Dr Sloane Secretary of the Society. Dr Sloane sent me a discourse of Mr Keil where he justified what he said after a manner wch reflected even upon my integrity. I took this for a private animosity peculiar to that person without having the least suspicion that the Society & even Mr Newton himself took part therein. And not judging it worthy the while to enter into a dispute wth a man ill instructed in former affairs & supposing also that Mr Newton himself being better informed of that wch had passed would do me justice, I continued only to demand that satisfaction wch was due to me. But I know not by what chichan[e]ry & foul play some brought it about that this matter was taken as if I was pleading before the Society, & submitted my self to their jurisdiction, wch I never thought of. And according to justice they should have let me know that the Society would examin the bottom of this affair & have given me opportunity to declare if I would propose my reasons & if I did not hold any of the judges for suspected. So they have given sentence, one side only being heard, in such a manner that the nullity is visible. Also I do not at all beleive that the judgment wch is given can be taken for a final judgment of the Society. Yet Mr Newton has caused it to be published to the world by a book printed expresly for discrediting me, & sent it into Germany, into France, & into Italy as in the name of the Society. This pretended judgment, & this affront done without cause to one of the most ancient members of the Society it self & who has done it no dishonnour will find but few approvers in the world. And in the Society it self I hope that all the members will not agree to it. The able men among the French, Italians, & others disapprove highly of this proceeding & are astonished at it, & I have several Letters upon it in my hands. The proofs produced against me appear to them very short.

As for me I have always carried my self with the greatest respect that could be towards Mr Newton. And tho it appears now that there is great room to doubt whether he knew my invention before he had it from me; yet I have spoken ~ ~ [2] as if he had of himself found something like my method; but being abused by some flatterers ill advised, he has taken the liberty to attaque me in a manner very sensible. Judge now Sr, from what side that should principally come wch is requisite to terminate this controversy. I have not yet seen the book published against me, being in Vienna which is in the furthest part of Germany where such books come very slowly, & I have not thought it worth the while to send for it by the Post. So I have not yet been able to make such an Apology as the affair requires. But others have already took care of my reputation. I abhorr disobliging disputes among men of Letters & have always avoyded them: but at present all means possible have been taken to engage me in them. If the evil could be redressed, Sr, by your interposition which you offer so obligingly, I shall be very glad, & I am already very much obliged to you for it.

NOTES

(1) Add. 3968(30), fo. 443. The translation seems certainly to be Newton's own work, as there is considerable emendation throughout. It extends as far as the portion we have printed in French. The translation was read before the Royal Society at its meeting on 20 May 1714. As Chamberlayne could not attend this meeting (see Letter 1079) he asked Keill to send him an account of it. Keill accordingly furnished him with an extract from the Journal book of the Royal Society; Chamberlayne sent both this extract and a copy of Keill's accompanying letter to Leibniz on 27 July 1714 (see Letters 1089, 1092 and 1092*a*).

(2) *Sic* in MS., but there is no omission.

1063 KEILL TO NEWTON
[19 APRIL 1714]
From the original in the University Library, Cambridge.[1]
Reply to Letter 1053; for the answer see Letter 1064

Honoured Sr

I have read both the peices concerning the Commercium inserted in the Journal Literaire [2] and I think I never saw any thing writ with so much impudence falshood and slander as they are both, I am of opinion that they must be immediatly answered, and I am now drawing up an answer wch I will finish as soon as I hear from you, the Author of the Remarks says that when Mr Leibnits had published his method in 1684 you had never communicated any thing like it either to the Publick or to any particular man till a long time after, to confute this I designe to insert your demonstration of the method wch is at the end of the Analysis, wch that author must have seen and have read, [3] I only want to know what he means by saying that their are things in your principia contrary to the Doctrine of Fluxions if it be that where you say the terms of the series for the binome are the first 2d and 3d Fluxions, &c But this is not in the Principia but in the book of Quadratures and it is plain you only mean that they are Proportional.[4] I would likewise know who that great Mathematician is who has long agoe observed it, and where it is I doe not find it in the Acta Lipsiæ, Did ever any Geometer before you give the quadrature of a curve in finit terms whose ordinate was expressed by several quantities connected under a Vinculum, I cannot find that any of them did. If you would let me know your opinion about these things, and what you think should be inserted in the Answer you will for ever oblige

Sr

your most Humble and obedient

Faithfull servant

JOHN KEILL

For

Sr Isaack Newton
at his house in
St Martins Street
near Leicester Feilds
  Westminster

NOTES

(1) Add. 3985, no. 5. As the letter is postmarked 20 April (on the day of delivery in London) we have assumed that it was written on the nineteenth. Newton answered on the day of receipt.

(2) That is, the *Charta Volans* (Number 1009) and Leibniz's 'Remarques'; see Letter 1018.

(3) It is true that Leibniz read *De analysi* in 1676—his notes on it are printed in Whiteside, *Mathematical Papers*, II, pp. 248–59 (for discussion see Hofmann, pp. 181–94, and Whiteside, *op. cit*, pp. 169–71). However, Whiteside points out that Leibniz was interested only in Newton's work on series, and paid no attention to fluxions—the notes emphasize the independence rather than the derivative nature of Leibniz's thought. Whiteside also prints Leibniz's February 1712 review of *De analysi* and Newton's draft rebuttals of it (pp. 259–73). These were probably composed in late 1713 or early 1714, before the period of this (surviving and resumed) correspondence with Keill.

(4) See Letter 1004.

1064 NEWTON TO KEILL
20 APRIL 1714

From the original in Trinity College Library, Cambridge.[1]
Reply to Letter 1063; for the answer see Letter 1069

Sr

I am glad you have read both the pieces concerning the Commercium inserted in the Journal Literaire & are of opinion that they must be immediately answered & are thinking of an Answer. As to what you want to know

107

concerning things in the Principia contrary to the doctrine of fluxions or differences I take it to be this. In the Scholium of ye 10th Proposition of the second book of the Principia I have made use of ye method of Infinite Series for determining the Curves in wch Projectiles will move in a resisting Medium such as is air. John Bernoulli has published in the Acta Eruditorum for Febr. & March was a twelve month, a Paper upon that Scholium, in wch he represents that the Method there used is the Method of fluxions, & that it appears thereby that I did not understand ye 2d Fluxions when I wrote that Scholium because (as he thinks) I take the second terms of the series for the first fluxions, the third terms for the second fluxions & so on. But he is mightily mistaken when he thinks that I there make use of the method of fluxions. [2] Tis only a branch of ye method of converging series that I there make uses of. The Acta Eruditorum for the last year are but just come to London, & I find thereby that John Bernoulli is the great Mathematician who accuses me on this account. But I beleive it's better not to reflect upon him for it nor so much as to name him any otherwise then by the general name of the great Mathematician. They are seeking to pick a quarrell with me & its better to lett them begin it still more openly without a provocation.

There is another great Mathematician to whom Leibnitz referred the examination of the Commercium Epistolicum. [3] He makes use of two arguments against me. One that I made no use of the prickt letters till of late, the other that when I wrote the Principia I understood not the second fluxions as a certain great Mathematician (Bernoulli) has observed. The Answer is that I use any notation for fluents & any other notation for fluxions, & an unit for the fluxion of time or its exponent & the letter o for the moment of time or of its exponent, & the rectangles of the fluxions & the moment o for the moments of other fluent quantities. That in the Analysis per æquationes numero terminorum infinitas I represent fluents by the areas of figures, time by the Abscissa flowing uniformly, the fluxions of fluents by the Ordinates of curves, the moments of fluents by the rectangles under the Ordinates & o the moment of the Abscissa: but do not confine myself to any certain symbols for the Ordinates or fluxions. That I do the same in the book of Quadratures & even to this day. That where I use prickt letters they signify not moments or differences wch are infinite little quantities but fluxions or the Ordinates of curves as the exponents of fluxions wch are finite quantities, unless they be multiplied by the symbol o (either exprest or understood) to make them infinitely little: but it is not necessary that the Ordinates of curves should be represented by prickt letters. Such letters may be a convenient sort of notation but not necessary to the method. That prickt letters are older symbols for fluxions then any used by Mr Leibnitz: for he has no symbols for fluxions to this day. That the rectangles under the Ordinates of curves & the moment *o* are older symbols for moments or differences then any used by Mr Leibnits they being used by me in my Analysis above-mentioned communicated by Dr Barrow to Mr Collins in the year 1669,[4] & the symbols *dx* & *dy* being not used by Mr Leibnitz before the year 1677. And whereas Mr Leibnitz præfixes the letter ∫ to the Ordinate of a curve to denote the Summ of the Ordinates or area of the Curve, I did some years before represent the same thing by inscribing the Ordinate in a square as may be seen in the Analysis.[5] My Symbols therefore (so far as I have used any particular symbols) are the oldest in the kind.

The other argument used by the great Mathematician is that when I wrote my Principia I understood not the second differences, as a certain great Mathematician (vizt Bernoulli) has noted, meaning in the Scholium to ye 10th Proposition of ye second Book. But this great Mathematician is grosly mistaken in taking the method there made use of, wch is a branch of the method of converging series to be the method of fluxions. The Elements of the method of fluxions are set down in ye 2d Lemma of the second Book & are very different from ye method made use of in this Scholium.[2]

The author of the Remarks cites Dr Wallis as favouring Mr Leitnitz & yet Dr Wallis in the Preface to the first Volume of his works A.D. 1695 writes that in my two letters of June 13 & Octob. 24, 1676 I expounded my method of Fluxions to Mr Leibnitz found by me ten years before.[6]

In my letter of 10 Decem. 1672 sent to Mr Collins,[7] in writing of a method whereof the method of Tangents of Slusius was but a Corollary, & which stuck not at surds, & wch was therefore the method of fluxions, I represented that this method was very general & amon[g]st other things extended to the determining the curvature of Curves. Whence its manifest that I then understood the second fluxions or differences of differences.

I received your Letter this afternoon at three of the clock & have time to add no more but that I am

Your most humble Servant

London 20 *April*   Is NEWTON
1714

In the book of Quadratures where I use prickt letters for fluxions I solve some Problems in the Introduction to ye book without making use of such Letters & therefore did not then confine the method of fluxions to such Letters.[8]

For the Rnd Dr John Keill Professor
of Astronomy in the University of
Oxford

NOTES

(1) R.16.38, fos. 424–5, printed in Edleston, *Correspondence*, pp. 170–4.

(2) See Letter 951, note (3) (vol. v, p. 348) and Edleston, *Correspondence*, p. 171, footnote.

(3) Newton was not yet aware that this was the same mathematician, that is, Bernoulli.

(4) See vol. i, pp. 12–14. The *De analysi* was probably written in the early summer of 1669.

(5) See (for example) Whiteside, *Mathematical Papers*, ii, p. 226 and p. 227, note (78).

(6) In the Preface to Volume i of his *Opera mathematica* (Oxford, 1695) after referring to these letters Wallis added 'ubi methodum hanc [*sc.* de Fluxionibus, consimilis naturae cum Leibnitii calculo differentiali] Leibnitio exponit tum ante decem annos, nedum plures, ab ipso excogitatam' (compare vol. iv, p. 102, note (7)). For Wallis's unequivocal defence of Newton's priority see the same volume, pp. 100, 115 and 117.

(7) Letter 98, vol. i, p. 247.

(8) Newton's standard dot notation first appeared in print in Wallis's *Opera mathematica*, ii, pp. 392–6 (Oxford, 1693; the first of three to be published), in an excerpt from the still unpublished first (1691) version of Newton's *De quadratura curvarum*.

1065 NEWTON & PHELIPPS TO THE TREASURY
28 APRIL 1714

From the copy in the Public Record Office.[1]
Reply to Letter 1060

To the most Honorable the Lord High Treasurer of Great Brittain

May it please Yr Lordp.

In Obedience to your Lordps. Order of Reference upon the two Memorials of Hen. Smithson relating to prosecutions against false Coyners since Jan. 1710, together with an account of his charges in carrying on the same, we have considered the matters conteined therein & heard him thereupon & find that he had no Order from the Warden of the Mint for performing the Services expresst in his Bill except the prosecution of Horsman & Robinson for which he was allowed in Mr. Weddals Account.[2] That upon Accounts Stated between him & Mr. Weddal a little before Mr. Weddel's death, he was indebted to Mr. Weddel above five pounds as appears by his Note under his hand and as he acknowledged in a Letter to Mrs. Weddel since the death of her Husband. And in general it did not appear to us that any moneys were due to him for prosecuting under the Warden, except a bill of £10.18.2 wch. is incerted into Mr. Barrows account and was the only charge demanded by him at that time or before.

And we further humbly represent to your Lordp. that John Thompson [3] being examin'd by Sr. Isaac Newton about what Services he was able to do, could think of no Services but such as were looked upon as too Stale to be regarded in Courts of Justice and thereupon was dismist by the Warden for the present & we are satisfied that what Smithson represents against the Warden in relation to the Said Thompson is groundless false & scandalous.

Is. N.

Edw. Ph.[4]

28. *April* 1714

NOTES

(1) Mint/1, 7, p. 66.
(2) See Letters 982 and 990, vol. v.
(3) See Letter 1060, note (3), p. 102.
(4) Edward Phelipps, Comptroller of the Mint.

1066 LOWNDES TO THE MINT
29 APRIL 1714
From the copy in the Public Record Office[1]

Gent[lemen]

My Lord Treasurer having received from Mr. Bertie the inclosed Memorial about Assaying of Copper [2] His Lordp commands Me to transmit the same to You for Your Consideration I am &c. 29th Aprill 1714

Wm Lowndes

NOTES

(1) T/27, 21, p. 193. For the content, compare Letter 1047.
(2) James Bertie's memorial to Oxford about the purity of copper (dated 26 April 1714) was carefully preserved by Newton (Mint Papers, II, fos. 414–15). Copper cannot be assayed by cupellation like gold and silver, writes Bertie; it may be examined by grain and colour 'especially by a microscope' (is this the first introduction of the microscope to metallurgy?) but there is no certainty in such an inspection. Again, the tests of hammering with the smith's hammer when hot or of bending when cold are good but not infallible. The best test is provided by hammering the metal under the great hammer—the water-driven hammer—when red-hot, for no alloy of copper will withstand this. Bertie also points out that pure copper cannot be cast into bars on any scale, for tin must be added to prevent sponginess in the casting; hence, he argues, hammered copper must be used even though it is more costly than the fused metal, and if this is done the copper coins will be as good as those of Charles II's reign.

1067 LOWNDES TO THE MINT
29 APRIL 1714
From the copy in the Public Record Office[1]

Gentlemen

My Lord Trea[su]rer having received a Letter from an unknown hand proposing to make a Discovery of False Coining provided he has protection & Encouragement His Lordship Commands Me to transmit the same to you, that you may take Care such Publication may be made in the Gazette or other wise that the Writer of the said Letter may Attend You for that Purpose. I Am

Gentlemen

Your Most Humble Servt.

WM. LOWNDES

Tre[asu]ry Ch[ambe]rs.
Officers of the Mint

NOTE

(1) Mint/1, 7, p. 62. There is another copy in T/27, 21, p. 193. No reply has been found. The anonymous letter from 'P.B.' follows promising a complete but unspecific disclosure of all the counterfeiters' tricks if an indemnity is granted to the writer; this letter is dated 22 April. The attempt to contact 'P.B.' through the *London Gazette* was successful; see Letter 1071.

1068 LOWNDES TO THE MINT
29 APRIL 1714
From the copy in the Public Record Office.[1]
For the answer see Letter 1085

Gentlemen

I send you by my Lord Treasurers Order a Letter from the Lords of the Privy Council in Ireland representing that it will be a great prejudice to the Trade of that Kingdom if several species of French Gold and Silver Coines therein mentioned be not made current there,[2] And desiring the same may be laid before the Queen, that Her Majtys. Order may be obtained for that purpose, together with the Report of Mr. Vincent Kidder Assay Master there concerning the Intrinsick Value of the said Coines;[3] which Letter & Report his Lordship is pleased to direct you to Consider, And let him have your Opinion what may be fit for her Majty. to do therein. I am

Gentlemen

Your most Humble Servant

WM LOWNDES

Trea[su]ry Chambers
29th. April 1714
Officers of the Mint

NOTES

(1) Mint/1, 7, p. 67.

(2) On 23 April 1714 Bollingbroke had sent to Oxford a letter from the Irish Privy Council (dated 17 April), which follows immediately the letter printed here in the Minute-book, pointing out that the older French Louis d'or, double Louis and silver crowns formerly and still circulating in Ireland and recognized as legal tender there, had been withdrawn in France and replaced by new issues (of the same names) which were now in turn beginning to circulate in Ireland although not yet recognized by proclamation. The Lords desire recognition of these coins (see also *Cal. Treas. Papers, 1708–14*, p. 578).

(3) The Assaymaster's report (dated 27 February 1713/14) draws attention to discrepancies in weight and fineness between samples of the new French coins submitted to him for examination (with details), discrepancies arising 'thro haste or neglect of the Assay Master of the Mint or Comptroller . . . as it often does in the Mint in London' (a sentence hardly welcome to Newton). He suggested that the Louis d'or should pass as equivalent to a guinea at twenty-three shillings, and a silver crown for five shillings and seven pence. Newton recommended slightly lower values (Letter 1085).

1069   KEILL TO NEWTON
2 MAY 1714

From the holograph original in the University Library, Cambridge.[1]
Reply to Letter 1064; for the answer see Letter 1074

Oxford May 2d 1714

Honoured Sr

I here send you a part of the Answer [2] I have made to the tuo Anonymous Authors who have attacked the Commercium, [3] I never saw a bad cause defended with so much face and impudence before. There second Argument about second Fluxions, I have not yet considered, because I was willing first to see what Mr Bernoulli had said on that matter, and I had the Acta for the year 1713 (where it is) but yesterday; I remember you told me that Mr Leibnits had made a blunder in his Tentamen de motuum Celestium causis by not

113

understanding 2d Fluxions I was [4] you would take the pains to mark it down and write it to me. [5] I beleive my Answer will contain a sheet more, for I am to take him to Task for filshing of series. [6] I am for putting my name to it, for I have said nothing but what is fully made out, and they have on the contrary have thrown all the dirt, and scandal they could without proving anything they have said. and therefore they thought it best to conceal their names, I beleive Wolfius is the author of the Latin letter for it is exactly agreable to his candour and honesty who is inferior to no body but Mr Leibnits in prevarications. Dr. Halley and I doe often drink your health, he and I are both of opinion that there should be 50 copies of the commercium sent over to Johnson and that there should be advertisements in the forreigne Gazetts that the original letters of the Commercium are in such a mans hands to be viewed by [any] Gentleman that is to travel to England. [7] and particularly the letter with Gregories Quadrature of the Circle [8] I wish my paper were translated into French before it were sent over Dr Halley beleives that Mr DeMoivre will doe it I doe not care to trust them with the translating it I am

Sr

your

most obedient and Faithfull servant

JOHN KEILL

For
Sr Isaack Newton
at his house in
St Martins Street near
Leicester Feilds
    Westminster

NOTES

(1) Add. 3985, no. 6, postmarked 5 May.

(2) The paper—as the letter makes plain, not yet finished nor translated into French—is not now with the letter or extant. Presumably it was returned by Newton to Keill and then went to De Moivre for translation. Keill's 'Answer' was eventually to appear in the *Journal Literaire de la Haye* for July and August 1714, pp. 319–58; see Letter 1053a, note (1), p. 90.

(3) Presumably Keill means the writer of the *Charta Volans* (Number 1009) and the 'Remarques' in the *Journal Literaire de la Haye* (see Letter 1018) as one person, and the 'eminent mathematician' (unknown to Keill as yet, Johann Bernoulli) as the second. Farther on he suggests that the 'eminent mathematician' is Christian Wolf.

(4) Read: 'wish'; or perhaps Keill omitted 'hoping'.

(5) Leibniz had published his 'Tentamen de motuum cœlestium causis' in the *Acta Eruditorum* for February 1689 (pp. 82–96). In it he developed a vortical hypothesis of planetary

motion (owing something to the earlier ideas of Kepler, Descartes and Borelli) which, he claimed, preceded his first-hand knowledge of the *Principia*; see the articles on the celestial mechanics of Leibniz by E. J. Aiton in *Annals of Science*, **16** (1960), 65–82; **18** (1962), 31–41; **20** (1964), 111–23; and **21** (1965), 169–73. In the second and third of these articles Dr Aiton discusses Newton's criticisms of the 'Tentamen' which were taken up and amplified by Keill in his 'Answer' (see Letter 1053*a*, note (1)).

The dates of Newton's reading of Leibniz's articles in the *Acta* have not been firmly settled. One of his reactions to the 'Tentamen', the curiously titled *Ex Epistola cujusdam ad Amicum*, has already been published by Edleston (*Correspondence*, pp. 308–14) from the autograph by Newton and now in the *Lucasian Papers* (U.C.L., Res. 1893 (*a*), Packet 8); there are other partial versions in Add. 3968, no. 4 and elsewhere. Edleston supposed this document to belong to the *Commercium Epistolicum* period (1711–12), but it is more likely of somewhat later date. In any event it was written after Newton's *Notæ in Acta Eruditorum An.* [16]89, which we print, following this letter, where further comment may be found; it may be remarked here that in both his papers of comments Newton recorded Leibniz's inability (in 1689) to apply second differentials successfully to the more difficult problems of physics.

Leibniz's 'Tentamen' together with additional introductory matter and a later re-working of the whole paper is printed in Gerhardt, *Leibniz: Mathematische Schriften*, VI, pp. 144–93.

(6) Since the *Journal Literaire de la Haye* was printed in 16mo, the text of Keill's 'Answer' (see Letter 1053*a*, note (1)) in 39 pages extended to nearly two and a half printed sheets. It is clear from Newton's Letters 1074 and 1077 that Keill sent him with this letter more than eleven draft pages, and that these went beyond a point corresponding to p. 335 of the printed 'Answer', for in Letter 1074 Newton rewrites a passage on the eleventh of Keill's draft pages, which appears in print on p. 335; that is, Keill sent more than enough copy for a single sheet. So far his 'Answer' is essentially concerned with the *Commercium Epistolicum* material and the intended inference from it of Newton's priority in the discovery of the new infinitesimal calculus. Keill only tackled Leibniz's 'Tentamen' ten pages from the end of the printed letter (p. 348) whereas the 'filshing of [Gregory's] series' (see Letter 1008, note (5)) is noted on p. 335, the draft of which was now in Newton's hands. In fact, perhaps influenced by Newton's Letter 1077, Keill added no more about Leibniz's 'theft' of series but went on to discuss the development of notation in relation to the priority issue, and then the question of errors in second differentials (see also Letter 1076, note (1), and Letter 1078 note (6)).

(7) Virtually all the original documents printed (in whole or in part) in the *Commercium Epistolicum* were the property of William Jones. They were at this moment, presumably, on the deposit with the Royal Society where they were open to inspection—or so said Johnson's note at the end of Keill's printed 'Answer' in the *Journal Literaire de la Haye* for July and August, p. 358. For a time later Jones took them back, but they have now (bound together in a separate volume) long been preserved in the archives of the Royal Society. Conti inspected the MS. letters later to see if anything to Leibniz's advantage had been suppressed in the *Commercium Epistolicum*.

(8) In James Gregory to Collins, 15 February 1670/71, in vol. I, p. 61 and *Commercium Epistolicum* p. 25; see Turnbull, *Gregory*, pp. 168–72.

1069a NEWTON'S NOTES ON LEIBNIZ'S 'TENTAMEN'
From the original in private possession

Notæ in Acta Eruditorum an. 89 p. 84 & sequ [1]

1. Hypoth. 1. Omnia corpora, quæ in fluido lineam curvam describunt, ab ipsius fluidi motu agi. Nam conatus recedendi a centro non coercetur nisi a contiguo & moto

Absurd. 1. Ergo corpus non movetur nisi ab agente corporeo, non a mente humano nisi corporeo, non a Deo nisi corporeo.

Absurd. 2. Deus non regit mundum proindeque non est Dominus Deus.

Absurd. 3. Vortex gravitatem causatur, & gravitas propterea tendit ad axem vorticis, non ad centrum Solis.

2. Hypoth. 2. Ergo Planetæ moventur a suo æthere.

Absurd. 1. Ergo Cometæ moventur a suo æthere.

Absurd. 2. Ergo Cometarum et Planetarum æqualiter distantium a sole æquales sunt velocitates. [2]

7. Hypoth. Orbem fluidum cujusque Planetæ moveri circulatione harmonica, id est in circulis concentricis, & cum velocitatibus reciproce proportionalibus distantiarum a centro. [3]

Absurd. Ergo tempora periodica Planetarum sunt ut quadrata distantiarum a centro. [4]

9. Sollicitatio gravitatis male dicitur pro pondere. Nam solicitatio est actio rei intelligentis. [5]

Leibnitz affirms that all bodies wch describe a curve are carried in a fluid (Acta 1689 Feb. p.84 l.8) & thence concludes that the Vortex circulates harmonically (ib p.86 l.19,22,26.) But after Dr Gregory had objected that the tempora periodica of the Planets were in a sesquialterate proportion of the mean distances of ye Planets from ye center & therefore the circulation of the Vortex, could not be harmonical: Leibnits restrains the circulatio harmonica to ye narrow latitude of the orb of every single Planet & in the intervals of the orbs allows the circulation to be not harmonical (Acta 1706 Octob. pag.86) And yet by his Principles it follows from ye motion of Comets that the circulation must be harmonical throughout. [6]

10. *Hunc conatum* [7] *metiri licebit perpendiculari ex puncto sequenti in tangentem puncti præcedentis insignabiliter distantis.* Negatur ubi differentiæ considerandæ veniunt

11. *Conatus centrifugus exprimi potest per sinum versum anguli circulationis.* [8]

116

Negatur ubi circulatio non fit in circulo. Negatur etiam ubi differentiæ secundæ considerandæ veniunt.

12.[9] Negatur in iisdem casibus.

14. Negatur ubi differentiæ secundæ considerandæ veniunt.

15. *In omni circulatione harmonica elementum impetus paracentrici est differentia solicitationis paracentricæ et dupli conatus centrifugi.* Negatur. Nam 1 si circulatio circa centrum solicitationis paracentricæ, peragatur in circulo concentrico, elementum impetus paracentrici nullum erit, ideoque solicitatio paracentrica (seu vis centripeta) æqualis erit duplo conatui centrifugo, cum tamen revera æqualis conatui centrifugo. 2 Si circulatio sit in curva quacunque excentrica, differentia sollicitationis paracentricæ et dupli conatus centrifugi erit conatus centrifugus, et differentia sollicitationis paracentricæ et conatus centrifugi erit nihil (nam sollicitatio paracentrica et cenatus centrifugus semper sunt inter se æquales:) sed elementum impetus centrifugi nec nihil erit nec æquale conatui centrifugo. 3 Si curvatura Curvæ minuatur donec curva cum tangente coincidat, solicitatio paracentrica et conatus centrifugus cessabunt, sed motus paracentricus et ejus elementum non cessabunt. 4 In calculo hujus Propositionis Leibnitius errat ponendo $N_2 M$ æqu[ale] $G_2 D$ et NP æqu. $_2 D_2 T$. Nam hæ quantitates habent differentias secundas, et differentiæ secundæ in investigatione differentiæ differentiarum negligi non debuerant. Leibnitius igitur calculum differentialem ubi differentiæ secundæ considerandæ veniunt nondum intellexerat.[10]

16. Negatur.

19. Colligitur ex duabus falsis Propositionibus 12mam et 15mam: nam 12ma non valet nisi in circulo viribus concentrico; & Prop. 15 falsissima est. Leibnitius igitur non invenit Prop. 19 per calculum differentialem sed inventam computare conatus est ut suam faceret, & computando bis erravit in hac Propositione.[11]

Translation

Notes on the *Acta Eruditorum* 1689, pp. 84 ff.[1]

1. Hypothesis 1. All bodies describing a curved line in a fluid [medium] are impelled by the motion of that very fluid. For the endeavour to recede from the centre is only overcome by a contiguous moving [body].

1st Absurdity. Then a body is only moved by a corporeal agent, not by the human mind (unless it be corporeal) nor by God (unless he be corporeal).

2nd Absurdity. God does not govern the world and so he is not the Lord God.

3rd Absurdity. The vortex is the cause of gravity and therefore gravity tends towards the axis of the vortex, not to the centre of the Sun.

2. Hypothesis 2. Therefore the planets are moved by their æther.

1st Absurdity. Therefore comets are moved by their æther.

2nd Absurdity. Therefore the velocities of comets and planets equally distant from the sun are equal.[2]

7. Hypothesis. The fluid orb of each planet is moved in a harmonic circulation, that is, in concentric circles with velocities reciprocally proportional to the distances from the centre.[3]

Absurdity. Therefore the periodic times of the planets are as the square of the distances from the centre.[4]

9. Gravitational 'urging' is mistakenly said in place of [gravitational] 'weight'. For 'urging' is the action of an intelligent being.[5]

[A passage in English follows; see p. 116]

10. *This [force of] endeavour* [7] *to the perpendicular will properly be measured from the following point to the tangent to the preceding point, insignificantly distant from it.* This is not the case where second differences enter into consideration.

11. *The centrifugal [force of] endeavour may be expressed by the versed sine of the angle of revolution.*[8] This is not the case when the revolution does not take place in a circle. It is not the case either when second differences enter into consideration.

12.[9] Not true under the same conditions.

14. Not true where the second differences enter into consideration.

15. *In every harmonic circulation the element of the paracentric velocity is the difference between the paracentric urging and twice the centrifugal [force of] endeavour.* Not true. For 1. if the revolution about the centre of the paracentric urging continues in a concentric circle, the element of the paracentric velocity will be zero and so the paracentric urging (or centripetal force) will be equal to twice the centrifugal [force of] endeavour, whereas it is really equal to this endeavour. 2. If the revolution takes place in an excentric curve of any kind the difference between the paracentric urging and twice the centrifugal endeavour will be the centrifugal endeavour, and the difference between the paracentric urging and the centrifugal endeavour will be zero (since the paracentric urging and the centrifugal endeavour are always equal to one another); but the element of the centrifugal velocity will neither be zero nor equal to the centrifugal endeavour. 3. If the curvature of the curve be diminished until the curve coincides with the tangent, the paracentric urging and the centrifugal endeavour will cease, but the paracentric speed and its element will not vanish. 4. In his computation in this proposition Leibniz makes a mistake, taking N_2M as equal to G_2D and NP as equal to $_2D_2T$. For these quantities have second differences, and second differences ought not to be regarded as negligible in the investigation of the differences of differences. Accordingly Leibniz did not yet understand the differential calculus in so far as second differences come under consideration.[10]

16. Not true.

19. This is inferred from two false propositions, the twelfth and fifteenth; for the twelfth is only valid in the case of the circle whose centre is that of the forces; and Prop. 15 is completely false. Therefore Leibniz did not discover Prop. 19 by the differential calculus but attempted to compute what was [already] discovered in order to make it his own, and in computing it he twice made a mistake in this proposition.[11]

NOTES

(1) Apparently Keill's allusion to the matter in Letter 1069 is the earliest dated evidence for Newton's detection of Leibniz's 'ignorance' of second differentials. This may suggest, but it does not prove, that the *Notæ* were written earlier than the date of Keill's letter. Further, there is in the Lucasian papers a copy of the *Notæ* in Keill's hand which must presumably have been made in the early summer of 1714. However, Letter 1080 of 21 May 1714 makes it plain that Keill had not at that time seen the *Notæ* since he shows himself unaware of Newton's own detailed onslaught upon the fifteenth proposition of Leibniz's 'Tentamen' (see Letter 1069, note (5)). In preparing his 'Answer' for the *Journal Literaire de la Haye* Keill was no mere scribe for Newton, who himself was to return to the defects of Leibniz's 'Tentamen' in the *Recensio* or 'Account' of the *Commercium Epistolicum* in *Phil. Trans.* **29**, no. 342 (January and February 1714/15), 173–224 (see Letter 1162, note (2)).

Considering Leibniz's 'Tentamen' as offering a dynamical theory of the motion of a body in an orbit about some centre (*not* as an account of the solar system) and setting aside all the defects of his 1689 paper, it may be said that (though neither Newton nor Leibniz realized the fact) this theory is congruent mathematically with that of Newton in the *Principia*. The difference between the two treatments lies in the difference between the *physical* interpretations of Newton and Leibniz. When a body M moves freely along a path oblique to some centre O with uniform inertial motion, as seen by an observer at O rotating with the radius vector OM it appears to accelerate away from him, with a non-uniform radial (or in Leibniz's terminology, paracentric) acceleration, c^2/r^3 (c being a constant and r being the radius). The puzzle is that this acceleration is not seen to be produced by a real force; Leibniz described it as the result of a *conatus centrifugus* in the vortex which drives the body freely circulating in harmonic motion (that is, in accord with Kepler's area law) out of its force-free path to move along the straight line MM'. (Thus whereas for Newton rectilinear motion is force-free *in vacuo* apart from the generating *vis inertiæ*, for Leibniz such motion in the vortex is force-free apart from the *conatus centrifugus*; the difference is that between the physics of free space and the physics of the plenum.)

To explain how a body moves in a circle or a conic section, Newton introduced a centripetal force, Leibniz a *sollicitatio paracentrica* (paracentric urging). These are equivalent forces in that both cause the accelerative deviation of the moving body M from the tangent into the orbit and are alike measured by the difference between the radial acceleration in the tangent and the radial acceleration in the orbit, that is by $(c^2/r^3) - (d^2r/dt^2)$. Hence the second differential of the radius with respect to time does certainly enter the argument. For Newton the equation $(c^2/r^3) - (d^2r/dt^2) = 0$ gives the condition for uniform rectilinear motion. For Leibniz's rotating observer, $d^2r/dt^2 = 0$ gives the condition for uniform, acceleration-free motion in a circular orbit, but when $d^2r/dt^2 = c^2/r^3$, the *conatus centrifugus*, uniform rectilinear motion is obtained. Or again, Newton's system starts from the postulate of the First Law of Motion and *derives* Kepler's area-law as a consequence; Leibniz starts from Kepler's area-law (the harmonic vortex) and *derives* inertial rectilinear motion along the tangent as a consequence.

Certainly such differences, and the existence in Newton's thought of a single active accelerative force (the centripetal) as contrasted with the existence in Leibniz's scheme of *two* active forces—the *conatus centrifugus* and the *sollicitatio paracentrica*—helped to conceal the mathematical consistency of the two analyses. Moreover, Newton naturally compared with his *Principia* of 1687 Leibniz's 'Tentamen' of 1689, not the later, corrected version of 1706. This was certainly fair, since Leibniz enhanced his somewhat exaggerated opinion of the 'Tentamen' by claiming—probably truthfully—that he had written it after coming across the account of the *Principia* in the *Acta Eruditorum* and knowing no more of Newton's ideas and propositions. (Leibniz's annotations on the *Principia* itself have recently been published in *G. W. Leibniz, Marginalia in Newtoni Principia Mathematica*, ed. E. A. Fellmann (Paris, 1973).) However, the 1689 'Tentamen' was all but unintelligible. There were numerous printing errors, particularly in the references to the figure and in the figure itself. Secondly, in the important Proposition 15, arguing from the congruence of triangles, Leibniz went wrong in setting N_2M as equal to G_2D (as Newton points out; see also E. J. Aiton, *Annals of Science*, **16** (1960), 73–5; **18** (1962), 35–9). However, this geometrical error in first-order differentials can easily be annulled, as Newton and Keill showed. Thirdly, Leibniz measured the *conatus centrifugus* not by c^2/r^3 but by $\frac{1}{2}c^2/r^3$ (or $\frac{1}{2}a^2\theta^2/r^3$ in Leibniz's notation). Again, both Newton and Keill discovered this mistake and attributed it to Leibniz's inadequate handling of second differentials. Dr Aiton (*Annals of Science*, **20** (1964), 111–23) has argued that their interpretation involved a basic misunderstanding of the 'Tentamen' (see note (7) below) and that the error derived simply from a geometrical slip. However, it is certain that the second differential with respect to time of the radius vector of the orbital curve (a focal ellipse) is to be computed; that Leibniz (rightly) related the differential to the tangential motion and the central force; and that (in 1689) his argument was faulty in assigning the factors of proportionality therein. Thus, though Leibniz's indubitable error does not establish (as Newton and Keill claimed) his ignorance (in 1689) of the general nature of second differences, it does demonstrate his fallibility in a particular case of handling second-order quantities. In short, this mistake (seized upon by the English with malicious enthusiasm) certainly related more genuinely to the use of the calculus than did that which the Bernoullis had seized upon in *Principia*, Book II, Prop. 10.

Although the 'Tentamen' was severely criticized by Huygens (1690) in his correspondence with Leibniz, it was Pierre Varignon (in a letter of 6 December 1704) who pointed out the definite mistake in Proposition 15; this, and some printing errors, Leibniz amended in the *Acta Eruditorum* for October 1706, pp. 446–51 ('Excerptum ex Epistola G.G.L. quam pro sua Hypothesi physica motus planetarii . . . ad amicum scripsit').

(2) This is a powerful argument against any theory of an aetherial vortex transporting the planets about the sun; a comet ought only to move at the vortical speed appropriate to its solar distance, that is all cometary orbits would be alike, but this is not so. Newton returns to this point in U.L.C. Add. 3968 (12), fo. 176r.

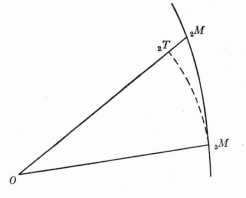

(3) This is a correct rendering of Leibniz in *Acta Eruditorum* for 1689, p. 84 (see E. J. Aiton, *Annals of Science*, **21** (1965), 172). Leibniz resolved the orbital motion of the planet (say a portion of an ellipse) about the Sun O at a point $_3M$ into a concentric *harmonic* component $_3M_2T$ and

a rectilinear *paracentric* component $_2T_2M$. If the speed of the harmonic circulation be further defined in the manner of Leibniz (following Kepler) as $_2T_3M/dt \propto 1/O\,_2M$ (where dt is the orbital time from $_2M$ to $_3M$) then Kepler's area law is an immediate consequence.

(4) Compare *Principia*, Book I, Prop. 4. Since $T = 2\pi R/V$ and $V \propto 1/R$, $T \propto R^2$; only $V \propto 1/\sqrt{R}$ gives $T^2 \propto R^3$, the true condition for the solar system. Thus Leibniz's vortex theory of the planets necessarily fails; it can only be redeemed by postulating that the vortex divides into discontinuous layers ('deferents'), the motion within each 'orb' being harmonic but the orbs moving relative to each other according to Kepler's Third Law.

(5) The term *sollicitatio* was, of course, Leibniz's and more or less equivalent to *conatus*. The second sentence of Newton's comment seems to say, pejoratively, that *sollicitatio* is a term properly applied only to the acts of intelligent beings and was therefore inconsistent with Leibniz's hypothesis. Newton himself was deeply inclined to believe that gravity *is* the act of an intelligent being, God, though not an arbitrary act since it would be a miraculous event should God decide to breach his own physical laws by which the universe is governed. See A. Koyré, *Newtonian Studies* (London, 1965), pp. 90–107, 154–8 and R. S. Westfall, *Force in Newton's Physics* (London, 1971), pp. 395–400 and note 184, with references there cited.

Perhaps the further comment should be added here that though Leibniz in the 'Tentamen' identifies the 'paracentric urging' with the body's gravity, and even speaks of the body's attraction towards the centre, he was far from abandoning the plenist explanation of gravity which he originally derived from Huygens: 'tho' he call this Force Attraction', as David Gregory put it, 'he does not question but tis derived from the impulse of the ambient Fluid, like the actions of the Magnet' (David Gregory, *Astronomiæ physicæ et geometricæ elementa* (Oxford, 1702); we quote from the English translation of 1715, I, p. 176). Thus Leibnizian physics required at least two dynamic æthers—one to move the planets, the other to effect gravitation—as well as, possibly, other subtle fluids appropriate to the phenomena of light, electricity, magnetism and so forth.

(6) In Book I, Prop. 77 of David Gregory's *Astronomiæ physicæ et geometricæ elementa* (Oxford, 1702, pp. 99–104; English translation, 1715, I, pp. 172–82) is an account and critique of the 'Tentamen', which contains the objection concerning comets (note (2) above) and a statement of the 'deferent' fallacy (*ibid.*, Latin edition, pp. 102–3; English translation, p. 179) just discussed in note (4) above.

(7) In translation it seems advisable to distinguish (as seventeenth-century Latin writers did) between *conatus* and *vis*, rather than render both by the (loaded) term 'force'. The italicized words are taken from Leibniz, here and below.

(8) See E. J. Aiton, *Annals of Science*, **20** (1964), 115–16. Leibniz conceived of the tangent not as the limit of the chord but as the line 'joining two [consecutive] points on a curve separated by an infinitely small distance, or the produced side of the polygon [of infinitely many sides] which for us is equivalent to the curve'. However, according to Aiton's analysis of the 'Tentamen', Leibniz confused the tangent thus conceived with the Euclidean tangent and so occasioned an error (by a factor of two) in his measure of *conatus centrifugus*; this was the error first pointed out to him by Varignon, and corrected in Leibniz's 1706 revision of the 'Tentamen'.

(9) Leibniz's twelfth proposition was 'Conatus centrifugi mobilis harmonice circulantis sunt in ratione radiorum reciproca triplicata.' ('The centrifugal [forces of] endeavour of a moving [body] circulating harmonically are reciprocally proportional to the radius cubed.') Newton's criticism of this and the following proposition could again be rooted in Leibniz's

treatment of second differences (see note (8) above, and note (1), p. 119—it is in Proposition 12 that Leibniz wrongly introduces an additional factor of two and states the *conatus centrifugus* to be $a^2\theta^2/2r^3$). But a rough draft by Newton, on an unfoliated scrap in U.L.C., MS. Add. 3972(2) indicates that his objections were probably, at least at first, more a result of his misconceptions about the fundamentals of Leibniz's dynamical system. He wrote, '[Leibniz deals with] the center & angle of harmonical circulation, without considering the curvity of the Orb described. And by confounding these two centers with one another in the fundamental Propositions upon wch he grounds his calculus & taking the vis centrifuga for a force sometimes bigger sometimes less then that with wch the body tends from either of the centers, he has perplexed his calculations & sometimes erred for want of it.'

(10) The draft has many deleted variants not in general recorded here, but the following is too important to omit:

In calculo hujus Propositionis D. Leibnitius negligit differentias secundas usque donec ad differentiam differentiarum pervenitur: quod fieri non debuit. Et propterea calculum differentialem ubi differentiæ secundæ considerandæ veniunt nondum intellexit. Differentiam differentia[ru]m sic investigat.

Jam P_2M *æqu.* (N_2M) *seu* $G_2D + NP$. *Et* $_2T_2M$ *æqu.* $_2MG + G_2D - _2D_2T$. *Ergo* $P_2M - _2T_2M$ *(differentia differentiarum) erit* $NP + _2D_2T - _2MG$ *hoc est (quia* NP *et* $_2D_2T$ *sinus versi duorum angulorum & radiorum incomparabiliter differentium coincidunt) bis* $_2D_2T - _2MG$. Quod falsum est. Errat ponendo N_2M æq. G_2D et rursus ponendo NP æqu. $_2D_2T$ Nam hæ quantitates habent differentias secundas quæ hic negligi non debuerant. Rectius dixisset. Jam P_2M æq. N_2M $+PN$, et $_2T_2M$ æqu. $_2MG + G_2T$ Et $P_2M - _2T_2M$ æq $N_2M - G_2T + PN - _2MG$.

The quotation from the 'Tentamen' is exact.

Newton's second criticism was mistaken, for $NP = _2D_2T$ to the third order. Given that $_2ML$ is the true tangent, his first criticism is exact, and the reasoning is more fully given in the variant just quoted. From $N_2M - G_2T + PN - _2MG$ the line equivalencies in Leibniz's figure give $(PG - _2MG) - (G_2T + _2MG)$ whence $2(_2D_2T - _2MG)$, not $2_2D_2T - _2MG$, is the difference of P_2M and $_2T_2M$. Now in Leibniz's mind (1689), if o represents the time of motion from $_2M$ to $_3M$, then $_2D_2T$ represents (*conatus centrifugus* $\times o^2$) and $_2MG$ represents (*sollicitatio paracentrica* $\times o^2$). Hence he stated that the former was twice the latter. His error may be amended in the manner suggested by Newton (whereby $_2D_2T = _2MG$); or if, with Dr Aiton, Leibniz's argument making $_2MG = _3ML$ represent (*sollicitatio paracentrica* $\times o^2$) be accepted as correct (see note (7) above), then—since the *conatus* must be made equal to the *sollicitatio* and the factor of two eliminated—the correction adopted by Leibniz in 1706 whereby $2_2D_2T = $ (*conatus* $\times o^2$) must be adopted. So that whether or not justified in proclaiming *two* errors of Leibniz relating to second differentials, Newton was certainly justified in proclaiming *one*.

(11) The meaning becomes very plain from the *Recensio* (see Letter 1162, note (2)) where the same charge is levelled against Leibniz: '. . . the 19th Proposition has an erroneous Demonstration adapted to it. It lies upon him [Leibniz] either to satisfy the World that the Demonstration is not erroneous, or to acknowledge that he did not find that and the 20th Proposition thereby, but tried to adapt a Demonstration to Mr. Newton's Proposition to make it his own.' (p. 208). In Proposition 19 of the 'Tentamen' Leibniz derived the 'equivalent' to Newton's inverse-square law of gravitation, with some ostentation (Huygens, *Oeuvres complètes*, IX, p. 52); its argument is analysed by E. J. Aiton, *Annals of Science*, **20** (1964), 120–2. (For Leibniz, the gravitational impulse is inversely proportional to the square of the distance in an elliptical orbit; but this is *not* identical with Newton's gravitation.)

1070 J. BERNOULLI TO BURNET
4 MAY 1714
Extracts from the copy in the University Library, Basel.[1]
Reply to Letter 1055

à Bâle ce 15. *May* 1714 [N.S.]

Monsieur.

Ce que vous me mandez dans l'honneur de vôtre derniere du huitieme Avril
(apparement vieux style) touchant les 2 exemplaires du *Commercium epistolicum*,
destinés pour mon Neveu et pour moi, avoit deja ete notifié par Mr. de Moivre
dans une lettre, qu'il écrivit, non pas à moi, comme vous penséz mais à mon
Neveu.[2] J'ai enfin reçu, mais un peu mal conditioné par la voiture, le livre
de Mr. Newton,[3] que vous avez eu la bonté de m'envoyer en present, dont je
vous repete les remercimens, que j'ai fait dans ma dernière:[4] Mr. de Moivre
m'en avoit promis un exemplaire il y a long temps de la part de l'Auteur, mais
apparement il ne s'est pas trouvé en etat de tenir sa parole, puisque je n'en ai
point reçu de luy; Vous dites que Mr. Newton n'en a fait aucun present qu'a
Mr. Anisson ce pendant je sçay, que Mr. Varignon et d'autres Mathematiciens
de France, en ont reçu,[5] si bien que Mr. Varignon lui même me dit dans une
de ses lettres, qu'il ne doutoit nullement que Mr. Newton ne me fit aussi ce
même present comme il luy en avoit fait: mais je ne suis pas surpris qu'il
menage ses exemplaires et qu'il n'en envoye pas à tout le monde car c'est un
livre veritablement preçieux et il lui en couteroit trop de les prodiguer
j'usqu'en suisse; je ne laisse pas de lui être redevable de quelques uns de ses
autres livres, dont il m'a regalé autre fois: Pour la nouvelle edition des *Principia
Phil. Mat.* je l'aurois acheté à tout prix, si vous n'aviez pas eu la generosité de
m'en rendre possesseur. Pour ce qui est de mes remarques sur l'anciene
edition, j'espere que Mr. Newton en aura toute la satisfaction, qu'il peut
souhaiter, il est vray, qu'entre autres de ces remarques mon Neveu luy a
indiqué une meprise avec la correction quoique sans la methode dont je me
suis servi pour trouver cette correction ce qu'il verra dans les actes de Leipsic;
Il est vray aussi que là dessus Mr Newton a corrigé son erreur [6] dans sa Nouvelle edition et il y a inseré la correction de sa façon en cassant quelques
feuilles deja imprimees et en y substituant d'autres sans pourtant faire
mention avec un seul mot, par qui il avoit eté averti de sa meprise, ni à qui il
en etoit redevable demarche dont tout autre plus glorieux que moi se plain-
droit publiquement sur tout dans cette conjoncture, ou vos Mathematiciens
Anglois se mettent en campagne contre les Etrangers et se donnent tant de
mouvement pour assurer á leur nation la gloire des decouvertes, qu'on a

faites en Geometrie depuis quelques temps : pour vous dire la verité, je croyois que la candeur de Mr. Newton lui arracheroit quelques mots de confession de la verité à l'honneur d'un Suisse sans blesser par là la reputation de vôtre Illustre Nation ; mais quelque puisse être la cause de ce silence, j'ai trop bonne opinion de l'amitié de Mr Newton par les marques qu'il m'en a données pour croire, que ce soit un effet de mepris en me jugeant indigne, que mon Nom paroisse dans son livre. Quant au reste, j'ai remarqué en le parcourant jusqu'ici legerement, qu'il y a inseré plusieurs belles choses et d'autres corrections, il trouvera pourtant, quand il les confrontera avec mes remarques, qu'il a laissé quelques endroits defectueux sans correction. J'ai fait autre fois sur l'anciene edition d'autres remarques, que je n'ai pas encore communiquées, mais que je communiquerai peutêtre avec le temps. Je voi aussi que Mr. Newton a profitè des observations de Mr Huguens, qu'un certain [homme] nommé Groningius aprez les avoir achetés à la vente de la biblioteque de Mr Huguens fit ensuite imprimer, il y a environ 15 ou 16 ans avec un petit traité sur la Cycloide, que Mr Newton peut avoir vu. [7] Vous me demandez Monsieur, mon sentiment sur le *commerci[um Episto]licum* mais je n'ai garde de decider d'une querelle qui est entre deux grands hommes, que j'éstime egalement et qui tout deux m'honorent de leur affection ; cependant il y auroit du pour et du contre sans qu'il me prenne envie de m'en meler, de peur que je n'offense l'un ou l'autre ou peutetre tous les deux. Il servit à souhaiter, que les partisans de Mr Newton eussent embrassé sa cause avec moins d'emportement et avec d'autant plus de moderation : selon la maniere derisive dont ils parlent avec un air de superiorité on diroit que tout ce qu'on a invente dans le mathematiques depuis prés d'un demi siecle etoit dû uniquement à la Methode des Fluxions ou au Calcul differentiel qu'on pretend être tiré de cette methode, comme si nous autres petits compagnons n'avions rien fait que commenter sur elle et tout au plus enformer quelques legers corollaires suivant l'expression de vôtre Cousin Mr. Cheynes. [8] Mais si j'avois l'humeur querelleuse, je pourrois peutêtre montrer que nous n'avons pas peu contribué à la perfection de la methode des infiniment petis, que c'est moi le premier qui ai inventé et communiqué la maniere de reduire en regles d'algorithme pour remonter des fluxions aux fluentes ou comme nous parlons des differentielles aux integrales, dont il est resulté le calcul integral, qui est aussi le nom dont je l'ai baptisé, n'en sçachant alors point de plus convenable ; au lieu, qu'auparavant on n'avoit que des regles particulieres sans Algorithme, que l'on employoit suivant l'état des questions, et le plus souvent on étoit obligé de recourir aux suites ou progressions infinies, qui faisoient la principale application de vos Mess[ieurs]. Si de plus mon naturel me le permettoit, je pourois bien faire valoir mes decouvertes et les faire sonner bien haut par quelques Partisans

zelés et outrès en les prenant par tout et en montrant, qu'elles son indepen-
dantes de la methode des fluxions et du Calcul differentiel; mais je n'aime pas
à me rendre celebre par des cricries des esprits serviles et des esclaves, ni à
employer des plumes mercenaires, qui se piquent de dire des flatteries sans
mesure quand il s'agit de loues quelqu'un de leur nation. Je me contente que
la verité parle elle même en ma faveur et se developpe á qui voudra séclaircir
du fait par la lecture des livres et de journaux qui contiennent nos productions
car aprez la lecture on jugera, qui en a eté le premier inventeur. Pour ajouter
un mot sur le *commercium epistolicum*. Je ne sçai si vous sçavez que Mr Leibnits,
qui se trouve à Vienne depuis plus d'un an, y a publie une petite response sur
une feuille de papier,[9] qui doit servir d'avant coureur à une reponse plus
ample quand il sera de retour chez luy; si je n'avois eu peur de charger trop
cette lettre, ou que cela ne Vous fut pas agreable, je vous aurois envoyé cette
reponse, . . .

NOTES

(1) LIa 654 no. 13. No more letters between Bernoulli and Burnet are extant in the University Library, Basel.

(2) We have not traced this letter.

(3) The second edition of the *Principia*; see Letter 1040.

(4) Letter 1040.

(5) Burnet had in fact mentioned that three copies were sent to the Académie Royale.
Fontenelle wrote to Newton on behalf of the Academy to thank him for the gift; see Letter
1036.

(6) See Letter 1040, note (3), p. 65; Bernoulli's tone is becoming increasingly testy.

(7) Johann Grœning, *Historia cycloeidis . . . Accedunt Christiani Hugenii annotata posthuma in
Isaaci Newtoni* Philosophia naturalis principia mathematica (Hamburg, 1701). As is described
in Cohen, *Introduction*, pp. 186–7, this was a garbled jumble of notes made by Huygens with
drafts and corrections made by Newton himself and Fatio de Duillier. It was *not*, as Grœning
seemed to suppose, a list of mistakes detected by Huygens.

(8) George Cheyne (1671–1743); see Letter 1008, note (1), p. 14.

(9) The *Charta Volans* (Number 1009).

1071 'P.B.' TO NEWTON
8 MAY 1714
From the copy in the Public Record Office[1]

May 8th: 1714

May it Please Your Honour

This day in the Gazette from Tuesday April the 27 to Saturday May 1st
1714 I had the good fortune to meet with an Answer of a Letter I directed to
the Most Honble. the Lord High Treasurer of Great Britaine Relating to

Matters of certain Discoverys for the preservation of the Curr[en]t Coin of this Kingdom, and therein I am Ordered to Attend Your Honours for that Purpose.

But in Regard I have Run great Hazards by Conversing with Some persons concerned in that Sort of Evil practice though only to find the Depths of that Matter purely to Arme Myself with Such Knowledge that may be Serviceable the better to make out and Compleate what I have proposed, & if Encouraged it will putt an End to all those Things for Ever.

Therefore I hope it will not be thought unreasonable that I shall Desire my Pardon for all things that I have both seen and known in those Matters, before I give My Attendance since I have ventured hard for no other End then to Serve my Native Country and that without Charge if I shall be Encouraged so to Doe.

Att the sight of which pardon I shall be ready to give My Attendance but not till Some days after by Reason of Distance and will make out that which will be Satisfactory to Your Honrs. and to all Good & Wellmeaning people.

Yours Honrs: Most Obedient Servt.

P: B.

To the Honorable Sr. Is. Newton.
at her Majes. Mint in the Tower
of London
    These

NOTE

(1) Mint/1, 7, p. 63. Compare Letter 1067. It is evident that the advice to insert a notice in the *London Gazette* was followed.

1072 NEWTON TO CHAMBERLAYNE
11 MAY 1714
From Des Maizeaux, *Recueil*.[1]
Reply to Letter 1062

Lettre de Mr. Newton à Mr. Chamberlayne

Le 11. *de Mai* 1714.*V.St.*

Monsieur,

Je n'entends pas assez à fond la Langue Françoise,[2] pour sentir toute la force des termes de la Lettre de Mr. Leibniz;[3] mais je comprends qu'il croit que la Societé Royale & moi, ne lui avons pas rendu justice.

126

Ce que Mr. Fatio a écrit contre lui, il l'a fait sans que j'y aye eu la moindre part.

Il y a environ neuf ans que Mr. Leibniz attaqua ma réputation, en donnant à entendre que j'avois emprunté de lui la Méthode des Fluxions. Mr. Keill m'a défendu; & je n'ai rien su de ce que Mr. Leibniz avoit fait imprimer dans le Journal de Leipsic, jusqu'à l'arrivée de sa premiére Réponse à Mr. Keill, où il demandoit, en effet, que je rétractasse ce que j'avois publié.

Si vous pouvez me marquer quelque chose en quoi je lui aye fait tort, je tâcherai de lui donner satisfaction; mais je ne veux pas rétracter ce que je sai être véritable, & je crois que le Commité de la Societé Royale ne lui a fait aucun tort.

Je suis

Votre très-humble Serviteur

Is. NEWTON.

NOTES

(1) II, pp. 126–7. We have modified the printing conventions and removed the notes, without altering the spelling. The translation into French was presumably made by Des Maizeaux himself.

(2) As the beginning of Letter 1074 reveals, Newton was insensitive to the subtleties of French idiom but had a perfectly adequate grasp of a piece of French prose.

(3) Letter 1079 confirms that Chamberlayne had submitted Letter 1062 to Newton, presumably meeting him for the purpose soon after Leibniz's letter arrived in London, Chamberlayne hoped that Newton would write Leibniz a private and conciliatory reply but as this Letter and Letter 1079 make plain, Newton had no such intention.

1073 DERHAM TO NEWTON
11 MAY 1714
From the original in King's College Library, Cambridge[1]

Upminster. May. 11 1714♂ [2]

Sr

I have just been reviewing my *Physico-Theology* for a 3d Impression,[3] & shall be troubled to have it to go to the Press without your castigations: & am therefore forced to give you the trouble of this L[ette]r to beg the favour of what you promised me in that matter. When I was last in town I called at your house, but it proved an unseasonable time. I am sorry I could not again wait on you, as I earnestly desired.

I am obliged to make a great chasm in my *Survey of the Heavens*,[4] for want of knowing the centrifugal force of ♃'s [5] Rotation, wch I intreat you to

127

compute for me: for I dare not trust to my own calculations, & it is not difficult to you. I am satisfied you have much other business of your own lying upon your hands, & am sorry I am forced to press you thus: but these two matters are of such great consequence to me, & I hope I may say to God's service also, yt I can't but beg you to undertake so much trouble for me, I hope you have no other opinion of either my fidelity, friend[shi]p, or discretion to think I shall make any ill use of your favour, such as may in the least prejudice your credit, or disserve you, but shall strictly observe any directions you shall give me in concealing, or gratefully acknowledging any favours to

Sr Your obliged & affectionate

humble servant

WM DERHAM

To Sr Isaac Newton

NOTES

(1) Keynes MS. 95B; microfilm 931.4. On the cover Newton has made notes on chronology and the calculus dispute. For William Derham see Letter 973, vol. v.

(2) Tuesday.

(3) First published in 1713, as the substance of Derham's Boyle Lectures delivered in 1711 and 1712, it was a highly popular book of which the sixth edition was published in 1723.

(4) The work published in the following year as *Astro-Theology: Or a Demonstration of the Being and Attributes of God, From a survey of the Heavens* (London, 1715); *Physico-Theology* is almost entirely concerned with living things.

(5) Jupiter's. We have found no reply to this letter, and a footnote in Derham's *Astro-Theology* (p. 144, note (5)) implies that Derham eventually obtained the required information from Halley.

1074 NEWTON TO KEILL
11 MAY 1714

From the original in Trinity College Library, Cambridge.[1]
Reply to Letter 1069; continued in Letter 1077; for the answer see Letter 1076

Sr

I have read over your Letter & find it right. The Marquess de l'Hospital in his Treatise de Infinitement Petits [2] teaches that if the Ordinates *AB, CD, EF* be at Equal distances, & the chord *BD* be produced till it cuts the Ordinate *EF* produced in *N*, the line *FN* shall be the second difference of the three Ordinates. And the points *B, D* being infinitely neare, perhaps Burnoulli may take *BD* for a tangent of the Curve at *D* & so reccon that the distance between the

128

Curve & ye Tangent is the second difference: whereas *BDN* is not a tangent but cuts the Curve at *D*, & the tangent at *D* is parallel to the chord *BF* & bisects the second difference *FN*, suppose in *G*. So that the line *FG* wch lies

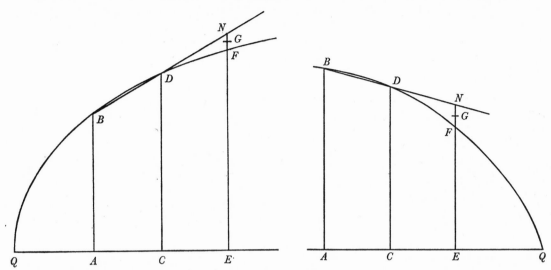

between the Curve & the tangent, & is equal to the third term of the series, is but half the second Difference, as I have put it. Mr Burnoulli therefore is mistaken in affirming that I put the third terme of the series equal to the second difference, & I am in the right in putting it equal to ye line between the Curve & the Tangent & by consequence to half the second difference as you observe. And I think your Demonstration is good.

I have corrected a paragraph in ye 11th page of ye papers you sent me & put it thus, [3dly We do not dispute about the antiquity of the symbols of Fluents Fluxions & Moments, Summs & Differences used by Mr Newton & Mr Leibnitz, they being not necessary to the method, but liable to change. And yet the symbol $\boxed{\dfrac{aa}{64x}}$ used by Mr Newton in his Analysis for fluents or summs is much older then the symbol $\int \dfrac{aa}{64x}$ used by Mr Leibnitz in the same sense.[3] And some of the symbols of fluxions used by Mr Newton are as old as his said Analysis, whilst Mr Leibnitz has no symbols of fluxions to this day. And the rectangles under the fluxions & the letter *o* used by Mr Newton for moments are much older then the symbols *dx* & *dy* used by Mr Leibnitz for the same quantities. But these are only ways of Notation & signify nothing to ye method it self wch may be without them.] [4] I have made this alteration to avoyd quoting my Manuscripts wch are not upon record. And for the same reason the last leaf of the papers you sent me must be altered. But I have time to add no more at present but that I am

Sr

Your most humble Servant

Is. NEWTON

London May 11*th.*
　1714.

For the Rnd Dr John Keill
Professor of Astronomy in the
University of
　　　　Oxford

NOTES

(1) R.16.38, fo. 426, printed in Edleston, *Correspondence*, pp. 174–5.
(2) See Letter 1008, note (4), p. 15, and Letter 1053*a*, note (24), p. 92. The definition of

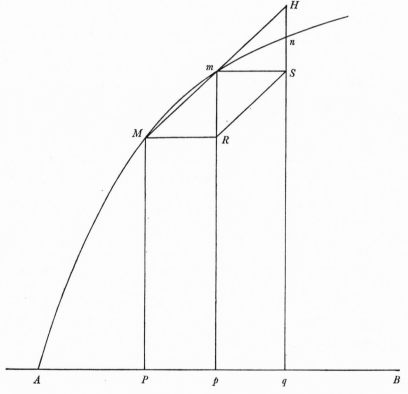

the second difference appears as Definition I in Section IV (*Analyse des infiniment petits* (Paris, 1696), p. 55):

La portion infiniment petite dont la différence d'une quantité variable augmente ou

diminue continuellement, est appellée la *différence de la différence* de cette quantité, ou bien sa *différence seconde*. Ainsi si l'on imagine une troisième appliquée *nq* infiniment proche de la seconde *mp*, & qu'on mene *mS* parallèle à *AB*, & *mH* parallèle à *RS*; on appellera *Hn* la *différence de la différence Rm*, ou bien la *différence seconde* de *PM*.
Newton paraphrases this definition: *mH* is his chord *BD*, *Hn* Newton's *FN*.

(3) There is a very similar passage in Newton's third observation on Leibniz's 1712 review of *De analysi* (see below, Letter 1077, note (5)) reproduced in Whiteside, *Mathematical Papers*, II, p. 273.

(4) The alteration was made in Keill's 'Answer' as printed in the *Journal Literaire de la Haye* for July and August 1714, p. 335 (see Letter 1053*a*, note (1)). The square brackets are Newton's.

1075 J. BERNOULLI TO LEIBNIZ
12 MAY 1714

Extract from Gerhardt, *Leibniz: Mathematische Schriften*, III/2, pp. 931–3.[1]
For the answer see Letter 1122

. . Cl. Wolfius mihi misit . . . plura exemplaria schediasmatis continentis responsionem Tuam (nam Tuam esse dixit Wolfius et publice extat in Diario illo Germanico: Büchersaal, quod Lipsiæ imprimitur [2]) meque rogavit, ut ea inter Mathematicos mihi notos distribuerem, quod equidem jam feci, præsertim in Galliam non pauca misi; sed in Angliam mittere nulla volui, ne me Angli pro Auctore suspicentur hujus Responsionis, aut saltem illius Epistolæ, quam inseruisti. Si Tibi ad manus fuisset scripta et literæ inter Te Anglosque olim reciprocatæ, invenisses fortasse plura et fortiora argumenta in usum Tuum facientia: quales enim hæ Literæ extant in Commercio Epistolico, quædam earum valde suspectæ videntur, si non omnino confictionis, saltem alterationis et falsificationis. En aliquod exemplum: 'Pag. 25 Commercii Epistolici habetur Epistola Jacobi Gregorii ad Collinsium 15. Februarii 1670 data, ubi hæc leguntur:[3] Quod attinet Newtoni Methodum universalem, aliqua ex parte, ut opinor, mihi innotuit, tam quoad Geometricas quam Mechanicas curvas Nihilo tamen minus ob series ad me missas gratias habeo, quas ut remunerem, mitto quæ sequuntur. Sit Radius $= r$, Arcus $= a$, Tangens $= t$, Secans $= s$, et erit

$$a = t - \frac{t^3}{3r^2} + \frac{t^5}{5r^4} - \frac{t^7}{7r^6} + \frac{t^9}{9r^8} \text{ etc. eritque}$$

$$t = a + \frac{a^3}{3r^2} + \frac{2a^5}{15r^4} + \frac{17a^7}{315r^6} + \frac{62a^9}{2835r^8} \text{ etc. et}$$

$$s = r + \frac{a^2}{2r} + \frac{5a^4}{24r^3} + \frac{61a^6}{720r^5} + \frac{277a^8}{8064r^7} \text{ etc.}$$

Sit nunc tangens artificialis etc. [']

Jam quod spectat ad primam seriem $a = t - \dfrac{t^3}{3r^2} + \dfrac{t^5}{5r^4} - \dfrac{t^7}{7r^6}$ etc. potest esse,

ut ea nunc demum ab Editore Commercii Epistolici huic Epistolæ Gregorianæ fuerit callide inserta, ut Tua quadratura Arithmetica Circuli = $1 - \dfrac{1}{3} + \dfrac{1}{5} - \dfrac{1}{7} +$ etc. jam ante a Jacobo Gregorio communicata adeoque inventi gloria non Tibi, sed Gregorio deferenda dici posset: probabilem enim reddit hanc conjecturam locus quidam, in quem nuperrime incidi, legens in *Davidis Gregorii Exercitatione Geometrica de Dimensione Figurarum*, impressa Edinburgi anno 1684, ubi hæc diserta habentur verba: 'Fortassis notatu erit dignum, hinc consequi Præstantissimi Geometræ Gothofredi Gilberti (Guilielmi) Leibnitii Circuli Quadraturam Transact. Phil. Mensis Aprilis anni 1682 editam, si nempe ponatur $DA = \dfrac{1}{2}$, arcus AF 45 grad. erit AC etiam $\dfrac{1}{2}$; unde erit sector

$$DAF = \frac{1}{2} DA \times AF \quad \text{arc.} = \frac{1}{2} \times \frac{1}{2} + 1 \times \frac{1}{2} - \frac{1}{3} \times \frac{1}{2} + \frac{1}{5} \times \frac{1}{2} - \frac{1}{7} \times \frac{1}{2} + \text{etc.} \quad \text{ejusque}$$

octuplum, nimirum Circulus, cujus quadratum circumscriptum est 1, erit $1 - \dfrac{1}{3} + \dfrac{1}{5} - \dfrac{1}{7} + \dfrac{1}{9} -$ etc. in infinitum. [']

Ergo David Gregorius asserit Tibi primæ inventionis gloriam. Quis autem tam vecors esset, ut putaret, Davidem hoc fecisse in præjudicium Jacobi, Patrui sui, cujus inventum, si Inventor fuisset, ipse ignorare non potuisset, deprædicans alioquin singulis fere paginis ejus Methodos et Series, fatensque ab initio suæ Exercitationis se nactum fuisse post mortem Patrui sui ejusdem adversaria, in quibus utique reperturus fuisset seriem Tibi ab Anglis controversam; si vel maxime Jacobus vivens eam celare voluisset Davidem; quod tamen minime verisimile est, nam ne umbram quidem verisimilitudinis haberet, si quis diceret Jacobum Gregorium studio abscondidisse Davidi, sanguine sibi proximo, quod cum Collinsio, homine ad se nihil pertinente, tam liberaliter communicaverit? Quæ cum ita sint, nescio quid sentiam de eo, quod habetur in Epistola quadam Oldenburgi ad Te scripta d. 15. Aprilis 1675, et Commercio Epistolico inserta pag. 39, in qua nempe hæc verba reperio: 'Et conversim ex tangente invenire arcum ejus

$$a = t - \frac{t^3}{3rr} + \frac{t^5}{5r^4} - \frac{t^7}{7r^6} + \frac{t^9}{9r^8} - \text{etc.'}\ ^{(4)}$$

Forsitan hæc ab Editore Commercii Epistolici nunc sunt per fraudem intrusa; quare operæ pretium esset, ut autographum ipsum inspiceres; talem enim si dolum detegere posses, actum esset de candore Adversariorum Tuorum, et nullam fidem amplius inveniret narratio alias satis speciosa, quam contexunt et per epistolas ubique probant. Incumbit igitur Tibi, ut montres publice, has epistolas fuisse corruptas et adulteratas, adeoque reliqua omnia esse sublestæ fidei. Hæc me judice brevissima est via confundendi Antagonistas eosque ad turpe silentium redigendi . . .

Translation

Mr Wolf has sent me many copies of the sheets containing your reply (for Wolf has said it is yours, and the statement appears publicly in the German journal, *Büchersaal*, which is printed in Leipzig [2]); and has asked me to distribute it amongst the mathematicians known to me; of course I have already done so, and I have especially sent quite a number into France; but I was reluctant to send any to England, lest the English suspect that I am author of that reply, or at least of that letter which you have inserted [there]. If you had had the papers and letters formerly exchanged between you and the English to hand, you would perhaps have discovered more and stronger arguments prepared for your use; for of those letters set out in the *Commercium Epistolicum* some seem strongly suspect; if not wholly fabricated, then at least altered and falsified. I give you an example: On page 25 of the *Commercium Epistolicum* is James Gregory's letter to Collins dated 15 February 1670[/71] where the following words appear:[3] 'As for Mr Newton his universal method, I imagine I have some knowledge of it, both as to geometrick & mechanick curves, however I thank you for the serieses ye sent me, and send you these following in requital. Let the radius = r, the arc = a, the tangent = t, the secant = s, then

$$a = t - \frac{t^3}{3r^2} + \frac{t^5}{5r^4} - \frac{t^7}{7r^6} + \frac{t^9}{9r^8} \text{ etc, and}$$

$$t = a + \frac{a^3}{3r^2} + \frac{2a^5}{15r^4} + \frac{17a^7}{315r^6} + \frac{62a^9}{2835r^8} \text{ etc, and}$$

$$s = r + \frac{a^2}{2r} + \frac{5a^4}{24r^3} + \frac{61a^6}{720r^5} + \frac{277a^8}{8064r^7} \text{ etc.}$$

Now let the artificial tangent be &c'

Now as regards the first series

$$a = t - \frac{t^3}{3r^2} + \frac{t^5}{5r^4} - \frac{t^7}{7r^6} \text{ etc.}$$

it may be that it has been cunningly inserted into this letter of Gregory by the editor of the *Commercium Epistolicum*, so that your arithmetical quadrature of the circle = $1 - \frac{1}{3} + \frac{1}{5} - \frac{1}{7} +$ etc. can be said to have been communicated before by James Gregory, thus giving the glory of its discovery not to you but to Gregory; a certain passage which I came across very recently when reading David Gregory's *Exercitatio Geometrica de Dimensione Figurarum*, printed in Edinburgh in the year 1684, lends probability to this conjecture; for there the following is affirmed: 'Perhaps it will be worthwhile to note that there follows from this the quadrature of the circle of that outstanding geometer Gottfried Gilbert (William) Leibniz published in the *Philosophical Transactions* for April 1682; for if we put $DA = \frac{1}{2}$, arc $AF = 45°$, then AC is also $\frac{1}{2}$, whence sector DAF will be

$$\frac{1}{2}DA \times \text{arc } AF = \left(\frac{1}{2} \times \frac{1}{2}\right) + \left(1 \times \frac{1}{2}\right) - \left(\frac{1}{3} \times \frac{1}{2}\right) + \left(\frac{1}{5} \times \frac{1}{2}\right) - \left(\frac{1}{7} \times \frac{1}{2}\right) + \text{etc.},$$

and eight times this, which is, of course, the circle whose circumscribed square is unity, will be $1 - \frac{1}{3} + \frac{1}{5} - \frac{1}{7} + \frac{1}{9} -$ etc. to infinity.'

Thus David Gregory gives to you the glory of the discovery. But who would be so foolish as to believe that David made this statement prejudicial to his father's brother James, of whose discovery he could not be ignorant, if it was James's discovery. Otherwise he plunders whole pages of James' methods and series, confessing at the beginning of his *Exercitatio* that after the death of his uncle he acquired possession of his notebooks, in which (surely) he could have found the series in dispute between you and the English. Or else James, when living, must have been very anxious to keep it secret from David; which is nevertheless extremely unlikely. For does it have the least shadow of probability, to say that James Gregory would zealously conceal from David, his closest relative, what he so freely communicated to Collins, a man having no connection with him? Whatever the case, I do not know how to judge what is found in a certain letter that Oldenburg wrote to you on 15 April 1675, printed in the *Commercium Epistolicum* on page 39, in which I find these words: 'and conversely from the tangent to obtain its arc,

$$a = t - \frac{t^3}{3r^2} + \frac{t^5}{5r^4} - \frac{t^7}{7r^6} + \frac{t^9}{9r^8} - \text{ etc.'} \,^{(4)}$$

Perhaps these words are now fraudulently inserted by the Editor of the *Commercium Epistolicum*; so it would be worthwhile for you to inspect the original letter itself; for if you could discover such a deceit, it would detract from the candour of your adversaries, and no one would have any further confidence in a narrative that is otherwise specious enough, which they have stitched together and substantiate at every point by the letters. It is therefore incumbent upon you to show publicly that these letters have been tampered with and altered, and so all the rest are not to be relied upon. This in my opinion is the shortest way of confounding these antagonists and reducing them to a shameful silence.

NOTES

(1) This extraordinary letter, which aroused alarm and indignation in its recipient, is really the strongest testimony to Newton's case; for Bernoulli was in fact saying that the *Commercium Epistolicum* evidence was so damaging to Leibniz that it must have been falsified; Leibniz must surely be able to convince the world that it *had* been falsified. Leibniz knew,

however, as we know, that however unjust to him the construction placed by the *Commercium* on its evidence might be, the evidence itself was perfectly authentic and beyond challenge. In the summer of 1714, Leibniz wrote to Wolf giving a very fair summary of the matter; see Gerhardt, *Briefwechsel zwischen Leibiz und Wolf*, p. 158.

(2) *Neuer Büchersaal der Gelehrten Welt* (xxvii Oefnung), 1713, p. 224. Here in a brief and anonymous article it is stated that Leibniz 'hat nur neuligst wieder eine Schrifft von Wein aus in der gelehrten Welt bekannt machen lassen, worinne er sich nicht nur hefftig über die Englischen Widerfacher beschweret, sondern auch ihm alle beygemessene Beschuldigung von sich ablehnet.' The paper referred to is of course the *Charta Volans* (Number 1009).

(3) We take the few lines of English here from Gregory's letter as printed in vol. i, p. 62; Gregory's subsequent mathematical Latin was, of course, faithfully transcribed in the *Commercium*.

(4) See vol. ii, p. 340; Collins supplied the material for this answer to Leibniz's letter of 24 March 1674/5 (received in London on 15 April), in which he acquainted Leibniz with four series of Newton and five of James Gregory, all related to the circle, including the last from Gregory which so struck Bernoulli (r = radius, t = tangent, a = arc, hence $t = r\tan(a/r)$). No proofs were given.

1076 KEILL TO NEWTON
14 MAY 1714

From the original in the University Library, Cambridge.[1]
Reply to Letter 1074

May 14*th* 1714

Honoured Sr

I have here sent you up what I have said in Answer to the 2d impudent Argument of these Gentlemen and I have said as I beleive so much to show Mr Bernoulli to be in the wrong, that he will repent his forwardness in finding of faults, I am almost confident he wrote that paper on purpose to show that you did not understand 2d Fluxions. I wish you would let me know if any thing is to be said about Mr Leibnits not understanding [2]d Differences.[2] I being in hast at present I have just [t]ime to conclude that I am

Sr

Your

most Humble and faithfull servant

JOHN KEILL

For
Sr Isaac Newton
at his house in St Martins Street
near Leicester Feilds
Westminster

NOTES

(1) Add. 3985, no. 7. Keill now submits the second part of the proposed rejoinder to be sent to the *Journal Literaire de la Haye*, in continuation of that sent with Letter 1069.

The first 'impudent argument' was that Newton's priority was refuted by the late occurrence of the dot-notation; the second, that he did not understand second differentials. Keill's rejoinder to this 'autre preuve' begins on p. 343 of the printed 'Answer' and concludes on p. 348, where the subject of the 'Tentamen' is introduced. The draft of this part of the text, then, accompanied the present letter.

(2) See Letter 1069. Keill proceeded—from p. 348—to attack the 'Tentamen' and so Newton's agreement must be presumed.

1077 NEWTON TO KEILL
15 MAY 1714

From the original in Trinity College Library, Cambridge.[1]
Continuation of Letter 1074; for the answer see Letter 1078

Sr

I wrote to you on Tuesday[2] that the last leafe of the papers you sent me should be altered because it refers to a Manuscript in my private custody & not yet upon Record. For setting right this leafe it is to be considered that altho I use prickt Letters in the first Proposition of the book of Quadratures, yet I do not there make them necessary to the method. For in the Introduction to that book I describe the method at large & illustrate it wth various examples without making any use of such letters. And it cannot be said that when I wrote that Preface I did not understand the method of fluxions because I did not there make use of prickt letters in solving of Problems. The book of Quadratures is ancient, many things being cited out of it by me in my Letter of 24 Octob 1676.[3] A copy of the first Proposition where letters with pricks are used, was at the request of Dr Wallis sent to him in the year 1692 & the next year published in the second Volume of his works.[4] And in the Principia Pholosophiae [*sic*] pag 254 the Notarum formulae used in those days in explaining this Proposition are referred unto.

Fluxions & moments are quantities of a different kind. Fluxions are finite motions, moments are infinitely little parts. I put letters with pricks for fluxions, & multiply fluxions by the letter o to make them become infinitely little & the rectangles I put for moments.[5] And wherever prickt letters represent moments & are without the letter o this letter is always understood. Wherever $\dot{x}$, $\dot{y}$, $\ddot{y}$, $\dddot{y}$ &c are put for moments they are put for $\dot{x}o$, $\dot{y}o$, $\ddot{y}oo$, $\dddot{y}o^3$. In demonstrating Propositions I always write down the letter o & proceed by the Geometry of Euclide & Apollonius without any approximation. In resolving Questions or investigating truths I use all sorts of approximations wch I think will create no error in the conclusion & neglect to write down the letter o, & this [I] do for making dispatch. But where $\dot{x}$, $\dot{y}$, $\ddot{y}$, $\dddot{y}$ are put for fluxions without the letter o understood to make them infinitely little quantities, they never signify differences. The great Mathematician therefore acts unskilfully in comparing prickt letters with the marks dx & dy, those being quantities of a different kind. Mr Leibnitz has no mark for fluxions & therefore prickt letters are older marks for fluxions then any used by him & so are others [*sic*] marks used by me for fluxions. The rectangles under fluxions & the moment o being my marks for moments are to be compared with the marks dx & dy of Mr Leibnitz & are much the older being used by me in the Analysis communicated by Dr Barrow to Mr Collins in the year 1669.

The Author of the Remarks represents that Dr Wallis was for Mr Leibnitz & yet the Dr in the Preface to the first Volume of his works represents that I in my Letters of June 13 & Octob 24, 1676 explained to Mr Leibnitz this method found out by me ten years before or above, that is in the year 1666 or 1665.

I am

Your most humble Servant

Is. NEWTON

For the Rnd Dr John Keill
Professor of Astronomy in the
University of Oxford

NOTES

(1) R.16.38, fo. 427, printed in Edleston, *Correspondence*, pp. 176–7. This letter like the preceding one from Keill was postmarked 15 May. It is the third and last of the consecutive letters from Newton to Keill given to Trinity College by Richard Watson (1737–1816), Professor of Chemistry and Divinity and later Bishop of Llandaff, a Fellow of the College. No further letters from Newton to Keill are extant for this period.

(2) Letter 1074.

(3) The *Epistola posterior* to Leibniz, see Letter 188, vol. II, pp. 110–29, and Whiteside, *Mathematical Papers*, IV, pp. 618–33 and 671–4. D. T. Whiteside finds the material of this letter rather in the 1671 'Tractatus de Methodis serierum et fluxionum' (*ibid.*, III, pp. 32–352), itself a revised expansion of the 1669 *De analysi* (*ibid.*, II, pp. 206–46); the reference to *De quadratura curvarum* was hardly appropriate. Compare Letter 1064.

(4) See above, Letter 1064, note (8), p. 110.

(5) Neglecting the matter of symbolism, Newton had certainly laid his 'moment' ($\dot{x}o$) before the public as equivalent to the differential in the *Principia* 'fluxions' Lemma 2 of Book II (Whiteside, *Mathematical Papers*, IV, pp. 521–5). It was of course not in dispute that the symbol o had long been used to denote the increment of a variable; Newton had employed it

since 1664 (*ibid.*, I, p. 55, note (21); II, p. 262, note (14)). Newton's attention had perhaps first been drawn to this miscomprehension of the fluxion/moment/differential relationship on the part of Leibniz and Bernoulli by Leibniz's anonymous review of *De analysi* (as published by William Jones in *Analysis per quantitatum series* etc. (London, 1711)) printed in the *Acta Eruditorum* for February 1712 (partially reprinted in Whiteside, *Mathematical Papers*, II, pp. 259–62). When Newton first read this review and prepared three draft sets of observations upon it is not known, but they seem to belong to successive periods between the end of 1712 and the early part of 1714; some part of these drafts (all in *ibid.*, II, 263–73) passed over into Newton's anonymous 'Recensio' of the *Commercium Epistolicum* (*Phil. Trans.* **29**, no. 341 (January–February 1713/14, pp. 173–224)). For passages very similar to this paragraph see Whiteside, *Mathematical Papers*, II, pp. 264 and 266.

1078   KEILL TO NEWTON
17 MAY 1714
From the original in the University Library, Cambridge.[1]
Reply to Letter 1077

Oxford May 17th 1714

Honoured Sr

The papers I have sent I intirely submitt to you, and shall be glad you'l take any pains about them to change add or leave out anything; tho I am of opinion that, the manuscript treatise wch you composed in 1671 [2] tho it has been in your own hands, is a sufficient testimony of the Antiquity of points as marks for fluxions, this manuscript having been seen by several of your freinds, non of Mr Leibnits party will have the impudence to denye it, since it contains all those things wch you describe when you gave an Account of it, in your letter 1676.

If you think fitting I would add to what I have said about Mr Leibnits takeing from Dr Barrow, A quotation of Mr James Bernoulli, so after these words, *only Dr Barrow expresses the difference of the Absciss by* e *and of the ordinate by* a *wch Mr Leibnits changes into* dx *and* dy. Add [3] Mr James Bernoulli who has advanced the use of the calculus differentialis further than ever Mr Leibnits did, in [his] first specimen of the calculus differentialis printed in the Acta Lipsiæ 1691 pag 14, after having complained and not without reason of the obscurity and brevity of Mr Leibnits has these words. [4] Quamquam ut verum fateor qui calculum Barovianum (quem decennio ante in Lectionibus suis Geometricis, adumbravit Auctor, cujusque specimena sunt tota illa propositionum inibi contentarum farago) intellexerit, alterum a Dno Leibnitio intentum ignorare vix poterit, utpote in priore illo fundatus est et nisi forte in differentialium notatione et operationis aliquo compendio ab eo non differt.

I sent you up by the Fridays post my Answer to their 2d Argument. [5] But because you mention nothing of it in your Letter I am affrayed it had not then come to your hands, wch it should have done by saturday at tuo of the clock, if you have not received it pray let me know that I may send you another copy.

There is one sentence wherein I would have Mr Bernoulli make publick acknowledgement that he has wronged you changed into this: Since therefore it is evident that Mr Newton has not given in his principles any false Rule for the collecting of 2d Fluxions and these of a superior degree, as also that he has not supposed the Terms of the series $\frac{ao}{e} + \frac{n^2o^2}{2e^3} + \frac{an^2o^3}{2e^5}$ &c each of them to be equal to a correspond[ing] difference;[6] but on the contrary that he expressly makes the 2d [diffe]rence double of the term $\frac{n^2o^2}{2e^3}$, the world will expect that Mr Bernoulli should doe justice to Mr Newton, and publickly acknowledge that he has mistaken and misrepresented him. wch he is more particularly obliged to doe because our tuo Authors (who doe not seem to understand much of that matter themselves) have made use of his great name to defame Mr Newton on that score. The speedy doeing of this will be very agreeable to the character I have always had of Mr Bernoullis natural candor and sincerity.

In a day or tuo I will send you up all the rest of my Answer I wish you could get it translated into French to be sent over, for I am of opinion that there is no time to be lost. I am

Sr your

most faithfull and obedient servant

JOHN KEILL

For
Sr Isaack Newton
at his house in St Martins
Street near Leicister feilds
Westminster

NOTES

(1) Add. 3985, no. 8.

(2) See Letter 1077, note (3).

(3) The addition was not made; see Keill's 'Answer' in *Journal Literaire de la Haye* for July and August 1714, p. 325.

(4) 'Although, to speak the truth, whoever has comprehended the calculus of Barrow (sketched by that author in his *Lectiones geometricæ* more than a decade earlier, and of which there are examples in all that mass of propositions contained therein) can scarcely fail to understand this other one conceived by Mr. Leibniz, as being founded upon that earlier one and in no way differing from it except perhaps in the notation of differentials and a certain convenience in operation.'

(5) Letter 1076, which (as Keill might have guessed) crossed with Letter 1077, which Keill must have received on the day he wrote this reply.

(6) The series relates to *Principia*, Book II, Prop. 10, so often discussed before. In the 'fluxions' lemma Newton had said nothing of the fluxion of a fluxion. The question of Newton's error and its source is fully canvassed by Keill on pp. 343–6 of the printed 'Answer'. The following passage demanding an apology from Johann Bernoulli (essentially as drafted here in the letter, but without the final sentence) appears on pp. 345–6. All this, of course, had been sent to Newton with Letter 1076.

1079 CHAMBERLAYNE TO NEWTON
20 MAY 1714
From the original in King's College Library, Cambridge[1]

Petty France
20 *May* 1714

Honored Sr

I am very sorry I can't wait on you this Afternoon when you are to consider of the Letter from Mr Leibnitz to me,[2] concerning you, which Letter I did not intend to have expos'd to anybody's view, but your own, because I am not sure it wil be agreable to the writer, but since you have desired it, & to shew my Respect to you (who ought to have no Enemy in the world, but such as are Enemys to Philosophy & Truth[)] I am content you should make what use you please thereof, only submitting it as a matter of Prudence how far you, in your private Capacity, may think it adviseable to keep some Measures with a Gentleman that is in the Highest Esteem at the Court of Hanover &c [3] and if I could be so Happy as to be Instrumental in beginning a friendship between Two Men of very great Merit, it would be an exceeding great Pleasure to Hon[oure]d Sr

your most Faithful Humble servent

JOHN CHAMBERLAYNE

For the Honble
Sr Isaac Newton &c

NOTES

(1) Keynes MS. 141(C), microfilm 1011.25.

(2) Compare Letter 1072. As Letters 1089 and 1092 further reveal, it was Newton's intention to lay Letter 1062 before the Royal Society at its meeting on 20 May, treating it not so much as a private communication as a formal challenge to the *Commercium Epistolicum*; indeed, he had evidently done so at the preceding meeting. The Society declined to concern itself with the letter.

(3) Chamberlayne has an eye to the future political situation.

1080 KEILL TO NEWTON
21 MAY 1714
From the original in the University Library, Cambridge[1]

Oxford May 21*st* 1714

Honoured Sr

You will find by the enclosed that I have been studying Mr Leibnits Tenta-men, and I think (tho it was with much a doe) I have found out both his meaning and his mistakes he is designedly obscure, and I beleive he had purposely many faults in the print, with lines and letters left out of his figure on purpose to perplex; so that I beleive very few have read his demonstrations.[2] By centrifugal force he means not the centrifugal force the body has in the Curve, but that wch a body has, that revolves in a circle at the same distance with an Harmonical velocity; he has been told by somebody that there was a mistake in his computation, and that the force of Gravity was to the conatus centrifugus as $\dfrac{2a\theta\theta}{r^2}$ to $\dfrac{aa\theta\theta}{r^3}$ and therefore in the year 1706 he layes his mistake on a wrong estimation as he sayes of the Conatus Centrifugus without knowing where the blunder truely lay.[3] Next post I will send you all the rest of my Answer wch does not consist of above a leaf more, mean time I am

Honoured Sr

Your

most obedient faithfull servant

JOHN KEILL

For
Sr Isaack Newton
at his house in St Martins Street
near Leicester Feilds
    Westminster

NOTES

(1) Add. 3985 no. 9.

(2) On Leibniz's 'Tentamen', see Number 1069*a*. From Huygens to the present time this work has baffled commentators. Huygens, a devotee of plenum physics, compelled Leibniz to give way on some points as early as 1690. To Keill it was 'le morceau de Philosophie le plus incomprehensible qui ait jamais paru' (*Journal Literaire de la Haye* for July and August 1714, p. 348).

(3) The blunder concerned a factor of two (see Number 1069*a*). In 1704 Leibniz sent a revised 'Tentamen' to Christian Wolf, intended for publication in the *Acta Eruditorum*. Wolf

passed the paper on to Otto Mencke (then the editor of the journal), who published only an extract; see *Acta Eruditorum*, for October 1706, pp. 446–51. Both extract and unpublished paper are printed in Gerhardt, *Leibniz: Mathematische Schriften*, vi, pp. 247–80.

1081 KEILL TO NEWTON
25 MAY 1714
From the original in the University Library, Cambridge [1]

Oxon. May [2]5*th* 1714 [2]

Honoured Sr

I have here inclosed all the rest of my Answer, I had a great deal more to say, but I am affrayed I have already made it too long. I find Mr Leibnits understood nothing of centripetal forces when he published his Tentamen he makes the centripetal force of a body moving in a circle equal to twice its conatus centrifugus this blunder he endeavours to correct in the year 1706, but he again blunders in excusing himself, I leave my whole paper to You and Dr Halley to change or take away what your please, I only desire that it may be done quickly and sent over, if there be any necessity for my coming to town I will wait upon you, I desire that it may be advertised in the foreigne Gazets where the Original letters and books are to be seen. I am

> Sr
>
> Your
>
> Most faithfull and obedient
>
> servant
>
> JOHN KEILL

For
Sr Isaack Newton
at his house in
St Martin's Street
near Leicester Feilds
    Westminster

[Postscript] [3]

I would insert the following Paragraph just before the Paragraph wch begins thus. *Before we proceed to answer the Arguments wch &c*

The Author of the Remarks further tells us that M. Newton never troubled Mr Leibnits in his possession of the honour of inventing the Calculus till after

142

Mr Hugens and Mr Wallis's deaths who were well informed and could have
been impartial Judges in this cause. To wch I answer that Mr Newton was
very easy notwithstanding the injustice done him by Mr Leibnits and the
compilers of the Acta Lipsiæ, he did not love to trouble himself about the
matter. It was his freinds who saw and complained of the wrong that was done
him. In his 2d letter that was sent to Mr Leibnits he told him that he had a
designe of publishing both his Optical and Geometrical discourses, but that
after he had near finished his treatise in 1671. He was detterred from [printing]
it by the troubles he dayly met with upon the account of what he had already
published and accused himself of imprudence for following a shadow so far as
to lose his own quiet wch was a substantial good. [4] Mr Leibnits soon perceived
his humor in this point and made as we see his own use of it. Mr Gregory likewise when he was pressed by Mr Collins to publish his series, he writes back to
him in these words. [5] Non est quod metuas cuiquam quicquod miserim
communicare parum enim sollicitus sum utrumne meo an alieno nomine in
publicum prodeat.

It were indeed to be wished that the two great men our author mentions
were alive to give their opinions in this controversy. Mr Hugens knew nothing
of the correspondence Mr Leibnits had with Mr Collins nor of any letters he
had seen of Mr Newtons Mr Leibnits was wise enough to keep that a secret
from him, and without the knowledge of these letters he could not be well
enough informed so as to judge of the of the [sic] matter. if he were now alive
and had the whole evidence layed before him, there is no doubt to be made,
but that he would pronounce in our favour. I am supprised our Author should
be so unwarry as to mention Dr Wallis and to think that he would have agreed
with him he indeed knew more of the matter than Mr Hugens, and upon that
account has given his opinion already against him in the Preface to the the
first volum of his works where he says that Mr Newton in a Letter written in
the year 1676 had expounded to Mr Leibnits his method wch was found by
him ten years before, to wch he adds these words Quod moveo ne quis causetur
de hoc calculo differentiali nihil a nobis dictum esse. [6]

NOTES

(1) Add. 3985, no. 10. This letter repeats the preceding letter. The copy sent with it has
disappeared; it contained the last ten pages of the printed 'Answer' but these were to be much
modified in Letter 1086.

(2) The date has been partially torn away in breaking the seal; the postmark is May 26.

(3) This addendum is written on the same sheet as the letter; an equivalent passage in
French was inserted in the printed version, pp. 332–4.

(4) The phrase is adapted from the *Epistola Posterior* (vol. II, p. 114).

(5) In James Gregory's own English: 'Ye need not be so close handed of anything I send

you: ye may communicate them to whom ye will, for I am little concerned if they be published under others' name or not.' (To Collins, 15 February 1671; Turnbull, *Gregory*, p. 171).

(6) 'And this I bring forward lest someone should argue that we had said nothing at all about this differential calculus.' For Wallis's attempts to print Newton's mathematical letters to Leibniz (which finally appeared in Wallis's *Opera mathematica*, III (Oxford, 1699), pp. 622–9 and 634–45) see Whiteside, *Mathematical Papers*, IV, p. 672, note (54).

1082   NEWTON TO THE DUKE D'AUMONT
27 MAY 1714
From a copy in private possession.
Reply to Letter 1042

May it please yr Grace

The Letter you were pleased to honour the Royal Society with came so late to their hands, that I could not sooner return you their Thankes for the great humanity and Civility wherewith you have treated them. Your Graces Letter was read in a full Meeting of the Society,[1] to the great satisfaction and Pleasure of all the Members present.

Whenever anything comes to their knowledge which they may thinke acceptable to your Grace, they will take care to communicate it, and in the meantime desired me to signify to yr Grace, how exceedingly you have have obliged them

I am Yr Graces most humble

and most obedient servant

ISAAC NEWTON

A Letter from
The President
to the Duke D'Aumont
Ordered may ye 27th
1714

NOTE
(1) On 13 May 1714; see Letter 1042, note (1), p. 66.

1083   W. J. 'sGRAVESANDE TO NEWTON
28 MAY 1714
From the original in the University Library, Cambridge[1]

Monsieur

Je me donne l'honneur de vous ecrire à loccasion de ce que M. Burnet m'a marqué dans une de ses lettres [2] que vous n'aviez pas encore vû ce que M.

Bernoulli a fait imprimer l'année passée dans les actes de leipsic,[3] et que vous souhaitiez d'avoir les mois de ces actes ou ces pieces se trouvent, je n'aurois pas attendu jusques a present a vous les envoyer s'il m'avoit été possible de les avoir plustot. J'aurois fort souhaité Monsieur d'avoir pu vous marquér par ma promtitude a executer cette commission, avec quel soin je rechercherai toujours les occasions de vous faire voir l'estime que j'ai pour vous. Vous m'obligerez sensiblement Monsieur de ne me point epargner dans toutes les occasions que je pourai vous estre de quelque utilité dans ce pais et d'estre persuadé que je suis avec respet

Monsieur

Vostres tres Humble

et tres obeisant serviteur

G. J. 'sGRAVESANDE

de la Haie ce 8 *de Juin* 1714 [N.S.]

NOTES

(1) Add. 3968(42), fos. 594–5. Willem Jacob 'sGravesande (1688–1742) was, as already noted, one of the editors of the *Journal Literaire de la Haye*. He was at this time practising law at the Hague. Early in 1715 he was to travel to England, where he would meet Newton and become F.R.S. (on 9 June 1715); this was on the proposal of William Burnet, mentioned in the letter, with whom 'sGravesande was already acquainted.

(2) Compare Letter 1055.

(3) See Letter 1055, note (2), p. 96.

1084 FONTENELLE TO NEWTON
29 MAY 1714
From the original in the Royal Society, London[1]

Monsieur

Je suis chargé par l'Academie Royal des Sciences de vous remercier d'un Recueil de differentes piéces de vous,[2] qu'elle a reçü des mains de M. le Chevalier de Louiville.[3] ie vous rends aussi en mon particulier très humbles graces de votre nouvelle édition des Principes, que i'ai reçüe de M. Taylor.[4] ie vous suis d'autant plus obligé de m'avoir honoré d'un si beau present, que ie n'avois aucun droit de m'y attendre. ie ne le pouvois meriter tout au plus que par l'admiration que i'ai pour tous vos Ouvrages, mais elle m'est commune avec tout ce qu'il y a de gens au monde, qui ont quelque teinture de Geometrie, et il s'en faut méme beaucoup que ie ne sois assès habiles pour vous admirer comme il faudroit. i'espere que ma reconnoissance suppliera a tout, et ie vous supplie d'étre bien persüadé que ie suis avec une veneration singuliere

Monsieur

Votre très humble et très

obeissant serviteur

FONTENELLE

sec. perp. de l'ac. Roy. des Sc

de Paris ce 9 *Juin* 1714 [N.S.]

NOTES

(1) MS. MM 5, no. 47; printed in *Modern Language Notes*, **54** (1939), 188–90. Bernard Le Bovier de Fontenelle (February 1657–January 1757) had been *secrétaire-perpétuel* of the Académie Royale des Sciences since 1697; his main accomplishment was the edition of forty volumes of its *Histoires et Mémoires* and the writing of *Eloges* of its members. His highly successful *Entretiens sur la pluralité des mondes* had appeared in 1686. According to his own statement Fontenelle had composed (but did Newton know this?) the Preface to the Marquis de l'Hospital's *Analyse des infiniment petits* (1696—see Letter 1008, note (4)) in which after much encomium of the achievements of Leibniz and Bernoulli there is added the bald statement that Newton 'avoit trouvé quelque chose de semblable' as could be seen from the *Principia*, 'lequel est presque tout de ce calcul'.

(2) When the Académie des Sciences was reorganized in 1699, places were provided for eight *associés étrangers*. Three foreigners (Leibniz, Tschirnhaus and Guglielmini) were already members and continued as *associés*; Jakob and Johann Bernoulli, with Hartsoeker, were elected on 4 February 1699 O.S. and, finally, Roemer and Newton in that order a week later. Perhaps the present here mentioned was of William Jones, *Analysis per Quantitatum Series* (London, 1711), though possibly the *Commercium Epistolicum* could be meant.

(3) Jacques-Eugène d'Allonville, chevalier de Louville (1671–1732), an astonomer, elected to the Académie as *associé* on 24 February 1714. He visited England in 1715, when he was elected F.R.S. on 9 June; it is to be presumed that he had been in communication with Halley or other friends of Newton earlier. In his *Construction et théorie des tables du soleil* (*Mémoires de l'Académie Royale des Sciences* for 1720 (Paris, 1722) pp. 35–84) the Chevalier proved himself a precocious, if discreet, French adherent to Newtonian celestial mechanics (see P. Brunet, *L'introduction des théories de Newton en France au XVIIIe siècle* (Paris, 1931), pp. 85–6).

(4) The mathematician Brook Taylor (1685–1731), Secretary of the Royal Society since January in the present year.

146

1085 NEWTON TO OXFORD
[MAY 1714]
From the copy in the Public Record Office.[1]
Reply to Letter 1068

To the most Honble the Earl of Oxford & Earl Mortimer
Lord High Treasurer of Great Britain

May it please your Lordp.

According to your Lordps. Order of Reference signified to us by Mr. Lowndes his Letter dated 29th of April last We have considered the annexed Letter of the Lords of the Privy Council in Ireland concerning the making current the new French moneys of Gold & Silver in that Kingdom together with the annexed Report of Mr Vincent Kidder Assay master there concerning the intrinsic values of those moneys. And by the weight of 55000 Louid'ors of the new species amounting to 1197£wt. 1oz. 11dwt we find that singly they are in weight one with another 5dwt. 5gr. $7\frac{1}{2}$ mites. And by the Assays of several Ingots melted out of new Louid'ors compared with the Assays of many single peices, We find that they are at a medium one grain & one twelft part of a grain worse than standard. And therefore by the weight & assay together they are singly worth but twenty shillings & six pence & three farthings in England. And in Ireland where a Guinea passes for 1£. 3s. they are singly worth 1£. 2s. At which rate the half Louid'or may pass for 11sh. & the Quarter for 5s. 6d.

Fifteen hundred silver Louises of the new species weighed 1470 Ounces Troy, & therefore one with another they weigh singly an ounce wanting 9gr. 12 mites. They are an half-penny weight worse than standard one with another, & therefore their standard weight at a medium is an ounce wanting $10\frac{1}{3}$ grains. And so they are worth 5s. & three farthings a piece in England at present. And in Ireland where a Crown piece English passes for 5s. 5d these Louises singly are worth 5s. $5d\frac{3}{4}\frac{1}{16}$ & in the nearest round numbers may pass for 5s. 6d, & the half Louises for 2s. 9d, & the Quarter pieces for sixteen pence half penny.

All wch is most humbly submitted to your Lordps. great Wisdome

Mint Office May 1714   Is. NEWTON

E. PHILIPPS

NOTE

(1) Mint/1, 7, p. 69. This follows very closely Newton's holograph draft in Mint Papers II, fo. 223 (without date). The copy bears the name of the Comptroller as well as Newton's.

1086   KEILL TO NEWTON
2 JUNE 1714
From the original in the University Library, Cambridge [1]

Oxford June 2d

Honoured Sr

I hear by Dr Halley that you have gote my papers translated into French, I could wish that the last part concerning Mr Leibnits mistake in his Tentamen [2] were altered according to what I here send you wch lays open his blunder more evidently Where I talke about the nature of converging series I would change the example and make it fitter for the present purpose and I would say. [3] And therefore the the series $1-x+x^2-x^3+x^4$ & in infinitum is the quotient of 1 div[id]ed by $1+x$ where 1 is greater than x yet it is not that quotient if 1 be less than x &c

But because my Paper is to[o] long already perhaps it would be better to leave the whole out. I am

Sr

your

most obedient Fathfull servant

JOHN KEILL

For
Sr Isaack Newton
at his house in St Martins Street
near Leicester Feilds
     Westminster

NOTES

(1) Add. 3985, no. 11.
(2) Sent with Letter 1081.
(3) Apparently this was not done, and perhaps there was some deletion from the original draft; but see the printed 'Answer', p 347 (Letter 1053 *a*, note (1)).

1086*a*   KEILL'S REVISION [1]

Whereas in the Apsides of the Ellipse it is easy to demonstrate that the one force is to the other only as *a* is to $2r$[.] All wch follows these words concerning Mr Leibnits mistake I would have altered as follows.

As Mr Leibnits is generally very obscure in his philosophical notions, so he seems here not to be very clear about the centrifugal force, wch in reality is nothing but the Reaction to the Centripetal force; or the Resistance arising from the Vis Inertiæ, that a body has to be turned out of its direction. and therefore the centripetal and centrifugal forces are always equal and contrary to each other. But so litle did he then understand of this, or of the way of explaining the celestial motions by the laws of Gravity that in pag. 96 of the Tentamen, he expresly asserts that a circle is described when the Attraction of Gravity (he should have said the force of Gravity) is equal to twice the centrifugal force[.] Tis true this consequence follows rightly from the conclusion he had made, but I have showen that this conclusion is not true, being founded on a mistake of his for not rightly apprehending the nature of 2d Differences. for from the true conclusion that should have been made viz that the element of the paracentrick impetus is equal to twice the difference of the sollicitation of Gravity and the conatus centrifugus It follows that in a circle the vis centripeta and centrifuga are just equal. [2]

In the year 1706 he perceived that his former assertion was false, and therefore in the Acta Lipsiæ that year he publishes another paper [3] where to correct his former error he says that the centrifugal force should be estimated by a perpendiculare not upon the Tangent but upon the chord of the Arch that was described before. and there he gives a new distinction of the conatus centrifugus Arcualis and Tangentialis wch is founded on a mistake that the direction of the body should be conceived to be in the chord and not in the Tangent of the Arch. Which is not true for in describing an Arch even infinitely small the body constantly changes its direction wch never coincides with the chord[.] But granting this he likewise committs another mistake in not estimating the conatus centrifugus and the sollicitation of Gravity by the same Rule, for if the one ought to be estimated by a perpendiculare on the chord certainly the other ought to be so likewise, and if a body in the one case may be supposed to move in the chord of the Arch why should not the same supposition hold in the other. Thus endeavouring to mend one mistake he committs tuo, wch proceeds from his not rightly knowing the source from wch his former Error flowed, for even in that paper he stands to his first computation, and tho he mends all the Typographical errors committed in his first Tentamen (wch by the way are not a few) yet he takes no nottice of this mistake about 2d Differences, but expressly asserts that the element of the Paracentrick Impetus is the conflict (by wch I suppose he means the difference) of the solicitation of Gravity and double the conatus centrifugus meaning the Tangential conatus, as he calls it; wch assertion I have showed to be false and his mistake proceeds from his not rightly computing the 2d Diff. Whence it is plain that Mr Leibnits does not yet fully comprehend the nature of 2d Differences.

*It [4] were easy to show several other mistakes and Absurdities Mr Leibnits has committed in that Tentamen, But I think it scarcely worth while and it would take me from the present purpose wch is to answer our tuo Authors.

*Tho the Compilers of the Acta Lipsiae take all occasions to proclaim the supposed mistakes of others, yet I beleive they will scarcely be so impartial as to mention this palpable error of Mr Leibnits.

NOTES

(1) This revision is written on the same sheet as Keill's letter, immediately preceding. A similar passage appears in the final French version of Keill's 'Answer'; see *Journal Literaire de la Haye* for July and August 1714, pp. 350–2.

(2) See Letter 1080. The acceleration ('impetus') is proportional to the difference of the forces, so the 'element' (distance) is proportional to twice this difference.

(3) The 'Excerptum ex Epistola' in the *Acta Eruditorum* for October 1706, pp. 446–51; see Letter 1080, note (3), p. 141.

(4) This and the following paragraph (the asterisks are Keill's) are not part of the revision and were not, of course, printed.

1087 LOWNDES TO NEWTON
7 JUNE 1714
From the original in the Mint Papers [1]

Sr

My Lord Treasurer having received from an unknown hand the Enclosed Paper relating to Copper Farthings His Lordp commands me to transmit the same to you for your consideration. [2]

I am

Sr

Your most humble servt.

WM LOWNDES

Trea[su]ry Chambrs
7th June 1714
Sr Isaac Newton

NOTES

(1) II, fo. 388; there is a copy in P.R.O. T/27, 21, p. 219.

(2) The letter to Oxford, signed simply 'L.W.', follows immediately afterwards; it was considered at the Treasury on 31 May. The writer proposed to divide sixpence into 25 farthings, making the shilling £0·05 etc. exactly as was done in the recent decimalization of the British money.

1088 DE MOIVRE TO J. BERNOULLI
28 JUNE 1714
Extracts from the original in Basel University Library.[1]
Reply to Letter 1046; for the answer see Letter 1090

. . . J'ay eté tenté plusieurs fois de vous ecrire simplement pour vous assurer de la continuation de la tres profonde estime que j'ay pour Vous; Mais le desir que j'avois de voir Vos remarques[2] sur le livre de Mr Newton, me retenoit, et je mourois d'envie de pouvoir vous en parler; j'ay veu vos remarques Monsieur, mais j'ay eté dans une impossibilité absolue de les aprofondir, je ne les [ai] encore lues qu'historiquement, mais autant que j'en puis juger, ou plustost que je puis l'entrevoir, elles sont remplies de tout ce qu'il y a de plus fin dans le calcul, et en même temps de beaucoup d'elegance; ce sont deux choses qui jointes ensemble rendent un ouvrage infiniment recommandable, j'ay dessein presentement que nous sommes dans la belle saison d'essayer l'air de la campagne pour quelques semaines; j'espere qu'il me retablira, et alors je consacreray constamment une partie du jour a etudier vos remarques. Il ne m'est jamais venu dans l'esprit que vous ayiez pris parti pour Mr Leibnitz sur le differens qui est entre luy et Mr Newton, mais Monsieur, quand cela vous seroit arrivé, ce n'auroit pas eté une raison pour moi d'en marquer du mecontentement par mon silence, il ne seroit aucunement raisonnable que la reconnoissance que je Vous dois pour la bienveillance dont Vous m'honorez, fut sujette a Varier sur un incident aussy leger; il ne paroit pas que Mr Newton ait eu la moindre inquietude sur le sujet de la gloire due au premier inventeur des fluxions ou du calcul differentiel; il a eté à la verité un peu piqué de ce qu'on a voulu insinuer dans quelques pieces volantes que c'etoit lui qui avoit appris le calcul des differences de Mr Leibnitz & qu'on avoit voulu le faire passer pour plagiaire; cependant je ne croy pas qu'il agisse luy même pour se justifier. On dit que M. Keil en fait son affaire, et qu'il paroitra au premier jour une piece de luy qui doit servir de reponse aux papiers ecrits depuis peu en faveur de Mr Leibnitz contre Mr Newton.[3] Je crains que vous n'ayiez formé un jugement des suites de Mr Newton qui puisse donner quelque petite prise contre Vous, . . .

. . . Monsieur Newton m'a chargé de vous dire que ce n'a eté que par un pur effet du hazard qu'il ne vous a pas envoyé la seconde edition de ses principes; il avoit oublié qu'il vous l'avoit promis;[4] cependant comme il n'a point de Mathematicien dont il estime plus la science, il se seroit determiné naturellement a vous en envoyer un exemplaire, si un accident ne l'avoit empeché, c'est que tous les exemplaires sont demeurez dans la possession du Dr Bentley qui avoit voulu se charger du soin de l'impression, et Mr Newton a l'exception de deux ou trois a eté obligé lui-meme de payer les autres fort cher, et de n'en distribuer qu'a un petit nombre de personnes, je vous diray que Mr Halley et moy avons eté obligés d'achetter les notres, et la même chose est arrivée a plusieurs autres personnes à qui Mr Newton avoit fait present de ses autres livres; mais je croy que nous en aurons encore une autre impression avant qu'il soit peu, plus correcte que la seconde, mieux imprimée et sur de meilleur papier; elle sera imprimée aux depens de Mr Newton et sous ses yeux, les figures doivent etre gravées et inserées dans le corps du livre; je suis chargé par Mr Newton de vous en promettre un exemplaire en cas que cela s'execute, et il n'y a presque point de doute; mais en attendant et pour reparer en quelque sorte son oubly, Mr Newton vous en enverra un exemplaire de la seconde edition aussytost qu'il scaura la voye de vous le faire tenir. Je luy ay deja indiqué la voye de Mr Holmius, mais nous attendrons que vous m'ayiez fait l'honneur de m'ecrire un mot . . .

[P.S.] Apres ma lettre finie et signée, j'ay veu M. Newton qui m'a dit qu'il avoit lu avec beaucoup de plaisir votre methode de resoudre le Probleme de la resistance;[5] il vous rend justice, en homme qui n'est nullement offensé, il dit qu'elle est admirablement belle & meme qu'elle est commode pour des expressions finies.

NOTES

(1) L.I.a 664, 10*; printed in Wollenschläger, *De Moivre* pp. 289–93.

(2) That is, Bernoulli's article in the *Acta Eruditorum* for February and March 1713. See Letter 1055, note (2), p. 96.

(3) De Moivre refers to Keill's projected 'Answer' eventually printed in the *Journal Literaire de la Haye* for July and August 1714; see Letter 1053*a*, note (1), p. 90. De Moivre clearly did not know, or at least did not want Bernoulli to know, the extent of Newton's involvement in the matter.

(4) See Letter 1028, note (2), p. 47.

(5) *Mémoires de l'Académie Royale des Sciences* for 1711 (Paris, 1714), pp. 47–54. See also Letter 951*a*, note (2) (vol. v, p. 349).

1089   CHAMBERLAYNE TO LEIBNIZ
30 JUNE 1714
Extract from the original in the Niedersächsische Landesbibliothek, Hanover.[1]
Reply to Letter 1062; continued in Letter 1092

Honored & Dear Sr

I am asham'd I have been so long in acknowledging the great Favour of your last dated from vienna 28 Apr. 1714 [N.S.], but the time that has been

152

necessarily spent in communicating the same to Sr Isaac Newton,[2] & by him to the Royal Society (for I did presume your consent to show it to all whom it concern'd) [3] wil I hope Justify my Delay, . . . But I am sorry to tel you Sr that my Negotiations have not met with the desired success, & that our Society has been prevail'd upon to vote that what you writ was insufficient & that it was not fit for them to concern themselves any further in that Affair &c, as will more fully appear by the Resolution enter'd into their Books of which I shal shortly send you a Copy,[4] & in the mean time the inclos'd printed pamphlet which Dr Keill gave me last Thursday [5] wil stay your Stomach till you can hear further from

[etc.]

NOTES

(1) Leibniz Letters and Papers; see Bodemann's catalogue, no. 149, fo. 14.
(2) See Letter 1072.
(3) See Letter 1079.
(4) See Letter 1092.
(5) The long 'Answer' from Keill in the *Journal Literaire de la Haye* for July and August 1714, pp. 319–58, discussed in the preceding correspondence; Keill had a few copies of the article printed separately in pamphlet form, and it is one of these that Chamberlayne here sends to Leibniz.

1090 J. BERNOULLI TO DE MOIVRE
24 JULY 1714
Extracts from the copy in the University Library, Basel.[1]
Reply to Letter 1088

. . . J'ai été ravi d'entendre que mes remarques sur le livre de Mr. Newton a eu le bonheur de Lui plaire, et qu'il a temoigné beaucoup de satisfaction de ma methode de resoudre le probleme de la resistance; voila rempli le desir que j'avois en composant cette piece là d'êcrire quelque chose qui pût meriter l'approbation de ce grand Geometre, et en même temps vôtre applaudissement, ce que Vous temoignez assez dans Votre Lettre, quoique Vous n'ayez encore lû mon écrit qu'historiquement comme Vous dites: peutetre qu'une lecture plus attentive ne Vous fera pas repentir de la partie du jour que Vous y aurez consacrée. Je suis bien aise que Vous ne me soupçonniez pas d'avoir pris parti pour Mr. Leibnits sur le differend qui est entre Lui et Mr Newton; je souhaiterois que la paix fût retablie entre ces deux Messieurs ou plûtôt que cette querelle n'eut jamais été suscitée. Pendant qu'on dispute à qui d'eux appartient la gloire de quelques inventions, je croi que nous autres pourions bien pretendre aussi de participer à cette gloire; car si Eux, pour qui leurs partisans se sont mutuellement declaré la guerre, ont jetté la premiere pierre, nous pouvons nous vanter d'avoir bâti la plus grande partie de cet Edifice, ce que Mr. Leibnits Lui même reconnoit quelque part dans les actes de Leipsic; cependant les partisans le dissimulent et ne nous veulent point admettre à la participation de la moindre partie de la gloire disputée. *Quæ ipsius magni Newtoni reperta* (c'est ainsi que finit Mr. Cheynes son discours sur la methode des fluxions pag. 59 et 60.) [2] *cum mecum animo perpendo, non possum abstinere me quin dicam, omnia in hisce vel per hasce (aut non absimiles methodos) ab aliis (intra hosce viginti quatuor annos proxime elapsos) edita, esse solum eorundem ab Ipso diu antea cum Amicis vel Publico communicatorum Repetitiones, aut non difficilia Corollaria:* Vous avez eprouvé la plume empoisonnée de cet Ecrivain, mais je ne croi pas que son poison Vous ait tant pu déplaire que cet encens impertinent, et cette fade flatterie aura deplû a Mr. Newton. Si j'etois d'une humeur querelleuse, je trouverois bien de la matiere à me plaindre de quelquesuns des habitans de vôtre Isle, car outre ce Mr. Cheynes qui sous le titre de *Addenda et Adnotanda in Libro fluxionum* [3] a fait imprimer mot pour mot mes remarques sur son livre, comme si ç'eussent été ses propres remarques ne me laissant pour toute gloire que celle de Lui avoir communiqué quelques fautes d'impression ou de calcul, aprez avoir injurié furieusement ceux qui *humillimam in numeris contexendis diligentiam adhibent,* outre ce Monsieur Cheynes, dis-je, il y en a d'autres, dont je n'ai pas trop sujet d'etre content, s'étant servi dans leurs livres de nos decouvertes sans faire mention d'un seul mot de qui ils les ont empruntées ou plutot ravies; Mr. Keil lui meme qui crie tant aux voleurs! n'en est pas exemt. Mais je cherche et j'aime la tranquillité tant necessaire a ceux qui par leurs meditations tachent d'enrichir les sciences; car assurement la bile echaufée et l'esprit querelleux ne sont point propres pour faire des decouvertes. Que si neantmoins Mr. Keil trouve a propos de m'attaquer (puisqu'il est plus facile de critiquer que de produire quelque chose de meilleur, et que je n'ai encore rien vû de ses propres productions, tout ce qu'il a écrit n'étant que commentaire ou compilation) je le souffrirai sans etre persuadé pour cela que ses insultes puissent faire le moindre tort à ce que j'ai écrit ou pour l'amour de la verité, ou pour l'avancement des sciences, d'autant plus que le grand Newton Lui meme dont il entreprend la defense, l'a jugé digne de son approbation. Mais Vous craignez Monsieur (peutetre en vain) que je n'aye formé un jugement des suites de Mr. Newton qui puisse donner quelque petite prise contre moi, et puis Vous dites qu'il y a plus à craindre d'un autre coté, c'est qu'on *s'imagine* chez Vous, que par la remarque sur les secondes differences, j'ai voulu fournir des armes aux ennemis de Mr. Newton contre Lui; imagination ridicule et plaisante! car il est facile de croire que mes remarques qui parurent aux mois de fevrier et de mais 1713, doivent avoir été ecrites 3 ou 4 mois pour le moins auparavant, vu le grand voyage que mon êcrit à fait d'ici à Leipsic, et la lenteur avec laquelle on insere ces pieces qui sont un peu longues, en sorte que la mienne doit avoir été achevée avant la publication du *Commercium Epistolicum* par consequent avant que de sçavoir de quoi il s'agissoit entre Messrs. Newton et Leibnits: peutetre me diraton que quelqu'un m'avoit donné avis d'Angleterre, qu'on alloit commencer une telle dispute par la publication de ce livre, mais que me repondrez vous Monsieur si je Vous dis que cette remarque sur les secondes differences, qui est proprement de mon Neveu, se trouve deja sous son nom inserée dans les memoires de l'acad. des sciences de Paris de l'annee 1711 pag. 54 edit. Paris. en forme d'addition à une de mes lettres ecrite à cette Academie le 28 Janvier 1711 [4] pres de deux ans avant la publication du *Commercium Epistolicum* et partant longtemps avant qu'on ait seulment songé en Angleterre d'intenter procés à M. Leibnits. voyez donc je Vous prie avec quelle justice on m'accuse que j'ai voulu fournir des armes contre Mr. Newton! Que si nonobstant ces remontrances Mr. Keil s'opiniatre à me faire la guerre là dessus, je Vous ai dejà dit que je ne perdrai point mon temps à le refuter; mais il se trouvera peut etre quelqu'un qui fera mon apologie. Il suffit que je declare hautement, que je n'ai jamais eu le dessein de causer le moindre chagrin à Mr. Newton, mais bien au contraire que je cherche toutes les occasions pour faire paroitre la tres-profonde estime que j'ai pour Lui et le grand cas que je fai de son rare merite; c'est de quoi je Vous prie tres-instamment de l'assurer, en Lui faisant mes respectueux complimens. Pour ce qui est de la dispute que Vous eutes autre fois avec mon Neveu touchant les tangentes et les differences secondes, et sur laquelle Vous me demandez mon avis dans votre penultieme: Je vous y repondis amplement dans la derniere lettre que je me donnay l'honneur de Vous écrire en date du 18 fevrier 1713, [5] mais à laquelle je n'ai point reçu de reponse, et meme sans en faire mention dans l'honneur de vôtre derniere; Vous y aurez donc vû, que je suis tout à fait de votre sentiment, sçavoir que l'on peut considerer la courbe ou comme un polygone ou comme un assemblage d'une infinité de petits arcs ou circulaires ou parabolique pourvû que l'on raison toujours consequemment selon son hypothese et que l'on ne confonde pas l'un avec l'autre, on trouvera toujours la verité. Je me suis même servi de la seconde idee plutot que de la premiere, quand il s'agit de la determination des forces parce que je la trouve plus commode que la premiere, ce que Vous remarquerez assez dans la dite piece que je publiai dans les actes de l'annee passée, ou j'ai consideré les elemens des courbes comme de petits arcs des cercles osculateurs; c'est encore par l'explication de cette idée que je levai autrefois à Mr. Varignon il y a pres de dix ans une difficulté sur la force centrifuge, qu'il croyoit être deux fois plus grande que ne l'avoit determinée Mr. Huguens; Mais pour Vous asseurer pleinement que cette maniere de considerer les courbes ou l'idée des tangentes Geometriques dans les elemens des courbes ne m'est pas nouvelle, je Vous communiquerai à la fin de cette lettre un Extrait d'une de mes lettres écrite à Mr. Herman,[6] lorsqu'il etoit encore à Padoue; Vous en jugerez si je ne raisonne pas juste sur cette matière et si j'ai jamais confondu, comme Vous pensez, l'interceptée entre la tangente geometrique et la courbe avec l'interceptée entre le prolongement de la corde de l'arc adjacent infiniment petit et la courbe; je ne sçai donc ce que Vous voulez dire par la *petite inadvertance* que Vous me voulez imputer, si c'est parceque Vous pensiez que je n'avois pas pris garde que la premiere interceptée n'est que la moitié de l'autre, Vous voyez maintenant que Vous m'auriez fait tort: Mais si c'est parce que nôtre jugement des suites de Mr. Newton n'est pas conforme à l'explication qu'il leur donne présentement, ce ne sera pas nôtre faute; il falloit les prendre dans le sens qu'il leur donnoit Lui même, or il dit quelque part en termes expres, que les termes de ces suites etoient des differences 2des, 3mes etc. S'il s'est reservé quelque restriction, nous n'etions pas obligés de deviner cette reservation mentale: Voila tout ce que j'ai à dire pour appaiser les murmures qui s'elevent contre moi, si on n'est pas content, patience! j'attendrai l'arret de mort avec un entière resignation.

Vous me parlez Monsieur de quelques pieces volantes que l'on doit avoir publiées contre Mr. Newton; je n'en ai vu qu'une seule que l'on m'a envoyée de Saxe;[7] mais comme on y parle de certains faits de meme que dans le *Commercium Epistolicum*, nous autres qui n'avons point de connoissance de ces faits nous ne pouvons ajouter qu'une foi historique à toutes ces relations et à toutes ces Lettres dont on nous donne des copies pour fideles et sinceres de part et d'autre . . .

. . . Mr. Newton se met trop en peine de ce qu'il a oublié de m'envoyer un exemplaire de la nouvelle edition de son livre,[8] il me l'a promis, il est vrai, mais il n'etoit pas obligé de s'en acquitter contre la possibilité; J'accepte donc tres-volontiers son excuse, d'autant plus que je lui suis endetté pour plusieurs de ses autres livres dont il m'a fait present. J'ay un plaisir extraordinaire d'apprendre qu'il veut nous donner une troisieme edition de son excellent ouvrage, qui s'imprimera sous ses yeux et qui sera comme vous dites beaucoup meilleure et plus correcte que la seconde; je lui ai de l'obligation de sa promesse de m'en envoyer un exemplaire en cas que cela s'execute, je croy que la voye la plus seure et la plus courte de me le faire tenir, c'est de l'envoyer à Paris, en l'adressant à Mr. Varignon ou à quelque autre ami qui trouvera de frequentes occasions pour le faire transporter ici. . . . .

NOTES

(1) LIa 664, no. 9; printed in Wollenschläger, *De Moivre*, pp. 294–9.

(2) George Cheyne, *Fluxionum methodus inversa, sive quantitatum fluentium leges generaliores* (London, 1703); see Letter 1046, note (1), p. 74. 'When I reflect inwardly upon the discoveries made by the great Newton himself, I cannot refrain from remarking that everything published by others during the twenty-four years last past in these or through these (or not dissimilar methods) is nothing but a repetition of the same things long ago communicated by him to his friends or to the public, or easy corollaries.'

(3) See Letter 1046, note (1), p. 74.

(4) Compare Letters 951 and 951*a*, vol. v. Johann Bernoulli refers, of course, to his *Lettre . . . écrite de Basle le 10. Janvier 1711 touchant la maniere de trouver les forces centrales dans les milieux resistans . . .* printed in the *Mémoires de l'Académie Royale des Sciences* for 1711 (Paris, 1714), pp. 47–54. (The heading in the *Mémoires* gives 10 January; the marginal note gives 28 January as the date of reading; Wollenschläger accordingly (and silently) altered the date in the text of the letter to 10 January.) This *Lettre* is followed by the *Addition* (pp. 54–6) in which Nikolaus [II] explained how Newton's error in Prop. 10, Book II, was due to ignorance of differentiation: 'C'est cette méthode de changer les quantités indeterminées & variables en suites convergentes, & de prendre les termes de cette suite pour leurs différentielles respectives . . . qui a conduit M. Newton, à des solutions fausses dans l'exemple dont je viens de parler, & dans les suivans . . .' and further maintained that this was just the method taught by Newton in the Scholium at the end of *De quadratura curvarum*. (As already explained, Nikolaus' 'correction' gives the right answer but was not the source of Newton's mistake.) It seems that in writing to Leibniz Johann was eager to claim this 'correction' for himself, whereas in writing to Newton's friends he was happy to attribute it to his nephew Nikolaus.

(5) See Wollenschläger, *De Moivre*, pp. 281–2, Bernoulli answering De Moivre's letter of 17 December 1712 (*ibid.*, pp. 277–8). The point at issue was (again) the *geometrical* interpretation of a second differential. Compare (for what follows) the third sentence of Letter 1074.

(6) The extract is not now with the copy at Basel. For Hermann, see Letter 1005, note (11), p. 10.

(7) The *Charta Volans*; see Number 1009.

(8) See also Letter 1055.

1091 HANNAH BARTON TO NEWTON
24 JULY 1714
From the original in private possession

Dear Bro.

Just now I have Receved word of half a Buck to be Delive[re]d to you of Monday in ye after Noon.

I request yr Acceptance and to hear you say very fast if it comes safe & is good. in hast ye Bearer stays

Your Most Affeconate

Sister   H BARTON [1]

July ye 24 1714
   For
Sr Isaac Newton in
St Martins Street near
   Lecester fields
      London

NOTE

(1) Hannah Barton, Newton's half-sister. See Letter 2, note (2) (vol. I, p. 3) and Letter 878, note (1) (vol. v, p. 199).

1092 CHAMBERLAYNE TO LEIBNIZ
27 JULY 1714
From the original in the Niedersächsische Landesbibliothek.[1]
Continuation of Letter 1089; reply to Letter 1062

Petty France Westmr
27 July 1714

Most Honor'd Sr

I have already acknowledged the great Honor of your Letter of the 28 April 1714 N.S. from vienna,[2] but as I told you then, I did not pretend to have Answer'd it fully til I had procur'd the Resolution of the R. Society about it, which I send you here inclos'd, together with a copy of a Letter from Dr Keill, & a little book writ by that Gentleman relating to the Differences between Sr Isaac Newton & you Sr,[3] of which I am forced to say: *Non nostrum est Tantas*

Componere lites, & therefore humbly desiring you to accept of my Good will in that Affair, subscribe myself with the utmost Passion & Truth

Honor'd Sr

your most Devoted Humble Servant

JOHN CHAMBERLAYNE

NOTES

(1) Leibniz letters and papers; see Bodemann's catalogue, no. 149, fo. 15. The enclosures, which follow, are at fos. 27–8.

(2) Letter 1062.

(3) See Letter 1089, note (5), p. 153.

1092*a* THE ENCLOSURES

[*In a clerical hand*]

Extract from the Journal of
the Royal Society

May. 20*th*. 1714.

The Translation of Mons: Leibnitz Letter to Mr. Chamberlayne produced the last meeting was read.

It was not judged proper (since this Letter was not directed to them) for the Society to concern themselves therewith, nor were they desired to do so: But that if any Person had any Material Objecon to ye Commercium, or to the Report of the Com[mitt]ee, it might be considered at any time.

Copy of a Letter from Dr. Keill
to Mr. Chamberlayne dat. 20, July 1714

Sr.

I have sent you the Inclosed according to my Promise, you'l find by it how much reason I had for what I before asserted, there were but 18 of them printed here but they are to be reprinted in Holland, and if Mr Leibniz makes any more Noise I will stil give the world a greater knowledge of his Merits & Candor   I am

Sr

Your very humble Servt.

John Keill

[*Then in Chamberlayne's hand*]

P.S.

I am almost asham'd to send to the Honorable Mr Leibniz the Copy of such a Harsh Letter, had I not thought it my Duty to hide nothing from him in this Affair &c.

J.C.

1093 LOWNDES TO NEWTON
27 JULY 1714
From the copy in the Public Record Office[1]

Sir

I am Commanded by my Lord Trea[su]rer to Transmit to You for Your Consideracon the Enclosed Letter and paper Annext Concerning the Discovery of the Longitude at Sea.[2] 27 July 1714

WM LOWNDES

NOTES

(1) T/27, 21, p. 250.

(2) We presume that the enclosed documents are Numbers 1093 *b* and *c* that follow, from the Portsmouth Collection. On 25 May 1714 the House of Commons had received a petition from the Captains of several naval vessels and merchantmen, and from London merchants, setting out that the discovery of the longitude was so necessary for Great Britain and the safety of her ships and commerce, that an encouragement should be offered to persons to 'prove the same before the most proper judges'; a committee was set up to consider the petition, which reported on 7 June.

The Committee had heard the evidence of expert scientific witnesses—perhaps it was the first hearing of his kind in history—beginning with Humfrey Ditton and William Whiston (see Letter 1049 above), upon whose proposals Newton and others were then invited to give their opinion (see Number 1093*a* below). The Committee resolved that a reward should be offered for the 'discovery of longitude', the prize to be proportionate to the accuracy of the results obtained. Its resolution was adopted by the House of Commons, and a bill to put it into effect drawn up, which received the Royal Assent early in July, 1714.

By the Act (13 Anne, *c*. 14) Commissioners—among whom, besides the Lord High Admiral, The Master of Trinity House, and the Speaker of the House of Commons, were the President of the Royal Society, the Astronomer Royal, and the Savilian, Lucasian, and Plumian Professors of Oxford and Cambridge as well as William Lowndes—were appointed to judge proposals and, if necessary, organize trials of them. The reward was to be £20000 if the longitude was ascertained to half a degree, £15000 for two-thirds of a degree, and £10000 for a whole degree. Half of the award was to be paid when the Commissioners were satisfied that the 'Method extends to the Security of Ships within 80 Geographical Miles' of the most dangerous shores, and the whole of it when a voyage from Britain to the West Indies had been completed without loss of the longitude beyond the limit claimed.

1093*a* SCIENTIFIC EVIDENCE ON THE DETERMINATION OF LONGITUDE
[EARLY JUNE 1714]
From the *Journals of the House of Commons* [1]

Mr. Ditton and Mr. Wiston, being examined, did inform the Committee, That they had made a Discovery of the Longitude; and were very certain, that the same was true in the Theory; and did not doubt but that, upon due Trial made, it would prove certain and practicable at Sea:

That they had communicated the whole History of their Proceedings towards the said Discovery to Sir *Isaac Newton*, Dr. *Clark*, Mr. *Haley*, and Mr. *Cotes*; who all seemed to allow of the Truth of the Proposition, as to the Theory: but doubted of several Difficulties that would arise in the Practice.

Sir *Isaac Newton*, attending the Committee, said,[2] That, for determining the Longitude at Sea, there have been several Projects, true in the Theory, but difficult to execute:

One is, by a Watch to keep Time exactly: But, by reason of the Motion of a Ship, the Variation of Heat and Cold, Wet and Dry, and the Difference of Gravity in different Latitudes, such a Watch hath not yet been made:

Another is, by the Eclipses of *Jupiter's Satellites*: But, by reason of the Length of Telescopes requisite to observe them, and the Motion of a Ship at Sea, those Eclipses cannot yet be there observed:

A Third is, by the Place of the Moon: But her Theory is not yet exact enough for this Purpose: It is exact enough to determine her Longitude within Two or Three Degrees, but not within a Degree:

A Fourth is, Mr. *Ditton's* Project: And this is rather for keeping an Account of the Longitude at Sea, than for finding it, if at any time it should be lost, as it may easily be in cloudy Weather: How far this is practicable, and with what Charge, they that are skilled in Sea-affairs are best able to judge: In sailing by this Method, whenever they are to pass over very deep Seas, they must sail due East, or West, without varying their Latitude; and if their Way over such a Sea doth not lie due East, or West, they must first sail into the Latitude of the next Place to which they are going beyond it; and then keep due East, or West, till they come at that Place:

In the Three first Ways there must be a Watch regulated by a Spring, and rectified every visible Sun-rise and Sun-set, to tell the Hour of the Day, or Night: In the Fourth Way, such a Watch is not necessary: In the First Way, there must be Two Watches; this, and the other mentioned above:

In any of the Three first Ways, it may be of some Service to find the Longitude within a Degree; and of much more Service to find it within Forty Minutes, or Half a Degree, if it may be; and the Success may deserve Rewards accordingly:

In the Fourth Way, it is easier to enable Seamen to know their Distance, and Bearing from the Shore, 40, 60, or 80, Miles off, than to cross the Seas; and some Part of the Reward may be given when the First is performed on the Coast of *Great Britain*, for the Safety of Ships coming Home; and the rest, when Seamen shall be enabled to sail to an assigned remote Harbour without losing their Longitude, if it may be. [3]

Dr. *Clarke* said, That there could no Discredit arise to the Government, in promising a Reward in general, without respect to any particular Project, to such Person or Persons who should discover the Longitude at Sea.

Mr. *Haley* said, That Mr. *Ditton's* Method for finding the Longitude did seem to him to consist of many Particulars, which first ought to be experimented, before he could give his Opinion; and that it would cost a considerable Sum to make the Experiments; but what the Expence would amount to, he could not tell.

Mr. *Whiston* affirmed, That the undoubted Benefit which would arise on the Land, and near the Shore, would vastly surmount the Charges of Experiments.

Mr. *Cotes* said, That the Project was right in the Theory near the Shore; and the practical Part ought to be experimented.

<hr>

NOTES

(1) xvii, 1803, pp. 677–8; the date of the Committee's report to the House was 7 June. See also Brewster, *Memoirs*, ii, pp. 259–62. The hope of winning a large state reward for a method of determining longitude at sea goes back at least to Galileo's discovery of Jupiter's satellites, and it was widely believed in the late seventeenth century (though apparently mistakenly) that the Dutch government had offered a handsome prize for a successful solution of this problem. In 1715 the French government followed the British example.

(2) According to Whiston, writing many years later, Newton presented his evidence in writing, and failed (till prompted by himself) to make his views understood; see Edleston, *Correspondence*, p. lxxvi, note (167). Whiston's story is born out by an autograph version of Newton's evidence (U.L.C., Add. 3972, fo. 33), almost word for word identical with that printed above, but ending with the additional sentence—after '. . . Longitude, if it may be.'— 'But whether rewards shall be setled & what rewards I desire to be excused from giving any opinion.' It was doubtless this indecision which confused the Committee and provoked Whiston into making Newton declare verbally that he thought the proposed bill for a reward

ought to pass because the Ditton–Whiston method was useful near the shore. When Newton had been thus goaded into a positive declaration, the final indecisive sentence was naturally deleted from his written evidence.

(3) Newton's recommendation was adopted in the consequent Bill. The first part of it is clearly favourable to the proposal of Ditton and Whiston. A system of warning mariners when they pass from the deep oceans to the continental shelf, though highly useful (and to some extent realized today) does not possess the universality of a true method of determining longitude.

1093*b* JORDAN TO [?OXFORD]
11 JULY 1714
From the original in the University Library, Cambridge[1]

A Barleduc le 22e Juillet 1714 [N.S.]

Mylord

Sans avoir l'honneur d'estre connu de Vostre Grandeur, je prends la liberté de lui adresser le memoire ci joint dela part d'un de mes amis, qui s'est trouvé encouragé. de donner sa decouverte, touchant les Longitudes sur Mer, par l'esperance des recompenses, que le Parlement Britannique se propose de donner, à ceux qui feront cette decouverte.

Vous faites paroitre, Mylord, tant d'attachement & de zele, pour tout ce qui a du rapport aux avantages de la Nation Britannique, qu'il y a lieu de croire que vous voudrez bien ajouter un nouveau degré de gloire à celle qui vous est deja acquise, en contribuant a lui procurer le nouvel avantage qu'elle recherche sous vôtre Ministere.

M. Romual le Müet, Auteur du Memoire ce joint, remplit à Metz un Employ, qui ne lui permettroit pas d'en deplasser pour aller en Angleterre, donner des preuves de sa decouverte. C'est pour cela, Mylord, que Votre Grandeur est Suppleé, de proposer; qu'il plaise au Conseil de sa Majesté, la Reine de la Grande Bretagne, d'envoyer des Commres. à Metz, pour conferer avec l'Auteur de la decouverte des Longitudes sur Mer.

Si neantmoins la ville de Metz ne vous convenoit pas, Mylord, & qu'on aima mieux que la Conferance se tint à paris, en presence du Ministre d'Angleterre, M. Romuald le Muet offre de s'y rendre, en l'avertissant quinze jours ou trois semaines à l'avance, & alors il executera dans l'une ou l'autre ville, de Metz ou de Paris, tous ce à quoi il s'engage par son Memoire.

Je vous supplie tres humblement, Mylord, d'avoir la bonté de nous faire honnorer d'un mot de reponse, soit à mon adresse, soit à celle de *Mons[ieu]r. Romuald le Müet a Metz*; afin que sur icelle on puisse prendre les mesures convenables, & que la reponse soit écrite en langue françoise, ou en Latin. En attendant, j'ai l'honneur d'estre, avec un tres profond respect

Mylord

    De Vostre Grandeur

          Le tres humble & tres ob[eissant]
          serviteur JORDAN
          con[seill]er historiographe

            A Barleduc

To be sent to Sr Isaac Newton.[2]

NOTES

(1) Add 3972, fos. 7–10, enclosing Number 1093.
(2) Treasury endorsement.

1093*c* LE MUET TO THE LONGITUDE COMMISSION
29 JUNE 1714
From the original in the University Library, Cambridge [1]

    Aus Seigneurs du Parlement
    de la grande Bretagne et aux
    commissaires établis pour l'affaire
    concernant les Longitudes par mer.

La proposition faite en dernier lieu dans le Parlement Britannique, d'assurer une recompense proportionnée au merite de la chose, à celui qui fera la découverte des Longitudes, m'a déterminé à envoyer ce petit memoire, pour estre communiqué à qui il appartiendra.

    J'offre de donner une methode geometrique Nouvelle, pour connoitre surement la Longitude aussi facilement sur mer que sur terre, sans qu'on ait besoin d'autre connoissance que de la Latitude et du Rumb du Vent, ou dela direction du Vaisseau. ce qui n'empeschera pas que par cette même metode, on ne puisse aussi facilement resoudre les quatre autres problemes, ou Regles generales dela Navigation. il n'est besoin ni de plume ni d'Encre en cette metode pour faire les calculs: un compas suffit; quoiqu'elle indique tout ce qu'on peut desirer, j'entens la Latitude du depart, la Latitude de l'arrivée la difference en Latitude; la Longitude du depart, la Longitude de l'arriveé, la difference en Longitude; le Rumb de vent; les lieües Majeures, les lieües mineures, et les lieües de distance, le tout bien précisement exprimé et tres facile à pratiquer. Sans aller en mer, sans sortir de la chambre, et en moins d'une heure on peut connoitre la verité et la solidité de cette metode. Enfin elle est si simple, si naturelle que les sçavans auront lieu d'admirer comment elle a pû echaper à leurs reflexions depuis tant de siecles qu'ils la cherchent.

Je promets de satisfaire à ces Engagemens moyennant une recompense proportionnée au merite exquis d'une découverte si utile.
Ecrit à Metz ce dix juillet 1714 [N.S.]

S. ROMUALD LE MUET.

NOTE

(1) Add. 3972, fos. 7–10. It is obvious that this paper was written while the Longitudes Bill was passing through its stages in the House of Commons, before it passed in the Lords (8 July) and received the Royal Assent. In fact, an announcement of the proposed prize had appeared in the *Journal Literaire de la Haye* for May and June 1714, pp. 235–6.

1094 ANDRÉ D'ALESME TO [?NEWTON]
[?END OF JULY 1714]
From the original in the University Library, Cambridge[1]

Monsieur,

Ayant appris que le Parlement a promis trois cent mille livres ou aprochant a celuy qui trouvera les longitudes maritimes et que vous en estes l'éxaminateur,

Jay l'honneur de vous envoyer une idée de pendule que le mouvement et secousse du vaisseau ne fera jamais arrêter par ce moyen[2] si l'on na pas les longitudes maritimes dans toute la justesse du moins elle aidera a en aprocher bien plus que tout ce que lon a fait jusqués apresent.

Sile Parlement sur votre raport a la bonté de me faire une gratification telle que votre prudence jugera apropos,

ou pour la Cric[3] que j'ay l'honneur de vous envoyer j'esperequil ne la perdra pas, car cela me mettra en estat de continuer mes nouvelles decouvertes dont plusieurs pouroient estre utiles au service de son Etat.

Voila un memoire de quelques machines que j'ay inventées et executées.[4]

Je suis avec toute l'éstime possible

Monsieur

Votre tres humble et

tresobeissant serviteur

ANDRÉ DALESME

Si vous m'honorez d'une reponse dont je vous suplie Mon adresse est a M Dalesme de l'academie royale des Sciences rue St Denis proche la porte de parte a la ville danvers a Paris

NOTES

(1) Add. 3972, fos. 5–6. We place this letter arbitrarily here. The writer (1645–1727) was appointed *pensionnaire mécanicien* at the refoundation of the Académie Royale des Sciences in 1699.

(2) The letter transcribed here is a clerical copy signed by D'Alesme. With it (fos. 1–3) is an autograph letter in very similar terms (also without date). This continues with a sketch and brief description of a 'pendule que le mouvement du vaisseau ne faira iamais arrester'—in fact, a proposal to gear two pendulums together by toothed segments so that they swing 180° out of phase.

(3) Screw-jack. This phrase is not in the autograph version; no sketch survives.

(4) This also is missing (the autograph has 'une liste d'une partie des machines . . .').

1095 LOWNDES TO THE MINT
27 JULY 1714
From the copy in the Public Record Office.[1]
For the answer see Letter 1098

Gentlemen

My Lord Treasurer having received the Enclosed Representation Subscribed by Several Traders in the Woolen Manufactures in and about Taunton relating to the Current price of Moydors in those parts, His Lordship Commands me to transmit the same to you for your Consideration, and desires you will transmit hither Your Opinion what may be proper to do therein. I am

Gentlemen

Your Most Humble Servt.

Wm. Lowndes

Tre[asu]ry Chambers
27 *July* 1714
Officers of the Mint

NOTE

(1) Mint/1, 7, p. 70; the memorial (Number 1095*a*) precedes this letter on the same page.

1095a THE TAUNTON PETITION
21 JULY 1714
From the copy in the Public Record Office [1]

Taunton July 21*st*: 1714

May it please your Lordship

We whose Names are Subscribed being the principal Traders in the Woollen Manufactury of this Town and Country adjacent, and drawing Bills for Vast Summs of Money for the said Manufactures, Crave leave humbly to represent to Your Lordship, that Moydors for some years past have been Currant with us in all payments at 28*s* each and at that price never refused to take them on our Bills of Exchange from any of Her Majestys Collectors whether they be of the Taxes, Excise Customs &c. And we have promised allways to Supply them with good Bills of Exchange for what Moydors they shall receive for her Majestys Revenue at the above said price. Notwithstanding which Severall of the Collectors of the Taxes, Customs and Excise or Agents under them for Some private Interest to themselves design'd have lately attempted and are attempting to lower the price of Moyders to 27*s*. 6 each, which if Effected would prove so very discouraging to the Trade in these parts that the poor could not be Employed for the Moyders would soon then be sent hence to London as the Guineas in great measures have already been, so Consequently not half Money enough left in these parts to Carry on our Trade, Wherefore as we are willing for the Encouragement of Trade and Employment of the poor at all times to Supply the Collectors with good Bills of Exchange for what Moydors they shall receive at 28*s* each, So we hope and humbly pray That Your Lordship will give Orders that the Several Collectors in these parts do not refuse any in the Said price And if they can Object against the Bills of Us, or any of Us the Subscribers they may refuse them and the rest will Supply them.

NOTE

(1) The original memorial, bearing numerous signatures, is also in T/1, 180, no. 30.

1096 NEWTON TO THE DUKE OF SHREWSBURY
2 AUGUST 1714
From the holograph original in the Public Record Office. [1]
For the answer see Number 1097

May it please your Grace

The Mint being at a stop for want of authority to proceed with the Dyes & Puncheons last in use untill new ones can be made, & great quantities of gold

167

Bullion being in the Mint to be coyned, & more Bullion being dayly expected; I have hereunto annexed a copy of the Warrant signed by her late Majesty upon the like occasion, & a Draught of a new Warrant suitable to the present occasion,[2] & most humbly pray your Grace to lay the matter before their Excellencies the Lords Justices that the Coynage may proceede. I am

May it please your Grace

Your most humble and

most obedient Servant

ISAAC NEWTON Master & Worker

of his Maties Mint

Mint Office
Aug. 2. 1714
To his Grace the Duke of Shrewsbury
Lord High Treasurer of Great
Britain

NOTES

(1) T/1, 180, no. 14. This trivial note reflects one of the dramatic moments of English history: as she lay mortally ill, Queen Anne on 30 July 1714 dismissed Oxford from the office of Lord High Treasurer and appointed in his stead Charles Talbot (1660–1718), Duke of Shrewsbury (1694), who had long before accompanied William III on his voyage to Torbay; she thereby ensured (a few days later) the peaceful succession of the House of Hanover. Shrewsbury resigned in October and the Treasurer's office has ever since been in commission.

(2) Neither copy nor draft is now with the letter.

1097 WARRANT FOR CONTINUING THE EXISTING COINAGE
3 AUGUST 1714
From the copy in the Public Record Office.[1]
Reply to Letter 1096

Harcourt C [2]	Devonshire	Orford
Buckingham P	Argyll	Halifax
Dartmouthe C. P. S.	Roxburghe	Cowper
Somerset	Kent	T. Parker
Bolton	Abington	
Shrewsbury	Townshend	

Whereas we are informed That there are in his Majties. Mint in the Tower of London, great Quantities of Gold, and some Silver, which if it should remain uncoined would be to the dissatisfaction and prejudice of the Importers, and a Discouragement to the Coynage. And that the making of new Puncheons and Dyes with his Majties Effigies and Armes will take up some considerable time, before they can be finished, We do therefore hereby Authorize and require you the Master and Worker of his Majties. said Mint, to proceed in coining the Gold & Silver Bullion already imported, as also all such other Gold and Silver Bullion as shall be brought into the said Mint, with Dyes make from the Puncheons used in her late Majties. Reign; untill new Puncheons & Dyes, with his Majties. Effigies and Arms shall be made and finished as aforesaid. And for so doing this shall be your Warrant. Dated at his Majties. Palace at St James's the third day of August 1714 in the first year of his Majties. Reign

To Sr. Isaac Newton Knt. Master
and Worker of his Majties. Mint

NOTES

(1) Mint/1, 8, p. 100.

(2) From his accession on 1 August George I was absent from Great Britain until 18 September; during the interval the Lords Justices fulfilled the functions of Head of State. Those named are: Simon Harcourt (?1661–1727), first viscount, Lord Chancellor; John Sheffield (1648–1721), first Duke of Buckingham, Lord President of the Council; William Legge (1672–1750), first Earl of Dartmouth, Keeper of the Privy Seal; Charles Seymour (1662–1748), sixth Duke of Somerset; Charles Paulet (1661–1722), second Duke of Bolton; the Duke of Shrewsbury, Lord Treasurer; William Cavendish (c. 1673–1729), second Duke of Devonshire; John Campbell (1680–1743), second Duke of Argyll; John Ker (c. 1680–1741), first Duke of Roxburghe; Henry Grey (1671–1740), first Duke of Kent; Montagu Venables-Bertie (d. 1743), Earl of Abingdon; Charles Townshend (1674–1738), second viscount, to be appointed Secretary of State on 17 September; Edward Russell (1652–1727), Earl of Orford, first Lord of the Admiralty; Charles Montagu (1661–1715), first Earl of Halifax of the second creation, appointed First Lord of the Treasury on 11 October 1714; William Cowper (d. 1723), Baron Cowper, to be appointed Lord Chancellor (for the second time) on 21 September 1714; Sir Thomas Parker (?1666–1732), created Baron (1716) and Earl of Macclesfield, Lord Chief Justice.

1098 NEWTON & PEYTON TO SHREWSBURY
5 AUGUST 1714
From the original in the Public Record Office.[1]
Reply to Letter 1095

To his Grace the Duke of Shrewsbury
Lord High Treasurer of Great Britain

May it please Your Grace

Upon an Order from my late Lord Treasurer. Dated 27. July 1714. that we should give our Opinion upon the Inclosed Petition of Several Traders in the Woollen Manufactures in and about Taunton that the Moyders of Portugall may there be received by the Collectors of the Revenues at 28*s* apiece.

Wee humbly represent that by their Weight and assay they are worth about 27*s*. 7$\frac{1}{4}$*d*. apiece at a Medium but are frequently culled in London, and the lighter pieces sent to other places to be put away where people will receive them, and those lighter pieces are scarce worth above 27*s*. 6*d*. one with another wch is the price that the Collectors of the Revenues are willing to take them at.

All which is most humbly Submitted to your
Graces great Wisdom

Mint Office   CRAV: PEYTON
5th August 1714   Is. NEWTON

NOTE

(1) T/1, 180, no. 9, signed. There is a copy in Mint/1, 7, p. 70.

1099 KEILL TO NEWTON
6 AUGUST 1714
From the original in the University Library, Cambridge[1]

Oxford August 6th 1714

Honoured Sr

I would have written to you sooner, If I had [not] been resolved to wait on you my self before now, but my housekeeper falling ill of the Small pox I cannot leave this place yet I had a letter from Mr Johnson, who writes to me in these words. Sr I received with a great deal of pleasure some days agoe your Packet[2] from Mr Ayeest [?] what you have sent shall be inserted in the

Journal of July and August, and every thing to your mind, if it had come a litle sooner it should have been in that of may and June, there is a curious peice in it about Air pumps wch I hope you will not dislike.[3] the 24 Copies of the Commercium I have received some tuo days before your letter, I shall give notice in the Journal that they may be had of me, and that the original letters are to be seen at London where you direct.[4] These are Mr Johnsons words and in a post[s]cript he writes I received this moment your letter by post wch I suppose to be of tuesday last, the faults shall be corrected and every thing else to your satisfaction. This letter is dated Hague August 3d. By wch Sr you will perceive that Mr Johnson deals fairly by us. I hope Mr Leibnits after this will not have the impudence to show his face in England. if he does I am persuaded that he will find but few freinds I am Sr with all respect

Your most obliged Humble servant

JOHN KEILL

For
Sr Isaac Newton
at his house in
St Martins Street
  near Leicester Feilds
    Westminster

NOTES

(1) Add. 3985, no. 13.

(2) Keill's 'Answer', see Letter 1053a, note (1), p. 90.

(3) W.J. 'sGravesande, 'Remarques sur la Construction des Machines Pneumatiques', *Journal Literaire de la Haye* for May and June 1714, pp. 182–208; an attempt at a mathematical theory.

(4) See the announcement at the end of Keill's 'Answer' (*ibid.* for July and August 1714, p. 358) that the original letters in the *Commercium Epistolicum* might be seen at the Royal Society.

1100   ALEXANDER MENSHIKOV TO NEWTON

12 AUGUST 1714

From the original in Babson College.[1]
For the answer see Letter 1111

Monsieur,

L'Inclination particuliere, que j'ay toujours eu pour la Nation Angloise jointe à la juste admiration, ou Elle a mis tout le monde par Sa Sagesse, par sa Valeur & par tant d'autres excellentes qualités & rares talents, m'a fait chercher avec Empressement non seulement les occasions de la Servir, mais m'a aussi donné l'envie de m'unir plus etroitement avec Elle.

Vous avez moyen, Monsieur, de me donner cette satisfaction, si Vous voulez bien me faire l'honneur de me recevoir dans l'Illustre Societé, dans la quelle Vous occupez si dignement la premiere place, honneur que je sai forte bien qu'on ne doit pas pretendre àla legere & qui doit etre attaché à beaucoup de merites, aussi je Vous assure, que j'en aurais toujours la reconnoissance düe, & que je tacheray par toutes sorts de moyens de ne vous être pas un membre inutile, c'est de quoi je vous prie de vouloir être persuadé & que je suis

Monsieur,

Votre tres-obeissante

St Petersbourg
ce. 23 *d'Aouste*
1714

Меншикофф[2]
[Menshikoff]

NOTES

(1) MS. 445, printed in Boss, p. 46. Alexander Danilovich Menshikov (*c.* 1660–1729), a man of humble birth who (as the signature to the present letter shows) could scarcely sign his own name, became, as a principal exponent of the Petrine reforms, a Prince, and after Peter's death, during the regency of his widow Catherine (February 1725–May 1727), virtual dictator of Russia. He was an able, forceful and grasping man. He had followed Peter the Great on his tour of Western Europe, worked with him in the Dutch shipyards and presumably accompanied the Czar on his visits to the Tower, where he may have met Newton (see *ibid.*, pp. 13–15, and vol. IV, p. 265).

On 29 July 1714 the Council of the Royal Society had assembled specially in order to consider a letter written by two English merchants in 'Peterburgh', Spilman and Hodgkin, on 25 June to a London merchant, Samuel Shepherd, intimating that the Prince wished to be elected as a Fellow (this letter considerately stated his official style which Newton was able to use in his answer). It was agreed that Prince Menshikov be proposed as a candidate for election, and he was elected by the Society on the same day along with four other Fellows. A Diploma and accompanying letter were drawn up on 9 October, and these on the 25th (possibly with Newton's letter of the same date) were 'delivered to Mr Sam. Shepherd in Bishopsgate Street to be transmitted to Muscovy'.

(2) The Prince has seemingly written 'Alexander' and 'Menshikoff' in Cyrillic characters one name above the other; the Christian name is really indecipherable but the characters of the family name can be made out as printed.

1101 LEIBNIZ TO CHAMBERLAYNE
14 AUGUST 1714
Extract from the printed version in Des Maizeaux.[1]
Reply to Letter 1092

Vienne ce 25. d'Aout 1714.

Monsieur,

Je vous suis obligé de la tentative que vous avez faite à la Societé Royale. L'Extrait de son Journal du 20. de Mai fait connoître qu'elle ne prétend pas que le rapport de ses Commissaires passe pour une décision de la Societé. Ainsi je ne me suis point trompé en croyant qu'elle n'y prenoit point de part. Quant à la Lettre peu polie, dont vous m'avez envoyé la Copie, je la tiens *pro non scripta*, aussi-bien que l'Imprimé François. Je ne suis pas d'humeur de vouloir me mettre en colére contre de telles gens.

Puisqu'il semble qu'on a encore des Lettres qui me regardent, parmi celles de Mr. *Oldenbourg* & de Mr. *Collins*, qui n'ont pas été publiées, je souhaiterois que la Societé Royale voulût donner ordre de me les communiquer. Car quand je serai de retour à Hanover, je pourrai publier aussi un *Commercium Epistolicum*,[2] qui pourra servir à l'Histoire Littéraire. Je serai disposé de ne pas moins publier les Lettres qu'on peut alléguer contre moi, que celles qui me favorisent; & j'en laisserai le jugement au Public . . .

Je suis avec passion, Monsieur, Votre &c.

NOTES

(1) II, pp. 128–9. Newton made a copy of this letter for Des Maizeaux's use (British Museum, MS. Birch 4284, fo. 214) differing only trivially in spelling from the printed version as far as the end of the second paragraph. Newton then placed dots to indicate an omission before 'Je suis avec passion.' However, Des Maizeaux had access to the complete letter, and printed it. It was read at a meeting of the Royal Society on 11 November 1714.

(2) Leibniz made frequent mention (see for example Letter 1109) in his correspondence of his intention of publishing a rival *Commercium Epistolicum* based on papers and letters in his own possession. Moreover, possibly shortly before his death he composed a manuscript *Historia et origo calculi differentialis*, in which he attempted to reconstruct the development of his own thoughts upon the calculus. (This was not published until the nineteenth century; see Gerhardt, *Leibniz: Mathematische Schriften*, v, pp. 392–410 and, for an English translation, J. M. Child, *The Early Mathematical Manuscripts of Leibniz* (Chicago and London, 1920); however, Child's version should be used with caution and his annotations disregarded.) In the manuscript Leibniz explains why he never in fact carried out his plan of publishing his own *Commercium*. He writes, referring to himself in the third person, (see Child, *loc. cit.*, p. 27),

. . . he was absent from home when these reports [the *Commercium Epistolicum*] were circulated by his opponents, and returning home after an interval of two years and being occupied with other business, it was then too late to find and consult the remains of his own past correspondence from which he might refresh his memory about matters that had happened so long ago as forty years previously. For transcripts of very many of the letters once written by him had not been kept; besides those that Wallis found in England and published with his consent in the third volume of his works, Leibniz himself had not very many.

1102　EDWARD SOUTHWELL TO NEWTON
1 SEPTEMBER 1714

From the copy in the Public Record Office.[1]
For the answer see Letter 1103

At the Council Chamber Whitehall
The 1st. of Septr. 1714

By the Rt. Honble. the Lords of the Committee appointed
to Consider of his Majties: Coronation

It is this day Ordered by the Committee, that the Master of his Majties. Mint, do prepare the Drafts of Gold and Silver Medals, of the Kings Majty. to be distributed on the day of his Majties. Coronation, and that He present the same to this Board on Friday Morning next at ten of the Clock.

EDWARD SOUTHWELL [2]

NOTES

(1) Mint/1, 8, p. 100.
(2) Edward (1671–1730), son of Sir Robert Southwell P.R.S., clerk to the Privy Council. He had also sat in the House of Commons and held or was to hold various other offices.

1103　NEWTON TO THE LORDS OF THE CORONATION COMMITTEE
3 SEPTEMBER 1714

From the copy in the Public Record Office.[1]
Reply to Letter 1102

To the Rt. Honble. the Lords of the Committee of Council
appointed to consider of his Majties. Coronation.

May it please yr. Lordps.

I have according to your Lordps. Order prepared Drafts of Gold & Silver Medals of the Kings Majties. to be distributed on the day of his Majties. Coronation, & herewith present the same to yr. Lordps. The form of his

Majties. face is taken from a Medall made in Germany: but Medals made there by different Gravers are not like one another.[2] The designe on the Reverse relates to his Majtys. accession to the Throne on Account of Religion & represents him Defender of the faith conteined in the Scriptures, & that the Bible is still open to the people: & according to the manner of the ancient Medals of the Greeks & Romans which are accounted the best it is grave, proper simple & free from reflections & difficult to be reflected upon [3]

All which is most humbly submitted to Your

Lordps. greate Wisdom

ISAAC NEWTON

Mint Office
3d Septr. 1714

NOTES

(1) Mint/1, 8, p. 101. There is naturally no copy of the sketch with the copy.

(2) As already noted, George I had not yet arrived in England.

(3) The design Newton describes here was not adopted. The medal as struck is shown on Plate I (facing p. 40).

1104 HOPTON HAYNES TO NEWTON
3 SEPTEMBER 1714
From the holograph original in the Mint Papers [1]

Exchequr. friday 3d. Septr. 1714

Sr

I am so engagd in bringing forward our Entries that I cou'd not give myself ye pleasure of waiting upon You. And the Invention of a design proper for a Coronation Medal is an undertaking in which few can hope for success there being no one thing wherein the tast of Mankind is nicer, & there is so little agremt. in their Opinions. In complyance however wth yr desires wch will have always the force of commands upon me, I have sent 2 or 3 designs for a Reverse gr[aving] [2] wch may suggest somthing for You to improve, & I shall be glad if even they have so much use wth You.

Reverse for Coronatn. Medals

The Legends.—

Pro Aris et Focis

⎧ The King on a Throne Crown'd, & Extending his
⎪ Sceptre towards a Britannia kneeling & Smiling
⎨ Upon him, at ye Bottom, Inauguratus   die
⎪ Octobr MDCCXIII
⎩

Jure Divino et Voto Populi	The King on a Throne, the Sun beams breaking through Clouds, & Shining upon him, & Britannia kneeling & presenting an Imperial Crown. Inaugurat. ut Supra
Statori et Vindici	A Crown'd Eagle the Sovereign Bird, coming out of the Clouds, & flying downwards, at whose Sight the Cock, an Emblem of France, expresses his fear by endeavoring to fly away, a View of Europe at ye Bottom, Inscrib'd, Liberata. Inaugurat. ut prius.
Georgio vere Divo, Cæsari Britannico	A Virgin representing Britannia in a mantel powderd wth British Lyons, in danger of being ravished by a Satyr, denoting ye pretender, who rides upon a dragon, denoting ye assistance of ye French Kg. A Hero in armour comes to her rescue, upon which she presents him wth. an Imperial Crown. Inaugurat. &c.
Exultent Cæli Lætetur Britannia	The King on a Throne, Britannia behind him, wth one hand covering him wth a Shield, & wth ye other putting a Crown on his head. before him a building of ye Corinthian Order standing on a hill to represent ye Stability of Church & State, wth a small scroll flying over it wth these words, pro Aris et Focis, at the upper pt. of ye Legend a Choir of Angels sounding trumpets on Each side of the Sun rising in glory & directing its rays on ye Kg. at ye bottom over against him ye people of Britn. etc
Jure Divino, et Voto Britanniæ	Britannia represented by a Lamb ready to become a prey to ye Wolf, ie. Rome, & to the Fox & Bear, ie. the Fr. Kg. & the Pretendr. who al retire wth fear upon the approach of the British Lyon Crown'd.

   I am afraid how these designs will succeed in Yr. good opinion but if I think of any more, I will take leave to acquaint You wth 'em.

I am

Honoured Sr.

Yr. most obliged &

most obednt. Servt.

HOPTON HAYNES.

176

NOTES

(1) III, fo. 309. Hopton Haynes (?1672–1749), had been appointed Weigher and Teller of the Mint upon Newton's very warm recommendation (see Letter 641, vol. IV, pp. 375–6) in 1701; he had worked there since about 1687. This post he doubled with another in the Tally Office of the Exchequer. He apparently did not seriously conceal (as Newton did) his unitarian beliefs, and published a number of theological writings. (Was he the 'Hopton Heinsius', *pauper puer*, matriculated at Exeter College, Oxford, in March 1684? If so, he must have been born earlier than 1672.)

(2) None of these designs was adopted; see Plate I (facing p. 40).

1105 NEWTON TO THE LORDS OF THE CORONATION COMMITTEE
6 SEPTEMBER 1714
From the original in the Public Record Office [1]

To the Rt. Honble: the Lords of the Committee of Council appointed to Consider of His Majestys Coronation.

May it please Your Lordships

In Obedience to Your Lordships Order that I should lay before Your Lordships an Acc[oun]t of the Medals made upon the last Coronation and of the time requisite to make Medals upon the present Occasion, I most humbly represent that twelve Hundred Medals of Silver and three hundred of Gold were then made by Order of Council and delivered to the Treasurer of the Household to be distributed at the Coronation, [2] and that upon her Majesty's Order Signifyed by the Lord Trea[su]rers Warrant, five Hundred and fifteen Medals of Gold were made afterwards for the House of Commons and delivered to their Speaker, and forty more were delivered to the Lord Chamberlaine for Foreign Ministers: A pound weight of fine Gold was then Cutt into Twenty Medals, and a pound weight of fine Silver into twenty and two Medals. But the Medals for foreign Ministers (except Agents & Consuls) were of double this Value. [3] At the Coronation of King William there were but two Hundred Medals of Gold made by Order of Council.

After the Form of the Medals and of His Majesty's Effigies is settled it will take up about a Calendar Month to make the puncheons, and three or four days more to make the Dyes, [4] and Coin ye Medalls by the Mill and Press. And if either of the puncheons should break as sometimes happens a fortnight more will be requisite to repair the loss. The Coinage Duty being appropriated, money should be advanced from the Civil List to buy Gold and Silver.

If the Impression is to rise high like that of the Medals made upon the late peace, they must be Coined in a Ring, and it will take up Six Weeks to make the puncheons and Dyes and Coin 1500 Medals of this Sort, or two Months if a puncheon should happen to break.[5] And the Medals must be weightier that there may be Substance to make the Impression rise high Sixteen Medals of this Sort will require a pound weight of fine Gold, and twenty a pound weight of fine Silver.

All which is most humbly Submitted to Your Lordps: great Wisdom

Is: NEWTON

Mint Office
6 *Sept.* 1714

NOTES

(1) Mint/1, 8, pp. 101–2. There is an almost identical draft in the Mint Papers, III, fo. 332, dated 4 September.

(2) See Letters 648 and 649, vol. IV.

(3) The same design was reproduced on two sizes of medal.

(4) To expedite the process, two or more pairs of dies were made from the puncheons, so that as many presses could be at work.

(5) The coronation took place six weeks and two days later, on 20 October.

1106 SHREWSBURY TO NEWTON
17 SEPTEMBER 1714

From a copy in the Public Record Office.[1]
Reply to Letters 1103 and 1105; for the answer see Letter 1108

At the Council Chamber St. James's the 10th *Septr.* 1714.

Present

Their Excellency's the Lords Justices in Council

The Right Honble: the Lords of the Committee Appointed to Consider of the Manner of His Majesty's Coronation, having this Day presented to the Board a Report from Sr. Isaac Newton Master of the Mint with the Drafts of Medals to be Distributed on the Day of His Majestys Coronation,[2] Which Drafts having been fully Considered Their Excellencys in Council were pleased to Approve of that Draft, Representing His Majestys Effigies, on one Side, with this Inscription, Georgius D.G. Mag: Britt: Fr: et Hib: Rex—And on the Reverse thereof, His Majesty in his Royal Robes, Seated in a Chair, with a Brittania Setting a Crown upon his Head; And no other Motto than in the Exergue, Inauguratus Oct MDCCXIV.[3] And to Order, as it is hereby Ordered, that the said Master of the Mint do Cause forthwith to be made and prepared such Numbers of the said Medals as Usual; And that particular Care be taken, that the Impression of the said Medals do rise high, like those Medals made upon the late Peace, And to be of such Value according to the said Draft; Which Medals are to be Distributed at His Majestys Coronation in such manner as His Majesty shall think fitt. And the Right Honble the Lord High Trea[su]rer of Great Brittain is to give the Necessary Directions herein accordingly.

EDWARD SOUTHWELL[4]

Let the Master and Worker of His Majestys Mint take Care to Cause the Number of Three Hundred Medals of Gold, and twelve Hundred Medals of Silver (being the like Number as was provided for the Coronation of Her late Majesty Queen Anne) to be forthwith provided, and made in the manner directed by the above written Order of Council. The Medals of Gold to be of the like Value respectively with those made and provided upon the late peace.[5] And the Medals of Silver to be of the like Weight with those made for the Coronation of Queen Anne Which Medals are to be delivered by the Said Master & Worker to the Treasurer of His Majestys Household, to be distributed by Him on the day of His Said Majestys Coronation, in Such manner as His Ma[jes]ty shall think fitt. And the Acquittance of the Said Treasurer of the Household Shall be a Sufficient Discharge to the Said Master and Worker. Whitehall Trea[su]ry Chambers. 17 Septr. 1714.

SHREWSBURY

NOTES

(1) Mint/1, 8, p. 103; the Treasury copy is in T/54, 22, p. 298.

(2) See Letter 1103.

(3) This was, indeed, the medal actually struck, engraved by Croker. See Plate I (facing p. 40).

(4) Compare Letter 1102.

(5) See Letter 1006, note (2), p. 11.

1107   CHRISTIAN WOLF TO LEIBNIZ
22 SEPTEMBER 1714
Excerpt from Gerhardt, *Briefwechsel zwischen Leibniz und Wolf*, p. 160[1]

Accepi (prout nuper monueram) epistolam Keilii,[2] quam opposuit Animadversionibus in recensionem controversiæ de inventore calculi differentialis ab E.T. publicatis: eam igitur ut statim communicarem, e re esse duxi. Miror hominis impudentiam, miror quoque jactantiam, cum tamen constet ex litteris a Moyvræo ad Varignonium datis (quemadmodum me certiorem

179

reddit Hermannus) ipsum non propriis, sed Newtoni armis instructum pugnare; ipsius tamen ingenio tribuenda esse judico, quæ pueriliter adversus argumentum a litteris punctatis sumtum excipit. Neque vero ego video, qui dici possit, in demonstratione quadraturæ curvarum Newtoniana calculum differentialem, qui per modum algorithmi exercetur, contineri; etsi concedam, principiis illius demonstrationis etiam locum esse in applicatione calculi differentialis ad quadraturas et alias quæstiones inde pendentes. Forsan non inutile foret, si E.T. responsionis loco veram calculi differentialis originem ostenderet et naturam ne quidem iis, qui eo quotidie cum successu utuntur, satis perspectam explicaret.[3]

Translation

As I told you recently, I have received Keill's letter,[2] which he opposes to the reflections in the account of the controversy over the invention of the differential calculus published by Your Excellency: I have concluded the matter is such that I should communicate it to you at once. I wonder at the impudence of the man; and also I wonder at his boasting, when it nevertheless appears from letters written by De Moivre to Varignon (as Hermann has informed me), that he fights not with his own weapons, but with Newton's; but I judge that that childish reasoning which he advances against the argument based upon the dotted notation is to be attributed to his own genius. Nor indeed do I see who can say that the differential calculus, which may be effected by means of an algorithm is contained in the Newtonian demonstration of the quadrature of curves; even if I should concede that there is a place for the principles of that demonstration in the application of the differential calculus to quadratures and to other questions depending upon them. Perhaps it would not be without use if Your Excellency were to show, by way of reply, the true origin and nature of the differential calculus, lest he should not explain it sufficiently clearly even to those who make a daily and successful use of it.[3]

NOTES

(1) Christian Wolf (1679–1754) had already been carrying on a considerable correspondence with Leibniz (see also Number 1009, note (1)) Wolf taught for some years at Leipzig, where he was enormously influenced by Leibniz's philosophy. In 1706 he became Professor of Mathematics at the University of Halle, where he remained until 1723 when, accused of atheistic ideas, he was deprived of his Chair.

(2) Keill's 'Answer' in the *Journal Literaire de la Haye* for July and August 1714; see Letter 1053a, note (1), p. 90.

(3) Leibniz did not publish a reply, unless we consider his brief article in the *Journal de Trévoux* as such (see Letter 1109). For Leibniz's writings on the calculus dispute see also Letter 1101, note (2), p. 173.

1108 NEWTON TO SHREWSBURY
25 SEPTEMBER 1714
From a copy in the Public Record Office.[1]
Reply to Letter 1106

To His Grace the Duke of Shrewsbury Lord High Trea[su]rer
of Great Britain

May it Please Your Grace

I most humbly beg leave to Acquaint Your Grace that fifty Medals of Gold made upon the late peace cost 168£ and at the same rate Three Hundred Medals of Gold now Ordered upon His Majestys Coronation will cost 1008£.

There were no Medals of Silver made upon the Peace for Her late Majestys Use, but the twelve Hundred made at Her Coronation took up 634 Ounces and a Quarter of fine Silver which was after the Rate of above ten pennyweight and a half to each Medal, and if there be now Allowed eleven pennyweight to each Medal, that the Impression may rise high, the Silver and Refining at 5s. 8d per ounce will amount unto 187£ and the Workmanship at 6d per Medal to 30£. In All the Charges of the Silver Medals will amount to 217£ wch being added, to the Charge of the Gold Medals, makes the whole Charge 1225£, besides the Charge of Receiving the Money which may amount unto about five pounds more. I humbly pray Your Grace that this Money may be Imprest to Me out of any Branch of the Civil List for buying the Gold and Silver and paying for the Coynage, the Coinage Duty being Appropriated.

All wch. is most humbly Submitted to Your Lordships [sic] great Wisdom

ISAAC NEWTON

Mint Office
Sept: 25: 1714

NOTE
(1) Mint/1, 8, p. 102. There is an identical draft in the Mint Papers, III, fo. 339.

1109 LEIBNIZ TO RENÉ-JOSEPH TOURNEMINE
17 OCTOBER 1714
From the extract printed in the *Journal de Trévoux*[1]

Ce n'est que depuis peu de semaines que j'ai vû le livre qu'on a publié contre moi en Angleterre; & afin que le Public tire quelque fruit de cette contestation, je pense à publier aussi un *Commercium Epistolicum*[2] sur ces matieres, & j'espere que Messieurs les Anglois me communiqueront ce qu'ils trouvent encore chez eux, & que je ne trouve point chez moi après tant d'années. Ce que je pourrai donner au Public à cette occasion servira à éclaircir les progrès de la science, & comment j'ai avancé peu à peu. Tout ce qu'on allegue en faveur de Mr. Newton ne peut valoir que pour lui attribuer l'invention de quelques *suites*; mais cela n'a rien de commun avec *le nouveau calcul* qu'il ne paroît pas avoir bien entendu, non plus que Monsieur David Gregory, & ses autres disciples en Angleterre. J'ai aussi quelque part aux *suites* mais après lui.

Autant que je me souviens de la lettre où Mr. Descartes veut rendre raison de la méthode *de maximis & minimis* de Mr. de Fermat, il n'y dit rien qui n'ait déja été connu de Mr. de Fermat & d'autres. Ce ne sont que quelques commencemens fort éloignez du nouveau calcul, qui est tout autre chose. Au lieu de prendre les grandeurs infinitesimales pour *o*, comme Monsieur de Fermat, Monsieur Descartes, & même Mr. Newton & tous les autres ont fait, avant que mon *Algorithme des incomparables* ait paru dans les actes de Leipsic; il faut supposer que ces grandeurs sont quelque chose, qu'elles different entre elles, & qu'elles soient marquées de differente maniere dans l'Analyse nouvelle. Car elles y seroient confondües, si elles étoient prises pour des zero. Je les prens donc, non pas comme des riens, ni meme pour des infiniment petits à la rigueur; mais pour des quantitez incomparablement ou indéfiniment petites, & plus que d'une grandeur donnée, ou assignable, inférieures à d'autres dont elles font les differences ce qui rend l'erreur moindre qu'aucune erreur assignable, ou donnée, & par conséquent elle est nulle.[3]

...

[*Brunsvik, le* 28. *Octobre* 1714 N.S.] [4]

NOTES

(1) Printed in the *Mémoires pour l'Histoire des Sciences et des beaux Arts* for January 1715 (Trévoux, 1715), pp. 154–5; this is often referred to in contemporary literature as the *Journal de Trévoux*. The original is in the Niedersächsische Landesbibliothek, Hanover (see Bodemann's catalogue, no. 937). René-Joseph Tournemine (1661–1739), a French Jesuit, spent much of his life teaching philosophy, theology and the humanities at Jesuit colleges. From 1701 to 1718 he took over the direction of the *Journal de Trévoux*.

(2) See Letter 1101, note (2), p. 137.

(3) Both Newton and Leibniz were groping towards a formal concept of a limit, Leibniz by the use of arbitrarily small indivisible quantities, with the geometrical tangent as their ratio (as in his first publication of his method in the *Acta Eruditorum* for 1684), and Newton by considering fluxions as instantaneous speeds. Both accused the other of an unallowable use of infinitely small quantities (cf. Letter 1053*a*, note (10)).

(4) The original letter bears this date.

1110 KATHERINE RASTALL TO NEWTON
19 OCTOBER 1714
From the original in the University Library, Cambridge[1]

Sr

I writ to you about a month a goe and fearing that my letter might not come safe to your hands makes me so bold as to troubel you with this to lett you know of my misfortun[e]s for I haveing had a great deal of illness which has been very chargable to me humbley disires that you will plleas [*sic*] to asist me in this my great nesesety for my wants is very great makes me so bold to troubel you about a fortnight sence my husband had all his goods seased by the landlord so Sr I humbley disire you that you will be pleased to give the bearer sumthing for me and she will take care to send it to me my son Thomas marryed her mother Sr Humbley beging the faver that you will be pleased to Answer this I remain Sr your humble sarvant

KATTHERN RASTALL

basingthorp [2]
oct 19 1714

    For
Sr Isaac newton in
German street near
St James Church
    London

NOTES

(1) Add. 3968(41), fo. 120v. The reverse of the sheet has been used for an anti-Leibniz draft. The writer was probably Newton's cousin Katherine Rastall, the second daughter of William Ayscough, his mother Hannah's elder brother. Her knowledge of his address was five years and two moves out of date.

(2) Bassingthorpe, in Kesteven, Lincolnshire.

1111 NEWTON TO MENSHIKOV
25 OCTOBER 1714
From the holograph draft in Babson College.[1]
Reply to Letter 1100

Potentissimo et maxime colendo Domino, Dno Alexandro Menzicoff, Romani et Russi Imperii Principi, Dno de Oranienbourgh,[2] Czarianæ Majestati

Primo a Concilijs, Equitum Stratego, Devictarum Provinciarum Dynasti, Ordinis Elephantis, Aquilæque Albæ [3] ac Nigræ Equiti &c.

Isaacus Newton salutem.

Cum Societati Regiæ dudum innotuerit Imperatorem vestrum, Czariensem Majestatem, Artes et Scientias in Regnis suis quam maxime promovere, eumque ministerio vestro, non solum in rebus bellicis et civilibus administrandis, sed etiam in bonis literis et scientijs propagandis, apprime adjuvari: maximo omnes suffusi fuimus gaudio, quando Mercatores Angli nobis significarent, Excellentiam vestram pro humanitate summa, & singulari suo in scientias affectu, & erga Gentem nostram amore, in corpus Societatis nostræ cooptari, se dignari. Atque eo quidem tempore cætibus nostris finem, pro more, donec tempestas æstiva et autumnalis præteriret, imposituri eramus. Sed hoc audito semel adhuc [4] convenimus ut Excellentiam vestram suffragijs nostris eligeremus: id quod fecimus unanimi [5] consensu. [6] Et jam, ut primum cætus nostros prorogatos renovare licuit, Electionem Diplomate sub Sigillo nostro communi ratam fecimus. Societas autem Secretario suo in mandatis dedit, ut transmisso ad vos Diplomate, Electionem vobis notam faceret.

Vale.
Dabam Londini.
xxv *Octob. Anno*
MDCCXIV

Translation

Isaac Newton greets the most powerful and honourable Mr Alexander Menshikov, Prince of the Roman and Russian Empire, Lord of Oranienburg,[2] Chief Councillor of his Caesarian Majesty, Master of the Horse, Ruler of the Conquered Provinces, Knight of the Order of the Elephant, of the White and Black Eagle, etc.

Whereas it has long been known to the Royal Society that your Emperor, his Caesarian Majesty, has furthered very great advances in the arts and sciences in his Kingdom, and that he has been particularly aided by your administration not only in military and civil affairs but also in the dissemination of literature and science, we were all filled with the greatest joy when the English merchants informed us that Your Excellency (out of his high courtesy, singular regard for the sciences, and love of our nation) designs to join the body of our Society. At that time we had concluded our meetings until the summer and autumn seasons should be past, as is our custom. But hearing of this we at once assembled, so that by our votes we might elect Your Excellency, which we unanimously did.[6] And now, as soon as it is possible to renew our postponed meetings, we have confirmed the election by a diploma under our common seal. The Society, however, has instructed its Secretary that when he has sent the Diploma off to you, he should advise you of the election. Farewell.

London, 25 October 1714

NOTES

(1) The original of this celebrated letter is not known. Three drafts were sold, together with Letter 1100, in 1936 (*Sotheby Catalogue*, Lot 145); of these (i) Babson 446, reproduced in facsimile in Boss, p. 241, is a rough draft dated 21 October 1714 differing very little from the others in substance; (ii) Babson 447 is the draft printed here—there is a facsimile of it in Boss, figure 19, and a transcription (not quite accurate) on pp. 46–7; (iii) a third fair copy was presented by the Royal Society to the Academy of Sciences of the USSR in 1942, where it is now preserved (see Boss, p. 242). This last draft has been printed (with Latin, Russian, and English texts) in a separate leaflet [?1942].

(2) Perhaps an error for Oranienbaum (now Lomonosov) near St Petersburg, though Boss, p. 46, note (2), mentions that Menshikov owned an estate of this name in the Ukraine.

(3) Previously printed 'Altæ'.

(4) 'Semel amplius' in drafts (i) and (iii).

(5) Previously printed 'unanimo'.

(6) On 29 July. It is curious that notice of this was not sent to Menshikov at once, but only when the diploma was ready.

1112 CHAMBERLAYNE TO NEWTON
28 OCTOBER 1714
From the original in King's College Library, Cambridge[1]

Petty France
28 *Octob*. 1714

Sr

I was in hopes I should have receiv'd your Commands in some manner or other by this time, I am sure I have taken Pains to do it, & now you refer me to ye R. Society where I fear I can not meet you,[2] being oblig'd to wait on his Ma[jes]ty with an Adress at Court,[3] wch with Dining with my Friends it wil I fear prevent my going to Crane Court this day again; besides I must frankly own to you, that I don't love to make a Cypher anywhere,[4] being grown a little too Bulky, not to Great for that; but I shall always be in particular to you

Sr

a most Humble Servant

JOHN CHAMBERLAYNE

Be pleas'd to return me Mr Leib:s letter by the Bearer[5]

NOTES

(1) Keynes MS. 141(D), microfilm 1011.25.

(2) Presumably Newton had suggested that Chamberlayne should meet him at Crane Court on Thursday, 28 October, the Society's regular meeting day.

(3) A Loyal Address following George I's Coronation.

(4) Chamberlayne means that the influence of Keill and others of like mind prevailed with Newton.

(5) Letter 1062.

1113 LOWNDES TO THE MINT
28 OCTOBER 1714
From the minute in the Public Record Office.[1]
For the answer see Letter 1118

Gentlemen,

The Lords Com[missione]rs of his Mats. Treasury are pleased to direct you to lay before their Lordps any [such] propositions as have been made or You shall think fit to make for Coyning halfe pence & farthings of English Copper [2]

I am &c 28th Octr. 1714

WM LOWNDES

NOTES

(1) T/27, 21, p. 296.
(2) See Letter 1034, note (2), p. 57, and Letter 1047.

1114 ANTONI VAN LEEUWENHOEK TO NEWTON
29 OCTOBER 1714

This letter was sold in a lot of 38 items, 'mostly [letters] of compliment from foreign Scholars (in Latin)', by Messrs Sotheby in 1936 (Lot 129). All of these items have vanished for the present.

This appears to be the only evidence of correspondence between Newton and the Dutch microscopist; the letter, written in Dutch, may have some connection with another letter of the same date, but written in Latin and addressed to the members of the Royal Society (see Letter book of the Royal Society of London, xv (copy), p. 182). This was one of the many letters on microscopical observations sent by Leeuwenhoek to the Royal Society.

1115 VARIGNON TO NEWTON
7 NOVEMBER 1714
From the original in King's College Library, Cambridge [1]

Equiti Nobilissimo,
Viroque Doctissimo D.D. Newtono,
Regiæ Societatis Anglicanæ Præsidi Dignissimo
S.P.D.
Petrus Varignonius.

Rure Parisios, quibus per tres menses aberam, adventanti mihi die 10. mensis hujus Novemb. N.S. redditæ sunt Clarissimi D. Moivræi Literæ 2. August. V.S. exaratæ, quibus summa cum Lætitia intellexi mei paucis antea diebus in Regiam ac Celeberrimam vestram Societatem fuisse cooptatam; quod cum te Fautore, præpotentique tua commendatione sit factam, non reperio quibus verbis gratias agam tibi pro tanto honore mihi a te curato. Video exinde quam promptus tibi desiderio meo satisfaciendi fuerit animus, quantumque tua valeat commendatio quæ (ut monet D. Moivræus) omnia mihi suffragia conciliavit. Ecquod de Geometra testimonium plus possit quam summi Geometræ D. Newtoni, cujus Ingenij profunditatem & acumen omnes, quotquot sunt in orbe, mathematici jure mecum suspiciunt & mirantur eo magis quo sunt oculatiores. Quantacunque autem sit de tua sublimi doctrina existimatio apud omnes, hac sane nemo me vincit, qui dum tua Lego, magis ac magis miror Ingenij tui Solertiam Singularem. Admirationi tui jam addo memorem animum quod benevole mihi Locum obtinuerit inter doctos ac Celebros viros quibus constat Regia Societas cui tam digne præes. Gratias igitur pro tanto Beneficio tibi quas possum maximas ago, & eo majores quod hæc tua erga me officia demerendi nunquam mihi data fuerit occasio. Vale generose Fautor, mihique Celeberrimi Nominis tui cultori devotissimo favere perge.

Dabam Parisijs die 18 Novemb. 1714. n.s.

Aliis Literis ad Clarissimum D. Moivræum missis gratias etiam ago Regiæ Societati.

Translation

Pierre Varignon sends a grand salute to the very
noble and learned Knight, Sir Isaac Newton,
most worthy President of the English Royal Society

After my return to Paris, whence I have been absent in the country for three months, the letter of the famous De Moivre written to me on 2 August O.S. was delivered to me on the tenth of this present November, by which I learned to my very great delight that a few days before I had been elected into your Royal and celebrated Society. I cannot find words with which to return you thanks for so great an honour which you have obtained for me, because it was done with yourself as my patron and upon your preponderant recommendation. From this I see how prompt was your inclination to satisfy my wish, and how powerful was your commendation which (as De Moivre advises me) won me a unanimous vote. Can any testimony concerning a geometer be superior to that of the supreme geometer, [Sir Isaac] Newton, the profundity and sharpness of whose genius all mathematicians in the world, do rightly (with myself) esteem and admire, and more in proportion to the keenness of their vision? Yet, however great the repute of your sublime learning in the eyes of all, no one outdoes me in this surely, who as I read your writings marvel more and more at the extraordinary skill of your genius. Now I add to admiration for you recollection of the kindly purpose which has secured a place for me among those learned and famous men of whom the Royal Society is composed, over which you preside so honourably. Thank you therefore for so great a kindness, with all my heart, and all the more because I have never yet been offered an opportunity of deserving these services of yours towards myself. Farewell, generous patron, and continue to bestow favour on me, a most devoted admirer of your famous reputation.

Paris, 18 *November* 1714 *N.S.*

I return thanks to the Royal Society also in another letter sent to the famous De Moivre.

NOTE

(1) Keynes MS. 142(B); microfilm 1011.26 Varignon was elected at the special meeting on 29 July, with Menshikov. This letter of thanks was read to the Society at its meeting on 2 December 1714.

1116 C. R. REYNEAU TO JONES
12 NOVEMBER 1714
Extract from the original in private possession [1]

Monsieur

. . . Jay vu avec plaisir dans ce recueil les premieres decouvertes de cet Excellent Autheur [Newton], qui seroient des chef d'oeuvre pour les autres, et comment il les a portées a leur entiere perfection. J'avois vû dans la première édition du sçavant et profond ouvrage des principes de la philosophie naturelle les belles applications qu'il fait de ses Methodes à decouvrir tout ce qu'il y a de plus caché dans la nature. La seconde édition en est encore si rare ici que je n'ay pû la voir que des instans par le moyen de ceux qui ont enlevé ce qui en étoit venu d'abord; Mais j'y ay vu avec plaisir, que ceque l'on avoit repris dans quelques journaux de lypsik [2] qu'on m'avoit montré sur la resistance des milieux n'etoit que des erreurs de calcul, comme je le jugeais en les lisant, par ce que je scavois que sa Methode de prendre les 2es differences, les 3es &c n'étoit pas celle qu'on luy attribuoit, mais celle qu'il a un peu plus expliquée dans l'endroit de la nouvelle édition où il retablit ce qui n'etoit qu'erreur de calcul dans la premiere . . .

a paris le 23e novembre 1714

NOTES

(1) Printed in Rigaud, *Correspondence*, ı, pp. 265–6. In the first part of the letter Reyneau thanks Jones for a gift of a copy of his *Analysis*, which he praises highly, and in the last part he apologizes for delaying so very long in his acknowledgement of this present.

Charles René Reyneau (1656–1728), an Oratorian, had been a successful professor of mathematics at Anger (1682–1705) before retiring to Paris, where he became (1716) an *associé libre* of the Académie Royale des Sciences. He was a particular friend of Malebranche, whose philosophical doctrines he adopted. Reyneau had already published *L'analyse démontrée, où la méthode de résoudre les problèmes de mathématiques est expliquée* (Paris, 1708), also influenced by Malebranche; his other book, *La science du calcul des grandeurs en général* appeared in this same year, 1714.

(2) *Acta Eruditorum* for February 1713, pp. 77–95; March 1713, pp 115–32, where Johann Bernoulli discusses Newton's error in Prop. 10, Book ıı of the *Principia*; see Letter 951*a*, note (2) (vol. v, pp. 349–50).

1117 TAYLOUR TO THE MINT
13 NOVEMBER 1714
From the copy in the Public Record Office [1]

Gentlemen

Sr Lambert Blackwell Knt & Edwd Gibbon Esqr. having in the Years 1710 & 1711 Advanced 60000 £ for the Service of her late Ma[jes]ty [2] there was delivered to their Order at hambrô [3] 960 Tuns of Tynn for Securing the repaymt: of the Same which said Sum being now near repaid they have by their Memoriall to the Lords Comm[issione]rs of his Mats. Treasury proposed to be Accountable for any part of the said Tinn which their Lordps shall Order them to dispose of at the rate of 4 £ per annum [4] and in the same manner as Mr Beranger is Obliged by Contract with the late Lord Treasurer I am Commanded by their Lordps to Transmitt the said Memo[ria]l to You for Your Consideracon pticularly with respect to the price the said Tynn should be sold for. I am &c 13 Novr. 1714

JO: TAYLOUR

NOTES

(1) T/27, 21, p. 307.

(2) See Letters 867 and 870, vol. v. The first-named is presumably to be identified with the 'Sir John Lambert' of vol. v, as Sir John Lambert Blackwell, grandson of the regicide.

(3) Hamburg.

(4) That is, four per cent per annum interest on the value of the tin. The memorial is not, of course, with this minute.

1118 NEWTON AND PEYTON TO THE TREASURY
15 NOVEMBER 1714

From the holograph draft in the Mint Papers.[1]
Reply to Letter 1113; continued in Letter 1119

To the Rt. Honble the Lords Commissioners
of his Majesties Treasury

May it please your Lordps.

According to your Lordps Order signified to us by Mr Lowndes his Letter of 28 October last, We humbly lay before your Lordps the following method of coyning copper money, vizt

That it be made of fine English Copper malleable under the hammer without cracking when red hot. For such Copper is free from mixture & is of about the same degree of fineness with the Swedish copper money & with copper vessels made at the battering mills.

That such copper be made into Fillets or Barrs of a due breadth & thickness either at the battering mills or at the drawing Mills & be received at the Mint upon the Master & Workers Note expressing the weight thereof: & that the Master & Worker upon delivering back to the Importer the same weight of copper in Scissel & Money together be discharged of his Receipt; the Importer at the same time paying the Master a certain seigniorage for bearing the charges of the Mint & Coynage, & the Master & Worker being accountable for the Seigniorage alone.

The Fillets imported may be assayed by heating a few of them red hot at one end & trying if they will beare the hammer without cracking. The Assays may be made by the Kings Assay-master or his Clerk or by the Smith, & all persons concerned may be present if they please.

The moneys may be assayed before delivery in the following manner. Let a Tunn of copper money (more or less) be very well mixed together, & at each of the four sides of the heap let so much copper money be counted out for a trial as should make a pound weight. And if each of the parcells counted out, makes a pound weight without the error of the weight of an halfpenny, & one or two pieces taken out of each parcel endures the assay by the hammer, then the money to be deliverable; otherwise to be returned to the Moneyers.

If the said four parcells differ not in weight from one another, above the weight of a farthing, the tale of the whole Tunn to be estimated in proportion to its weight, as the tale of all the four parcells is to their weight. And these four assays with the weight & tale of every Tunn of copper money to be entred in books. And if the money prove at any time too light or too heavy, the weight may be corrected in the coynage of the next copper imported, so as to make the whole tale of all the copper money beare a just proportion to the weight.

Two or more pieces of money may be taken out of every Tunn & put into a Pix & tried yearly by such person or persons as the Lord High Treasurer or Lords Commissioners of the Treasury shall appoint.

Mr Thomas Eyres [2] a Refiner of Copper proposes to make & size the fillets by a drawing Mill for fifteen pence per pound weight of the blanks cut out of them, whenever the price of fine copper in the market is no higher then at present, vizt 100£ per Tunn. And if a penny more be allowed to him for putting away the copper money, & four pence be added for seigniorage the whole will be answered by cutting a pound weight of copper into twenty pence. If the price of fine copper in the market rises or falls, then the price of the fillets to rise or fall as much.

Out of the Seigniorage the Master & Worker may have for himself the Graver & Smith one penny per pound weight & for the Moneyers two pence, & the remaining penny may be for bearing the charges of weighing, assaying, entring in books, making a Controllment Roll repairing the buildings buying coyning tools & putting them into repairs & buying barrells boxes & baggs to put the money into, &c.

After the coyning tools are once put into repairs the Moneyers are to keep them in repairs.

Mr Eyres hath not yet erected a drawing Mill but proposes that he can do it & be ready to deliver Fillets of Copper within the space of two months.

The charge of making the Fillets at the Battering Mills & sizing them will be more then at the drawing Mills by three half pence per pound weight of the blancks besides the charges of erecting sizing Mills: wch charges make us prefer the other method.

The buildings in the Mint where the coynage is to be performed are out of repairs.

The Proposalls of Mr Eyres & the Moneyers are hereunto annexed. The Moneyers demand 1½d per Lwt for coyining the Blanks, but in the reign of King Charles ye 2d had only 1d per Lwt. Mr Eyres demands 7 per cent for

191

putting off the copper money, but is willing to abate some thing; & we think 5 per cent, or 1*d* per Lwt sufficient.

All wch is most humbly submitted to your Lordships great wisdome [3]

[*Mint Office*   [CRA: PEYTON

15 *Nov.* 1714]   Is: NEWTON]

NOTES

(1) II, fo. 363. There are preliminary drafts at fos. 337 and 359. A minute copy, dated, and bearing the names of Peyton and Newton, is in P.R.O. T/29, 22, pp. 26–8. This is reproduced in *Cal. Treas. Books*, XXIX (Part II), 1714–15, pp. 28–9, and is virtually identical with the present text.

(2) Possibly a small manufacturer—but compare Letter 1119. The first contract for the supply of bars was awarded to Hines.

(3) When this report was considered their Lordships approved of its proposals and its preference for drawing rather than hammering the bars, but thought there was no need to allow a percentage for 'putting off' the money into circulation. They desired Newton to ask Eyres to hasten the erection of his draw-mills.

1119   NEWTON TO HALIFAX
19 NOVEMBER 1714

From the holograph original in the Public Record Office.[1]
Continuation of Letter 1118

Novem. 19. 1714

May it please your Lordp

I have spoke wth Mr Eyres, & his Answer is, that he can make the experiment in a horse-mill, & within a week will deliver into the Mint an hundred weight of barrs drawn after the manner proposed by him. But a water-mill is cheaper & goes with more strength for making dispatch with fewer draughts & less wast of the metal & so enables him to perform the undertaking at a cheaper rate.

I am

My Lord

Your Lo[rdshi]ps most humble

& most obedt servant

Is: NEWTON

Earl of Halifax [2]

NOTES

(1) T/1, 182, no. 9. The letter is endorsed 'Sir Isaac Newton—read 19th Novr. 1714. My Lords desire Mr. Eyres to go on wth all the Expedicon that may be.'

(2) Halifax had been appointed First Lord of the Treasury on 11 October 1714, and so remained until his death on 19 May 1715.

1120 SIR CHRISTOPHER WREN TO NEWTON
30 NOVEMBER 1714
From a copy in the British Museum[1]

Sr

I Present to the Royal-Society, a Description of three distinct Instruments, proper (as I conceive) for Discovering the Longitude at Sea: They are describ'd in Cypher,[2] and I desire you would, for Ascertaining the Inventions to the Rightfull Author, Preserve them among the Memorials of the Society, which in due time shall be fully explain'd by

yr Obet. Humble Servt.

CHR: WREN

NOTES

(1) Add. 25071, p. 115. The original letter is lost. This copy was found by J. A. Bennett in the MS. notes for *Parentalia; or Memoirs of the Family of the Wrens* (London, 1750) compiled by Christopher Wren Junior, who has added the heading 'On the 30th: Nov: 1714. He [Sir Christopher Wren] Presented to the *Royal-Society* a Packet seal'd up; together with this Letter to the President.' Halley confirms that it was the younger Christopher who delivered the parcel and letter to the society, although it was his father, then in his eighty-third year, who was author of the inventions. The Royal Society does not now possess a copy of the letter.

(2) They are printed by Brewster (*Memoirs*, II, p. 263) from a copy by Halley once among Newton's papers (cf. *Sotheby Catalogue*, Lot 314). A Mr Francis Williams of Chigwell, Essex, first pointed out that the message could be deciphered by reading each line backwards and omitting every third letter; thus:

(i)   WACH MAGNETIC BALANCE WOUND IN VACUO

(ii) FIX HEAD HIPPES HANDES POISE TUBE ON EYE

(iii) PIPE SCREW MOVING WHEELS FROM BEAKE

In each line the omitted letters read: CHR WREN MDCCXIV (or MDCCXIIII). (See D. Brewster, *Report of the British Association*, 1859 (London, 1860), second pagination, p. 34.) Perhaps the third invention describes a log or hodometer worked by a propeller ('screw') mounted in the bows ('beak') of the vessel.

1121 JOHN FRENCH TO THE ROYAL SOCIETY
9 DECEMBER 1714
From the original in the University Library, Cambridge[1]

To the most Hond: and Worthy Gentlemen belonging to the Royal Society: wth Humble Submission a most wonderful & Useful Secret in Nature is hereby communicated; by Experiments made and Discovered by John French.

That there is a Natural Sympathy and Concordance between the Magnetick Powers of ye Poles of ye Earth, and ye Magnetick Needle is known; though a Variation of ye Needle happens according to ye Place Situated on ye Terrestrial & Aqueous Globe.

That he has invented such an Instrument, yt with ye Use of Fire therein; it will cause a Disunion between the Magnetick Powers of the Poles of ye Earth, and that of the Needle; whereby ye Needle will swerve from ye Magnetick Meridian, in East Variation; to ye Right hand: but in West Variation, it will swerve to ye left hand of the Magnetick Meridian; so as to make an Angle wth ye Magnetick Meridian equal to the Complement of ye Latitude of ye Place then in: & will there stand while ye Fire is continued But so soon as ye Fire is Extinct; ye Needle will recurr to its Magnetick Meridian again. wch sd: angle he by a New Method Measures to a greater Acuracy than can be done by ye Divisions of a Circle of Ten times its Radius.

He has also Invented another Instrument most useful for Celestial Observations, aboundantly beyond all hitherto Done; both for Land and Sea Service. And he does also purpose; if God Gives him Life, and Ability; to perfect ye Theorax [2] of ye Moon by his constant continued Observ'ns &c. so as to render it as fit for Practice; both by Sea and Land: as that of the Sun, is at present: wch. will prove of Infinite Use and Service to Mankind. But being a Poor Man; and Reduc't; by his Great Charges he has been at; together wth his Great Loss of time for ye Publick Good; he can go no farther: having spun himself out; and no manner of Encouragement given him. He therefore Humbly prayes You would please to make ye Experiment of ye Discovery of the Latitude: and to Subscribe something toward the Building his Instruments

And Your Most Humble and Devoted Servant; &c. will ever Pray.

JOHN FRENCH

Decembr: ye 9th 1714

   To

the Honble: and Worthy Gentlemen
    of the Royal Society

(1) Add. 3972, fo. 11. The writer is stated by E. G. R. Taylor, *Mathematical Practitioners* (Cambridge, 1954), p. 431, to have printed in 1715 a *Perfect Discovery of the Longitude at Sea*, of which no copy survives. The present letter was not only kept by Newton, but mentioned by him later.

(2) Perhaps a variant of 'theorics' (=theory), though *Oxford English Dictionary* states that this word was obsolete long before 1714.

1122   LEIBNIZ TO J. BERNOULLI
19 DECEMBER 1714
Extract from Gerhardt, *Leibniz: Mathematische Schriften*, III/2, pp. 934–5.
Reply to Letter 1075; for the answer see Letter 1129

Multi illic Viri præclari adulatorum Newtoni temeritatem parum probant. Cl. Camberlanus (celebris Statu Magnæ Britanniæ novissime edito) cum literas quasdam ad se meas in conventu Societatis Regiæ ostendisset, quibus male mecum actum esse querebar, misit ad me excerptum ex Protocollo Societatis ea die habito, quibus negant litem a se decisam, neque viam esse præclusam ajunt ei, qui aliquid relationi Commissariorum (male proinde pro sententia Societatis editæ) opponere velit. [1]

At Keilius interea Gallicum quemdam libellum edidit, quibus inter alia errorem, quem in Newtono deprehenderas, removere conatur frustra. [2]

Mihi consilium est, edere aliquod Commercium literarium meum, unde apparebit, quam in aliis quoque Newtonus olim tenuis fuerit. [3] Et quia video, ex Collinsianis homines partium studio deditos ea suppressisse, quæ Newtono minus placere debere judicabant, sollicitabo ut ea ipsa quoque producant. Dabo etiam operam, ut quædam edam, in quibus Newtono aquam hærere scio.

Fieri potest, ut Gregorius etiam invenerit seriem, qua arcus circuli exprimitur per tangentem, quamquam meas schedas (quarum pars vel periit vel latet) nondum omnes excutere licuerit: nec putem .... homines quantumvis iniquos falsificare literas Gregorii vel Collinsii, sed arbitror mihi non fuisse communicatam hanc Gregorii seriem, nisi tunc cum meam jam in Angliam misissem. Itaque cum varias series Gregorianas et Newtonianas acciperem, credo me judicasse hanc inter cæteras post meum initium relatam. [4] Certe Newtonus ipse in suis literis ejus inventionem tum mihi tribuit. [5] Itaque de Gregorio inventionis participe nulla ad me suspicio pervenit. Ubi vacavit hæc minutatim excutere, inspiciam etiam, an Keilius in sua scheda olim per Secretarium Societatis jam correctum dederit errorem Newtoni. [6]

Translation

Many distinguished men there [in England] do not at all approve the boldness of Newton's toadies. Mr Chamberlayne (celebrated for the recently published *State of Great Britain*), after circulating some of my letters to him at a meeting of the Royal Society (in which letters I complained of their ill usage towards me), sent me an extract from the Minutes of the Society for that day in which they deny that they had taken a decision on the dispute or that the way was barred (so they say) to any person wishing to oppose the report of the commissioners (which is thus wrongly published as the judgement of the Society).[1]

But Keill has meanwhile published a certain pamphlet, in French, in which among other things he endeavours vainly to remove the error you detected in Newton.[2]

I am resolved to publish some correspondence of my own, from which it will appear how weak Newton once was in other respects.[3] And as I see from that of Collins that men, given over to partisan zeal, have suppressed those points which they thought less agreeable to Newton, I shall urge them to bring these matters out too. I shall also make it my business to publish certain things which I know look bad for Newton.

It may be that Gregory also discovered the series in which the arc of a circle is expressed by the tangent although it has not yet been possible to go through all my papers (of which a part is destroyed or hidden), nor can I think that . . . [these] men, however iniquitous, would falsify the letters of Collins and Gregory, but I conclude that this series of Gregory was only communicated to me after I had already sent mine to England. Thus as I received various series of Gregory and Newton, I believe that I judged this one among the rest to have been reported after my own beginning.[4] Certainly Newton himself then assigned the discovery of it to me in his letters.[5] And so no suspicion of Gregory's share in the discovery came to me. When I have leisure to go into this thoroughly I will examine also, whether Keill in his paper formerly [transmitted] by the Society's Secretary gave Newton's mistake as already corrected.[6]

NOTES

(1) See Letter 1092.

(2) This is, of course, Keill's 'Answer' in the *Journal Literaire de la Haye* for July and August 1714, pp. 319–58, which Leibniz as yet knows only from Chamberlayne's letter.

(3) See Letter 1101, note (2), p. 173.

(4) Leibniz seems determined to confuse the issue of Gregory's prior knowledge of the series (which is dependent solely on the authenticity of Gregory's letter to Collins of 15 February 1670/71) with the question of his awareness of Gregory's series before sending his own to London. In fact Collins *had* sent over Gregory's series (in Oldenburg to Leibniz of 15 April 1675) before Leibniz had made his own series for $\pi/4$ known, though after he had discovered it in late 1673 (see Hofmann, pp. 34–5).

(5) In the *Epistola Posterior* (24 October 1676) Newton praises the mathematical ingenuity displayed in Leibniz's letter of 17 August in which he expressed the $\pi/4$ series for the first time (see vol. II, pp. 110–11); later (p. 122) he refers to it as 'Seriem Leibnitij' and shows no inclination to ascribe its prior discovery to Gregory.

(6) Presumably Letter 843*a*, vol. v.

1123 NEWTON TO THE TREASURY
[END OF 1714]
From a holograph draft in the Mint Papers.[1]
For the answer see Letter 1127

To the Rt Honble the Lds Comm[issione]rs of his
Mats Trea[su]ry

May it please your Lordps

It being usual to place the Kings Arms upon the the [*sic*] reverses of the
larger species of the moneys: I humbly pray your Lordps that so soon as his
Mats Arms are setled, I may have such Orders or Directions as are proper, for
preparing Draughts of the several species of the moneys to be affixed to a
Warrant under the sign manual for coyning the same. All wch is most humbly
submitted &c.

NOTE

(1) III, fo. 276. Presumably this letter preceded both Letter 1127 (the warrant in question)
and Newton's Letter 1129; it may have been written in the autumn of 1714, but we place it
here before related and dated documents. Compare Letter 1096.

1124 NEWTON TO ——
[c. END OF 1714]
From a draft in the University Library, Cambridge[1]

Sr

I received your Letter with the Paper wch Mons[ieu]r de Bernstorf[2]
ordered you to send to me, & desire you to acquaint him that the Longitude is
already found at Land by Astronomy & good Clockwork together, & the
method is practised every-day with success for rectifying Geography: & I have
been many years of opinion that there is no other way of finding it by sea than
by pursuing & improving this method,[3] [This has been my opinion above
these thirty years & I reported it to the Committee of the House of Commons
who summoned me to attend them about this matter & I can make no other
report then what I have made already for I am too old to change my opinions
or to take these matters into fresh consideration.] And having reported this
opinion I am incapable of meddling with the Paper which you sent me it
being contrary to my report. But if other Gentlemen have a mind that it
should be examined, I will not oppose them.

NOTES

(1) Add. 3968(34) fo. 477v. This is the second of three considerably differing drafts on the same sheet. All three are discouraging, and the third is particularly terse:

Sr

I received the Proposals wch Mon[sieu]r de Bernstorff desired you to send to me, & beg the favour of you to acquaint him that I do not approve of them & desire that the person who makes them may not be sent to me. I am

Material on the verso was written after July 1716, but this proves nothing as to the date of this draft. We place it here arbitrarily.

(2) Possibly Andreas Gottlieb, Baron von Bernstorff (1649–1726), first minister of the Elector of Hanover, who accompanied him to England in October 1714; or perhaps his cousin (?) Joachim Engelke, Baron von Bernstorff, chamberlain to the Elector-King.

(3) This is a most interesting declaration which would seem, effectively, to restrict Newton's hopes to the method of lunar distances later developed by Nevil Maskelyne.

1125 NEWTON TO THE TREASURY
31 DECEMBER 1714
From the copy in the Public Record Office[1]

To the Right the Lords Comm[issione]rs of His Majesty's Treasury.

May it Please Your Lordships

I understand that the Money is Intended to be Coined according to the Forms expressd in the Annexed Draughts. The Inscriptions, the Armes of Hannover, and the Crown above that Escutcheon, are drawn only in black Lead, that if any thing be amiss, it may be wiped out & amended without Spoiling the Draughts. And because these parts of the Draught may be wiped out and changed, it will be convenient that they be described in Words in the Warrant for Coyning the Moneys: To wch Warrant these Draughts are to be annexed if they be approved. If these Draughts are to be amended, or others made to be laid before the King in Council it shall be speedily done. The Warrant uses to be upon Order of Council and may express that the five pound peices, the forty Shilling peices, the twenty Shilling peices, and the Ten Shilling peices of Gold, be Coined after the Forms depicted in the two uppermost Figures, the Crowns, Half Crowns, Shillings and Sixpences after the Forms depicted in the two figures next below, and the Groats, threepences, twopences and pence after the Forme of the larger Silver Moneys on the Head Side with this Inscription GEORGIUS DEI GRATIA, and on the Reverse after the Forms depicted in the four Figures below wth the Numbers 4: 3. 2. & 1. crowned, and this Inscription: MAG: BRI: FR: ET HIB: REX &c 1715.[2] If in any of these Draughts the Work prove too much Crowded, it may be remedied hereafter by a new Warrant. For I fear that the Arms of Hannover will scarce be distinct upon the Half Guineas and Sixpences.[3]

All wch: is most humbly submitted to Your Lo[rdshi]ps: great Wisdom.

Is: NEWTON

Mint Office Dec: 31: 1714

NOTES

(1) Mint/1, 8, 105. There is a holograph signed draft in the Mint Papers, III, fo. 285 differing only trivially from this record copy, but dated 30 December. We have not traced the 'draughts' of the proposed coinage, which presumably went to the engravers.

(2) Newton's proposals were approved in an Order of Council of 5 January 1714/15, transmitted from the Treasury to the Mint on the seventh (P.R.O. T/54, 22, p. 429) with the difference that 'the Letters FD for Fidei Defensor' were to be added to the inscription.

(3) See Letter 1128.

1126 LOWNDES TO THE MINT
6 JANUARY 1715
From the copy in the Public Record Office[1]

Gentln

By Order of the Lords Comm[issione]rs of His Mats Treasury I send You the inclosed Account wch their Lordps have received from Mr Anstis[2] Shewing what money hath arisen by ye sale of Tin between Xmas 1713 & Xmas 1714 And what Tin hath been made in Devon and Cornwal in the four last Quarterly Coynages and ye Charges thereupon Their Lordps are pleased to direct You to consider ye said Acco[un]t & let them know if ye same be truly stated or not I am &c

6 *Jany* 1714

WM LOWNDES

NOTES
(1) T/27, 21, 331. There seems to be no written reply extant.
(2) For John Anstis see Letter 907, vol. v.

1127 THE EARL OF SUFFOLK TO THE MINT
10 JANUARY 1715
From the copy in the Public Record Office.[1]
Reply to Letter 1123

These are to Signyfye unto You His Majestys Pleasure That the Royal Armes to be Impressed on His Majestys Coin or Money as well Gold as Silver are to be Marshalled as the Same are Delineated in the Draughts hereunto Annexed which were approved of by the King in Council, the 6th: of December last, Since which His Majesty having been pleased to Direct the Words F.D. for Fidei Defensor to be added at the End of the Circumscription on the Head Side, and 1715 instead of 1714 on the Reverse, [2] You and the Other Officers of His Majesty's Mint now, and for the time being, are to take due Notice thereof and Conform to His Majesty's pleasure therein Accordingly; And for so doing this Shall be Your Warrant. Given under my Hand and Seal the 10th. day of January in the First Year of the Reign of Our Sovereign Lord, George by the Grace of God King of Great Britain &c. Annoque Domini 17$\frac{14}{15}$.

SUFFOLK. M.

To the Master, Warden Comptr
and other Officers of His Mats: Mint.

NOTES

(1) Mint/1, 8, p. 106. Henry Howard (1670–1718), Earl of Suffolk, had been appointed a Privy Councillor in October 1714. He was also Deputy Earl Marshal from 1706 and held a Court of Chivalry at Westminster on 26 April 1707.

(2) The last Order of Council relating to the coinage had, in fact, been made on 5 January and conveyed to the Mint by Halifax as First Lord of the Treasury (Mint/1, 8, pp. 104–5), with the usual instruction 'Let his Majesty's pleasure be duly complied with'. Two designs for the gold coins and two for the silver had been approved. The Order goes on to specify that pieces of one penny and twopence be made: 'after the Form of the larger Silver Moneys. on the Head Side with this Inscription; Georgius DEI GRATIA And on the Reverse after the Forms in the four Figures with the Numbers 4. 3. 2. 1. Crowned, and this Inscription. MAG. Bri: Fr: ET Hib: REX &c with the Date of the Year: And that there be added the Letters F:D. for Fidei Defensor.' The Warrant for the Coinage is Number 1135 below.

1128 NEWTON TO THE TREASURY
17 JANUARY 1715
From a holograph draft in the Mint Papers.[1]
Continuation of Letters 1123 and 1125

To the Rt Honble the Lords Comm[issione]rs of
His Ma[jes]ties Treasury

May it please your Lordps

Having in a late Memorial signified to your Lordps that the Arms in the fourth Escutcheon on the Reverses of the half Guineas & sixpences may prove too much crouded: for removing the inconvenience I humbly present your Lordps with the Designes hereunto annexed for those Reverses, [2] the same being the chief parts of his Majties Arms proposed by the Earl Marshal & approved of. And I humbly propose that the same may be laid before his Ma[jes]ty in Council for his approbation if it be thought fit. All wch is most humbly submitted to your Lordps great wisdome

Mint Office
17 *Jan.* 171$\frac{4}{5}$

NOTES

(1) III, fo. 286; there are rough drafts at fos. 275 and 280.

(2) The coin designs (see Plate III, facing p. 216, top) are at III, fo. 286*a*. The newly-designed George I silver coins were all to bear the arms of the Electorate of the House of Brunswick in the fourth (centre left-hand) escutcheon. Newton's suggestion that these arms be split up on the smaller coins to reduce congestion was not adopted (see Plate III, bottom). The inscription on Newton's design—BRUN. ET. L. DUX. S.R.I.A. TH. ET. PR. EL [Brunsvicensis et Lunenburgensis Dux, Sacri Romani Imperii Archi-Thesaurius et Princeps Elector] appeared only at the 1714 minting; thereafter PR[inceps] was omitted. (See also Letter 1135.) 1714 guineas are thus known as Prince Elector Guineas. (See H. A. Grueber, *Handbook of the Coins of Great Britain and Ireland in the British Museum* (London, 1899; reprinted 1970), pp. 142–4 and plate XXXV.)

1129 J. BERNOULLI TO LEIBNIZ
26 JANUARY 1715
Extract from Gerhardt, *Leibniz: Mathematische Schriften*, III/2, pp. 936–7.
Reply to Letter 1122; for the answer see Letter 1138

Facile credam, quod postquam Serenissimus Tuus Princeps Magnæ Britanniæ thronum conscendit, Societas Anglicana nolit jam haberi pro sua sententia, quod tamen ejus auctoritate et nomine prodiit in Commercio Epistolico, tanquam Iudicis sententia decisoria pro Newtono. Fortassis etiam Keilius suum libellum Gallicum [1] (quem ante Reginæ obitum publicavit) non fuisset publicaturus, si præsensisset optatam illam, quæ paulo post contigit, conversionem rerum Britannicarum.

Consultissimum erit, quod formasti consilium, edere aliquod Commercium Literarium alteri illi Anglico opponendum. [2] Cum enim Angli omnia per literas et rerum gestarum historias evincere contendant, de quibus autem nobis non constat qua fide referantur, æquum est, ut et Tua producas ad Commercium illud pertinentia, ab adulatoribus Newtonianis omissa aut dolose suppressa, quod Newtono minus favere vel placere posse judicaverint. Optarim libenter videre libellum illum Gallicum Keilii; nihil ex eo vidi, quam quod nuper noster Hermannus perscripsit, excerptum ex Diario Hagiensi mensis Julii et Augusti 1714 maximam partem ad me spectans, nisi quod

dicat, Keilium Te parum honorifice tractare, pariter ac Auctorem vel Auctores illius Epistolæ, quæ Apologiæ loco pro Te in lucem prodiit. Tuum
Schediasma quoque circa motus Planetarum, Actis Lipsiensibus mensis
Februarii 1689 insertum, sub examen revocari, in quo Keilius duos paralogismos notare moliatur.[3] Ad me vero quod attinet, agnoscit quidem errorem,
quem detexi in Newtono circa determinationem resistentiæ commissum (vide
Schediasma meum in Actis Lipsiensibus mensium Februarii et Martii 1713)
sed multis persuadere conatur Keilius, me multo gravius lapsum esse, quod
erroris Newtoniani originem rejicere voluerim in series ipsius, cujus terminos
a Newtono pro fluxionibus vel differentialibus superioribus adhibitos esse
Keilius contra me negat, licet Newtonus alicubi disertis verbis hoc dicat, quem
locum, si opus fuerit, ostendere possum,[4] quidquid nunc dicat Keilius, vel
Newtonus ipse aliive Cultores ejus, ad dissimulandum quod veram differentiandi continuationem ignoraverit. Sed de ipso, quem indicavi, errore Newtoni
mollissime loquitur Keilius, dicendo eum ex accidente irrepsisse,[5] producta
aliqua tangente ad partem oppositam ei, ad quam produci oportuisset, et ideo
facile fuisse errorem ejus indolis committere. Interim non addit, quod hic
error, quantumvis facilis, mansisset incorrectus in nova Editione Principiorum
Phil. si Newtonus de eo non fuisset opportune monitus ab Agnato meo in
Anglia tum degente, cum liber jam esset prælo evasurus. De reliquis quibus-
dam erroribus, quos pariter notavi, nihil dicit Keilius. Cæterum mihi lepida
videtur excusatio Keilii, dicentis errorem commissum esse per accidens,
producendo lineam in plagam non debitam, sed debitæ omnino contrariam;
quid hoc aliud est, quam dicere errorem, qui diametraliter pugnat cum
veritate et ab ea quam longissime abest, esse errorem accidentalem? Tandem
me hortatur Keilius, ut justitiam faciam Newtono circa series, et ut publice
agnoscam me in erroris Newtoniani origine indicanda errasse, quanquam
Newtonus ipse tantum æquitatem non habuerit, ut in nova Libri sui editione
fateretur, a quo erroris sui commonefactus fuerit, eum nempe ita corrigens,
quasi a nemine monitus sponte lapsum suum animadvertisset.

Recte facies, si quædam edas, in quibus Newtono aquam hærere scis.[6]
Suppetunt haud dubie multa eorum, quæ olim inter nos agitata fuere, et quæ
per communem differentialium methodum non facile obvia sunt: qualia sunt
quæ de transitu ex curva in curvam habuimus, quæ peraguntur singulari
quadam differentiatione adhibita. Proposui olim, si meministi, publice
quædam problemata hujus generis, sed quæ pro curvis dissimilibus a nemine
hactenus fuerunt soluta; ea ipsi de novo possent proponi. Ex. gr.[7] inter
infinitas Ellipses super communi axe AC descriptas quæritur illa ABC, quæ
secans rectam positione datam, vel quamcunque etiam curvam datam LBN,
relinquat arcum AB inter omnes AR, AS etc. minimum longitudine. Aliud

problema foret, sed difficilius: Iisdem positis, determinare Ellipsin *ABC*, cujus arcus *AB* a gravi ex *A* descendente brevissimo tempore percurratur. Hujusmodi multa alia sunt, quæ Anglis forte plus negotii facessent, quam sibi statim imaginabuntur. Ex eorum numero, quæ singularem differentiandi et integrandi methodum requirunt, etiam hoc esset, ubi petitur modus complanandi superficies conoideas obliquas, vel quasvis alias datas superficies curvas, etsi non

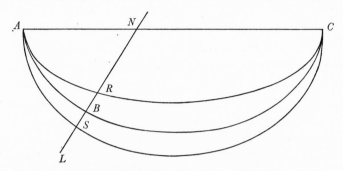

conoideas. Intelligo per superficiem curvam datam, cujus singula puncta determinantur (sic ut lineæ curvæ datæ puncta) per ordinatas tres *x*, *y*, *z*, quarum relatio data æquatione exprimeretur: sunt autem tres illæ coordinatæ nihil aliud, quam tres rectæ ex quolibet superficiei curvæ puncto perpendiculariter ductæ in tria plana positione data, et se mutuo ad angulos rectos secantia. Sit æquatio inter coordinatas ex. gr. hæc $xyz = a^3$, quaeritur hujus superficiei dimensio, vel saltem reductio ad figuram aliquam planam. Quod attinet ad solidum ipsum inter superficiem datam et plana positione data comprehensum, ejus quidem dimensio facilius habetur, quam superficiei, non tamen omni difficultate caret. Vale etc.

Basileæ a. d. 6. Februar. 1715, [N.S.]

Translation

I can easily believe that, after Your Most Serene Prince [George I] ascended the throne of Great Britain, the English [Royal] Society no longer wished that to be taken as its opinion which it nevertheless published in the *Commercium Epistolicum* by its own authority and in its own name as a decisive opinion of a judge in favour of Newton. Perhaps even Keill would not have wanted the French booklet [1] (which was published before the Queen's death) to be made public, had he had a presentiment of that welcome change in British affairs, which happened a little while after.

It will be wisest if you plan to edit another *Commercium Literarium* opposing that of the English. [2] For as the English try to prove everything by letters and narratives of events, of which, however, we are not certain how faithfully they are recounted, it is fair that you should disclose yours pertaining to that *Commercium*, left out by Newton's toadies or craftily suppressed, because they judged them possibly less favourable and pleasing to

Newton. I very much want to see Keill's French booklet: I have seen nothing of it, except what our countryman Hermann recently wrote about it, [being] an extract from the Hague *Journal* [*Literaire*] for July and August 1714, the greatest part relating to me, except that he says that Keill uses little respect in his treatment of yourself and also of the writer or writers of that letter which has been published by way of an apologia on your behalf. He [Keill] also scrutinizes that paper of yours on the circular motion of planets, inserted in the *Acta* [*Eruditorum*] of Leipzig for February 1689, in which Keill exerts himself to point out two errors.[3] As far as I am concerned he does indeed admit the error which I have detected as committed by Newton when dealing with the determination of resistance (see my paper in the *Acta* [*Eruditorum*] of Leipzig for February and March 1713), but Keill endeavours to persuade [his reader] by many considerations that mine is much the greater mistake, because I sought to refer the origin of Newton's error back to his series, whose terms Keill denies (contrary to myself) to have been used by Newton for fluxions or higher order differentials, although Newton says this somewhere in express terms; if it were necessary I could point out the place,[4] whatever Keill, or Newton himself or perhaps others amongst his supporters, says to hide the fact that he was ignorant of the true method of successive differentiation. But Keill speaks very softly concerning that same error of Newton's which I pointed out,[5] saying that it crept in accidentally, a certain tangent having been produced in the direction opposite to that in which it ought to have been produced, and so it was easy to make a mistake of that kind. However, he does not add that this error, be it ever so simple, would have remained uncorrected in the new edition of the *Principia Philosophiæ* if Newton had not been given timely warning of it by my nephew, then staying in England, when the book was about to come off the press. Keill says nothing about some other errors which I likewise noted. For the rest, Keill's defence seems amusing to me, saying as he does that the error was committed by accident, a result of producing a line in the wrong direction, indeed in quite the opposite direction. What is this, other than to say that an error, which diametrically contradicts the truth, and is as far as possible from it, is an accidental error? Finally, Keill exhorts me to do Newton justice concerning series, and to acknowledge publicly that I was mistaken in announcing the source of the Newtonian error; even though Newton himself has not been so candid as to admit, in the new edition of his book, by whom he was advised of this error, correcting it (I mean) in such a way as if (without any warning from anyone) he had perceived his mistake independently.

You would do well to publicize some [problems] where Newton would, as you know, find himself in difficulties.[6] Doubtless there are many of them to hand, which were once discussed between us, and which are not easily dealt with by the ordinary differential method: of this sort are those we considered concerning the transformation of one curve into another, problems which are dealt with by employing a certain particular method of differentiation. I once proposed publicly, if you remember, certain problems of this type, which have not hitherto been solved by anyone for dissimilar curves; so that they can be proposed anew to him. For example,[7] amongst an infinite number of ellipses described upon a common axis *AC*, it is required to find the one *ABC*, which, cutting a straight line whose position is given or even any given curve *LBN*, leaves an arc *AB* of minimum length among all [possible arcs] *AR, AS* etc. Another problem, more difficult, might be: the same things being assumed, determine the ellipse *ABC*, the arc *AB* of which is traced out in the least time by a heavy body descending from *A*. There are many others of this type which would keep the English much more busy than they would at first imagine. Among these [problems] which require a particular method

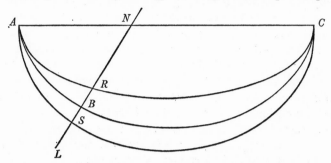

of differentiation and integration there would also be that in which a means of finding the area of oblique conoidal surfaces (or of any given curved surface, not conoidal) is sought. I understand by a given curved surface one of which individual points are determined (in the same way as are points of a given curved line) by three ordinates, *x, y, z*, the relationship between which would be expressed by a given equation: however, these three ordinates are nothing other than three straight lines drawn perpendicularly from any point of the curved surface to three given planes, intersecting mutually at right angles. For example, if the equation between the co-ordinates be $xyz = a^3$, it is required to find the area of this surface, or at least the area reduced to some plane figure. As far as a solid which is contained between a given surface and a given plane is concerned, its volume is more easily determined than its surface area, but still not entirely without difficulty. Farewell, etc.

Basel, 6 February 1715 [N.S.]

NOTES

(1) The separately published form of Keill's 'Answer'; see Letter 1092.

(2) Leibniz, despite Bernoulli's and Wolf's continued encouragement, never in fact published his own *Commercium*. See Letter 1101, note (2), p. 173.

(3) The paper referred to is Leibniz's 'Tentamen'; see Number 1069a. In a passage in his 'Answer' (see *Journal Literaire de la Haye* for July and August 1714, pp. 350–2) Keill turned Leibniz's accusation that Newton did not understand second differences back upon Leibniz himself, claiming that an error in the 'Tentamen' was a result of Leibniz's own misunderstanding of second differences. He added that in Leibniz's second version (*Acta Eruditorum* for October 1706, pp. 446–51) whilst correcting this mistake he committed others. See also Number 1086a, Keill's draft of the passage.

(4) Bernoulli refers to a passage in Newton's *De quadratura curvarum* where in the scholium following Proposition 11 the type-setter, (or perhaps Newton) had left out a vital 'ut'. (See Number 1053a, note (13)).

(5) Bernoulli refers, yet again, to the error in Book II, Prop. 10 of the *Principia*, and his own inapposite criticism of it; (see Letter 951*a*, vol. v).

(6) Leibniz eventually acted on Bernoulli's suggestion, although he did not use any of the problems Bernoulli mentions here (see Letter 1170, note (3)).

(7) The problems Bernoulli mentions here are largely drawn from those which we would now treat using the calculus of variations—they involve the choosing from amongst a given set of functions the one which satisfies certain given constraints. Three major problems of this type had been discussed by the Bernoullis and others at the end of the seventeenth century; (i) the geodesic problem of finding the curve of minimum length passing between two given points on a curved surface—the set of available functions is determined by the shape of the surface, and the constraint imposed is the minimization of the length of the curve; (ii) the brachistochrone problem, that is the determination of the line of quickest descent between two given points; (iii) isoperimetrical problems, that is the finding of the maximum area, on any surface, which may be enclosed by a line of given length. The first two problems Bernoulli mentions here clearly share many characteristics with these problems, and he had indeed set a number of problems of this type in 1697; see Letter 1138, note (3), p. 214. Their difficulty is of two kinds: first that of finding a suitable method of approach, and, second, that of completing the necessary integrations. (See also M. Cantor, *Vorlesungen über Geschichte der Mathematik*, III, Leipzig 1898, pp. 224–36.) The problems of surface integration which Bernoulli then mentions were probably inspired by his own development of the representation of a surface by an equation in three co-ordinates, and he rightly points out that surface integrals are, in general, more difficult than volume integrals.

1130 WOLF TO LEIBNIZ
FEBRUARY 1715

Extract from Gerhardt, *Briefwechsel zwischen Leibniz und Wolf*, pp. 161–2.
For the answer see Letter 1136

Cum Keilius in epistola responsoria Diario Hagiensi inserta, quam mature circa festum D. Michælis [1] communicavi, autores schedarum Commercio epistolico oppositarum ignorantiæ arguat, quasi nescirent inter rem et signa distinguere, nec argumenta Anglorum discutere valerent; vereor sane ne, qui argumenta ipsa examinare valeant, criminationibus hisce fidem habeant et ex silentio concludant, argumenta Anglorum revera invincibilia esse. Unde novi optare etiam alios, ut plenius respondeatur et expressius, præsertim cum etiam nonnulli, quorum non postrema est in Mathematicis auctoritas, argumenta Anglorum ad speciem composita esse videantur.

Halæ Saxonum Febr. 1715.

Translation

Since Keill, in the letter of reply printed in the Hague *Journal* [*Literaire*], which I at once communicated [to you] about Michaelmas, [1] charges the authors of the papers against the *Commercium Epistolicum* with ignorance, as if they did not know how to distinguish between realities and symbols nor were equipped to discuss the arguments of the English; I fear lest those who are able to examine the arguments themselves may have credence in these charges, and may conclude from [your] silence that the arguments of the English are truly insurmountable. Whence I have learned that others too wish that it should be more fully and directly answered, especially as some things which are not of the least authority in mathematics seem to have been compounded with the arguments of the English for appearance' sake.

Halle of the Saxons, February 1715

NOTE

(1) That is, 29 September N.S. = 18 September O.S.; see Letter 1107.

1131 TAYLOUR TO THE MINT
9 FEBRUARY 1715
From the copy in the Public Record Office [1]

Gentlemen

The Lords Commissioners of His Mats Treasury Command me to Acquaint You that the East India Company have agreed to buy 150 Tons of the Tyn in Your Custody at the Current Price in Order to dispose thereof in the East Indies, And their Lordps are pleased to direct, that You give all possible dispatch to the delivery thereof when it shall be required I am. &c 9th Febry 1714

J TAYLOUR

NOTE

(1) T/27, 21, p. 343. We include this note for completeness' sake although Newton seems not now to have been much concerned with the tin affair (but see Letter 1152); the rights in tin now belonged to the new Prince of Wales (as Duke of Cornwall) and on 1 June 1717 the Crown's system of purchase was abolished (Craig, *Newton*, p. 60).

1132 LOWNDES TO THE MINT
8 MARCH 1715
From the original in the Public Record Office. [1]
For the answer see Letter 1133

Whitehal Trea[su]ry Chambers 8th March 171$\frac{4}{5}$

Honble the Lords Comm[issione]rs of his Mats. Trea[su]ry are pleased to Referr this Petition to the Warden, Master and Worker, and Comptroller of his Mats Mint in the Tower of London who are to consider the matter therein contained and report to their Lordships what they think proper to be done therein. [2]

WM LOWNDES

NOTES

(1) T/1, 188, no. 52.

(2) The petitioner, Richard Barrow, claims in the following petition that he was deputed by Craven Peyton, late Warden of the Mint, 'to prosecute all Coyners and Utterers of False Money'; further details are sufficiently apparent from Letter 1133. Barrow was successor to Robert Weddell, deceased (compare Letter 990, vol. v, and Letters 1060 and 1065).

Peyton ceased to be Warden in December 1714, when Sir Richard Sandford was appointed his successor (*Cal. Treas. Books*, XXIX (Part II), 1714–15, pp. 208–9).

1133 NEWTON AND BLADEN TO THE TREASURY
9 MARCH 1715
From the original in the Public Record Office.[1]
Reply to Letter 1132

To the Rt: Hono[ura]ble the Lords Commiss[ione]rs
of His Majestie's Treasury

May it Please Your Lordshp's:

According to your Lordshp's. Order of Reference of the 8th instant, We have considered the Matter contained in the Petition of Mr. Richd. Barrow who about three years ago was by Mr. Peyton then Warden of the Mint Deputed to prosecute Clippers Coyners and Utterers of False Money, and that a Bill of Charges for such Services for about a year & three Quarters ending at Mich[ael]mas 1713 was in December following Referred to the Officers of the Mint which after some Abatements made by them came to the Sum of 387£. 14s 7d as they then represented in their Report. And We are humbly of Opinion that the said Bill be paid off out of the 400£ per Annum made applicable to this Service by Act of Parliament.

And We are further of Opinion that Mr. Barrow continue to Act in prosecuting Sarah Harris (mentioned in his Petition) and other Offenders untill the present Warden of the Mint return to London,[2] and that his Charges in prosecuting since Mich[ael]mas was a Twelve month be then Stated and paid off, and the Method of prosecuting and paying the Charges thereof be then settled upon the same Foot as in the Lord Treasurer Godolphins time or upon a better

All which is most humbly Submitted
to Your Lordshps. great Wisdom

Mint Office   Is. NEWTON
9th March 1714⅘
                                                MARTIN BLADEN

NOTES

(1) T/1, 188, no. 52.

(2) Sir Richard Sandford (1675–1723), third baronet, who, with Martin Bladen the new Comptroller (and signatory to this letter) took the oath of office on 30 March 1715 (*Cal. Treas. Books*, XXIX (Part II), 1714–15, p. 258). Sandford came from one of the oldest families in Westmorland and sat in parliament from 1695 to 1723. He was turned out of his office in 1717 for voting against the government. Martin Bladen (?1680–1746) had served in the war, selling his commission in 1710. Appointed Comptroller in December 1714 he was to retain this office until 1728, serving also as a highly energetic member of the Board of Trade and as a frequent defender of the government in the Commons. He was prominent in the West Indies sugar interest and a director of the Royal Africa Company (*History of Parliament*, I, pp. 465–6; II, p. 406).

1134 TAYLOUR TO THE MINT
10 MARCH 1715
From the original in the Public Record Office.[1]
For the answer see Letter 1158

Whitehall Treasury Chambers 10: *March* 17$\frac{14}{15}$

The Rt: Honble: the Lords Commissioners of His Mats. Treasury are pleased to Referr this petition to the Warden, Master and Worker, & Comptroller of His Mats. Mint, who are directed hereby to Consider the same, & Report to their Lordps their Opinion as to what they think fitt to be done therein

JO: TAYLOUR

NOTE

(1) T/1, 192, no. 13 a. This was written on the foot of the petition, referred to the Mint, from John Croker Chief Engraver and Samuel Bull his assistant. They had both worked for the Mint for eighteen years, and held their posts by Letters Patents for ten; these Letters Patents were due to expire in a few days and they were seeking renewal so that the business of the engravers could continue.

For the two engravers, see vol. IV, pp. 350–2, 395–6 and 416–18.

1135 WARRANT TO NEWTON
21 MARCH 1715
From a copy in the Public Record Office[1]

George R.

Our Will and Pleasure is and We do hereby Authorize & Command You to place upon Our moneys of Gold and Silver not less in Value then a Six pence,

the Inscription Georgius D.G.M. BR. FR. Et Hib. Rex. F.D. about Our
Effigies. And the Inscription Brun. Et. L. Dux. S.R.I.A. TH: Et El [2] 1715
about the Reverse, Suiting the Date unto the Year Currant, And upon the
Reverse of Our Moneys of Silver Extracted from English Lead in the Vacan-
cies between the four Escutcheons to place the Rose and Feathers alternately
as on the Silver Moneys of Our late Dear Sister Queen Anne Extracted from
Such Lead [3] And Our further Will and Pleasure is and We do hereby
Authorize and Command You that You do Coin Our Moneys of Gold and
Silver according to the Rules and Directions Set downe in the Indenture made
between Our Said Dear Sister and Your Selfe in the first Year of Her Reign,
until a New Indenture Shall be made And that You and all the Officers
Monyers and Ministers of Our said Mint do So long Conforme Your Selves
in all Things to the Rules and Directions contained in the Said Indenture.
And for so Doing, this shall be Your Sufficient Warrant. Given at Our Court
at St. James's the 21st. day of March $17\frac{14}{15}$. In the first Year of Our Reign

By His Mats. Command

HALIFAX

RID: ONSLOW [4]

WILL: ST. QUINTIN [5]

To Our Trusty and Wellbeloved Sr. Isaac Newton Knt.
Master and Worker of Our Mint within the Tower
of London.

NOTES

(1) Mint/1, 8, p. 106. There is another copy in T/52, 26, p. 424 (see *Cal. Treas. Books*,
XXIX (Part II), 1714–15, p. 432).

(2) Brunsvicensis et Lunenburgensis Dux, Sacri Romani Imperii Archi-Thesaurius et
Elector (see Letter 1128, note (2)).

(3) See Plate III (facing p. 216). Silver coins bore a variety of different design on the reverse between the escutcheons, often indicating the source of the metal used.

(4) Sir Richard Onslow (1654–1717), parliamentarian, Speaker of the House of Commons
1708–1710, was Chancellor of the Exchequer and one of the Treasury Lords from 13 October
1714 to 11 October 1715. He became a baron in 1716.

(5) William St Quintin (1662–1723), third baronet, of Harpham, Yorkshire, was M.P.
for Hull from 1695 to his death, and a Treasury Lord from 13 October 1714 to the resignation
of Walpole in April 1717 (*History of Parliament*, II, p. 405).

1136   LEIBNIZ TO WOLF
22 MARCH 1715

Extract from Gerhardt, *Briefwechsel zwischen Leibniz und Wolf*, pp. 162–3.
Reply to Letter 1130; for the answer see Letter 1140

Keilio homini impolito ut respondeam, a me impetrare non possum: vix
lectu digna habui ac ne vix quidem quæ effudit. Si quid notasti quod responsionem mereri aut Tibi aut aliis videatur, fac quæso ut sciam. Ita enim
dabo operam ut amicis satisfaciam. Ne speciem quidem argumenti notavi,
unde appareat notitiam inventi Calculi infinitesimalis a Newtono ad me pervenisse. Quin potius est cur judicemus, Newtono ipsi ante mea edita non satis
cognitum fuisse.

Translation

I cannot bring myself to make a reply to that crude man Keill. I have held what he
has put forward hardly worth reading. If you have noted something that seems to
you and to others to deserve a reply, I ask you to let me know [it]. For I would do my
best to satisfy friends. I have indeed observed no shadow of an argument whence it
appears that notice of the infinitesimal calculus invented [by him] came from Newton
to me. For which reason we judge rather that it was not well enough known to Newton
himself before I published [on the subject].

1137   NEWTON TO ——
[22 MARCH 1715]

From the draft in the University Library, Cambridge[1]

Sr

I have received your Letter (dated yesterday) [2] by the hands of Mr John
Vat & can acquaint you that his Project for the Longitude is as impracticable
as to make a perpetual motion like that of the heart but much more uniform
or to observe the Sun's meridional altitude to a second or to deduce the
Longitude from the complement of the Latitude, or to find that complement
by burning brandy. [3]

Upon my representing that the Longitude might be found by the motion of
the Moon without the error of above two or three degrees, & that if it could be
found to a degree it would begin to be useful if to $\frac{2}{3}$ of a degree it would be
more useful, if to $\frac{1}{2}$ a degree it would be as much as could be desired the
Committee of Parliament set Premiums upon finding it to these degrees of
exactness, & thereby the Act of Parliament points at the finding it by the

Moons motion. And I have told you oftner then once that it is not to be found by Clock-work alone. Clockwork may be subservient to Astronomy but without Astronomy the longitude is not to be found. Exact instruments for keeping of time can be usefull only for keeping the Longitude while you have it. If it be on[c]e lost it cannot be found again by such Instruments. Nothing but Astronomy is sufficient for this purpose. [4] But if you are unwilling to meddle with Astronomy (the only right method & the method pointed at by the Act of Parliament) I am unwilling to meddle with any other methods then the right one.

I had a few days ago a Letter from you to examin an instrument for performing plane Trigonometry at Sea. It is now above 20 years since I left of Mathematicks & I never applied my self much to practical mathematicks, so they are now much out of my mind. But I made the Judgement upon the Instrument that it is safer for the Kings fleet to keep to the Trigonometrical operations by Tables which are sufficiently expedite then in stead of them to use instruments wch are less exact & tend only to make dispatch by slubbering over the operations. [5]

NOTES

(1) Add. 3972, fo. 39. This is the longest of three drafts. One of these refers (adversely) to 'Mr French' (Letter 1121), hence the earliest date of writing (see the next note) would be 22 March 1715; but a later year is possible.

(2) Newton at first wrote (and deleted) 'of 21 March.'

(3) This is an all but unique example of Newton's making a joke.

(4) Obviously these statements by Newton are not literally correct; he must have known that in principle a perfect clock makes a perfect longitude calculation possible—always supposing, of course, that circumstances permit (by a noon observation or other means) astronomical determination of the ship's *local* time. Presumably he means that any practical clock was likely to be adequate 'for keeping a recconing at sea' as another draft puts it for only a few days; after a short period, due to mechanical uncertainties, it would be necessary to find the longitude again or to correct the clock by astronomical means. Perhaps Newton's opinion had something to do with the long distrust of Harrison's chronometers.

(5) The words '& to fill the Kings Fleet with ignorant seamen' were added but deleted. In another draft Newton begs his correspondent not to 'endeavour to concern me any further with Projectors.'

1138 LEIBNIZ TO J. BERNOULLI
29 MARCH 1715
Extract from Gerhardt, *Leibniz: Mathematische Schriften*, III/2, p. 939.
Reply to Letter 1129

Cum accepissem Newtonum mira quædam de Deo dicere in Optices suæ editione latina, quam hactenus nondum videram, inspexi et risi, spatium esse sensorium Dei, quasi Deus, a quo cuncta procedunt, sensorio opus habeat.[1] Præterea spatium nihil aliud est, quam ordo coexistendi, ut tempus ordo mutationum generalis seu ordo existendi incompatibilium; unde spatium abstractum a rebus non magis est res vel substantia, quam tempus. Atque ita Metaphysica huic Viro parum succedunt. Notavi etiam quædam, unde apparet Dynamicen seu virium leges non esse ipsi penitus exploratas. Vacui demonstratio, quam cum asseclis molitur, paralogistica est.[2]

Perplacent quæ Anglis proponi posse judicas, velut de tangentibus curvæ per magnitudines arcuum dissimilium curvarum determinatæ. Ni fallor jam alicubi in Actis vel Diariis tale quid publice proposuisti. Quæ etiam cum Fratre egisti, et minima a puncto ad punctum superficiei ducenda,[3] aliaque hujusmodi multa ex Methodis, quas jam didicere, non facile derivabunt. Doctrinam de æquationibus localibus trium coordinatarum, seu de Locis vere solidis, olim aggredi cœpi, eorumque intersectiones seu curvas etiam non planas, sed prosequi non vacavit. Operæ pretium faceret, qui studium impenderet. Viderisque in eo argumento nonnihil laborasse, quæ velim ne supprimas.

Paralogismus, quem mihi Keilius imputat, nihil est et redit ad modum loquendi. Cum scribat inciviliter et indecenter, a me responsum non habebit. De re ipsa agam, hominem non curabo. Quod superest, vale et fave etc.

Dabam Hanoveræ 9 *Aprilis* 1715 [N.S.].

Translation

When I was told that Newton says something extraordinary about God in the Latin edition of his *Opticks*, which until then I had not seen, I examined it and laughed at the idea that space is the sensorium of God, as if God, from whom everything comes, should have need of a sensorium.[2] Besides, space is nothing other than the order of coexisting, just as time is the general order of changes or the order of the existence of incompatible things; whence space, abstracted from things, is no more a substantial entity than is time. And so this man has little success with Metaphysics. I have even noticed certain things from which it appears that Dynamics, or the laws of forces, are not deeply explored by him. The demonstration of a vacuum, which he, like his followers, strives after, is a paralogism.[2]

The things which in your opinion can be proposed to the English please [me] exceedingly; that is, concerning the tangents of a curve determined by the magnitudes of arcs of dissimilar curves. If I am not mistaken you have already proposed such a thing publicly somewhere in the proceedings or journals. They [the English] will not easily derive the things which you, with your brother, have done and the drawing of shortest [lines] from point to point on a surface,[3] and many other things of this type from the methods which they have now learned. I once began an onslaught upon the theory of the equations of loci of triple co-ordinates, that is, of truly solid loci, and of the intersections of these, that is, of non-planar curves; but I did not have leisure to pursue it. He who took trouble over its study, would do something worthwhile. You seem to have done something in that business which I wish you not to conceal.

The paralogism which Keill imputes to me is nothing and amounts to a way of speaking. As he writes uncivilly and rudely, he will have no answer from me. I will deal with the matter of fact; to the man I will pay no attention.

For the rest, farewell and thrive, etc.

Hanover 9 April 1715 [N.S.]

NOTES

(1) This is Leibniz's first allusion to the idea he detected in Newton's *Opticks* that space is God's organ of sensation; he took up this metaphysical absurdity in his letters to Dr Samuel Clarke, whence it became public knowledge (see Letter 1173, note (3) and A. Koyré and I. Bernard Cohen in *Isis*, **52** (1961), 555–66, especially note (22)).

In the additional queries added to the 1706 *Optice*, *Quæst.* 20 asks (see *Optice* p. 315; we quote the later English versions, where this appears in *Quæry* 28):

Is not the Sensory of Animals that place in which the sensitive Substance is present, and into which the sensible Species of Things are carried through the Nerves and Brain, that there they may be perceived by their immediate presence to that Substance? And these things being rightly dispatch'd, does it not appear from Phænomena that there is a Being incorporeal, living, intelligent, omnipresent, who in infinite Space, as it were in his Sensory, sees the things themselves intimately, and throughly perceives them, and comprehends them wholly by their immediate presence to himself . . .

The critical words as first written and printed by Newton read (in Latin): 'Annon Spatium Universum, Sensorium est Entis Incorporei . . .' (Is not space the universal sensorium of an incorporeal Being . . .) but these were altered in a revise of p. 315 to read (in Latin) 'tanquam Sensorio suo' with the rest of the sentence as in the English above. Leibniz claimed that the normal meaning of *sensorium* is *organ of sensation* but Clarke claimed that it meant to Newton *place of sensation* (or rather, the place where the sense-organs meet the sentient soul).

Whether or not Leibniz examined one of the earlier printed copies of *Optice* (1706) containing the first, emphatic state of p. 315, he would have observed in *Quæst.* 23, p. 346, another passage denying that God has need of organs of sensation but attributing to God a 'boundless uniform Sensorium' (near the end of *Quæry* 31 in the later English editions).

(2) Presumably Leibniz again has in mind *Quæst. 20*, pp. 310–15. (*Quæry* 28) where Newton attacks the plenist conception of the Universe: 'And against filling the Heavens with fluid mediums, a great Objection arises from the regular and very lasting Motions of the Planets and Comets in all manner of Courses through the Heavens.' This argument would be false by the principles of Leibniz's 'Tentamen'.

(3) Leibniz seems to refer to a set of problems which Bernoulli proposed in the *Journal des Sçavans*, **33** (Paris, 1697), 394. See G. Eneström, 'Sur la découverte de l'équation générale des lignes géodésiques', *Bibliotheca Mathematica*, series 2, **13** (1889), 19–24.

1139 ANTONIO-SCHINELLA CONTI TO LEIBNIZ
[c. APRIL 1715]
Extract from Gerhardt, *Briefwechsel*, pp. 258–62.
For the answer see Letter 1170

[The letter begins with comments on F. M. Nigrisoli's *Considerazioni intorno alla generazione de' viventi, e particolarmente de' mostri* (Ferrara 1712) which Conti had previously criticized in the *Giornali de' letterati d'Italia*] [1]

M. Newton parle d'un certain esprit universel, et qui est repandu par toute la matiere. [2] Si par cet esprit M. Newton entend la Nature plastique, il ne dit rien de clair, ni de nouveau, mais ie suis tenté de croire, que l'esprit, dont parle M. Newton dans sa derniere edition de son Livre, n'est que le Concours des Loix Naturelles, et que par consequent M. Newton a son ordinaire, il dit la meme chose que vous, mais sous des termes un peu obscurs . . .

Mais en voilà assez pour la premiere fois que i'ay l'honneur de vous entretenir. Je vous prie de dire librement ce que vous pensez sur ma question, car ie ferai imprimer votre lettre, si vous le permettez. [3] Je parts demain pour l'Angleterre, [4] et ie ne manquerai pas d'y soutenir votre cause, comme i'ay fait à Paris. M. Remond [5] et M. l'Abbé Fraguier en sont temoins. Ils m'ont chargé tous les deux de vous faire leurs compliments. Je suis avec tout le respect etc.

NOTES

(1) In an attempt to collect information for a general history of philosophy, Antonio-Schinella Conti (1677–1748), born in Padua, travelled first to France, in 1713, where he met Malebranche, Fontenelle and others. In 1715, becoming interested in the work of Newton and his circle, he visited England, and was elected a Fellow of the Royal Society on 10 November 1715. The present letter was the beginning of an attempt to reconcile Newton and Leibniz, an attempt which Leibniz at least used only to stir up the controversy between the two men still further (see Letter 1170, note (3)).

The changing fortunes of Conti's relations with Leibniz and Newton may be traced in some detail in Leibniz's correspondence with the Princess of Wales (published in Onno Klopp, *Correspondenz von Leibniz mit Caroline* (Hildesheim, 1973); reprinted from Onno Klopp, *Die Werke von Leibniz*, erste Reihe: historisch-politische und staatswissenschaftliche Schriften, elfter Band, Hannover, 1884). Early in his correspondence with the Princess (10 May 1715) Leibniz laid his dispute with Newton before her, plainly suggesting that as the Court of Hanover was now settled in England, and the Newtonian party was inclined towards the Stuart house, the new monarch should show at least as great favour towards his old servant Leibniz as towards Newton (pp. 37–40). Later he sought to introduce Conti, then in London, to the Princess although he already feared that his loyalty to Leibnizian philosophy was weakening (pp. 62–3); by the end of the year Caroline was expressing appreciation of Conti's wit, and seeing herself and Conti as mediators between Leibniz and Newton, healing their differences (pp. 71–2). However, Leibniz soon became assured of Conti's infirmity of principle: he had now, doubtless, gone over to Newton though 'Quand il repassera en France, on le fera retourner du vuide au plein'. (p. 100, May 1716). Besides, he lost some of Leibniz's manuscripts.

Conti eventually managed to alienate himself from Newton as well as Leibniz, over the affair of the publication by Conti of *Abrégé de Chronologie de M. Le Chevalier Newton* (Paris, 1725); see vol. VII.

(2) Conti presumably alludes to the last paragraph of the second edition of the *Principia*, beginning: 'Adjicere jam liceret nonnulla de spiritu quodam subtillissimo corpora crassa pervadente, & in iisdem latente . . .' though it is strange that Conti should interpret 'spirit' in so metaphysical a sense.

(3) Conti's letter is a very general one, asking Leibniz's opinion of what Nigrisoli called the 'Lumiere seminale', and comparing it with various concepts to be found not only in the work of Newton and Leibniz, but of many other philosophers and scientists. Leibniz, in his reply (see Letter 1170) hardly answered the question posed, but added a lengthy postscript about the calculus disputes, clearly expecting that the letter might be printed.

(4) The exact date of Conti's departure is not known, but he intended observing the total eclipse of the sun visible in England on 22 April 1715, so the letter must have been composed some time before this date. The letter did not, however, reach Leibniz until considerably later; Conti had given it to Nicolas Rémond (see note (5) below) who forwarded it to Leibniz on 7 October 1715, together with excerpts from several of Conti's letters to himself (see also Letter 1170, note (2)).

(5) It is not clear whether this refers to Nicolas Rémond, or his elder brother, Pierre Rémond de Monmort. Nicolas kept up a considerable correspondence with both Leibniz and Conti, largely on philosophical matters; (for the letters between Leibniz and Nicolas, see Gerhardt, *Leibniz: Philosophische Schriften*, III). Pierre Rémond de Monmort (1678–1719), who travelled with Conti to England and was elected F.R.S. before him on 9 June 1715, had greater interest in mathematics, his most important work being *Essay d'analyse sur les Jeux de Hazard* (Paris, 1708); see also vol. IV, p. 534, note (1). He was greatly interested in the calculus disputes, and Leibniz used him as intermediary between Conti and himself.

1140 WOLF TO LEIBNIZ
24 APRIL 1715

Extract from Gerhardt, *Briefwechsel zwischen Leibniz und Wolf*, pp. 164–7.
For the answer see Letter 1142; reply to Letter 1136

Ex ejus litteris cognovi, Gallos fortiorem judicare epistolam Keilianam in Diario Hagiensi, quam ut a Keilio proficisci potuerit: arbitrantur itaque Newtonum ipsum argumenta suppeditasse. Iidem optant ut E.V. distinctius ad eadem respondeat, et Hermannus arbitratur, quod E.V. in necessitate aliqua reponendi nunc constituatur. Sed quænam tum illi, tum hic responsione potissimum digna judicent, non constat. Mihi quidem necessarium videretur ostendere, quod in demonstratione quadraturarum Newtoniana, quæ etiam in Actis A. 1712 p.75 legitur, [1] ipse calculus differentialis nondum contineatur,

PLATE III. Coin designs submitted by Newton for the reverses of a 1714 'Prince Elector' half-guinea (*top left*) and of a silver sixpence (*top right*). The designs were drawn by an engraver; but are annotated in Newton's hand. Both show a simplified heraldry in the fourth (left-hand) escutcheon. This was not adopted; compare the 1715 shilling shown below. See Letters 1128 and 1135. (Newton's designs are reproduced by courtesy of the Controller of Her Majesty's Stationery Office and the 1715 shilling by courtesy of the Trustees of the British Museum.)

etsi ea et hic communi quodam fundamento nitantur. Deinde scire velim, quid E.V. reponat ad accusationem, [2] quod in Tentamine de Causis physicis [3] motuum cœlestium differentias secundas rite æstimare non noverit, cumque propterea in errorem inciderit, errorem A. 1706 in Actis correctura ob ejus fontem non animadversum duplicem alium commiserit. Sane cum Keilius adeo audacter provocet Autores Schediasmatum, [4] contra quos calamum stringit, et argumenta Anglorum tanta evidentia niti asseveret, ut nec ipsa E.V. aliqua reponere ausura sit; plurimi sane ex silentio concludent bonam Anglorum causam. Videtur autem speciem aliquam habere hoc argumentum, quod Newtonus in epistola descripserit eos methodi suæ characteres, qui nulli alii nisi calculo differentiali conveniunt, et quod exempla dederit, unde ingenium mediocre adhibita illa descriptione ipsam methodum hariolari potuerit, præsertim cum una exhibita fuisse dicatur series omnes differentias possibiles quantitatis variabilis cujuscunque complexa. Bernoullium prorsus silere miror, qui ad retractationem parum honorifice provocatur a Keilio. [5]

Translation

I understood from his [Hermann's] letter, that in the opinion of the French Keill's letter in the Hague *Journal* [*Literaire*] was more powerful than Keill himself could have made it; therefore they think that Newton himself provided the arguments. They desire that Your Excellency should answer it in a more definite manner, and Hermann thinks that Your Excellency is now placed under some necessity of replying [to it]. But it is not clear whether they judge this point or that most worthy of a reply. It would seem to me necessary to show that the actual differential calculus is still not contained in the Newtonian demonstration of quadratures which is given in the *Acta* [*Eruditorum*] for 1712, p. 75, [1] although both spring from a certain common foundation. Next I would like to know what Your Excellency would say in answer to the accusation [2] that in the *Tentamen de causis* [3] *physicis motuum cœlestium* you did not know how to reckon second differences correctly, and on that account, moreover, lapsed into an error, and when that error was to be corrected in the *Acta* [*Eruditorum*] for 1706, because its source was not observed you have committed a double error. Surely, as Keill so audaciously challenges the authors of the [fly-]sheets[4] against whom he wields his pen, and asserts that the arguments of the English are founded upon so much evidence that Your Excellency himself would not dare to make any reply, most people may deduce from [your] silence that the English case is a good one. However, this argument seems to have a certain speciousness, because Newton in a letter has described those characteristics of his method which fit nothing other than the differential calculus, and because he gave examples whence even a moderately clever man furnished with that description could divine that very method, especially when one [of those] presented was said to have been a series embracing all possible differences of any variable quantity whatever. I wonder at Bernoulli's continued silence, after being dishonourably challenged by Keill [to make] a retraction. [5]

NOTES

(1) 'Analysis per Quantitatum Series, fluxiones ac differentias cum enumeratione linearum tertii Ordinis', *Acta Eruditorum* for February 1712, pp. 74–7. This was an unsigned review [by Leibniz] of William Jones' *Analysis* (London, 1711), which contained Newton's *De analysi* and his *De quadratura curvarum*. The review reproduces the Newtonian algorithm for integrating $ax^{\frac{m}{n}}$, and discusses the difference between Leibnizian differentials and Newtonian fluxions; see Whiteside, *Mathematical Papers*, II, pp. 259–62.

(2) In Keill's 'Answer' in the *Journal Literaire de la Haye*; see Number 1086*a*.

(3) This word does not properly appear in Leibniz's title.

(4) That is, the *Charta Volans*; see Number 1009.

(5) Bernoulli, of course, at this stage was still nominally an anonymous participant in the controversy, only Wolf and Leibniz sharing the knowledge of his identity as the writer of the letter quoted in the *Charta Volans* (see Number 1009, note (1)). But of course Keill and Newton had their suspicions. It may be that Wolf refers here to Keill's article on the inverse problem of central forces (see Letter 1153).

1141   COTES TO NEWTON
29 APRIL 1715

From the original in the University Library, Cambridge.[1]
Continued in Letter 1144

Cambridge Apr. 29*th* 1715

Sr.

I think it my Duty to send You what Observations I could make of the late Eclipse.[2]

Times by ye P.Clock	Times Corrected	
h ′ ″	h ′ ″	About 30 degrees in the Suns Limb covered by the Moon, estimated by memory after the end of the Eclipse.
8.07.35	8.10.09	
8.31.38	8.34.11	The first edge of the greater Spot touch'd by the Moon
8.32.42	8.35.15	The first edge of the middle Spot touch'd
8.34.22	8.36.55	The first edge of the third Spot touch'd
9.12.04	9.14.37	The first recovery of the Sun's Light
10.19.25	10.21.57	The end of the Eclipse.

The Times were corrected by an Instrument,[3] which I ordered to be made for the purpose about a day or two before the Eclipse: the form of which is as follows. *AB* is a strong wooden Axis of about six feet in length; *CD* & *DF* on one side, *FE* & *EG* on the other are peices fram'd to each other & to ye Axis as firmly as was possible. Into the peice *CD* & at the angle *E* were fix'd strong wooden pins nearly parallel to each other & perpendicular to ye plain *CDFEG. PQ* is the Cylindrical Brass Tube of a Five-foot Telescope,[4] this was well fastened with Iron staples & screws to ye peice of wood *IKML* whose under plain surface is here represented as objected to view. Into this surface there was

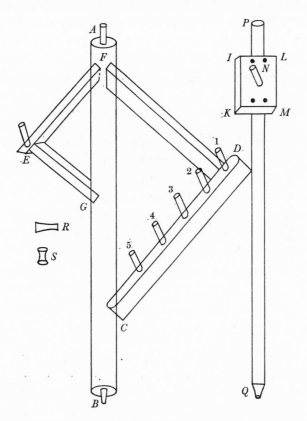

perpendicularly fix'd a strong wooden Pin *N* which was design'd to hang the upper end of the Telescope upon the Pin *E* whilst its lower end rested upon any of the Pins *CD*. Now that ye Telescope might be taken off, & yet afterwards be again plac'd accurately in the same position, I ordered the edges *IK* & *EF* which touch'd each other to be rounded like the surface of a Cylinder, as also the edge *CD* into which the row of Pins was fix'd & against which the Cylindrical Tube of the Telescope rested; so that the contact in both places might be made in a point. Upon the same account the Pin *E* was made a little hollow as is represented at *R*; the others were Frustums of Cones, that thereby the Telescope might more surely touch the edges *EF* & *CD*. Into the two ends of the wooden Axis were strongly fix'd two peices of well tempor'd steel: that at the upper end *A* was a Cylinder well turn'd which mov'd in a collar whose cavity (represented by *S*) was figured like two equal hollow & inverted Frustum's of Cones joyn'd togather: the lower at *B* was a Cone moving in a Conical Socket of a somewhat bigger angle. This socket had liberty to move horizontally & to be fix'd in any Position by three Screws which press'd against it sideways. The Instrument being thus prepared, I fix'd a Needle at the lower end of the Wooden Axis whose point stood out from it about an Inch; then suspending a fine Plumb-line from the upper end of the same Axis I altered the Position of the Instrument by the three screws, until the Plumb-line came to beat against the point of the Needle in the Whole revolution of the Instrument, & there I fix'd it as prepar'd for use: My Observations follow.

Day

1.	XXIpm	4.^{h} 01.′ 21″	the Sun's upper Limb observ'd at the 3d Pin	
2.	XXIIam	6.^{h} 48.′ 41″	Upper Limb ⎫	2d Pin
		6. 52. 09	Lower Limb ⎭	
3.	XXIIIam	6.^{h} 47.′ 29″	Upper Limb ⎫	2d Pin
		6. 50. 58	Lower Limb ⎭	
4.	XXIIIam	7.^{h} 51.′ 10″	Upper Limb.	3d Pin
5.	XXVam	6.^{h} 44.′ 53″	Upper Limb ⎫	2d Pin
		6. 48. 22	Lower Limb ⎭	
6.	XXVpm	5.^{h} 08.′ 18″	Lower Limb ⎫	2d Pin
		5. 11. 47	Upper Limb ⎭	

Allowing for the Variation of Declination [4] I find

By the 2d & 3d Observ: the length of ye Solar Day measured by the Clock was . . . . 24.^{h} 00.′ 18″

By the 3d & 5th Observ: the length of two Solar Days by the Clock was . . . . 48.^{h} 00.′ 18″

Which two Conclusions manifest a great inequality of the Clock's motion

By the 1st & 4th, the Meridian of the XXIId day was at   11.^{h} 57.′ 32″

By the 5th & 6th, the Meridian of the XXVth day was at   11.^{h} 58.′ 02″

And therefore the Meridian of the XXIId, allowing for the Clocks inequality . . . .   11.^{h} 57.′ 26″

I put the correct Meridian of the XXIId day at . . .   11.^{h} 57.′ 29″

I beg Your Pardon for troubling You with so large an Account of my Method for correcting the Pendulum. I confess to You I have a design in it for the advantage of our yet imperfect Observatory. [5] The Clock which I used was borrowed of a Clock-maker in this Town, who took it for a very good one. Not expecting so great inæquality in its motion, I was very much surpriz'd to find it by the Observations: & since I have found it I cannot think of making use of such ordinary workmanship again unless in case of necessity. To speak plainly, I beg of You to let that excellent Clock [6] be now sent down to Us which You ordered to be made for the Use of our Observatory. I cannot think of a more accurate Instrument for the setting of it, than such an one as I have been describing. Having it therefore by me I think I am prepar'd to receive Your Noble Gift. I have written to Mr Street [7] to wait upon You for Your resolution.

I am Sir

Your

Most Obliged Humble Servt

ROGER COTES

I will send You an account of what was observ'd at Cambridge during the total obscuration in another Letter.

NOTES

(1) Add. 3983, no. 39; Edleston, *Correspondence*, pp. 179–80, prints a partial draft (Trinity College, Cambridge, MS. R.16.38, no. 290).

(2) The eclipse took place on 22 April; this was one of those rare occasions when the line of totality passed right across the South of England, and the weather conspired to make observing conditions favourable. Halley, on behalf of the Royal Society, had a small map of England printed, showing the passage of the Moon's shadow, and distributed it all over England, in the hope of encouraging astronomers to make observations and to send them to him. Cotes' observations must eventually have fallen into Halley's hands, for they are included in the synopsis he made of the observations he received. (See *Phil. Trans.* 29, no. 343 (1715), 245–62.) There, too, he reports that Cotes 'had the misfortune to be oppressed by too much company, so that, though the heavens were very favourable, yet he missed both the time of the beginning of the eclipse and that of total darkness'. Cotes' somewhat confused account in Letter 1144 confirms this.

(3) This simple instrument allows the telescope to be placed at five distinct angular elevations, and to be rotated so that the time at which the sun reaches a given elevation during both rising and setting may be noted. Robert Smith, in his *Compleat System of Opticks* (Cambridge, 1738), II, p. 327, who quotes Cotes' description, describes the instrument as 'of small expence, but very accurate'.

(4) According to Smith, *ibid.*, p. 328, the telescope was the one belonging to the Observatory's sextant.

(5) Plans for an Observatory to be erected over the Great Gate of Trinity College were first made by Bentley, then Master of the College, in 1702 or 1703; by the end of 1703 a collection of instruments had been acquired. A sum of £1900 to be put towards the building of the Observatory was left to the University by Thomas Plume, who died at the end of 1704, but building began only in 1707, when Cotes became Plumian Professor of Astronomy. How-

221

ever financial difficulties prevented the completion of the building until 1739. (See D. J. Price, 'The early observatory instruments of Trinity College, Cambridge', *Annals of Science*, **8** (1952), 1–12.)

(6) Newton had decided to present a clock to the Cambridge observatory as early as 1708; he estimated that it would cost at least £50. See Cotes' letter to John Smith, 10 February 1708, printed in Edleston, *Correspondence*, pp. 197–201.

(7) Richard Street, of Shoe Lane, the maker of the clock, was, from 1715, one of the three wardens of the Clockmakers' Company of London. At least two of his clocks are still extant (see F. J. Britten, *Old Clocks and Watches and their Makers*, 7th edition (London, 1956)).

1142   LEIBNIZ TO WOLF
7 MAY 1715

Extract from Gerhardt, *Briefwechsel zwischen Leibniz und Wolf*, pp. 168–9.
Reply to Letter 1140

Cum Keilius scribat rustice, ego cum tali homine litigare nolim. Qui sola ejus asseveratione et jactatione moventur, iis frustra scribitur, neque enim rem examinant. Qui examinabunt, videbunt, nullum esse errorem in Schediasmate motuum cœlestium, sed phrasin postea a me redditam commodiorem circa vim centrifugam. Utique enim vera sunt et manent, quæ ostendi, ex Circulatione Harmonica cum gravitate conjuncta prodire Ellipses Keplerianas. Videbit etiam, qui examinabit, quadraturas Newtonianas jam ex Barrovii et similibus Notitiis derivari potuisse. Ego cogito aliquando rebus refellere hominem, non verbis. Cæterum si methodus differentialis eadem est cum his, quæ Newtonus olim dedit in Epistolis, cur suam Methodum Fluxionum, quæ utique præludium quoddam est Calculi differentialis, ænigmate tegere voluit? putavit ergo ipse, rem plane diversam. Dubito an Dn. Bernoullius Keilianam dissertationem viderit. Interim gratias ago, quod monuisti, in quo difficultas esse videatur, et si quid adhuc incidet cujus rationem haberi velles, indica quæso: alii talia sæpe melius pervident, quam nos ipsi. Et ego animadversa Keilii ruditate, nondum ejus schediasma accurata lectione sum dignatus.

Translation

Since Keill writes like a bumpkin, I wish to have no dealings with a man of that sort. It is pointless to write for those who respond only to his bold assertions and boasting, for they do not examine the substance. Those who look into the paper on celestial motions will see that there is no error there, only a phrase concerning centrifugal force which I later rendered more suitable. For what I have demonstrated, that the Keplerian ellipses results from a combination of the harmonic circulation with gravity, certainly is true and will remain so. He who examines it will also see that the Newtonian quadratures could have been derived already from the ideas of Barrow and similar [methods]. I think of knocking the man down, some time, with things rather than words. Moreover, if the differential method is the same as that which Newton formerly gave out in correspondence, why did he wish mysteriously to conceal his method of fluxions, which is surely a sort of precursor of the differential calculus? So he himself thought the thing quite different. I doubt that Mr Bernoulli has seen Keill's dissertation. Meanwhile, thank you for advising me where the trouble seems to lie, and if you wish to be given an explanation of what has happened up till now, I ask you to let me know: others often see such things better than we do ourselves. And I myself, considering Keill's rudeness, have not yet thought the paper worthy of thorough reading.

1143 TAYLOUR TO NEWTON
9 MAY 1715
From the copy in the Public Record Office.[1]
For the answer see Letters 1152 and 1164

Sir

The Lords Comm[issione]rs of his Mats: Treasury direct you to send hither forthwith An Acco[un]t of the Quantitys and value of the Tyn now in hand or paid for in the Country as also what remains at Amsterdam Hamburgh or elsewhere and if it stands with Your Conveniency to come here this Morning [2] I am &c 9 May 1715

J TAYLOUR

NOTES

(1) T/27, 21, p. 370.
(2) Compare Letter 1126.

1144 COTES TO NEWTON
13 MAY 1715
From the draft in Trinity College Library, Cambridge.[1]
Continuation of Letter 1141

Sr

Dr Bentley has told me, You have been pleas'd to give orders, that the Clock may be sent to Cambridge.[2] I take this oportunity of returning You my hearty thanks for it, & of giving You an account [3] of what was observ'd by Us during the time of the sun's total obscuration in the late Eclipse, so far as I judge it to be of any moment. The sky was perfectly clear all the Morning till about two or three minutes after the recovery of the suns light. It surpriz'd us to find to so great a quantity [of] Light remaining in the middle of the Eclipse: I think it did very much exceed the brightness of the clearest Moonlight nights. A Freind assur'd me He could very easily & distinctly read the smallest letters engrav'd about Mr Whistons Scheme of the Heavens, which he had in his hands at that time. We saw the Planets Jupiter, Mercury, & Venus, with some fix'd stars: but they appear'd with far less splendor & fewer in number than we expected, or than they might have done by Moon-light. I took the greatest part of this remaining light to proceed from the Ring which incompass'd the Moon at that time. As nearly as I could guess, the breadth of this Ring was about an eighth or rather a sixth part of the Moons Diameter, [4] the light of it was very dense where it was contiguous to the Moon but grew rarer continually as it was further distant, till it became insensible: its colour was a bright clear white. I saw this Ring begin to appear about five seconds before the total immersion of the suns body, & it remain'd visible to me as long after His emersion. I did not apply my self to observe whether it was of the same breadth in all its parts during the total Obscuration. Mr Walker [5] a Fellow of our College whom I can very well depend upon assur'd me He was very certain it was not. He says He took notice with a great deal of attention that at first the Eastern part was very sensibly broader & brighter than the western, afterwards they became equal, & some time before the emersion the Western side was manifestly broader & brighter then the Eastern. His design in attending so diligently to such an Observation was this; He thought, as he afterwards told me, that I might desire to note the Time of the middle of the Obscuration; & being in the same Room with me, He was willing to assist me in judging of that Time, & beleiv'd the method which He took to be the properest for it; accordingly I do remember that I heard him call out to Me, *Now's the Middle*, though I knew not at that time what he meant. I think this Observation of Mr Walkers is of moment, I have therefore been very particular in giving You the circumstances of it that You may Your self judge how far it may be depended upon, for my part I cannot see any reason to doubt of it. Besides this Ring there appear'd also Rays of a much fainter Light in the form of a rectangular Cross: I have drawn You a Figure [6] which represents it pretty exactly, as it appeard to Me. The longer & brighter branch of this Cross lay very nearly along the Ecliptick, the light of the shorter was so weak that I did not constantly see it. The colour of the Light of both was the same: I thought it was not so white as that of the Ring even in it's fainter parts, but verg'd a little towards the colour of very pale copper. You may observe, that in my Figure the branches of the Cross are represented as bounded by parallel lines, for so it was they appear'd to me. But there are others here, who saw a very different form. I have therefore sent You another Figure the most remote of any I have met with from my own, This was drawn by a very ingenious Gentleman representing the appearance as seen by himself. He differs also from me in this particular, vizt that he takes the Cross light to be only a

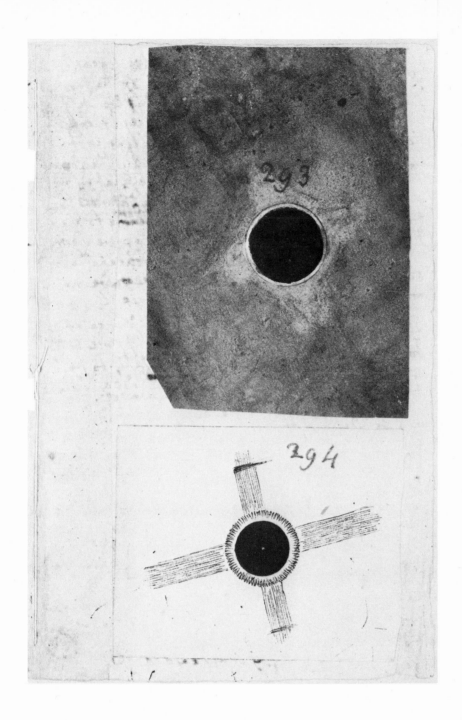

PLATE IV. Cotes' sketches of an eclipse of the sun seen on 22 April 1715; see Letter 1144. Our photograph reflects the poor quality of the original sketch (Trinity College, Cambridge, MS. R. 16.38, fos. 293 and 294). (Reproduced by courtesy of the Master and Fellows, Trinity College, Cambridge.)

continuation of the Ring whereas I make 'em to be intirely distinct from each other. I am Sir.

May 13. 1715.

NOTES

(1) R. 16.38, fos. 291–5; printed in Edleston, *Correspondence*, pp. 181–4.

(2) See Letter 1141, note (6), p. 222.

(3) Many descriptions of the eclipse appear in print; see for example the accounts by Halley in *Phil. Trans.* **29**, no. 343 (1715), 245–62 and 314–16. The qualitative descriptions of what these observers saw may be put down largely to the physiological effects of intense light upon the eye. Cotes' account corresponds fairly well with those of his contemporaries.

(4) The ring Halley saw was 'perhaps a 10th part' of the moon's diameter, and was 'of a pale whiteness, or rather pearl colour, seeming a little tinged with the colours of the Iris.'

(5) Richard Walker (1679–1764) graduated from Trinity College, Cambridge, in 1706/07, and became a Fellow in 1709. From 1734–64 he was Vice-master of Trinity, and from 1744–64 Knightbridge Professor of Moral Philosophy. Ordained in 1707 he was absent much of the time from Cambridge in the period 1707 to 1717, acting as Curate at Upwell in Norfolk. He was recalled by Richard Bentley, then Master of Trinity, in 1717 and became his intimate friend and right-hand man in his quarrels with the Fellows. He founded the Cambridge Botanic Garden in 1762.

(6) See Plate IV.

I 145   NEWTON TO SIR JOHN [NEWTON]
23 MAY 1715

From the original in Trinity College Library, Cambridge [1]

Leicester Fields. 23 *May* 1715

Sr John [2]

I am concerned that I must send an excuse for not waiting upon you before your journey into Lincolnshire. The concern I am in for the loss of my Lord Halifax [3] & the circumstances in wch I stand related to his family will not suffer me to go abroad till his funeral is over. And therefore I can only send this Letter to wish you & your Lady & family a good journey into Lincolnshire & all health & happiness during your stay there. And upon your first return to London I will wait upon you & endeavour by frequenter visits to make amends for the defect of them at present.

I am

Sr

Your most humble

& obedient Servant

ISAAC NEWTON.

NOTES

(1) R.16.38, fo. 438; first printed by A. de Morgan in *Newton: His Friend: And his Niece* (London, 1885), p. 49.

(2) Presumably Sir John Newton, a distant relation of Isaac (see Letter 1059, note (1)).

(3) Halifax died on 19 May 1715, and in a codicil to his will bequeathed considerable property to Catherine Barton, Newton's niece. For Halifax's relationship with Catherine Barton, and Newton's involvement in their affairs, see the Introduction to vol. v, pp. xliv–xlvi. See also Letter 1151.

1146 NEWTON TO THE TREASURY

[After 26 MAY 1715]

From the draft in the Mint Papers[1]

To the Rt Honble the Lords Comm[issione]rs of his
Majties Trea[su]ry.

May it please your Lordps

In obedience to your Lordps Order of Reference of May 26th last, upon the Petition of Mr William Hamilton for being restored to his Place in his Mats Mint at Edinburgh:[2] I humbly represent that he who had been Clerk of the Bullion for the Mint in Scotland before the Union of the two kingdoms, & who by this Petition appears to be the Petitioner, was upon the ceasing of that place by the Union, & in recompence for the loss thereof, made an Officer in the present Mint in North Britain with a salary of 50£ per annum, that salary having been allowed to him before when he was Clerk of the Bullion. And in the list of the Officers of that Mint annexed to the Indenture made between her late Ma[jes]ty Queen Ann & John Montgomery Esq Master & Worker of that Mint, I find him entered in these words:

To the Clerk of the Bullion, as assistent to the Weigher & Teller & Surveyor of the Meltings, fifty pounds.

But that Indenture becoming void by the death of her said Majesty, & no grant of that place from his present Ma[jes]ty being entred in the Books of the Treasury, & it not appearing lately that he was alive nor what was his name; he was omitted in the Warrant lately directed to the General of that Mint for continuing the said Indenture in force. But since it now appears that he is alive, if, in consideration of the services mentioned in his Petition, his Ma[jes]ty shall think fit to restore him to that place, it may be done by a Warrant under his Mats signe Mannual directed to the said General for supplying the omission in the Warrant above mentioned wch was directed to him.

All which is most humbly submitted to your Lo[rdshi]ps great wisdom.

(1) III, fo. 199; there are earlier drafts at fos. 196 and 197.

(2) He has not been previously mentioned in this *Correspondence*. All surviving posts in the Edinburgh Mint were, of course, complete sinecures. William Hamilton had been granted the gift of the clerkship of the bullion on 2 December 1706 (*Cal. Treas. Books*, XXII (Part II), 1708, p. 127).

1147 TAYLOUR TO THE MINT
16 JUNE 1715

From the copy in the Public Record Office.[1]
For the answer see Letter 1150

To the Rt. Honble: the Lords Comm[issione]rs: of His Mats: Trea[su]ry
The Memorial of Sr. Richard Sandford Bart. Warden of
His Mats: Mint.

Sheweth

That the Late Wardens of the Mint have been allowed a Clerk for assisting them in prosecuting Clippers and Counterfeiters of the Coine and the Charges of the Prosecutions not exceeding 400£ per annum being now by Act of Parliam[en]t made payable out of the Coinage Duty and a Person skilled in the methods of Executing Warrants and carrying on the Prosecutions at Assizes and at the Old Bailey having been allowed to former Wardens and very necessary now, I humbly Pray Your Lordships Order in this matter and about money for bearing the Charges of such prosecutions and the methods of Accounting for it so that this Service may be carried on in the cheapest and best manner [2]

All which is most humbly Submitted to Your Lordships great Wisd[om].

Whitehall Trea[su]ry Chambers 16 *June* 1715

The Right Honble: the Lords Commissioners of His Mats: Treasury are pleased to Refer this Memoriall to the Principal Officers of His Mats. Mint in the Tower of London who are to Consider the matter therein Contained and to Report to their Lordships a true State thereof together with Opinion what is fit to be done therein

J TAYLOUR

NOTES

(1) Mint/1, 8, p. 107.
(2) Compare Letters 1132 and 1133.

1148 FLAMSTEED TO NEWTON
30 JUNE 1715
From the copy in the Royal Greenwich Observatory[1]

June 30 1715

To Sr. Is. Newton
Sr

You were pleased when you were here last about two year's agone to tell me yt you would restore my M.S.S. in yr hands.[2] I have sent Mr. Hodgson to receive them of you and he will return ye Note you gave for them. I doubt not but that you will herein obliege Sr

Yr humble Servant

JOHN FLAMSTEED

NOTES

(1) Flamsteed MSS. vol. 33, fo. 107.
(2) Of the *Historia Cœlestis*. At least some of the manuscript was returned shortly afterwards; see Letter 1151.

1149 J. BERNOULLI TO LEIBNIZ
2 JULY 1715
From Gerhardt, *Leibniz: Mathematische Schriften*, III/2, pp. 942–3 [1]

Has literas jam ad Te exaro in eum præcipue finem, ut commendare possim hunc, cui eas ad Te perferendas tradidi, Virum[2]. Is quidem natione Anglus, sed minime Keilii aliorumque invidorum amicus, per aliquot annos meus in Mathematicis Discipulus, peculiare monstravit probavitque mihi suum ingenium, non in vulgaribus tantum, sed etiam in penitioribus nostris Analyticis. Certe in Calculo differentialium et integralium non parum profecit; habebis in eo, ubi Patriam redierit, Tuum defensorem contra invidos et malevolos, utpote, cui totius controversiæ, quæ Te inter et Keilium viget, statum exposui, et ostendi ipsum Keilium aliosque ex Anglis non semper bona fide et qua deceret sinceritate nobiscum agere; ac sæpe verborum sensum malevole detorquere et studio res ipsas confundere, ut eo melius Lectori imponere queant; quod imprimis factum video in Keilii responsione, vel potius in Libello illo famoso, Tibi maxime injuriose in Diario Literario Hagiensi mensibus Junio et Julio anni superioris edito,[3] cujus excerpta quædam mihi transmisit Cl. Hermannus[4] noster. Meo quidem judicio optime feceris, si injurias istas promtissime retundas nuda rei expositione, conquisitis hunc in finem ex scriptis Tuis literis omnibus Schedis et Actis; quo quid inter vos actum sit, et quantum unicuique sit tribuendum et quousque quilibet, ut par est, de Inventorum gloria participet, toti Mundo constet, et ita oblatranti Keilio os obturetur. De cætero ad præcedentes meas literas me refero. Vale, Vir Illustrissime, et fave etc.

Basileæ a. d. 13 *Julii* 1715. [N.S.]

Translation

I write these words to you now to this particular end, that I may commend this man [2] whom I have trusted with bringing them to you. He is indeed an Englishman, but not at all a friend of Keill and other envious men, and he was for several years my pupil in mathematics, and has shown and proved to me particularly his intelligence, not only in common matters but also in the higher branches of our analysis. Certainly, he makes great progress in the differential and integral calculus; you will have in him, when he returns to his country, your defender against envious and malicious men, since I have explained to him the state of the whole controversy which rages between you and Keill, and have shown him that Keill himself and others among the English do not always deal with us in good faith and with fitting sincerity; and also often maliciously twist the sense of words and confuse the facts themselves by partiality, so that they are able to impose the better upon the reader. This I see particularly has been done in Keill's reply or rather in that notorious pamphlet, most injurious to you, which was published in the Hague *Journal Literaire* for June and July of last year,[3] excerpts from which were sent to me by our fellow countryman, Mr Hermann.[4] Certainly you would do best in my opinion if you immediately blunt those insults by a bare exposition of fact, researching for this purpose among the letters written by yourself and all your papers and records; so that it may appear to the whole world what happened between you, and how much should be attributed to each of you, and how far it is fair that each should participate in the glory of the invention. Thus Keill's railing mouth may be closed. For the rest I refer back to my previous letters. Farewell, famous sir, and flourish, etc.

Basel, 13 *July* 1715 [N.S.]

NOTES

(1) Bernoulli had already replied to Leibniz's last letter (Letter 1138) and now writes again. Leibniz replied on 25 August and Bernoulli wrote again on 31 August (see Gerhardt, *Leibniz: Mathematische Schriften*, III/2, pp. 943–7) but these letters contain no reference to Keill. Leibniz returns to the subject of the calculus controversy in Letter 1163.

(2) John Arnold; see Letter 1014, note (1), p. 27.

(3) Keill's 'Answer'; see Letter 1053 a, note (1), p. 90.

(4) See Letter 1005, note (11), p. 10.

1150 NEWTON AND BLADEN TO THE TREASURY
5 JULY 1715
From a copy in the Public Record Office.[1]
Reply to Letter 1147

To the Rt. Honble. the Lords Comm[issione]rs of His Mats. Trea[su]ry

May it Please Your Lordships

In Obedience to Your Lordships Order of Reference of June 16 last upon the Memoriall of the Warden of His Mats: Mint concerning a Clerk to be allowed him for assisting him in the prosecution of Clippers and Counterfeiters of the Coine & concerning Moneys to bear the Charges of such prosecutions and the method of accounting for the Same We humbly represent to Your Lordships that the late Wardens have been allowed such a Clerk and have been Supplied with moneys partly by a Privy Seal impowering them to Seize the forfeited Estates of Convicted Criminals and partly out of the Civil List Moneys, untill an allowance not exceeding 400£ per annum was made out of the Coynage money by Act of Parliament for Carrying on this Service.

Wee are therefore humbly of Opinion that the Warden of the Mint do find out a fit person to attend him as Clerk and Sollicitor to draw up Informations and Warrants and enter them in Books & attend prosecutions at the Old Baily and Assizes, and that this Officer be allowed a Sallary of 60£ per annum as formerly.[2] And that upon being Sent into the Country to apprehend and prosecute Offenders he be further allowed 15 *sh.* per day travelling Charges for himself and his Horse And Such other persons as shall be sent upon the like Services be allowed for themselves & their Horses such travelling Charges as the Warden shall approve of, nott exceeding 10 *sh.* per day to the two principal of them, & 8 *sh* per day to the rest. And that in lieu of pocket expences, Coach hire fees of Court, charges of wittnesses and such other incident expences as cannott be ascertained by good Vouchers or would perplex the Accts. there be allowed for every house searched in Londo. the Sum of 10 *sh.* & in the Country the Sum of 25 *sh.* and for every person apprehended and brought before a Justice of the Peace and Committed to prison in Londo. the Sum of 10 *sh.* & in the Country the Sum of 25 *sh.* to be equally divided between the persons who Search or apprehend and Carry to prison; and for every person indicted & tried in Londo. the Sum of 50 *sh.* and in the Country the Sum of 4£ to him or them who by the Wardens Order shall Sollicite and manage the Prosecution, besides Councellours Fees; provided there be So many wittnesses as may justifye bringing the person accused to a Tryall.

And We are also humbly of Opinion that the Master & Worker of the Mint be impowered by a Sign Mannual to pay the Sallary of the Said Clerk or Sollicitor quarterly as it shall become due and to advance to him such Summs of money from time to time as by allowance or Order under the Wardens hand He the Said Sollicitor shall desire or demand for carrying on the Prosecutions not exceeding in the Whole 200£ per annum and to pay what shall be further due to the Said Sollicitor upon his Acc[oun]ts duly Stated and allowed. And that the Auditors of the Mint be directed & required in the said Sign Mannual to allow in the Acc[oun]ts of the Master & Worker all the Said payments not exceeding in the whole the Sum of 400£ per annum allowed by Act of Parliament for this Service.

And We are further humbly of Opinion that the Said Sollicitor do at the end of every half year lay before the Master & Comptroller of the Mint an Acc[oun]t of the Services performed by Him or his Assistants, upon which any moneys be due or have been paid to him or them with a Certificate from the Warden in writing that those Services were done by his Order. and also a Bill or Bills of moneys paid for Councellours Fees, extraordinary Habeas Corpus's, Lord Chief Justices Warrants &c. And that he incert nothing into those bills without Receipts or other good Vouchers for the Same, nothing for Country people assisting in Searching houses and in apprehending and Carrying to prison persons Suspected or accused. And that the Master and Worker and the Comptroller do Examine and Sign those Bills or Vouchers to the Annual Acct. of the Said Sollicitor. And that at the end of every Year the Said Sollicitor do Lay his Acc[oun]t for that Year first before the Warden of the Mint to be Examined and Signed & then before the Lord High Trea[su]rer or Lord Commrs of the Treasury for their approbation or direction how the Same shall be Audited stated allowed & discharged.

And if the Accomptant craves any augmentation of any allowance for Services nott within the Rules above mencond the Same be laid before the Lord High Trea[su]rer or Lords Comm[issione]rs of the Trea[su]ry at the End of his Accompt.

All which is most humbly Submitted to your Lordships great Wisdom

ISSAC NEWTON

Mint Office
July 5 1715

M BLADEN

NOTES

(1) Mint/1, 8, pp. 108–9. There is another copy in T/52, 27, pp. 135–7, and there are drafts (some holograph) in the Mint Papers, I, fos. 450 (440, 442, 444, 447–8).

(2) A Warrant for the payment of this salary by the Treasury Commissioners was issued on 29 July 1715; see *Cal. Treas. Books*, XXIX (Part II), 1714–15, p. 657.

1151 FLAMSTEED TO SHARP
9 JULY 1715
Extract from the original in the Royal Society [1]

I doubt not but you have heard yt ye Ld Halifax is dead of a violent fever. [2] if common fame speaks true hee died worth 150 thousand pounds out of which he gave *Mrs. Barton* S I N'[s] neice for her *excellent conversation* a curious house £5000 with lands jewells plate money & household furniture to ye value of 20M*L* [3] or more S.I.N. loses his support in him. & haveing been in with Ld Oxford Bollingbrock & Dr Arbuthnet is not now looked on as he was formerly.

I sent last week for My Manuscripts [4] My Man brought me but 2 of them ye 3d was in Dr Hallys hands who is loth to part with it but S.I.N. I doubt not will force him

After which Sr. I.N. Will have still in his hands all the Observations made wth ye Murall Arch from 1689 to 1713 compleat. which I shall recall as soone as I have got back ye book yt Dr Hally deteins.

I beleive I have now an interest in some of the prime officers at Court that will not suffer me to be used as I have been formerly I shall recall ye MS of my Observations from 1689 to 1705 compleat. & know how he has disposed of 1200*lb* of prince Georges mony whereof I never received but 125*lb*

I shall not deal proudly with him nor call him Names as he did me, God forgive him, but I shall use him gently & calmly till I have got what he has of mine in his power out of his hands God has blest me hitherto all that has hapned I doubt not was by the order of his good providence but for ye best. I will attend him patiently till my hopes are turned into certaintys.

NOTES

(1) Sharp Letters, fo. 94, printed in Baily, *Flamsteed*, p. 314.
(2) Compare Letter 1145, and see the Introduction to vol. v. This letter from Flamsteed makes it obvious that the terms of Halifax's will were widely known, and given an unflattering interpretation.
(3) £20000.
(4) See Letter 1148.

A LIST of the MEDALS struck in the Reign of her Late Majesty, Queen *Anne*; with their Price.

	In Gold.			Silver.			Copper.		
	l.	*s.*	*d.*	*l.*	*s.*	*d.*	*l.*	*s.*	*s.*
THE Coronation Medal	3	15	0	0	5	0	0	1	6
A large Medal, the Motto, *Novæ Palladium Trojæ*	30	0	0	1	17	0	0	17	0
— On the taking and destroying the Galeons at *Vigo*	4	5	0	0	6	0	0	2	0
— On taking *Keyserwaert, Venlo, Ruremond,* &c.	4	5	0	0	6	0	0	2	0
The Queen and Prince	7	0	0	0	12	0	0	4	0
On the Surrender of *Bonn, Huy* and *Limburg*	7	0	0	0	12	0	0	4	0
The Chain of Hearts	3	15	0	0	5	0	0	1	6
Entirely *English*	3	15	0	0	5	0	0	1	6
On the Battle of *Blenheim*	3	15	0	0	5	0	0	1	6
On the Sea-Fight and taking *Gibralter*	6	6	0	0	7	0	0	2	6
On the Relief of *Barcelona*	3	15	0	0	5	0	0	1	6
On the Battle of *Ramellies*	3	15	0	0	5	0	0	1	6
On the Queen's giving the First-Fruits and Tenths to the Clergy	8	0	0	0	12	0	0	5	0
The large Union Medal	8	0	0	0	13	0	0	4	0
A small Medal on the Union	3	15	0	0	5	0	0	1	6
On the Battle of *Oudenard*	8	0	0	0	12	0	0	4	0
On the taking *Lisle*	8	0	0	0	12	0	0	4	0
On the Pretender's Invasion of *Scotland,* in 1708	6	6	0	0	7	0	0	2	6
On the Taking *Sardinia* and *Minorca*	6	6	0	0	7	0	0	2	6
On the Surrender of *Mons*	6	6	0	0	7	0	0	2	6
On the Taking of *Tournay*	6	6	0	0	7	0	0	2	6
On the Battle of *Tannier*	8	0	0	0	13	0	0	4	0
On the Battle of *Saragossa*	10	0	0	0	15	0	0	5	0
On the Battle of *Almenara*	10	0	0	0	15	0	0	5	0
On the Taking *Douay*	10	0	0	0	15	0	0	5	0
On the Taking *Bethune, St. Venant,* &c.	10	0	0	0	15	0	0	5	0
On the Taking *Bouchain*	8	0	0	0	12	0	0	5	0
The Peace Medal	3	15	0	0	5	0	0	1	6
A large Medal on the Peace	20	0	0	1	5	0	0	10	0
	222	0	0	15	7	0	5	9	6

A LIST of the MEDALS struck from his Majesty King GEORGE's Accession to the Throne, to the Year 1718.

	In Gold.			Silver.			Copper.		
THE Coronation Medal	3	15	0	0	5	0	0	1	6
A large Medal on the King's first Arrival in *England*	30	0	0	1	15	0	0	15	0
On his Publick Entrance through the City	10	0	0	0	15	0	0	5	0
On the beating the Rebels at *Dumblain* in *Scotland*	8	0	0	0	12	0	0	5	0
On the Defeat of the Rebels at *Preston*	8	0	0	0	12	0	0	5	0
On the Victory gain'd over the *Spanish* Fleet by Sir *George Byng* in the Mediteranean	8	0	0	0	12	0	0	5	0
On the King's being Mediator of the Peace between the Emperor, *Turks* and *Venetians*	8	0	0	0	12	0	0	5	0

7 *l.* 10.

Note, *The Price of Gold Medals is according to their Weight, so may be two or three Shillings more or less than what is here set down.*

1152 NEWTON TO THE TREASURY
14 JULY 1715

From the original in the Public Record Office.[1]
Reply to Letter 1143

An Account of what Mony has arisen for Tin sold in London and elsewhere from 1st July 1714, to 1st July 1715, Viz

London.	from 1st July to Christmas following . . .	£74763 . 13 . 5
	from Christmas to Ladyday 1715	27179 . 19 . 5
	from Ladyday to 1st July	33858 . 12 . 10
		135802 . 5 . 8

Holland	Mr Beranger hath within that time certified the sale of 1493 Blocks, which being computed at 3 cwt each, makes 4479 cwt and being to pay four pound per hundred[weight] clear of all charges it will produce	17916 . . . .
	What has been sold at Hamborough is very inconsiderable as yet, and doth not amount to more then	300 . . . .
		£154018 . 5 . 8

The Sales of Cornwall and Devon are certified to the Receiver of the Monies arising by the Sale of Tin (to whom they are answerable) and are unknown to this office.

July 14th. 1715 This morning I received from Dr Francis Fauquier the Deputy agent for the sales of Tin in the Tower, the Account above written of the sales during the last year.

Is. NEWTON

NOTE

(1) T/1, 190, no. 16. The last three lines are in Newton's hand.

1153 WOLF TO LEIBNIZ
17 JULY 1715
Extract from Gerhardt, *Briefwechsel zwischen Leibniz und Wolf*, p. 174.[1]

Keilius consueta rusticate in Transactionibus Anglicanis [2] suggillat solutionem Bernoullianam problematis inversi de vi centrali, quam in Commentariis Academiæ Regiæ Scientiarum A. 1711 exhibuit,[3] narrante amico. Insolescere videtur, postquam sibi persuadet, ipsi responderi non posse:[4] responsionem enim postulaverunt a me Diarii Hagiensis Collectores. Unde consultum judicarem, si cui novitio tela suppeditarentur, quæ in hominem insulsum vibraret.

Translation

In the English [*Philosophical*] *Transactions* [2] Keill, with customary rudeness, inveighs against Bernoulli's solution of the inverse problem of central force which is set out in the *Mémoires de l'Académie Royale des Sciences* for 1711,[3] so a friend tells me. He seems to become haughty after persuading himself that he cannot be answered:[4] for the editors of the Hague *Journal* have asked me for an answer. Whence I would judge it wise to have darts supplied to some novice to thrust into that insolent man.

NOTES

(1) Leibniz has written the following comment at the end of the letter: 'Nescio an mihi conveniat respondere Keilio, qui scribit ruditer et inciviliter. Cum talibus conflictari meum non est. Volo Antagonistam ita scribere, ut disputatio inter nos sit cum voluptate conjuncta. Si qui ex silentio meo sinistrum judicium capiunt, eorum judicium parum moror.

Suppeditavi modum ostendendi demonstrationem Keilii pro vacuo esse inanem.'

('I do not know that it would be fitting for me to answer Keill, who writes in a rude and uncivil manner. It is not for me to wrangle with such men. Accordingly I wish my champion to write in such a way that the dispute between us be associated with enjoyment. If anyone places a bad construction upon my own silence, I care nothing for his opinion.

I have supplied a way of showing that Keill's demonstration in favour of a vacuum is ridiculous.')

(2) 'Joannis Keill M.D. & in Academia Oxoniensi Astronomiæ Professoris Saviliani, Observationes in ea quæ edidit Celeberrimus Geometra Johannes Bernoulli in Commentariis Physico Mathematicis Parisiensibus Anno 1710. de inverso Problemate Virium Centripetarum. Et ejusdem Problematis solutio nova': Keill's 'Observations' printed in *Phil. Trans.* **29**, no. 340 (1714), 91–111.

(3) See Letter 1023 and notes; for 1711 read 1710.

(4) There seems little doubt that Keill had been trying to stir up an argument between Bernoulli and Newton. He wrote in his 'Observations' (*Phil. Trans.* **29**, no. 340 (1714) 95): 'Impossibile est ut credat nullam Newtono notam fuisse hujus rei demonstrationem; Noverit enim eum primum & solum fuisse qui hanc omnem de vi centripeta doctrinam geometrice

tractavit, quique eam ad tantam perfectionem perduxit, ut post plures quam viginti annos, parum admodum a præstantissimis Geometris ei additum sit. Noverit etiam Bernoullius Newtonum, præter generalem problematis inversi solutionem, ostendisse modum quo formari possunt Curvæ, quæ vi centripeta decrescente in triplicata distantiæ ratione describuntur, adeoque alterum illum casum ignorare non potuisse.'

('It is impossible that he [Bernoulli] believes that no demonstration of this matter was known to Newton. Certainly he [Bernoulli] would have known that it was he [Newton] who first and alone treated this whole doctrine of centripetal force geometrically and who brought it to such perfection that after more than twenty years very little has been added to it by the most distinguished geometers. Bernoulli would also have known that Newton, besides the general solution of the inverse problem, has shown the way in which curves can be constructed which are described by a centripetal force decreasing in proportion to the cube of the distance, and so he could not have been ignorant of the other case.') Keill went on to give his own solutions of the inverse cube problem, using fluxional notation.

1154 NEWTON TO THE TREASURY
20 JULY 1715
From the holograph draft in the Mint Papers.[1]

To the Rt Honble the Lords Comm[issione]rs of the Trea[su]ry

May it please your Lordps

The Mint is a place not subject to any military power but is directly under the King & Council & the Ld Treasurer or Comm[issione]rs of the Treasury & its own Officers. And by the Indentures of the Mint no strangers can live or lodge in the Mint without the leave of the Officers of the Mint, & by an Order of K. Charles II all strangers were turned out of the Mint & prohibited to live there any more without the leave of the Ld Treasurer & Chancellour of the Exchequer. But notwithstanding these Constitution [2] General Compton the Leieutenant of the Tower [3] has brought the Earle of Oxford into the House of the Comptroller of the Mint, & there put a guard upon him, [4] as if that house & by consequence the whole Mint was under his jurisdiction.

My Lords, the safety of the Coynage depends upon keeping the Mint out of the hands of the Garrison, & the safety of Prisoners depends upon keeping them in a legal custody under the jurisdiction of their keepers. And I am humbly of opinion not only that the Prisoner be removed into a legal custody but also that something be done which may hinder this invasion of the Mint from being drawn into president hereafter.

All wch &

Mint Office   I. NEWTON
July 20. 1715

NOTES

(1) III, fo. 407.

(2) Newton first wrote 'Orders' then deleted this word in favour of 'Constitution' but forgot to alter 'these'. The draft seems to have been written under some excitement for the signature also is unusual.

(3) Hatton Compton (1661?–1741), nephew of the Earl of Northampton, was instrumental in saving William III from imprisonment during the retreat from Landen, 1693. He was Lieutenant-Governor of the Tower from 1713 to 1715, but retired from the Life Guards in 1718.

(4) Unlike Bolingbroke, who fled to France in March 1715, Oxford took his seat in the first House of Lords summoned after the Hanoverian accession (April 1715). As a result of an enquiry into his conduct as first Minister, particularly with regard to the Treaty of Utrecht, he was sent to the Tower under articles of impeachment on 16 July. After abandonment of the impeachment he was released two years later to live quietly for seven more years.

1155   THE TREASURY TO NEWTON
28 JULY 1715
From the copy in the Public Record Office [1]

After Our hearty Commendations. These are to Authorise and Require You out of any Money that is or shall be in your hands of the Coynage Duty to make payment unto the Chief Warden, or any other of the Wardens of the Jury of Goldmsiths Summoned by Order of Council for the Tryal of the Pix, [2] of the Mint in the Tower of London, on the Second of August next, or to his Assignes the Summ of Thirty pounds, the Same being to defray the Charges of Entertaining the Said Jury and Such other persons whose Attendances are required for the performance of the Said Service. And this shall be as well to You for payment as to all others herein Concerned in allowing thereof upon Your Accounts a Sufficient Warrant

Whitehall Trea[su]ry Chambers 28 July 1715.

CARLISLE [3]

RI: ONSLOW

WILL ST. QUINTIN

To Our very Loveing Freind Sr. Isaac Newton Knt.
Master & Worker of His Majestys Mint.

NOTES

(1) Mint/1, 8, p. 107.

(2) See Letter 771, note (3) (vol. v, p. 13).

(3) Charles Howard (1669–1738) succeeded as Earl of Carlisle in 1692. He had previously served as M.P., and as First Lord of the Treasury in 1701–2. He had been reappointed to this office after the death of Halifax (19 May 1715) on 23 May.

1156 NICOLAS RÉMOND TO LEIBNIZ
24 AUGUST 1715
Extract from Gerhardt, *Leibniz: Philosophische Schriften*, III, p. 650

Mons. l'abbé Conti [1] [qui est en Angleterre] qui m'ecrit tres souvent, est charmé de ce pays là; il me paroit que M. Newton n'a rien de caché pour lui; je suis donc presentement tres instruit des sentimens particuliers des Philosophes Anglois. Les dernieres pages du livre de M. Newton sont bien embrouillées, sa conversation est plus libre et plus degagée

NOTE

(1) See Letter 1139, and notes. At this time Conti was on fairly intimate terms with Newton, and in frequent correspondence with Nicolas Rémond (see Letter 1139, note (5)) whom he kept well informed of English scientific matters. Rémond later sent Leibniz some extracts of Conti's letters. (See Gerhardt, *Leibniz: Philosophische Schriften*, III, p. 651.)

1157 THE TREASURY TO NEWTON
12 SEPTEMBER 1715
From a copy in the Public Record Office [1]

After Our hearty Commendations, In pursuance of His Mats. Command Signifyed by his Warrant upon a Report made by yourself & the Comptr[oller] of His Majestys Mint upon a Peticon of Sr. Richard [2] Sandford Bart. Warden of the Said Mint all which are hereunto annexed. [3] These are to direct & require you out of the Summe nott exceeding Four Hundred pounds per annum Authorised by an Act 7mo: Annæ to be applied out of the Coynage Duty for the charge and expence of the Officers and others employed and to be employed in the prosecution of the Offences in counterfeiting diminishing or otherwise concerning the current coine of Great Britaine in that part thereof called England, to pay or cause to be paid to such person or persons as is or shall be employed by the Warden of the Said Mint as Clerk and Sollicitor for apprehending & prosecuting Clippers and Counterfeiters of the Coine a Sallary at the rate of Sixty pounds per annum to Commence from the 29th: day of July 1715 being the date of His Majestys Said Warrant, [4] And also to pay or cause to be paid all such other allowances for Travelling Charges and other Incidents relating to the Searching for, apprehending and prosecuting the Said Clippers and Counterfeiters in Such manner & under Such regulacons as are contained in the Said annexed Report approved by His Ma[jes]ty as aforesaid. And for So doing this and the Said annexed Warrant & Report being first Entered in the Offices of His Majestys Auditors of the Mint or one of them shall be Your Sufficient Warrant.

Whitehall Trea[su]ry Chambers 12 Sept. 1715.

CARLISLE

WILL ST QUINTIN

EDWD. WORTLEY [5]

To Our very loving freind Sr. Isaac Newton Knt.
Master and Worker of His Majestys Mint
And to the Master & Worker thereof for the time being.

NOTES

(1) Mint/1, 8, p. 110.
(2) The clerk actually wrote 'Riclland'.
(3) The report on which this action was based is Letter 1150.
(4) The warrant was attached to this order (Mint/1, 8, p. 109) and is printed in *Cal. Treas. Books*, XXIX (Part II), 1714–15, p. 657.
(5) Edward Wortley [Montagu] (1678–1761), educated at the Middle and Inner Temple, M.P. for Huntingdon, was made a Lord of the Treasury soon after the accession of George I, his cousin Halifax being First Lord. He had married Lady Mary, daughter of the Marquis of Dorchester, without her father's consent in 1712. He was dismissed from the Treasury when Walpole became First Lord (11 October 1715), and was sent as ambassador to Constantinople in June 1716. It was on this embassy that Lady Mary learnt of the Turkish practice of inoculating against smallpox.

1158 THE MINT TO THE TREASURY
14 SEPTEMBER 1715
From the original in the Public Record Office.[1]
Reply to Letter 1134

To the Right Honble: the Lords Comm[issione]rs of His
Majestys Treasury

May it Please Your Lordships

According to Your Lordships Order of 10th March 17$\frac{14}{15}$, Wee have Considered the Petition of the Gravers of His Majesty's Mint hereunto annexed, and are satisfyed of the Truth of the Things therein represented: And now an Act of Parliament is passed which makes Mr. Croker the head Graver (a Foreigner Naturalized many years ago) capable of being continued in his Place of Graver by a Patent or Constitution, Wee are humbly of Opinion that it is for his Majestys service to grant the said Petition.

All which is humbly submitted to Your Lordps great Wisdom

RICH SANDFORD

ISAAC NEWTON

Mint Office
Septr: 14: 1715

NOTE

(1) T/1, 192, no. 13.

1159   WOLF TO LEIBNIZ
20 SEPTEMBER 1715

Extract from Gerhardt, *Briefwechsel Zwischen Leibniz und Wolf*, pp. 174–5.
For the answer see Letter 1160

Litteræ E. T. recte mihi traditæ sunt, tum priores in quibus Keiliana temeritas notabatur, tum posteriores, quæ opus Hermannianum dignis encomiis prædicant.[1]

Postquam hisce diebus ex Anglo quodam, qui me inviserat, intellexi, Keilium ob mores sceleratos[2] (cum studiosis enim curæ ac fidei commissis cauponas et lupanaria frequentavit, lucrum insigne in ebrietate et fornicatione ponens) ab officio Professorio remotum id agere, ut controversiis inclarescat morum pravitate infamis, nec mihi consultum videtur cum istiusmodi homine congredi et litem, quæ plerisque videbitur, de lana caprina movere. Neque hoc rerum statu probare possem, si E. T. ad objectiones hominis insulsi responderet. Interim tamen e re Reipublicæ litterariæ mihi videretur, si Commercium aliquod epistolicum E. T. in publicum prostaret.[3]

Translation

Your Excellency's letters have been delivered to me safely, both the first, in which Keill's audacity is mentioned, and the second, which describes Hermann's work with deserving praise.[1]

A few days ago I learnt from someone from England who visited me that Keill had behaved so unlike the occupant of a professiorial chair because of his disgraceful morals[2] (for he has frequented drinkshops and bawdy-houses with the students entrusted to his care, spending heavily on wine and women) that he may become notorious for some infamous proceedings arising from his want of morals, and it does not seem wise to me to be involved with such a man, and to begin an argument which seems to most people to be about nothing. Nor in the present state of affairs can I approve Your

Excellency's replying to the objections of an idiot. Meanwhile it seems to me advantageous to learning that Your Excellency should present to the public a *Commercium Epistolicum*.[3]

NOTES

(1) Both the letters Wolf mentions are missing; but a rough note by Leibniz on Letter 1153 records his attitude to Keill.

(2) There is no other contemporary evidence to support the charge of immorality which Wolf's informant here brings against Keill. Leibniz in his reply takes a more moderate view of Keill's behaviour.

(3) See Letter 1101, note (2), p. 173.

1160 LEIBNIZ TO WOLF
[AUTUMN 1715]

Extract from Gerhardt, *Briefwechsel zwischen Leibniz und Wolf*, pp. 176–8.
Reply to Letter 1159

Audiveram ego quoque Keilii mores non admodum laudari, sed ignorabam eo rem processisse, ut ab officio fuerit depositus. Fortasse id contigit ante multos annos, et jam censetur expiatum;[1] idque ex eo suspicor, quod eum intelligo nunc in locum Walleri [2] demortui factum esse Secretarium Societatis Regiæ secundum et Hallejo adjunctum, quod suis in nos latratibus videtur apud Newtonum et alios ejus factionis meruisse.

Translation

I had heard myself that Keill's manners are certainly not to be admired, but I did not know that the thing had gone so far that he had been deprived of his position. Perhaps it happened many years ago, and he may be considered by now to have exonerated himself.[1] And I conjecture this from the fact that I understand him now to be made second secretary of the Royal Society, assisting Halley, in place of the late Waller,[2] which [position] he seems to have earned in the eyes of Newton and others of that faction by his railings at us.

NOTES

(1) There is no evidence that Keill was ever dismissed from his post, for although he left Oxford for a time (1709–12) it was his own choice to do so. In May 1712 he was unanimously elected Savilian Professor of Astronomy at Oxford.

(2) Leibniz was misinformed; in fact Brook Taylor was elected Secretary with Halley on 13 January 1715, in succession to Richard Waller, and continued till 1718.

1161 THE TREASURY TO NEWTON
21 SEPTEMBER 1715
From a copy in the Public Record Office [1]

After Our hearty Commendations By Virtue of a Warrant under His Mats.
Royal Sign Mannual dated the 15th Instant, These are to pray & Require You
to provide and deliver or cause to be provided and delivered unto Sir Clement
Cottrell Knt. (Master of Ceremonies to his Majestye) [2] Thirty such Medalls of
Gold as were distributed at his Majestys Coronation, And of the Same Weight
and fineness being intended for the Several Foreign Ministers and others
mentioned in the Copy of a List or Schedule hereunto Annexed, According to
the Directions of His Majestys said Sign Manual in that behalf And for So
doing this together with the Acquittance of the Said Sir Clement Cottrell shall
be Your Sufficient Warrant. [3] Whitehall Trea[su]ry Chambers 21 Sept 1715.

To Our very loving Freind Sr. Isaac Newton Knt.   RID: ONSLOW
Master and Worker of His Majesties Mint
                                                         WILL ST. QUINTIN

                                                         EDWD. WORTLEY

[There follows immediately afterwards the acquittance]

October 1715

Received then of Sr. Isaac Newton Knt. Master and Worker of His Mats:
Mint the Thirty Medals of Gold mentioned in the Warrant on the other Side
written, Weighing Twentyone Ounces Eighteen pennyweight and twenty
grains, I say received by Me

                                                         CLEMENT COTTRELL

oz dwt gr	£ s d	
21 : 18 : 20. of fine Gold at 4 : 07 : 06 per oz . .		£95 : 19 : 10½
making 30 Medals at 3s apeice		4 : 10 : 00
His Mats. Sign Manual to the Lords of the Treasury about the Said Medals		4 : 05 : 00
Their Lordships Warrt. to Sr. Is: Newton to deliver the Same		1 : 15 : 00
Warrt. to the Auditor of the Exchqr. for paying the Money out of the Civil List		1 : 15 : 00
Fees at the Exchequer in receiving the Money . . .		0 : 07 : 06
		£.108 : 12 : 04½

NOTES

(1) Mint/1, 8, p. 112. There is another copy in T/54, 23, p. 62. Compare Letter 1106.

(2) Sir Clement Cotterell [-Dormer] (d. 1758), Master of Ceremonies from 1710 to his death; he took the additional name Dormer on inheriting Rousham, Oxon.

(3) The Order is rehearsed at Mint/1, 8, p. 111 (there is a copy at T/52, 27, p. 152); it authorizes the distribution of two medals to each ambassador and one to each envoy or other official named on the list, which may be found in *Cal. Treas. Books*, xxix (Part ii), 1714–15, p. 740.

1162 HALLEY TO [KEILL]
3 OCTOBER 1715
Extract from the holograph original in Trinity College Library, Cambridge [1]

London Octob 3d 1715

Dear Sr

We have printed a French translation of ye account of the Commercium given in the Transactions, [2] in order to send it abroad: Sr Isaac is desirous it should be publisht in the Journal Literaire and Mr Gravesant [3] has promised to gett it done, but cares not to do it as of his own head; and therfore proposes that you would signifie to Mr Johnson at the Hague, by a letter enclosed either to Sr Isaac or me, [4] that you are desirous that the said French paper be inserted in his Journal, as containing the whole state of ye controversy between you and Mr Leibnitz. Sr Isaac is unwilling to appear in it himself, for reasons I need not tell you, and therfore has ordered me to write to you about it, who have been his avowed Champion in this quarrell; and he hopes you will gratifie him in this matter by the first opportunity.

NOTES

(1) R.16.38, fo. 429, printed in Edleston, *Correspondence*, pp. 184–5.

(2) The paper Halley refers to, commonly known as the *Recensio*, appeared in English, French and Latin: in English in *Phil. Trans.* 29, no. 342 (February 1715), 173–225 as 'An Account of the Book entitled Commercium Epistolicum . . .'; in French (the version discussed here) in the *Journal Literaire de la Haye*, 7 (1715), Part i, 114–58 and 344–65 as 'Extrait du Livre intitulé Commercium Epistolicum . . .'; and, much later, in Latin, as 'Recensio Libri qui inscriptis est Commercium Epistolicum . . .', prefixed to the 1722 edition of the *Commercium*, pp. 1–50.

The *Recensio* was published anonymously; James Wilson first mentioned that Newton himself was its author in Benjamin Robins's *Mathematical Tracts*, ii (London, 1761), p. 368—this is in a lengthy appendix by Wilson on the calculus dispute. The point was repeated by A. de Morgan in the *Philosophical Magazine*, 3, 4th series (June 1852), pp. 440–4. Numerous drafts in Newton's hand, are to be found in U.L.C. Add. 3968, *passim*, and in private possession. The paper reviews and puts into context all the letters which were previously printed virtually without comment in the first edition of the *Commercium Epistolicum*.

It is probably relevant to the *Recensio* that Newton, at a meeting of the Royal Society on 10 March 1715, ordered that 'the rest of the Extracts of Letters in the Commercium should be compared with the Originals and with the Copies in the Letter Books of the Society which was done accordingly; and the hands of Mr. Leibnitz Mr. Oldenburgh and Collins were very well known to several persons present.' The *Recensio* also brought to public light two new letters, which Newton deposited with the Royal Society at a meeting on 5 May 1715. These were a letter from Leibniz to Newton, dated 7/17 March 1693 (see Letter 407, vol. III, pp. 257–9) and a letter from Wallis to Newton dated 10 April 1695 (see Letter 498, vol. IV, pp. 100–1).

(3) W. J. 'sGravesande. See his letter to Newton (Letter 1083) where he offers to help as far as he can in settling the calculus dispute.

(4) Keill did this, and an extract of his letter to Johnson prefaces the articles in the *Journal Literaire de la Haye*. Keill explains that the *Philosophical Transactions* are difficult to obtain abroad, so that he is sending a French translation of the *Recensio* in the hope that the editor will see fit to publish it.

1163 LEIBNIZ TO J. BERNOULLI
24 OCTOBER 1715
Extract from Gerhardt, *Leibniz: Mathematische Schriften*, III/2, p. 948

Transiit hac nuper Dn. Johannes Arnoldus [1] Anglus, Vir, ut apparet, doctus et bonus, et a Te mihi literas gratissimas attulit. Dedi ipsi aliquot exempla Schedæ impressæ, [2] cui judicium Tuum est insertum, quod adversarios non parum urit. Keilius quidem responsione indignus est, sed rem ipsam brevi narratione complecti e re erit et adjicere problemata quædam, unde intelligamus, quid ipsi possint. Cum illam novam differentiandi rationem considerasses, cujus ope problemata solvuntur, quale illud de minimo arcu Elliptico intercepto, excogitaveras inde applicationem quandam sic satis generalem, qua, si bene memini, problemata quædam, quæ ad differentio-differentiales descendere solent, intra differentiales primi gradus coercentur. Ego nunc non bene memini, nec in literis antiquis quærere vacat. Te autem melius meminisse puto, itaque rogo ut si commodum est, iterum communices. [3] Inserviet enim fortasse ad aliquod problema proponendum, cujus non statim apparebunt fontes.

Translation

Mr John Arnold, [1] an Englishman, to all appearances a clever and good man, has passed through here lately, and has brought me a most welcome letter from you. I have also given him some examples of the printed sheet, [2] in which your judgement is given which much disturbs [our] adversaries. Keill, indeed, does not deserve a reply, but it will be advantageous to embrace the bare facts in a brief account and to add some problems, from which we may understand what they can do. Since you will have considered the new method of differentiating, by the help of which such problems are solved as that concerning the minimum arc cut off from an ellipse, you will have derived from it some application general to such a degree that, if I remember well, certain problems, which should reduce to differentio-differentials, are confined to differentials of the first order. Now I myself do not remember [that application] well, and I have not the leisure to look at old letters. I think, however, you will remember [it] better, so I ask you, if it is convenient, to impart [it] to me again.[3] For it may perhaps be useful in proposing some problem, of which the origins will not at once be apparent.

NOTES

(1) See Letter 1014, note (1), p. 27.

(2) The *Charta Volans*; see Number 1009. Arnold speaks of the copies Leibniz gave him in Letter 1181.

(3) Bernoulli sent excerpts from earlier letters, with permission for Leibniz to use the problems contained there, in his next letter, dated 23 November 1715. Gerhardt prints the letter, (see *Leibniz: Mathematische Schriften*, III/2, pp. 949–51) but the enclosed excerpts are missing. The excerpts apparently concerned problems involving the manipulation of variable parameters—for example the finding of the curve of maximum or minimum length amongst a set of curves, or for finding the curve cutting a second set of curves in a given angle, or in an angle varying according to a certain law. Bernoulli and Leibniz had discussed problems of this type in their correspondence of 1694; see Letter 1170, note (6), p. 254.

1164 TIN ACCOUNTS
26 OCTOBER 1715

From the holograph draft in the Mint Papers.[1]
Reply to Letter 1143

An Account of the Tin remaining unsold in the Tower, in Holland, & at Hamburgh, *Octob.* 25. 1715.

There remained in the Tower 1742 Tuns. 4 cwt. 0 Qr. 5 lb wt which at 76£ per Tunn (being the price it is now sold at) will make – – – – – – – – – –

£ s d

13247. 7. 4 [2]

There was in Holland in the hands of Mr Beranger 1600 Tuns, which at 80£ per Tun comes to   128000£
Out of wch summ must be deducted what he has paid to Mr Anstis upon acct being (as appears by his Accts) – – – – – – –   19500
And to Mr Nicoll – – – – – –   2000   75500
And what he has advanced by way of anticipation – – – – – – – –   54000

And there remains – –   52500. 0. 0

There was also at Hamburgh in the hands of Sr John
Lambert & Captain Gibbon 960 Tunns, which at 80£
per Tun comes to – – – – – – – – – 76800£
Out of wch must be deducted what they have
advanced by way of anticipation – – – –   14000
                                    And there remains   – –   62800. 0. 0

And the summ total of the three Accts. is   – – – –   247707. 7. 4 [3]

Mr Beranger & Sr John Lambert are to have an Interest at the rate of 5 per
cent to be recooped from what they sell wch must be deducted from this summ
as also the charges of the Management. And what the Agents have in Cornwall
must be deducted.

Is. NEWTON

Mint Office
Octob. 26. 1715

The Agents in Cornwall had in their hands at Christmas last about 397
Tuns, to which if the Christmas, Lady Day & Midsummer coynages be added,
& if from the summ be deducted 951 Tuns 15 cwt 3 Qr 20 lb sent from Cornwall to ye Tower; the remainder will be the quantity now in Cornwall. vizt [4]

NOTES

(1) III, fo. 541. The persons engaged in the 'tin affair' have been identified in vol. v.

(2) *Recte*, £132407. 7s. 4d; Newton has omitted the zero.

(3) The total is correct (obviously it was copied from an earlier computation) if the emendation noted above be made.

(4) The amount is not stated on the MS.

1165 KEILL TO NEWTON
[28 OCTOBER 1715]

From the original in the University Library, Cambridge.[1]
Continued in Letter 1169

Honoured Sr

Having had a great deall of other bussiness on my hand since I came here,
I have not had time to doe what I promised you till now, and I here send you
the first part wch I intirely submitt to your Judgement to change and alter as
you think fitt, If you doe not like the Title I desire you may give it another.[2]
I hope the next week to send you up the rest of it. I wish you could get it
translated into French and sent over to Mr Fontenel I think he is obliged to
doe us Justice and to receive a vindication since he accepted and published

245

the accusation. I would gladly have this paper inserted in the Memoires for this year and therefore wish it were done with all speed. My most humble service to my Brother [3] Halley and Dr Taylor. [4] and beleive me to be Sr

Your

Most affectionat Humble servant

JOHN KEILL

For
Sr Isaack Newton
at his house in St Martins
Street near
  Leicester Fields
    Westminster

NOTES

(1) Add. 3985, no. 16. The letter was clearly written from outside London, possibly from Oxford; it is postmarked 'OC 29' and was therefore written at latest on the 28th (since letters to London were postmarked at the Post Office on their arrival there). The year is determined by the contents; see note (2) below and Letter 1169.

(2) It was 'Apologie pour le Chevalier Newton, dans laquelle on repond aux remarques de Messieurs Jean et Nicolas Bernoully inserees dans les Mémoires de l'Academie Royale des Sciences pour les années, 1710 & 1711, par J. Keill . . .' There is an MS. copy amongst the Lucasian Papers (U.L.C., Res. 1893(a), no. 5. first manuscript). The article finally appeared under the title 'Defense du Chevalier Newton par M. Keil contre MM. Bernoully' in the *Journal Literaire de la Haye* for 1716 (Part II), pp. 418–33. This paper was an answer to Johann Bernoulli's 'letter' in the *Mémoires de l'Académie Royale des Sciences*, for 1710 (Paris, 1713), pp. 519–33 (see Letter 1023 and notes) on the inverse problem of central forces and Newton's incompetence therein—Keill had already produced one answer to this paper in *Phil. Trans.* **29**, no. 340 (1714), 91–111 (see Letter 1153, note (2)). The 'Apologie' Keill speaks of here was originally intended for publication in the *Mémoires de l'Académie Royale des Sciences*, and for this purpose it was sent by Halley to Fontenelle on 19 January 1716. The Académie, however, refused to print Keill's communication on the grounds that he was not a member. (See Edleston, *Correspondence*, p. 187 and Letters 1194 and 1206.)

Bernoulli's answer to Keill's article was the 'Epistola pro Eminente Mathematico', *Acta Eruditorum* for July 1716, pp. 296–315; see Letter 1196.

(3) Halley and Keill were both Savilian Professors at Oxford.

(4) Brook Taylor (1685–1731) graduated from St John's College, Cambridge in 1701. He became a Fellow of the Royal Society on 3 April 1712, and served as secretary from 14 January 1714 to 21 October 1718. *Methodus Incrementorum Directa et Inversa* was published in 1715 and contained the now-famous 'Taylor's Theorem'. His ardent support of Newton against Leibniz was partly fired by his own personal controversy with Johann Bernoulli (Letter 1217, note (2)); see Heinrich Auchter, *Brook Taylor, der Mathematiker und Philosoph; Beiträge zur Wissenschaftsgeschichte der Zeit des Newton–Leibniz–Streites* (Würtzburg, 1937).

246

1166 HENRI DU SAUZET TO LEIBNIZ
29 OCTOBER 1715
Extract from the original in the Niedersächsische Landesbibliothek, Hanover.[1]
For the answer see Letter 1167

Au reste on m'envoya il y a quelques semaines un Article[2] touchant la dispute que vous avez avec Mr Keill sur le droit d'invention de la Methode des Fluxion[s] qu'il pretend être düe à Mr Newton. J'inserai cet Article dans les Nouvelles du 21 Septembre [N.S.] tel qu'on me l'avoit envoyé. Si vous jugez à propos d'y faire quelque réponse,[3] ie vous offre de l'inserer telle que vous me la fournirez.

NOTES

(1) The letter is not listed in Bodemann's catalogue. In the microfilm of the Leibniz manuscripts in the University Library, London, it appears in section 13:59. Henri du Sauzet was editor of the *Nouvelles Litteraires*, a weekly journal containing book reviews, obituaries, scientific and literary news, etc. published at the Hague. The first issue appeared in January, 1715. Du Sauzet also later took responsibility for the publication of Des Maizeaux' *Recueil*, a task which proved extremely troublesome (see vol. VII).

(2) In the *Nouvelles Litteraires* for 21 September 1715 (N.S.), pp. 184–5, a brief review of the *Recensio* (see Letter 1162) appeared. It referred to the *Recensio* as 'une Histoire très éxacte de cette Dispute, & toute differente de ce qui en a paru dans les Journals étrangers, particuliére-ment dans le *Journal Literaire* de Juillet & Aout 1713. Article IV'.

(3) In this reply (Letter 1167) Leibniz wrote that he had no intention of answering the article. However in the *Nouvelles Litteraires* for 28 December 1715 [N.S.], pp. 413–14, an answer, presumably by Leibniz, was printed (see Letter 1172).

1167 LEIBNIZ TO DU SAUZET
4 NOVEMBER 1715
Extract from the draft in the Niedersächsische Landesbibliothek, Hanover.[1]
Reply to Letter 1166

je ne repondrois point à des gens qui m'attaquent d'une maniere incivile et grossiere.[2] je pourray un jour traiter la matiere meme sans aucun égard à de tels ecrivains; dont les connoiesseurs méprisent les petites raisons.

Vous m'obligerez en ne mettant rien qui me regarde avant que de m'en avoir fait part

NOTES
(1) The letter is not listed in Bodemann's catalogue. It appears in the microfilm of Leibniz's manuscripts at the University Library, London in section 13:59.
(2) But see Letter 1172.

1168 NEWTON TO COTES
8 NOVEMBER 1715
From the original in Trinity College Library, Cambridge[1]

Sr

By the death of Dr Burnet the Master of Charterhouse that Mastership is now voyd.[2] It is worth about 200*lb* per an in money besides lodgings meat drink & fire & the maintenance of one or two servants (I think a man & a maid servant) to attend him.[3] He puts in & turns out all the servants attending the Hospital at his pleasure. The place is accounted honourable & easy & there is no body yet puts in for it who is like to succeed. The Electors should have met a few days ago, but put of their meeting to a further day & there is some danger that they may let it lapse to the king. And this I take to be the preferment of wch Mr Herring[4] gave you a hint. You have been spoken of privately as a fit person & named to one or two of the Electors without their dislike. But on the other hand some question has been made of your firmness to the present government under King George. And if that objection be not removed there is danger that the king may be sollicited to speak to the Electors for some body else in opposition to you. If you like the Post & think to put in for it & can remove the Objection by a Letter wch may be shown to some of the Electors, the next step will be to propose you to them & sollicit them to meet & make an Election.[5] I am

Sr

Your most obedient

humble servant

Is. NEWTON

St Martin street
in Leicester Fields
8 *Novem.* 1715

For the Rnd Mr Cotes, Professor of
Astronomy, at his Chamber in
Trinity College in
    Cambridge.

NOTES

(1) R.4.42, no. 13. There is no record of an antecedent letter from Cotes to Newton.

(2) Thomas Burnet (b. 1635), of Christ's College, Cambridge became Master of the Charterhouse in 1685, and died on 27 September 1715. He was well known for his *Sacred Theory of the Earth* (London, 1681).

(3) The Charterhouse, in the City of London, originally a Carthusian monastery (founded in 1371), was endowed in 1611 by Thomas Sutton as a chapel, almshouse and school. The school, having become large and celebrated, removed to Surrey in 1872 but the brothers remain on the old site.

(4) It is not possible to identify the likely person among several 'possibles' of this name—including an elderly member of Cotes' College and a future Archbishop of Canterbury.

(5) What little is known of the remaining months of Cotes' life suggests that (unlike Newton) he chose to remain a professor at Cambridge. Newton himself had considered the Mastership of the Charterhouse a poor post in 1691 (vol. III, pp. 184–5).

1169   KEILL TO NEWTON
10 NOVEMBER 1715
From the original in the University Library, Cambridge.[1]
Continuation of Letter 1165

Oxford Novbr 10th 1715

Honoured Sr

I here send you up the rest of the paper[2] I have written against the Bernoulli[s] I beleive I am alitle to severe in my expressions tho they deserve to be more roughly used, yet I give you leave to alter them as you shall think fitt only I desire you would let me have a copy of what you send over to France. I am Sr

your

Most Humble and most obliged servant

JOHN KEILL

For
Sr Isaack Newton
at his house in St Martins
street near Licester fields
    Westminster

NOTES

(1) Add. 3985, no. 11.
(2) See Letter 1165, note (2), p. 246.

1170   LEIBNIZ TO CONTI
25 NOVEMBER 1715

Extract from Gerhardt, *Briefwechsel*, pp. 262–267.[1]
Reply to Letter 1139; for the answer see Letters 1187 and 1190

Hannover 6 *Decembr.* 1715 [N.S.]

On m'a dit tant de bien, Monsieur, de vostre penetration et de vos nobles desseins pour la recherche de la verité, que l'honneur de votre lettre ne m'a pû etre que tres agreable, et je souhaiterois de vous y pouvoir aider. Monsieur Negrisoli doit etre homme de merite et de reputation, puisque vous avés pris la peine d'entrer en dispute avec luy. La lumiere seminale est un beau mot, mais dont on ne connoist point le sens. Je m'imagine que ces Messieurs qui s'en servent, l'entendent dans un sens metaphorique. Que la lumiere leur signifie quelque matiere subtile douée de grandes perfections, comme la lumiere paroist le plus parfait fluide qui nous soit connu, ils logeront dans un corps si parfait un artifice assez grand pour former les animaux, mais cela n'explique rien; il faut qu'une matiere capable d'organiser soit organisée elle meme, mais un fluide tel que la lumiere ne dit pas cela. La lumiere prise dans le sens metaphorique, dont je viens de parler, conviendroit assez avec l'esprit de M. Newton . . .

P.S.[2]

Voila, Monsieur, la lettre dont vous pourrés faire usage si vous le jugés à propos. Je viens maintenant à ce qui nous regarde. Je suis ravi que vous estes en Angleterre, il y a de quoy profiter. Et il faut avouer qu'il y a là de tres habiles gens. Mais ils voudroient passer pour etre presque seuls inventeurs, et c'est en quoy apparremment ils ne reussiront pas. Il me paroist point que M. Newton ait eu avant moy la Caracteristique et l'Algorithme infinitesimal, suivant ce que M. Bernoulli a tres bien jugé, quoyqu'il luy auroit eté fort aisé d'y parvenir s'il s'en fut avisé. Comme il auroit esté fort aisé à Apollonius de parvenir à l'Analyse de Des Cartes sur les Courbes, s'il s'en etoit avisé. Ceux qui ont ecrit contre moy n'ayant pas fait difficulté d'attaquer ma candeur par des interpretations forcées et mal fondées, ils n'auront point le plaisir de me voir répondre à de petites raisons de gens qui en usent si mal, et qui d'ailleurs s'ecartent du fait. Il s'agit du Calcul des differences, et ils se jettent sur les Series, où M. Newton m'a precedé sans difficulté; mais je trouvay enfin une Methode generale pour les Series, et après cela je n'avois plus besoin de recourir à ses extractions. Ils auroient mieux fait de donner les Lettres entieres comme M. Wallis a fait avec mon consentement, et il n'a pas eu la moindre dispute avec moy, comme ces gens là voudroient persuader au public. Mes adversaires n'ont publié du Commercium Epistolicum de M. Collins que ce qu'ils ont crû capable de recevoir leur mauvaises interpretations. Je fis connoissance avec M. Collins dans mon second voyage d'Angleterre, car au premier (qui dura tres peu, parceque j'estois venu avec un ministre public) je n'avois pas encore la moindre connoissance de la Geometrie avancée, et n'avois rien vû ny entendu du commerce de M. Collins avec Mss. Gregory et Newton, comme mes lettres echangées avec M. Oldenbourg en ce temps là et quelque temps après feront assez voir. Ce n'est qu'en France que j'y suis entré, et M. Hugens m'en donna l'entrée. Mais à mon second voyage M. Collins me fit voir une partie de son commerce, et j'y remarquay que M. Newton avoua aussi son ignorance sur plusieurs choses, et dit entre autres qu'il n'avoit rien trouvé sur la dimension des Curvilignes celebres que la dimension de la Cissoide. Mais on a supprimé tout cela. Je suis faché qu'un aussi habile homme que M. Newton s'est attiré la censure des personnes intelligentes, en deferant trop aux suggestions de quelques flatteurs qui l'ont voulu brouiller avec moy.[3] La Societé Royale ne m'a point fait connoitre qu'elle vouloit examiner l'affaire, ainsi je n'ay point eté ouï et si l'on m'avoit fait savoir les noms de ceux qu'on avoit nommés comme Commissaires, j'aurois pû m'expliquer si je recusois quelques uns, et si j'en desirois. C'est pourquoy les formalités essentielles n'ayant point eté observées, la Societé a declaré qu'elle ne pretend point d'avoir jugé definitivement entre M. Newton et moy.

Sa Philosophie me paroist un peu étrange, et je ne crois pas qu'elle puisse s'etablir. Si tout corps est grave, il faut necessairement (quoyque disent ses defenseurs et quelque emportement qu'ils temoignent) que la gravité soit une qualité occulte Scholastique, ou l'effect d'un miracle. J'ay fait voir autresfois à M. Bayle que tout ce qui n'est pas explicable par la nature des creatures est miraculeux. Il ne suffit pas de dire, Dieu a fait une telle loy de Nature, donc la chose est naturelle. Il faut que la loy soit executable par les natures des creatures. Si Dieu donnoit cette loy, par exemple à un corps libre, de tourner à l'entour d'un certain centre, il faudroit ou qu'il y joignit d'autres corps qui par leur impulsion l'obligeassent de rester tousjours dans son orbite circulaire, ou qu'il mit un Ange à ses trousses, ou enfin il faudroit qu'il y concourut extraordinairement. Car le mobile s'ecartera par la tangente. Dieu agit continuellement sur les creatures par la conservation de leur Natures, et cette conservation est une production continuelle de ce qui est perfection en elles. Il est *intelligentia supramundana*, parcequ'il n'est pas l'Ame du Monde, et n'a pas besoin de *sensorium*.

Je ne trouve pas le vuide demonstré par les raisons de M. Newton ou de ses Sectateurs, non plus que la pretendue gravité universelle, ou que les Atomes. On ne peut donner dans le vuide et dans les Atomes, que par des vues trop bornées. M. Clark dispute contre le sentiment des Cartesiens qui croyent que Dieu ne sauroit destruire une partie de la matiere pour faire un vuide, mais je m'etonne qu'il ne voye point que si l'Espace est une substance differente de Dieu, la même difficulté s'y trouve. Or de dire que Dieu est l'Espace, c'est luy donner des parties.[4] L'Espace est quelque chose mais comme le temps, l'un et l'autre est un ordre des choses, l'espace est l'ordre des coexistences, et le temps est l'ordre des existences successives. Ce sont des choses veritables mais ideales, comme les Nombres.

La matiere même n'est pas une substance, mais seulement *substantiatum*, un Phenomene bien fondé, et qui ne trompe point quand on y procede en raisonnant suivant les loix ideales de l'Arithmetique, de la Geometrie et de la Dynamique etc. Tout ce que j'avance en cela paroist démontré. A propos de la Dynamique ou de la Doctrine des forces, je m'etonne que M. Newton et ses sectateurs croyent que Dieu a si mal fait sa machine, que s'il n'y mettoit la main extraordinairement, la montre cesseroit bientôt d'aller. C'est avoir des idées bien étroites de la sagesse et de la puissance de Dieu. J'appelle extraordinaire toute operation de Dieu, qui demande autre chose que la conservation des natures des creatures. Ainsi quoyque je croye la Metaphysique de ces Messieurs là, *a narrow one*, et leur Mathematique assez *arrivable*, je ne laisse pas d'estimer extremement les meditations physico-mathematiques de M. Newton, et vous obligeriés infiniment le public, Monsieur, si vous portiés cet habile homme à nous donner jusqu'à ses conjectures en physique. J'approuve fort sa methode de tirer des phenomenes ce qu'on en peut tirer sans rien supposer, quand meme ce ne seroit quelquefois que de tirer des consequences conjecturales. Cependant quand les *data* ne suffisent point, il est permis (comme on fait quelquefois en déchifrant) d'imaginer des hypotheses, et si elles sont heureuses, on s'y tient provisionellement, en attendant que des nouvelles experiences nous apportent *nova Data*, et ce que Bacon appelle *Experimenta crucis*, pour choisir entre les hypotheses. Comme j'apprends que certains Anglois ont mal representé ma Philosophie dans leur Transactions, je ne doute point qu'avec ce que je vous mande icy, Monsieur, je ne puisse estre justifié. Je suis fort pour la Philosophie experimentale, mais M. Newton s'en ecarte fort quand il pretend que toute la matiere est pesante (ou que chaque partie de la matiere en attire chaque autre partie) ce que les Experiences ne prouvent nullement, comme M. Hugens a deja fort bien jugé; la matiere gravifique ne sauroit avoir elle même cette pesanteur dont elle est la cause, et M. Newton n'apporte aucune experience ny raison suffisante pour le vuide et les Atomes ou pour l'attraction mutuelle generale. Et parcequ'on ne sait pas encor parfaitement et en detail comment se produit la gravité ou la force elastique, ou la magnetique etc., on n'a pas raison pour cela d'en faire des qualités occultes Scholastiques ou des miracles; mais on a encore moins raison de donner des bornes à la sagesse et à la puissance de Dieu, et de luy attribuer un *sensorium* et choses semblables. Au rest, je m'etonne que les Sectateurs de M. Newton ne donnent rien qui marque que leur maistre leur a communiqué une bonne Methode, j'ay eté plus heureux en disciples.[5] C'est dommage que M. le Chevalier Wren, de qui M. Newton et beaucoup d'autres ont appris quand il etoit jeune, n'a pas continué de regaler le public. Je crois qu'il est encor en vie. Il seroit bon de faire connoissance avec luy. Dans le temps qu'il estoit jeune, on se seroit moqué en Angleterre de la nouvelle philosophie de quelques Anglois. et on l'auroit renvoyée à l'école. Luy et M. Flamstead avec M. Newton sont presque le seul reste du siecle d'or d'Angleterre par rapport aux sciences. M. Whiston étoit en bon train. Mais un certain zele étrange l'a jetté d'un autre coté. Je plains le public de cette perte. Depuis quelque temps on s'y est jetté dans les *ghiribizzi politici*, ou dans les controverses Ecclesiastiques. Il y a un François en Angleterre, nommé M. Moyvre, dont j'estime les connoissances Mathematiques. Il y a sans doute d'autre habiles gens, mais qui ne font point de bruit, dont vous saurés sans doute des nouvelles, Monsieur, et vous m'obligerés de m'en apprendre. . . .

Billet.[6]

Pour tater un peu le pouls à nos Analystes Anglois, ayés le bonté, Monsieur, de leur proposer ce probleme comme de vous même ou d'un amis: Trouver une ligne *BCD* qui coupe à angles droits toutes les courbes d'une suite determinée d'un même genre, par exemple toutes les Hyperboles *AB*, *AC*, *AD*, qui ont le même sommet et le même centre, et cela par une voy generale. Car on marque ce probleme particulier seulement pour se faire entendre, car dans les sections coniques il a ses facilités particulieres, mais il s'agit de donner une methode generale. Et ce

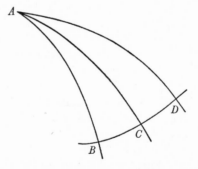

probleme general peut être conçû ainsi: Estant donnée la courbure des rayons de lumiere dans le milieu diaphane, changeant continuellement de refractivite, trouver l'onde de lumiere selon la maniere de parler de M. Hugens, ou selon la façon de parler de M. Bernoulli la synchrone, à la quelle les rayons ou les mobiles, pris convenablement, parviennent en même temps.

NOTES

(1) Gerhardt follows the original in the Niedersächsische Landesbibliothek, Hanover. The letter was sent via Nicolas Rémond (in France). See Letter 1170a.

(2) This postscript, and also Letters 1187 and 1197, and Number 1211, first appeared in print in Joseph Raphson's *History of Fluxions*, a book of which all copies have the same title page, dated London, 1715. As Augustus De Morgan first observed, the letters forming a short appendix to it (p. 97 onwards) do not appear in all copies; he surmised that a second, enlarged issue of it was made after Leibniz's death and called it a 'reprehensible proceeding' on Newton's part. However this may be, it is certain that Newton wrote the editorial matter in the appendix added under his direction (see the drafts in U.L.C. Add. 3968(37)). He wrote later to Des Maizeaux: 'As soon as I heard that he [Leibniz] was dead, I caused the letters and observations to be printed, lest they should at any time come abroad unperfectly in France.' The last sentence in Raphson's book (p. 123) makes it clear that this second issue took place in 1718: 'cum Author emortuus sit, & Liber annis abhinc plus tribus impressus fuit, . . .'; Raphson had died by 1715.

The letter was intended for Newton's eyes, and Conti indeed shewed it to him. Leibniz chose to write through Conti to Newton for two reasons; first, because of his growing irritation at Keill's treatment of him—he wrote to Nicolas Rémond on 11 June 1715:

'Comme Mr. *Keill* écrit d'une maniére un peu grossiére je ne lui répondrai pas. Pourquoi se chamailler avec de telles gens? Je pense à répondre à ces Messieurs par des réalitez, quand j'en aurai un peu plus de loisir. Je n'ai pas même encore lu le Livret de Mr. *Keill* [that is, Keill's 'Answer' in the *Journal Literaire de la Haye*, for July and August 1714, pp. 319–58; see Letter 1053a, notes] avec attention.' (see Des Maizeaux, p. 178 where the addressee is wrongly named as Monmort)—and second because he wanted to teach Conti a lesson. Nicolas Rémond sent him frequent reports of Conti's growing intimacy with Newton; (see his letters to Leibniz in Gerhardt, *Leibniz: Philosophische Schriften*, III, pp. 650ff. and also Letter 1170a).

What follows is, first, a brief review of the calculus dispute; then of the major points argued in the Leibniz–Clarke *Correspondence* (see Letter 1173, note (3)). Leibniz then gives his brief opinion of contemporary British scientists.

(3) In Raphson's text the remainder of the paragraph is omitted, thereby suppressing Leibniz's account of the Royal Society's ill proceedings.

(4) Raphson omits the next phrase: 'L'Espace est quelque chose mais comme le temps, l'un et l'autre est un ordre des choses.'

(5) The rest of the postscript is omitted from Raphson's text.

(6) Leibniz did not originally send the problem to Conti for Newton in the form we print here. Apparently the last two sentences, from 'Car on marque ce probleme particulier . . .' to the end were missing. The English mathematicians thus set about solving the problem for the *particular* case of a set of hyperbolas, which is much simpler to solve than the general case which Leibniz intended.

Leibniz, realizing his carelessness, hastily wrote to Nicolas Rémond on 16 January 1716 (see Letter 1178) asking him to add a sentence stressing that a general solution was required, presumably hoping that his letter would reach Rémond before he despatched the problem to Conti. Leibniz also wrote twice to Arnold (see Letter 1179) asking him to make the acquaintance of Conti and to suggest the addition.

The last sentence, where Leibniz gives as a physical model for the problem the determination of the wave-front produced by light travelling through a medium of varying optical density, (see Letter 1201, note (7)) may also be an addition.

The general problem of finding a curve cutting another, given series of curves at right angles was originally proposed privately by Johann Bernoulli to Leibniz in a letter of 2 Septemper 1694 N.S., and Leibniz gave an outline of a general solution in his reply dated 6 December 1694 N.S. (see Gerhardt, *Leibniz: Mathematische Schriften*, III/1, pp. 152 and 157). Leibniz's solution was later published in the *Acta Eruditorum*, October 1698, p. 471. Meanwhile Johann Bernoulli posed the problem to his brother Jakob, who attempted a solution in the *Acta Eruditorum* for May 1698, pp. 230–32. See Number 1187*a*, and Letter 1201, note (7), p. 320.

Leibniz's carelessness in setting the problem, implying that he required only a solution for one special case, caused considerable bad feeling amongst the English in the ensuing months. Before Leibniz had time to correct his mistake, the English had produced a number of solutions of the special case, which presented no real difficulty; one is given by Keill in his letter to Newton (Letter 1186). Newton's own attempted solution (see Number 1187*a*) is, however, remarkably feeble; he seems unable to construct either a general method (he introduces second order differentials quite unnecessarily) or a solution to the special case. In both Leibniz and Newton we seem to see signs of the failings resulting from advancing years.

It was Johann Bernoulli who provided a more suitable problem to pose to the English, and Leibniz transmitted it to Conti in Letter 1202. But the English rightly complained that this was hardly a special case of the problem originally posed by Leibniz, but was effectively entirely new; see Letter 1202, note (1), p. 323.

1170*a* LEIBNIZ TO NICOLAS RÉMOND
25 NOVEMBER 1715
Extract from Gerhardt, *Leibniz: Philosophische Schriften*, III, p. 662 [1]

Je ne suis pas faché que M. l'Abbé Conti se soit Anglisé un peu et pour un temps (pourveu, qu'il soit le même pour ses amis); il reviendra à nous par rapport aux sentimens, quand le charme des impressions presentes sera passé. En attendant, pendant qu'il donne dans le sens de ces Messieurs, il en profitera mieux. Cependant je luy envoye un petit preservatif, à fin que la contagion n'opere trop fortement sur luy.

NOTE
(1) Leibniz apparently enclosed Letter 1170 to Conti with this letter.

1171 WARRANT TO THE REFEREES
30 NOVEMBER 1715
From the copy in the Public Record Office. [1]
For the answer see Letter 1177

His Majesty being Informed that there are now remaining in the hands of Mr Churchill, Printer, three hundred and forty [2] Copies of the Astronomical

Observations made by the Revd. Mr. John Flamsteed, his Majesty's Astrono-
mer at the Observatory in Greenwich comprised in a Book entitled Historia
Cœlestis, which was printed at the Expence and Charges of his late Royall
Highness Prince George of Denmark.

These are to Signify his Majesty's Pleasure that You give Orders for
Delivering to the said Mr Flamsteed three hundred Copies of the said Book
entituled Historia Cœlestis as a Present from his Majesty,[3] and that You
likewise give Directions that the remaining forty Copies be sent to me, to be
disposed of as his Majesty shall think fit together with an account of how much
Money has been Recd. and Expended in that Service. And for so doing this
shall be Your Warrant. Given under my hand this 30th day of November 1715
in the 2d Year of his Majesty's Reign.

BOLTON[4]

To Sr Is Newton Knt the Honourable
Francis Robarts, Sr Christopher Wren Knt,
Dr Arbuthnot Referrees to his Royall
Highness Prince George of Denmark, and
to Mr Churchill Printer

NOTES

(1) T/1, 193, no. 43; T/1, 198, no. 56(a). Compare Letter 1148.

(2) T/1, 193, no. 43 reads 330 copies. The edition was of 400 copies.

(3) This was the successful conclusion of Flamsteed's struggle to maintain his opinion that
all copies of the *Historia Cœlestis*, though printed at Prince George's expense under the direction
of the Referees, should be given to himself 'as a part of a recompense for my great pains and
suffering from mind, and expenses' (see Baily, *Flamsteed*, pp. 317 and 319). He destroyed all
copies delivered to him; see *ibid*. pp. 101–2.

(4) Charles Paulet (1661–1722), second Duke of Bolton, Lord Chamberlain.

1172 [LEIBNIZ TO DU SAUZET
End of 1715]
From the *Nouvelles Litteraires*, 28 December 1715 [N.S.], pp. 413–14 [1]

Comme je suis persuadé, Monsieur, que vous n'adoptez pas tout ce qu'on
vous envoye pour être inseré dans vos Nouvelles Litteraires, & qu'on a mis
dans l'Article 5. du 7 Tom. du Journal Literaire [2] l'Extrait d'un Livre publié
en Angleterre contre M. *de Leibniz*, voici un Article dont vous serez bien aise
de faire part au Public. On trouve dans les Nouvelles du 7 [3] Septembre pag.
185 les paroles suivantes: *Tout le monde sait que M. Leibniz a voulu disputer à M.
Newton, l'Invention du Calcul des Differences*. Cette expression est renversée.
M. de Leibniz a toûjours passé pour en être l'Inventeur depuis l'an 1684,
qu'il l'a publié; ce n'est que depuis 3. ans ou environ que quelques Mathé-

maticiens d'Angleterre ont voulu lui contester cet avantage, & tout le monde en a été surpris. On ne connoit point de Savant hors de l'Angleterre, qui ait été persuadé par leurs raisons. Le Public a vû une Lettre écrite pour réfuter cette prétention, par M. Jean Bernoulli,[4] l'Auteur le plus profond dans ces matiéres, & qui a surpassé tous les autres dans l'usage du Calcul dont il s'agit. Comme sa Lettre est écrite en Latin, & qu'elle n'est pas assez connuë, il est nécessaire d'en donner ici la traduction. M. de Leibniz n'a point voulu jusques ici entrer dans cette dispute, sur tout parce que M. Newton n'a pas témoigné approuver tout ce que les Mathématiciens d'Angleterre avancent, & parce que la *Societé* dont ils font parade, a déclaré qu'elle n'a point prononcé définitivement sur cette matiére. En effet, la personne intéressée n'a point été ouïe, on ne lui a point fait connoître qu'on voulut en venir à une telle discussion, & on ne lui a pas nommé ceux qui ont été commis pour en prendre connoissance, afin de savoir si quelqu'un des Commissaires ne pourroit point être justement recusé.

Lettre de M. Jean Bernoulli de Bâle, du 7. de Juin 1713 [N.S.]

[*There follows a French version of the extract from the letter by an 'eminent mathematician', printed in Latin in the 'Charta Volans', Number* 1009.]

NOTES

(1) Although published anonymously, the article is clearly Leibniz's response to the review of the *Recensio* published in the *Nouvelles Litteraires* for 21 September 1715 N.S.; see Letters 1166 and 1167.

(2) That is, the volume of the *Journal Literaire de la Haye* for 1715, pp. 114–58, where the first part of the French translation of the *Recensio* was printed; see Letter 1162, note (2).

(3) This should read '21 Septembre pag. 185'—see Letter 1166, note (2), p. 247.

(4) Bernoulli's letter had originally been published anonymously in the *Charta Volans* (see Number 1009); but here Leibniz reveals its authorship. Bernoulli had specifically asked Leibniz to keep his name secret. This particular mention of Bernoulli as author seems to have received little notice; the real storm broke out as a result of Wolf's careless editing of the 'Epistola pro Eminente Mathematico' in 1716 (see Letter 1196, note (4)). Bernoulli much later, in 1719, persisted in denying authorship of the letter in the *Charta Volans* and Newton apparently believed him, at least at first (see vol. VII).

1173 LEIBNIZ TO WOLF
12 DECEMBER 1715

Extract from Gerhardt, *Briefwechsel zwischen Leibniz und Wolf*, pp. 180–1.
For the answer see Letter 1192

Isti homines alios ferre non possunt. Urit eos quod responsione ipsos non dignor. Itaque crambem commercii in Transactionibus[1] recoxerunt et versionem transactionis inseri curarunt Diario Hagiensi [2] literario. Et quo magis me ad respondendum permoverent, etiam mea principia Philosophica ibidem aggressi sunt, ut audio. Sed ibi quoque dentem solido illident. Serenissima Princeps Walliæ quæ Theodicæam meam legit cum attentione animi eaque delectata est, nuper pro ea cum quodam Anglo Ecclesiastici ordinis accessum in aula habente disputavit, ut Ipsa mihi significat. [3] Improbat illa, quod Newtonus cum suis vult, Deum subinde opus habere correctione suæ machinæ et reanimatione. Meam sententiam, qua omnia ex præstabilito bene procedunt nec opus est correctione, sed tantum sustentatione Divina, magis perfectionibus Dei congruere putat. Ille dedit Serenitati Suæ Regiæ schedam Anglico sermone a se conscriptam, [4] qua Newtoni sententiam tueri conatur meamque impugnare; libenter mihi imputaret Divinam gubernationem tolli, si omnia per se bene procedant, sed non considerat Divinam gubernationem circa naturalia in ipsa sustentatione consistere nec debere eam sumi ἀνθρωποπαθῶς. Respondi nuperrime et responsionem meam ad Principem misi. [5] Videbimus an ille sit replicaturus. [6] Gratum est quod materiam antagonista attigit, quæ non resolvitur in considerationes Mathematicas, sed de qua ipsa Princeps facile judicium ferre potest. Vale et fave.

Dabam Hanoveræ 23 *Decembr*. 1715. [N.S.]

Translation

Those men [the Newtonians] cannot tolerate others. It irritates them that I do not consider them worthy of a reply. Therefore they have dished up once more the same stale correspondence in the [*Philosophical*] *Transactions*, [1] and have taken care that a translation of that *Transaction* be inserted in the Hague *Journal Literaire*. [2] And, as I hear, they have also attacked my philosophical principles the better to persuade me to answer them. But here too they bite off more than they can chew. Her Highness the Princess of Wales, who read my *Théodicée* with an attentive mind and was delighted with it, not long ago disputed in its favour with a certain Englishman in Holy Orders, having access to Court, as she herself informed me. [3] She rejects what Newton and his [followers] maintain, that God repeatedly has need to correct and revivify his machine. My opinion, according to which everything proceeds rightly from its pre-established [state] without need for correction, but only divine maintenance, she considers more in accord with the perfection of God. [Clarke] has given Her Royal Highness a paper written by himself in English, [4] in which he tries to maintain Newton's opinion and attacks mine; he raises the facile charge against me that the divine governance [of the universe] would be destroyed if all things proceeded rightly on their own, but he has not considered that the divine governance consists, so far as natural things are concerned, in maintaining them, nor ought it to be judged according to 'human sentiments'. I have replied very recently [5] and have sent my answer to the Princess. We shall see whether he will reply. [6] It is welcome that my opponent has touched upon matters which are not to be resolved by mathematical considerations, but about which the Princess herself can easily form a judgement. Farewell and thrive.

Hanover 23 December 1715 [N.S.]

NOTES

(1) The *Recensio*; see Letter 1162, note (2).

(2) See Letter 1162, note (2) and Letter 1172.

(3) Extracts from the letters of Caroline, Princess of Wales, to the 'Englishman in Holy Orders,' Samuel Clarke (see Letter 1001, vol. v), are printed as an Appendix to *The Leibniz–Clarke Correspondence*, pp. 189–98. Caroline, for many years a correspondent of Leibniz, in 1715 approached Samuel Clarke in the hope of persuading him to translate Leibniz's *Essai de Théodicée* (Amsterdam, 1710, published anonymously; the second edition of 1712 bore his name). But Caroline was disappointed; on 26 November she wrote to Leibniz: 'Dr. Clarke is too opposed to your opinions to do it; he would certainly be the most suitable person of all; but he is too much of Sir Isaac Newton's opinion and I myself engaged in a dispute with him. I implore your help; he gilds the pill and is not willing to admit that Mr. Newton has the opinions which you ascribe to him, but you will see from the enclosed paper that it comes to the same thing.' The 'enclosed paper' was Clarke's reply to an extract of a letter to herself from Leibniz that Caroline had shown to Clarke in early November. The correspondence which ensued between Leibniz and Clarke, with the Princess as intermediary, terminated only on Leibniz's death in November, 1716.

We do not print the Leibniz–Clarke correspondence here for two reasons. First, it is readily available in English translation in H. G. Alexander's well-edited book, *The Leibniz–Clarke Correspondence*. Second, whilst it seems from Caroline's correspondence with Leibniz that Newton was to some extent consulted about Clarke's replies, his involvement was not nearly so great here as it was in Keill's letters to Leibniz about the calculus controversy. No correspondence between Newton and Clarke survives.

The theological and metaphysical questions which Leibniz and Clarke disputed were not new. Leibniz had attacked Newton's theory of gravitation and action at a distance both in the *Théodicée* and in correspondence with Hartsoeker (see Letter 918, vol. v, and notes). In both the *Opticks* (in the Quæries) and in the *Principia* (in the General Scholium) Newton had touched upon the theological implications of his physics, in particular the need for God's continued action in the Universe and the existence of a 'Sensorium' of God (see also Letter 1138, where Leibniz tells Bernoulli of his disagreement with Newton's ideas). Cotes (in his Preface to the second edition of the *Principia*) had attacked Leibniz more directly on behalf of Newton, and the arguments were also taken up in the *Recensio* (see Letter 1162). See also Koyré and Cohen, 'The case of the Missing Tanquam: Leibniz, Newton & Clarke' *Isis* **52**, part 4, 555–66.

(4) Clarke's first reply, transmitted 26 November 1715. See *The Leibniz–Clarke Correspondence*, pp. 12–14.

(5) Leibniz's second paper, undated [1715]; see *ibid.*, pp. 15–20.

(6) He did, of course. See Clarke's second reply, transmitted 10 January 1716, *ibid.*, pp. 20–4.

1174 JOHN ARNOLD TO LEIBNIZ
22 DECEMBER 1715
Extract from the original in the Niedersächsische Landesbibliothek, Hanover [1]

Monsieur

Monsieur le docteur Brandshagen [2] maiant fait savoir quil vous escriroit par cette Ordinaire. Je nai pas voulu manquer loccasion de vous offrir mes tres humbles services et de vous informer en même tems que Jai rendu les lettres des quels Vous M'aviès honorè a Hanover, Jai estè fort bien recu de son Excellence Monsieur de Bothmar [3] Qui entre ces autres Merites a celle de Plaire a mes Compatriotes, qui (entre nous) ne sont pas accoutumès de dire trop de bien des Etrangers. Monsieur le docteur Woodward [4] vous fait ses Compliments, il ma assurè que tout ce qui etè fait icy contre vous nest venu que de Monsieur Newton et quainsi il espere que vous ne lattribuerès pas a la Societe Roiale. Je souhaiterois de Vous voir dans ce pais pour convaincre nos Anglois quil y a des gens de lettres en Allemagne, Car les gens detude qui sont venus avec Sa Majeste [5] sont si peu chargè des Sciences, quil sont le sujet du mepris de nos scavans . . .

NOTES

(1) See Bodemann's catalogue, no. 17. For Arnold's connection with Leibniz and Bernoulli see Letter 1014, note (1), p. 27.

(2) See Letter 1200 for Brandshagen's connection with Newton's work at the Mint.

(3) See also Letter 1203, note (1), p. 329; Bothmer was chief Hanoverian agent in London.

(4) See Letter 785, vol. v. Woodward was never one of Newton's supporters.

(5) George I, who arrived in England on 18 September 1714 from Hanover. His retinue included the Baron and Baroness von Kilmansegge (see Letters 1187, note (1), and 1203), both of whom were to become marginally involved in the calculus controversy.

1175 LEIBNIZ TO J. BERNOULLI
[DECEMBER 1715]
Extract from Gerhardt, *Leibniz: Mathematische Schriften*, III/2, pp. 951–2

Dominus Abbas De Conti scripsit ad amicum Parisinum, [1] qui mihi significavit, Anglos longa recensione [2] Commercii Epistolici in Transactionum aliqua suas contra me argutationes iterasse, atque inter alia etiam Philosophiam meam impugnasse. De Analysi nostra dicunt, Newtonum originarium esse, nos tantum nomina adjecisse, eaque apta ad controversias in Mathesin introducendas. Philosophiam Newtoni esse mere experimentalem, meam conjecturalem: sed, ni fallor, harmonia præstabilita seu, quale nos statuimus,

Commericum Animæ et Corporis res demonstrata est; demonstratum etiam, ni fallor, firmitatis seu nexus in corporibus originem non posse desumi nisi a motibus conspirantibus, atomosque esse rem absurdam.[3] At Newtonus minime per sua experimenta demonstrat, materiam ubique esse gravem, seu quamvis partem a quavis attrahi, aut vacuum dari, ut ipse quidem jactat. De Deo etiam miras fovet sententias; extensum esse, sensorium habere,[4] et vereor ne revera inclinet in sententiam Averrhois et aliorum,[5] etiam Aristoteli tributam de Anima seu Intellectu agente generali in corpore quovis pro ratione organorum operante.[6] Illud etiam mihi plane absurdum videtur, quod putat machinæ mundanæ motum ex se desiturum, nisi a Deo subinde rursus animaretur. Itaque miraculis opus habet, nec sine perpetuis miraculis suam attractionem explicare poterit.

Dominus Abbas Contius Parisiis discedens Epistolam[7] ad me reliquit, quæ mihi nunc reddita est. Ipsi jam respondeo, et ut pulsum Anglorum Analystarum nonnihil tentemus, rogo ut, quasi suo proprio motu aut amici rogatu, Problema[8] hoc illis proponat: *Invenire lineam* BCD, *quæ ad angulos rectos secet omnes curvas determinati ordinis ejusdem generis, exempli causa, omnes hyperbolas ejusdem verticis et ejusdem centri* AB, AC, AD *etc. idque via generali.* Ita videbimus, quousque suis fluxionibus profecerint.

Translation

The Abbé Conti has written to a Parisian friend,[1] who told it to me, that the English have renewed their grumblings against me by a long review[2] of the *Commercium Epistolicum* in some issue or other of the [*Philosophical*] *Transactions*, and amongst other things attack even my philosophy. They say that Newton was the originator of our analysis, that we ourselves added only the names, and that these were appropriate for introducing controversies into mathematics; that Newton's philosophy is purely experimental, whilst mine is conjectural. But, if I am not mistaken, that pre-established harmony, or rather (as we postulate it to be) that correspondence between soul and matter, is a demonstrated thing; it is also demonstrated, if I am not mistaken, that the cause of cohesion or firmness in bodies can only be ascribed to a consort of motions, and that atoms are an absurdity.[3] But Newton in no way demonstrates by means of his experiments that matter is everywhere heavy, or that any part whatever is attracted by any other part, or that a vacuum exists, in accordance with his own boasts. He cherishes astonishing ideas about God; that He is extended, that He has a sensorium,[4] and I am afraid, indeed, that he shares the opinion of Averroes and others[5] which is indeed attributed to Aristotle, concerning a spirit or intellect acting universally in all bodies whatever, like the operative principle in machines.[6] What he thinks seems plainly absurd to me, namely that the motion of the world-machine will come to cease unless from time to time restored by God. Thus miracles are necessary to him, and he will prove unable to explain his attraction without perpetual miracles.

261

The Abbé Conti left a letter [7] for me when he departed from Paris, which has now been sent to me. I am now answering it, and in order to feel a little the pulse of the English analysts, I am asking [him] to propose this problem [8] to them as if it were suggested by himself or requested by a friend: To find the line *BCD*, which cuts at right angles all curves of a determinate order [and] of the same type, for example all hyperbolae *AB, AC, AD* etc. with the same vertex and the same centre and to do this by a general method. Then we shall see how much they will accomplish with their fluxions!

NOTES

(1) Presumably Nicolas Rémond; see Letter 1139, note (5).

(2) The *Recensio*, see Letter 1162, note (2); published in *Phil. Trans.* **29**, no. 342 (1715), 173–225.

(3) As a lifelong plenist, Leibniz had long maintained (in this resembling Descartes) the essential inertness of ordinary matter, attributing all phenomena to the motions of aetherial particles. He thus explained cohesion in his *Hypothesis physica nova* (Mainz and London, 1671), of which John Wallis wrote 'fateor mihi nondum satisfactum esse, ut, primis saltem cogita-tionibus, statim assentiar, Cohæsionem omnem ex continuo celerique sed inobservabili particularum motu fieri', ('I confess that I am not yet satisfied that I can at once agree, at least upon my initial reflections, that all cohesion is caused by a swift, continuous, but un-observable motion of particles'); see Hall & Hall, *Oldenburg*, VIII, p. 73. Newton's atomism was of course apparent to Leibniz from the final *Quæstio* of the 1706 *Optice*.

(4) Compare Letters 1138 and 1173.

(5) The Islamic philosophers, Averroes among them, certainly founded on Book XII of Aristotle's *Metaphysics* the ideas that Leibniz appears to castigate here, but they were not held by Aristotle himself. They are, briefly (i) that the heavenly bodies constitute an organic being, each orb possessing an intelligence which is its form, the first mover being also the first in-telligence created by God, that is, the first created thing; and (ii) the lowest and nearest to man of these celestial intelligences is the *active intelligence* in which every human mind partici-pates, and which is eternal though individual minds are mortal. As Renan pointed out, Is-lamic philosophy was wholly opposed to that tendency of seventeenth-century mechanism, maintained by Leibniz, to banish God from the finished creation; the Arab sought rather to create "une sorte de ministre [l'intelligence] à ce roi invisible pour le mettre en contact avec l'univers.' But Renan also points out that the conception of Averroes and his followers was utterly different from that of the Stoics and others in antiquity, who postulated a universal soul in the universe (and if Leibniz was attributing such an opinion to Averroes he was mis-taken), for in Averroist philosophy though the *active intelligence* was common to all minds, it was certainly human, not universal. (See Ernest Renan, *Averroès et l'Averroisme*, 3rd edition, Paris, 1866, p. 116ff.)

(6) Presumably Leibniz means that all parts of the *machina mundi* are inert unless inspired by an intellect.

(7) Letter 1139.

(8) See Letter 1170.

1176 NEWTON TO THE TREASURY
[End of 1715]
From the draft in the Mint Papers [1]

To the Rt Honble the Lords Comm[ission]ers of his
Mats Treasury

May it please your Lords

I have perused the Memorial concerning the Mint in Scotland & humbly represent that Dr Gregory was sent down into Scotland to assist in regulating that Mint during the time of the Recoinage only as appears by his Report annexed to the Memorial, & that the three new Clerks were proposed by me to assist the officers during that Service & no longer & are now superfluous, & in the copy of Mr Montgomeries Indenture in the Treasury are omitted.

That the estimate of the charge of Salaries Repairs & Incidents given in to Parliament by Mr Allardice then Master & Worker of that Mint, amounting to 1200 pounds, conteined the Salaries of the said three Clerks amounting to 120 pounds per annum, and the pension of 50 per annum to the Clerk of the Bullion wch is to cease & 50*li* per annum given to Mr Drummond the Warden of that Mint wch is also to cease, & the salary of the Generall of the Mint whose Office is useless. [2]

That the first Parliament of K. James VIIth allowed for salaries repairs new Tools & Incidents only 12000 pounds Scots wch answers to 900*l* English recconing a 60 pound peece Scots at 4*s* 6*d* English, & that the coinage then amounted only to about 300*l* per annum more, so that 1200*l* pounds per annum (wch is allready allowed) is sufficient for all the charges of that Mint, especially if the Generall's Office upon the next voidance of that place, should cease pursuant to ye Act of Union.

That the Coinage Acts before the Union ordered that all the moneys arising by those Acts should be issued to the Master & Worker of the Mint in England for the use & service of that Mint in general without distinguishing (in the Warrants) between the moneys issued for salaries & repairs & those issued for coinage: & by the Act of Union the coinage monies ought to be issued in the same manner to ye Master of ye Mint in Scotland, & therefore the clause annexed to ye Memorial & proposed to be added to the Bill is inconsistent with the Act of Union, wch ordeins that the Mint in Scotland be under the same Rules as the Mint in England. And for the same reason the clause in ye Act of the 7th of the late Queen concerning the Mint in Scotland ought not to be rest[r]ained to salaries repairs & necessaries of the Mint but to extend also to ye charge of coinage, & be issued out of the Exchequer of great Britain for the use & service of the Mint in general. And when the moneys are in the Exchequer of the Mint, the Officers of the Mint are to issue them out thence for particular services & see that no more be issued for salaries & repairs then are allowed by Act of Parliament.

NOTES

(1) III, fo. 56. This letter is in reply to an undated memorial (P.R.O. T/1, 185, no. 20; copy in the Mint Papers, III, fo. 23), to which was annexed a report on the Edinburgh Mint and the process of recoinage there written by David Gregory on 13 December 1707 (it is printed in vol. IV, pp. 503–4); in it the officers of the Edinburgh Mint desired a resumption of the Scottish coinage, with a higher allowance for that Mint than the £1200 already provided. Newton argued only against the financial recommendation in his reply, not the question of policy.

The date of this and the next document are highly uncertain. *The Calendar of Treasury Papers* supposes that the memorial was submitted very early in the reign of George I; Newton's reply is clearly written after Shrewsbury's resignation as Lord Treasurer in October. But in fact the question of the maintenance of the Scottish Mint may well have come up much later, for it was not until 10 March 1716 that the House of Commons resolved to authorize the Treasury to allow the Mint's running expenses of £15 000 a year out of general revenue, and to revive the former Coinage Act of 7 Anne for a further seven years. It seems likely that the maintenance of the Edinburgh Mint would not have been discussed much before the end of 1715.

(2) For George Allardes and William Drummond see vol. IV, p. 522, note (2) and Letter 1176a, note (5).

1176a NEWTON TO ——
[End of 1715]
From a draft in the Mint Papers[1]

Sr

I have consulted the Copy of Mr Montgomeries Indenture[2] entred in the Book of the Treasury & there are in it no salaries allowed for Clerks except to the Kings Clerk.[3] All the buisines of the other Clerks will not amount to above three or four good days work in a year, & therefore those Clerks are superfluous.[4] All the salaries in the Indenture amount only to 980*l* including Mr Drummonds[5] additional salary of 50*l* per annum & the pension of the late Clerk of the Bullion[6] of 50*l* per annum, both wch are extraordinary & annexed to their Persons, & ought not to affect the standing constitution of the Mint. If from the 1200*l* allowed to your Mint[7] by the last Coinage Act for Salaries the Fabric new tools & Incidents the 120*l* per annum reconed for three Clerks be deducted & the late Clerk of the Bullion be made Comptroller of the Mint so that his Pension of 50*l* per annum may cease, & about 30*l* per annum for new coining tools be also deducted there will remain 1000*l* per annum to fill up the blank in this clause AA. And when Mr Drummonds place becomes void there will be 50*l* per annum saved in the salaries.

The Parliament of Scotland in their Coinage Act made about thirty years ago allowed only 12000 pounds Scots for Salaries repairs & Incidents, which in those days amounted to 900*l* Sterling recconing a 60 shillings piece Scots at 4*s* 6*d* English,[8] & the coinage did not then amount to three hundred pounds more per annum one year with another as I understand by your discourse. And upon all these considerations I have proposed to Mr Lowndes that a summ not exceeding one thousand pounds sterling be allowed for Salaries the Fabric & Incidents & that a further summ not exceeding 400*l* be allowed for coinage, if you think fit to put a limit to the last allowance.

I leave it to you to offer to the House what clause you think fit for an allowance not exceeding 50*l* per annum for bearing the charges of prosecuting clippers & coiners.

I have sent you herewith a Copy[9] of the Bill so far as it relates to the Mint, with proposed amendments partly interlined, partly in the margin, & partly in the clause AA thereunto annexed. And in the margin of the clause AA I have set down one thousand pounds to fill up the blank in that clause.

NOTES

(1) III, fo. 57. This is the last of three partial holograph drafts without date or addressee. As it is closely related to the subject of the previous letter we reproduce it here.

(2) John Montgomery was Master of the (inactive) Edinburgh Mint.

(3) Robert Millar or Miller, Clerk and Bookkeeper since 1699 (Register of Appointments to the Mint, Scottish Record Office, E.105/2).

(4) Compare Letter 1176.

(5) Compare Letter 745, vol. IV. William Drummond of Blair Drummond had been appointed life Warden of the Edinbrugh Mint in 1705, at a salary of £150 per annum.

(6) See Letter 765, note (4) (vol. V, pp. 4–5).

(7) See Craig, *Newton*, p. 74; this phrase suggests that the addressee may have been Lauderdale, General of the Edinburgh Mint.

(8) James I had enacted that Scots money should pass at one-twelfth of the nominal value of the corresponding English coin, hence one shilling Scots equalled one penny English. Newton, relying on the bullion content, insisted that 10*s*. Scots pass only as 9*d*. English; Craig, *Newton*, pp. 22–3.

(9) This copy is, naturally, not with the draft.

1176B JOSIAH BURCHETT TO NEWTON
11 JANUARY 1716
From the minute in the Public Record Office.[1]
For the answer see Letter 1179A

11 *Jan[ua]ry* 1715/6

Sr

His Majesty having been pleased to referr unto the Lords Commissioners of the Adm[iral]ty the Petition of Mr Henery de Saumarez,[2] wherein he proposes the making some discoverys which may tend to the Benefit of Navigation,[3] I am commanded by their Lo[rdshi]ps to desire you will please to give youself the Trouble to examine into the Invention of the sd. Mr. de Saumarez who is the bearer hereof and report your Opinion of the same, that so their Lo[rdshi]ps may make a report thereof to his Majesty.

I am

Sir

Your &c

JB

Sr Isaac Newton

NOTES

(1) ADM/2, 449, p. 33; printed, together with the other documents concerning Saumarez's case (see below), in *An Account of the proceedings of Henry de Saumarez . ... Concerning His Discovery of an Invention, by which the course of a Ship at Sea may be better Ascertained than by the Logg-line* (London, 1717), p. 7. Josiah Burchett (?1666–1746) began his work in the administration of the Navy in 1680 as clerk to Samuel Pepys; he was secretary of the Admiralty from 1698 to 1742, and also M.P. for Sandwich 1703–13, and 1721–41.

(2) Viscount Townshend submitted Saumarez's petition for consideration by the Admiralty on 7 January 1716 (ADM/1, 4099, fo. 244, printed in Saumarez's *Account*, p. 6; see also note (3) below). Four days later the Admiralty Lords resolved to refer it to Newton for his opinion (ADM/3, 30, not foliated).

(3) For Saumarez, who was a member of the Guernsey family otherwise distinguished in naval history, see E. G. R. Taylor, *Mathematical Practitioners of Hanoverian England* (Cambridge, 1966), p. 141 and W. E. May and L. Holder, *History of Marine Navigation* (London, 1973), p. 112. He invented a form of 'patent log', described in a publication called the *Marine Surveyor* (1715) according to E. G. R. Taylor and also published in 1717 his *Account* (see note (1) above). Accounts of his invention and later modifications of it also appeared in the *Philosophical Transactions* (**33**, no. 391 (1725), 411–32; **36**, no. 408 (1729), 45–59).

The idea of causing the motion of the sea past a ship to rotate a dial was not new. Vitruvius (x, ix) described a primitive form of odometer, while the Byzantine general Belisarius had (it is said) devised at Rome (when the city was besieged by the Goths) the kind of water-mill known as a boat-mill. The two combined provide a machine for measuring the way of a ship; the combination (a paddle-wheel driving a wheel-work counting device) was shown by Jacques Besson (*Theatre des Instrumens mathematiques et mechaniques*, Lyon, 1579, no. 57). Robert Hooke in 1683 showed the Royal Society the vane of a 'waywiser for the sea' (invented twenty years earlier; Thomas Birch, *History of the Royal Society*, IV (London, 1757), p. 231), but Saumarez's simple device is the first of the sort positively known to have consisted of a propeller streamed behind the ship on a line, through which its revolutions were transmitted to an inboard counting mechanism. The propeller was quite large, consisting of two slightly inclined plates 8 ins. by $4\frac{1}{2}$ ins. mounted on the 15-ins. long arms of a Y-shaped rotor, the tail of the Y being joined to the line. Although several improvements of Saumarez's design were proposed by John Smeaton and others, the patent log only succeeded in the nineteenth century.

Saumarez had first petitioned the Royal Society, in a letter read on 20 October 1715, for recognition of his invention and of the possibility of using it for the determination of longitude at sea. The petition is printed in Saumarez's *Account* (see note (1) above); we have not traced the original document. It throws little light on the state of the invention at that time, but mentions that Saumarez had carried out trials in his laboratory on Guernsey and, later, at Greenwich and Woolwich. He confesses that he has not yet devised a means of transmitting the motion of the line through the vessel to the dial. Halley, as secretary of the Society replied that '1. [The Society] did not conceive [the invention] belonged to them. 2. That it was no new Invention, and that it had been discovered already. 3. That the Hindrance which the Ship should receive in her Sailing by the Application of the Machine, would not at all be compensed by the Advantage which the machine would give it.' (see *Account*, p. 5). Undaunted, Saumarez next petitioned the King on 5 December 1715 (see *Account*, pp. 5–6; again we have not traced the original document), once more giving scant detail of his invention. The reference to the Admiralty followed on 7 January 1716, and they in turn referred the petition to Newton, in the Letter we print here.

1177 THE REFEREES TO THE DUKE OF BOLTON
14 JANUARY 1716

From the original in private possession.[1]
Reply to Letter 1171

14 *Jan.* 171$\frac{5}{6}$

My Lord

Your Graces Order dated the 30th of November and directed to us as Referees to his Roy[a]ll Highness Prince George of Denmark came to our hands about ten days ago, We should have punctually obeyed his Majesties Commands, transmitted to us by Your Grace, were we not fully convinced that our Trust is now expired, and That we have no power to meddle further with the disposal of those Books.

At the same time, we think it our Duty to inform your Grace, that the Books are at present in the hands of Mr Aunsham Churchill Bookseller in Paternoster Row, where they were lodged by our Order for the use of his late Royall Highness the said Prince of Denmark.

The Account of the Receipts and Disbursements of the Mony will, with these Presents, be transmitted to your Grace by Sr Isaac Newton, who, besides what he has received, has disbursed a good deal of mony out of his own Pocket.

In discharge of our Trust, we think it Likewise our Duty to acquaint your Grace, That her late Majestie, in order to make the Observations as serviceable to future Generations as possible for the Improvement of Astronomy, Geography, and Navigation, intended to dispose them by making presents to such as were comprehended under the following heads,

1. Publick Libraries at home and abroad,
2. Professors and other eminent Mathematicians at home and abroad,
3. The Nobility of great Britain, who have Libraries of Note,
4. Those about the late Prince who had been instrumental in promoting the work (viz. the Rt Honble mr. Compton now Speaker of the House of Commons,[2] Mr Nicolas Treasurer,[3] and Mr. Clark[4] Secretary to the Said Prince) for themselves and their Friends.
5. The Referees for themselves and their Friends.

And tho Mr Flamsteed had a Gratuity[5] paid him of £125, and by the imperfect and uncorrect Copies which he delivered, and other breach of Articles, had given us a great deal of Trouble, yet we designed to have begged of her Majestie the Remainder of the Copies for him.

These things we humbly take the freedome to lay before your Grace, being with the utmost Respect

My Lord

Your Graces most humble

and most obedient Servants

F. ROBARTES

CHR. WREN

IS. NEWTON

JO: ARBUTHNOTT

NOTES

(1) This is formally written in a clerical hand with autograph signatures. The date has been added by Newton. There is a copy (printed by Baily, *Flamsteed*, p. 318) in *Flamsteed* MSS, vol. 55, fo. 133, annotated by Flamsteed. For the Duke of Bolton see Letter 1171, note (4).

(2) Spencer Compton (?1673–1743), younger son of the third Earl of Northampton, Speaker 1715–21, was created Baron (1728) and Earl of Wilmington (1730).

(3) Possibly Edward Nicholas either (i) the M.P. for Shaftesbury (b. *c*. 1662, d. 1726) or (ii) the commissioner of the Privy Seal, 1711 (b. 1682).

(4) George Clark (1661–1736); see vol. IV, p. 430, note (1).

(5) Flamsteed has added the annotation, 'not if a debt.'

1177ª ACCOUNT FOR PREPARING THE *HISTORIA CŒLESTIS*
13 JANUARY 1716
From the holograph original in private possession

An Accompt of the charges of printing Mr Flamsteeds Historia Cœlestis and preparing and correcting the Manuscript Copy thereof[1]

	£ s d
I received of Edward Nicolas Esqr Treasurer to the late Prince of Denmark 250£ at one time & 125£ at another. Total received — — — — — —	375.00.00
And upon recconing with the Princes Administrators, I paid back to Mr Compton 25£.3s.0d. & took his receipt in full of all Accompts 18 Apr. 1710 — — — — —	25.03.00
The total expence till that time — — — — — — —	349.17.00
Paid to Mr Churchill for paper & printing — — — — —	194.17.00
To Mr Flamsteed for preparing to [the] copy — — — —	125.00.00
To Mr Machin for correcting to [the] co[p]y by the Minute book & examining some calculations — — —	30.00.00
	349.17.00
Afterwards Dr Halley undertook to finish the Book & the Referees of the Prince acted no further & after the work was finished & the Accounts stated, moneys were imprest to me to pay them off, amounting to	364.15.00
Paid to Mr Churchill for paper & printing — — — — —	98.11.00
Paid for designing & graving the draughts and rolling of the Plates, above — — — — — — —	116.04.00
Paid Dr Halley for examining & correcting the catalogue of the fixt stars, adding about 500 new starrs to the Catalogue, drawing up the second book in due form from the Minute book & correcting ye Press — — — — — — — — — — —	150.00.00
	364.15.00

13 *Jan.* 171⅚

Is. NEWTON

NOTE

(1) Compare Letter 892, vol. v.

1178 LEIBNIZ TO NICOLAS RÉMOND
16 JANUARY 1716
Extract from Gerhardt, *Leibniz: Philosophische Schriften*, III, p. 669

27 Janvier 1716 [N.S.]

En revoyant ce que j'avois écrit à Monsieur l'Abbé Conti,[1] j'ay remarqué qu'on pourroit méprendre mon probleme, que je l'ay prié de proposer en Angleterre sans me nommer. C'est pourquoy je souhaiterois qu'il adjoutât ces mots: on allegue cet exemple des Hyperboles, non pas pour en demander la solution, car la question a des facilités particulieres dans les coniques, mais seulement pour se faire entendre, et l'on demande une methode generale.[2]

NOTES

(1) See Letter 1170.

(2) See Letter 1170, note (6), p. 254. Leibniz repeated his request to Nicolas to change the wording of the problem in another letter of 16 March 1716; see Gerhardt, *Leibniz: Philosophische Schriften*, III, p. 673.

1179 LEIBNIZ TO ARNOLD
27 JANUARY 1716
From the copy in the British Museum[1]

Hanover ce 7 Fevrier 1716 [N.S.]

Je serai bien aise d'apprendre Monsieur si vous aves fait Connoissance avec M. l'Abbè Conti Noble Venitien dont je Vous ai ecrit dans une de mes Precedentes. Un de Mes Amis a proposè un Probleme Analytique general a resoudre qui est *trouver les lignes qui coupent a angle droit une suite donnèes de lignes* comme une suite de lignes *AB, AB* &c estant donnèe trouver la suite des lignes *L,M,* qui coupent toutes les premieres a angle droit, par exemple les lignes AB pourroient etre toutes les Hyperboles d'un même Vertex et d'un même latus rectum. Il est vrai que dans ces cas des Coniques le Probleme n'est pas fort difficile et ce

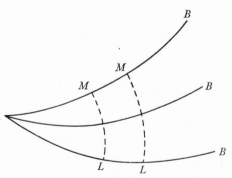

n'est pas dans ces Examples quon en demande la solution, mais par une methode generale. Ayès la bonte Monsieur d'en parler a M. l'Abbé Conti, que Jai priè de proposer ce Problême, mais il faut y adjouter que les Hyperboles [n']ont estè nommèes que pour mieux faire entendre la question.[2]

NOTES

(1) MS. Birch 4281, fos. 12–14. The extract, in Arnold's hand, was sent by Arnold to Chamberlayne on 16 May 1719 (see Letter 1318, vol. VII).

(2) For the correction to the problem see Letter 1170, note (6). Leibniz wrote again to Arnold on 6 March 1716, repeating his request to have the problem amended; see Letter 1193.

1179A   NEWTON TO [BURCHETT]
27 JANUARY 1716
From the holograph draft in the University Library, Cambridge.[1]
Reply to Letter 1176B

Sr

According to the Order of the Lords Commissioners of the Admiralty I have considered the inclosed Proposal of Mr Henry Saumarez & discoursed him upon the same & am humbly of opinion that by means of the Instrument wch he has invented a recconing of the distance sailed by a ship be kept wth less trouble then by the Log-line, but I am not yet satisfied, that the recconing will be so exact. Mr Savory who invented the raising of water by fire [2] told me about six years ago that he had invented an instrument to measure the distance sailed, [3] & by his description that Instrument was much like this, the sea water driving round the lowest & swiftest wheel thereof & that wheel driving round others the highest & slowest of which turned about an Index to show the length of the way sailed: but I have been since credibly informed that a Captain who tried that instrument found it less exact then the Log-line. The instrument now proposed will keep a recconing of the motion of the ship with respect to ye upper part of the sea water but not of the driving of the ship by currents, & tides, & by the motion of the upper surface of the sea caused by winds. And how far it will keep a true recconing of the motion of a ship sideways occasioned by a sidewind does not appear to me. [4] And therefore the Log-line is not to be laid aside. All the advantage of this Instrument will be only to save the labour of casting out the Log-line so often as is now the practise. Mr Savory is dead & the Petitioner tells me that he is willing to be at the charge of a Patent. And if a Patent be granted yet the Log-line alone should be depended upon untill it appear by experience how far this new instrument may be trusted.

I have no experience in Sea affairs nor ever was at Sea & therefore my opinion is not much to be relied on, without the opinion of Trinity House. [5] I am

NOTES

(1) Add. 3972, fo. 41. The date is given by Saumarez in his second petition (ADM/1, 4099, fos. 245–6) and confirmed by note (5) below; the letter is printed in Saumarez's *Account* (see Letter 1176 B, note (1)), pp. 7–8, with only trivial differences from the draft we print here.

(2) Thomas Savery (?1650–1715), whose patent for a steam-pump was taken out in 1698. He demonstrated the machine to the Royal Society in the following year. It was described in his *Miner's Friend* (1702) and in John Harris's *Lexicon Technicum* (1704; *s.v.* Engine). Savery was elected F.R.S. in 1705, and held various public appointments. See Rhys Jenkins in *Transactions of the Newcomen Society*, **3** (1924), 96–118.

(3) Newton's recollection may have been correct, and Saumarez himself reports having been told of Savery's device (see note (5) below), but we have found no independent or detailed description of it. It is possible there was some confusion between two highly similar devices—Savery had re-invented the paddle-boat, which he described in *Navigation improved* (1698). Saumarez affirmed later in his second petition that no one had appeared to assert a claim to the patent log on Savery's behalf 'which the sd. Sr. Isaac Newton endeavoured to set up in competition with ye petitioners' invention.' (ADM/1, 4099, fo. 245).

(4) Although Newton appears not to think so, the log-line was subject to the same defects.

(5) After receiving Newton's report, Burchett transmitted it with Saumarez's original petition to the King (see Letter 1176 B, note (3)) to Trinity House on 28 January 1716 (ADM/ 2, 449, p. 92). He informed the Brethren that Saumarez and an executor of Captain Savery would call upon them to make further report, and asked them to give their opinion of both inventions, returning the papers to him.

According to his *Account* (p. 9), Saumarez 'attended several times, and presented them with a Specimen of his Invention in Draughts upon Paper, and Models in Wood and Copper.' He also submitted two memorials to Trinity House, dated 7 March 1715 and 14 April 1716 (see *Account*, pp. 9–13). In the first of these, in which he reveals awareness of Newton's report, Saumarez mentions that a Captain Bennet had given him a verbal account of a device of Savery's on 24 February 1715, but gave no details of how the instrument worked (because Bennet was not allowed to examine it closely) except to emphasize its uselessness in heavy seas and the necessity of hauling it on board in order to take a reading. In the second memorial Saumarez replied to specific objections raised by the Brothers. First, to the objection that his instrument would work neither in the shallow water of rivers and ports, nor in very rough seas, when it would be damaged, he answered that it was not needed in port, and that in really rough seas it could be hauled in. Second, to the objection that in high seas it would record the up and down motion of the waves, thereby falsely augmenting the distance travelled, he replied the slow motion on riding up the waves would compensate the rapid motion riding down them, so that only forward motion would be measured. Third, to counteract the objection that the device measured only headway but not leeway, he suggested the use of a second, similar instrument set at right angles to the first.

Despite Saumarez's efforts, Trinity House, on 8 May 1716, returned answer to the Admiralty that his instrument was 'impractical' (*Account*, p. 13). Undefeated, he 'waited several times (but to no purpose) upon the Clarks of the Admiralty' (*Account*, p. 14) and then submitted his second petition, which he dates 5 July 1716 (*Account*, pp. 14–16; the manuscript version, ADM/1, 4099, fos. 245–6, is undated), where he complained of the unreasonable delays of Trinity House despite his efforts to evince the merits of his invention by the use of drawings and models. The admiralty did not respond, but merely approved, on the same day, the report from Trinity House (*Account*, p. 16). There the matter rested.

At the end of his *Account* (pp. 17–18) Saumarez briefly comments on the documents he prints there. He criticizes the Royal Society for showing no interest in a matter clearly, in his eyes, its concern (although he adds that Halley had treated him very civilly 'in several friendly conferences he had with him at his own house, before the presenting of the said Petition'). He objects to Newton's references to Savery's invention, which he considers irrelevant, since it had never proved useful. And he criticizes Trinity House for dismissing the invention as impractical without proper examination.

1180 FLAMSTEED TO JOHN LOWTHORP
3 FEBRUARY 1716
From the copy in the Royal Greenwich Observatory [1]

The Observatory Feb: 3 ♀ 1716

Sr

That I may not p[er]plex you I shall make some short remarks only on 2 or 3 p[ar]ticulars in ye representation Yours brought me yesterday,[2] after haveing first given You my thanks for it. The representation sayes, *Tho Mr Flamsteed had a gratuity of £125 and gave them a great deal of trouble yet they designed to have beggd the rest of ye Copies for him,* I told you in mine of October ye 24 last,[3] that when I dismist my Calculators at Midsummer 1706 I had been at £173 Charge for their Wages & accommodations, & some expense in attending the Work at ye Press; . . . hereby *You will see that this Money was so far from being a gratuity, that it was only p[ar]te of a just debt & that there is still £48 & above due to me on yt account*; As for trouble I gave *Them* (the pluralities of him-selfe) it was none but wt S[ir] I[saac] N[ewton] occasioned by his own ex-tremly p[er]verse behaviour, which some of his Referrees could not but take notice of at that very time. As to ye disposal of the books, ye Representation, you tell me sayes ye Queen designed to give ym

1. to Public librarys
2. to Professors and eminent Mathematicians at home & abroad.
3. to the Nobility who have Libraries of Note.
4. to the Princes friends, Mr Co[mpton], Mr. Cl[ark], & their friends.
5. to the Referrees & their friends.

When these are served Sir Is thinks he has made so many fast friends, but at whose cost does he do it? ye paines were mine, ye expense ye Prince's or ye Publick, Yet I must look upon it as a favour if he vouc[h]safe to beg ye rest for me, when all persons were furnishd yt would buy them. this is a spiteful & malitious contrivance; but there is a worse still remaines, of ye £714 12*s* 00*d* disburst on ye Edition he accounts wt he payd Halley & Machin for their pains bestowd in *spoyling it*, for ye Catalogue is absolutely spoyled the Abstracts of my Observations are very sorrily done, so yt it will be a shame to our Nation to have them seen in any *Publick Library. Whether Sr I. N[ewton] suffered these to be done out of Malice or Ignorance*, he ought to pay his Workmen and Tools out of his own pocket, & not to Charge his follies on ye Publick.

NOTES

(1) *Flamsteed* MSS., vol. 55, p. 135, printed in Baily, *Flamsteed*, pp. 319–20. John Lowthorp (*c.* 1659–1724) was admitted as a Sizar to St John's College, Cambridge in 1675, obtained his B.A. in 1679/80 and his M.A. in 1683. He was ordained Deacon at York in December 1680, and later Priest, 1683/84, He became Vicar of Tunstall in Holderness in 1683. He was elected F.R.S. on 30 November 1702, and is best known as the first abridger of the *Philosophical Transactions*. Evidently he served as an intermediaty between Flamsteed and the Duke of Bolton.

(2) See Letter 1177, obviously communicated to Flamsteed by Lowthorp.

(3) This typical Flamsteed letter is printed in Baily, *Flamsteed*, pp. 316–17, though without the name of the addressee.

1181 ARNOLD TO LEIBNIZ
5 FEBRUARY 1716
Extracts from the original in the Niedersächsische Landesbibliothek, Hanover.[1]
Continuation of Letter 1174

Monsieur

Il y a 8 Jours que Jai receu votre lettre du 21 de Janvier [N.S.] [2] et je Vous aurois respondu plutôt, Mais allant rendre Visite a M labbe Conti [3] pour lui faire vos compliments, Jy ai trouve M Newton et comme le discours rouloit sur linvention des calculs, il me demanda si Je navois pas vu de ces lettres [4] imprimès en Allemagne Je lui repondit que Jen avois apportè avec moi de Hall, et comme il marquoit quil seroit aise den Avoir Je lui promis de lui en apporter Jeudi dernier Je le fus voir chès lui avec Mr lAbbè Conti et Je lui portai une douzaine de ces lettres que Vous maviès donnès a Hanover; Il tacha de prouver que Vous estiès Autheur de ces lettres par cett expression dans le 2 page lin 4. *illaudibile laudis amore* et puis par ce qui ce trouve dans le lin. 27 & 28 du 3 pag quil pretend ne pouvoir estre connu que de vous. Il me dit quil estoit faché que cette dispute estoit survenu, quil y avoit du tems que Mr Keil lui avoit prie de lui permettre la permission descrire, mais quil Navoit Jamais consentit. Jusques a ce que Vous l'accusiès comme il disoit destre plagiere, et qualors il avoit donné la libertè a Mr Keil descrire tant ce quil lui plairoit. [5] Il me confessa pourtant que Mr Keil Vous avoit traitè

trop brusquement, Il sembloit aussi quil prenoit mal que vous avies envoie
une Probleme [6] a Mr lAbbè Conti pour estre resolu par les Mathematicien[s]
dicy, puis quil regardoit cela comme un defy fait a la Nation, Mr l'Abbe Conti
a plusieurs solutions de ce Problême quil vous enverra au premier Jour [7]
Mr Newton me fit beaucoup d'Honnetêtès et me retint a diner. Mais Je
crois que Vous trouverès tout ce quil me dit dans un brocheure [8] dont il ma
fait present et que Jai donnè hier a Mr Tollman [9] qui me fait esperer quil
vous lenverroit par un exprés qui devoit partir Je crois que vous y trouverès
quelque chose qui nest pas dans le Commercium epistolicum quoiquils lont
donnè le titre de Lextraict de commmercium [sic] epistolicum Monsieur
Tailour [10] Secretaire de la Societe duquel Je vous parlè dans ma derniere
lettre que Je Vous ai escrit il y a quelque tems Mais Je crains que les vents
contraires ne vous aient empeche de recevoir la lettre a publiè un traitè qu'il
l appelle Methodus incrementorum que Jai donnè aussi a Monsieur Tolman
pour vous lenvoier par le Moien de quelquun qui partira pour Hanover. Il y
resout plusieurs problemes sans citer personne que Mr Newton a la Mode d
Angleterre Je ne voi pas quel avantage il pretend tirer de ces increments. Je
me souviens dun Maxime que Jai appris autrefois a lecôle quod non debet
fieri per plura quod potest fieri per pauciorae [11] et je ne scois pas pour quoi
$\dot{x}$ ou dx nest pas si commode que $x + \chi$, il me semble quon en tire les mêmes
avantages sans y trouver tant d'embarras

Jai connu Mr lAbbè Conti autrefois a Paris, qui ma chargè de lettres pour
lItalie, Je le vois souvent icy, mais entre Nous Je doute que les honnêtetès
quil a recû de Mr Newton ne laient gagne de leur parti, peut estre vous seres
surpris quand vous scaurès quon tache de faire entrer les ministres estrangers
dans cette dispute. [12] Jai este assurè que Mr. Newton doit Montrer les
experiences sur les Couleurs dans la presence de Madame de Kilmansec [13]

. . . On me dit que Mr Keil va publier quelquechose [14] contre Mr Bernoulli
Je crois que cest a loccasion de ce que Mr Bernoulli a donne dans les Actes
de Leipsig de 1713 et dans les Memoirs de lAcademie des Science de 1711
contre Mr Newton, les Amis de Mr Keil disent quil va prouver que Mr
Bernoulli nentend pas le calcul differentiel, Je ne doute pas quune telle
demarche ne veille la paresse de Mr Bernoulli.

NOTES

(1) See Bodemann's catalogue, no. 17.

(2) Leibniz's letter to Arnold in reply to Letter 1174, mentioned here, is missing. Extracts
of subsequent correspondence taken from Arnold's letter to Chamberlayne of 16 May 1719
are given under their respective dates.

(3) Conti was on a visit to London, and had considerable contact with Newton. See Letter
1139.

(4) Presumably Arnold means the *Charta Volans* (Number 1009); compare Letter 1045, note (3), p. 72.

(5) Arnold refers to Leibniz's 'Remarques' in the *Journal Literaire de la Haye* for November and December 1713 (see Letter 1018) which Newton first saw in the spring of 1714 (see Letter 1053), and to Keill's 'Answer' to them in the same *Journal*, see Number 1053*a*, note (1).

(6) See Letter 1170, note (3), p. 254. Leibniz later took advantage of Arnold's friendship with Conti to send through him to Conti an addendum to the problem posed (that of finding the equation of the line intersecting a set of curves of the same type at right angles); see Letter 1179.

(7) Leibniz never received these solutions, as he explained to Arnold in Letter 1204.

(8) A copy of the *Recensio*; see Letter 1162, note (2), p. 242.

(9) Possibly a member of the diplomatic corps in London.

(10) Brook Taylor; see Letter 1165, note (4), p. 246.

(11) 'What can be done by means of a few things ought not to be done by many.'

(12) Newton had asked Conti to assemble foreign ministers at the Royal Society to examine the original letters included in the *Commericum Epistolicum*. Conti reported on the proceedings in a letter to Brook Taylor dated 22 May 1721 N.S. See vol. VII; it is also printed in Brewster, *Memoirs*, II, pp. 432–3.

(13) Baroness Kilmansegge, a mistress of George I; see Letter 1203.

(14) Presumably Arnold means Keill's 'Defense', originally intended for the *Mémoires de l'Académie Royale des Sciences*, but finally printed in the *Journal Literaire de la Haye* (see Letter 1165, note (2)). Leibniz took him to indicate a paper in the *Philosophical Transactions* and transmitted the same misapprehension to Bernoulli (see Letters 1201 and 1210).

1182 THE MINT TO THE TREASURY
9 FEBRUARY 1716
From the copy in the Public Record Office.[1]
For the answer see Letter 1184

To the Rt. Honble: the Lords Comm[issione]rs: of His Majestys Treasury

May it Please Your Lordships

The Act of Coinage being near expiring We have thought it our Duty humbly to represent to Your Lordships the present State of His Majesty's Mints in the Tower of London and at Edinburgh with Respect to that Act.

The Moneys leviable thereby have since the peace in the three Years between Christmas 1712 and Christmas last amounted unto 28707£; 19s : 05d which is after the rate of 9569 : 06 : 05½ per Annum at a medium.

In the Same three Years the charge of Coinage has been very great and encreased every Year so as in the last Year to amount unto 13430 : 06 : 01. between Christmas and Christmas and in all the three Years unto 28454:13:01.[2] To which if 3500£ per annum allow'd by Act of Parliamt. for Sallarys and Repairs of Houses and Offices in the Mint in the Tower and 1200£ per

Annum allowed to his Majestys Mint at Edinburgh and about 250£ per Annum for prosecuting Clippers and Coiners be added the whole Charge of the two Mints the last Year will amount unto 18380 : 06 : 01. and during the last three Years unto 43304 : 13 : 1 [3] And if this Expence continues to be so great as it has been the last two Years the Charge of the Mints will exceed the Income by five or six thousand pounds per Annum or above. This Charge has been Supplyed hitherto out of the Stock which accrued to the Mint in the time of War when the Coinage was very Small: and by the great Coinage since the Peace this Stock is now reduced to less then one Thousand pounds which will scarce serve to carry on the Coinage above a month longer.

Wherefore We humbly propose to Your Lordships that the House of Commons may be moved that the Act of Coinage be renewed this Session of Parliament with an Augmentation of the Duty from ten shillings to fifteen Shillings per Tun upon Wines and from 20 to 30 upon Brandy with such restrictions or applications of the money arising therefrom as the House shall think fitt [4]

Which is most humbly Submitted to Your Lordships great Wisdom.

Mint Office
9 *Feb* 1715.

NOTES

(1) Mint/1, 7, p. 71. There are several drafts in the Mint Papers, I, fos. 269, 313–14, 316, 324, 325, 328. The original was probably signed by Newton and Sandford.

(2) See Number 1182a.

(3) See Letter 1184, note (2).

(4) On the very day this letter was written, the House of Commons voted a Humble Address to George I, asking that the Mint officers be required to lay accounts for the last three years before the House (see Letter 1184). When these accounts showing the mounting annual deficiency in the Mints' operations had been examined, the House resolved (3 March) to provide the Mints with an income of £15000 per annum for seven years from 1 May 1715. On the 10th the Committee of Ways and Means proposed to achieve this object by continuing (for seven years) the Coinage Act of Anne, with supplementary provisions—including a tax on senna—to ensure that the Mints' income did not fall below £15000 p.a. The Act effecting these proposals received the Royal Assent on 26 June 1716.

1182a EXPENDITURE AND INCOME OF THE MINT, 1711–15

From the copy in the Mint Papers[1]

Abstract of Moneys Received out of the Exchequer out of the Coinage Duty, exclusive of what has been paid for the Mint in Scotland and for prosecuting Clippers and Coiners.

Charge of the Coinage of Gold and Silver Moneys Exclusive of Sallarys, and Repairs of Houses Offices and Buildings.

	[£	s.	d.]
1711	5865	13	$06\frac{1}{2}$
1712	4051	16	$09\frac{1}{2}$
1713	5055	05	$02\frac{1}{2}$
1714	10036	19	07
1715	9706	15	01
	34716	10	$02\frac{1}{2}$

	[£	s.	d.]
1711	5178	08	03
1712	1243	17	$01\frac{1}{2}$
1713	4796	03	09
1714	10228	03	03
1715	13430	06	01
	34876	18	$05\frac{1}{2}$

NOTE

(1) Mint Papers, I, fo. 314r. The method of computation here does not yield the averages quoted in the letter, but the figures illustrate the point the letter makes—the great increase of the cost of coinage in the last two years, and the deficiency of the revenue from the Coinage Duty.

1183 ALEXANDER CUNNINGHAM TO NEWTON
10 FEBRUARY 1716

From the original in the Library of King's College, Cambridge[1]

Sir Isaak Newton

Venice ye 21st Febry [N.S.]
1716

H[oure]d Sir

Having had good weather for traveling, I arrived, the end of Novbr in this place, just upon the change of the season, ever since we have had excessive colds wth a great quantity of snow, Our Lagune wer frozen over for a fortnight soe that carts came on the see from the main land to this town, wch has only hapned twice these 50 years past.

Since I came hither have recd infinite numbers of compl[i]m[en]ts from all places in Italy, The learned seldom omit asking about you Sir and assure

278

you of the high esteem they have for you, They are sorry you doe not write to them, I tell 'em of the multitude of your busines, and excuse you to them by a *Non si usa*, which tho it should be a good excuse evry where, I doe not find it more kindlie received any where, than here, Monsigr Bianchini [2] kindly salutes you, I have sent to him lately the short state of the Commercium, and as soon as he has red it I expect his judgmt of it; The Marqs Polenni was with me lately, [3] As he approves your Synthetique way in the search of nature, soe he admires your happyness in finding 'em out, I see he does not only understand your works, but alsoe knows your self as if he had been frequently wth you, by your way of writing. He tells me he thought the Adversarys of the Royal Soci[ety]: would [have] had something at Leest to have said for 'em themselvs. But since he saw that sheet they published two years agoe, [4] he thinks they have a poor cause indeed, and it has just had the effect here that you and I did forsee, for nothing has done 'em more harm, When some in France told me their opinions of that contraversie, I asked whey they did not publish it then and deal candidly, they have several measures to keep wth your adversarys and would not enter into a dispute in their present circumstances, In short Sr they oftn manage truth wth policy, soe they are like to doe at Rome, I have writ to some there, that this is not the Constitution Unigenitus, [5] soe they need not be so much on the reserve; they pretend they would be glad to hear what can be said on the otherside against the faith of the Commercium, I have told em, that all the doors of the societie, wer open to all strangers last winter, to examine the records referrd to, and alsoe more wer produced, that had not been mentioned;

Sr I'm ordered to tell you that there is one here, that has found out a way to come at the Longitude, by an instrument, I cant blame the inventor not to show it, but he's positive it will doe, He says its being eff[ec]tual, depends not on the nature of the mettale its made of, but on the form of the parts, wch in all Clims [6] are the same, I shall not now trouble you wth a longer acct of it, Only please to let me know if the perpetuum mobile would produce that effect.

Sr Isac I think tis about the 19 of March next, that thers du upon my 4 Excheqr Notes in your Custody, 12 pounds interest money, I desire you at that time to take up principal and interest and I shall order Mr Thomas Williams my Marcht to receive it from you, or if you find it too troublesome to you to receive the money I shall order him to receive the Exqr Notes and convert 'em into money for my use, I am

Sr
Your most humble servt
and friend
ALEXR CUNNINGHAM

Sr I was told this day that thers an old prophesie of St Bridget that says, the Turks are to overrun all Italy and burn Rome, than ar to become Christians, I know not But the Bi[sho]p of Woster [7] may have it from her for tis to be about this time

NOTES

(1) Keynes MS. 141(M); microfilm 1011.25. The writer must certainly be the historical author of this name (who was born in 1654 and died in 1737), whose acquaintance with Newton is mentioned in the *Dictionary of National Biography*. After having served as a tutor to various noblemen and as a secret agent acting for the Whigs, he was appointed envoy to Venice in 1715. Returning in 1720, he wrote *A History of Great Britain* from 1688 to 1715 which was published in 1787. In this work Cunningham refers to Newton several times: he joined with Newton in recommending the Great Recoinage (I, p. 182); he talked with Newton in London in 1705 (II, p. 103); he records the *Principia* as displaying 'such profundity of judgement as far surpassed both the genius and discoveries of antiquity, and the capacity of his own contemporaries' (I, p. 111).

(2) Probably Francesco Bianchini (1662–1729), celebrated Italian astronomer, antiquarian and man of letters.

(3) Giovanni Poleni (1683–1761), successively Professor of Astronomy (1708), Physics (1715) and Mathematics (1719) at Padua—in the latter Chair he succeeded Nicholas Bernoulli. He was elected F.R.S. in 1710.

(4) The *Charta Volans*, see Number 1009.

(5) The Papal Bull *Unigenitus* was issued by Clement XI in 1713 (under pressure from Louis XIV) against the Jansenists and the 'Gallican' claims of the French clergy.

(6) Climes or climates, presumably.

(7) William Lloyd (1627–1717), one of the seven bishops and partisan of Cunningham's *bête noire*, Gilbert Burnet (compare Letter 1021, note (1)).

1184 LOWNDES TO NEWTON
16 FEBRUARY 1716

From the copy in the Public Record Office.[1]
Reply to Letter 1182

Sir

His Ma[jes]ty having signifyed his pleasure that an account be forthwith layd before the House of Commons of the Clear produce of the Coynage duty for the three severall Years last past respectively And also of the charge of Coyning Gold & Silver money And the Established charge of the Mint in England & Scotland and Incident charges attending the same for the said three years respectively the Lords Commissioners of his Matys Treasury desire you to cause the said Account to be forthwith made up and laid before the House of Commons I am &c [2]

16 *Feby* 1715/16

WM LOWNDES

NOTES

(1) T/27, 22, p. 46.

(2) Compare Letter 1182, note (4), p. 277. The accounts (signed by Newton and Sandford and dated 16 February 1716) which were presented to the House of Commons showed a total yield from the Coinage Duties from Christmas 1712 to Christmas 1715 of £28654. 9s. 7½d; and a total expenditure upon the Mints of London and Edinburgh of £43216. 7s. 9d over the same period (increasing roughly as follows: 1713, £9912; 1714, £15171; 1715, £18132). Thus Newton had slightly revised the figures quoted in Letter 1182 (see *House of Commons Journal*, **18** (1716), 378).

1185 THE MINT TO THE TREASURY
22 FEBRUARY 1716

From a copy in the Public Record Office.[1]
For the answer see Letter 1195

To the Rt. Honble: the Lords Comm[issione]rs: of His Majestys Treasury.

May it Please Your Lordships

When the Gravers Pattent was Settled about ten Years since, it was thought necessary to take away from them the power they had of making Medalls without Controll.

Upon a Report made Some time after by the Officers of the Mint that good Graving was the best Security of the Coin of the Kingdom, and best acquired by Graving of Medalls, and that without Exercise in such Graving the Gravers might in time loose the Skill they had already acquired, Her late Majesty was then pleased to Grant a Warrant (a Copy whereof is hereunto annexed) impowering the Officers of the Mint to direct the Said Gravers to make Such Medalls as should be approved of by them [2]

As the Said Warrant is of no force at present, Wee humbly represent to Your Lordships that for the reasons therein expressed, Wee hope it may be thought necessary the Same should be renewed that the Gravers may improve their hands, when they are nott imployed about the Coins.

Which is most humbly Submitted to Your Lordships great Wisdom.

22 *Feb*. 1715

NOTES

(1) Mint/1, 7, p. 71.

(2) See Letters 676 (12 October 1704) and 679 (24 November 1704), vol. IV, pp. 419–20 and 429. The Warrant is not of course with the copy of the letter printed here, and we have not thought it useful to reproduce it. John Croker and Samuel Bull had in fact struck many medals to commemorate Marlborough's great victories (see Plate V, facing p. 233) under Newton's authority; and Croker was to strike a medal to commemorate Newton himself.

1186 KEILL TO NEWTON
[23 FEBRUARY 1716]
From the original in private possession[1]

Honoured Sr

I returned but last week to Oxford, but while I was in Northampton shire I began to think on Mr Leibnits Problem,[2] and I think I have hit upon the solution of it, it not being difficult to any that understand Fluxions, The method[3] I used was this

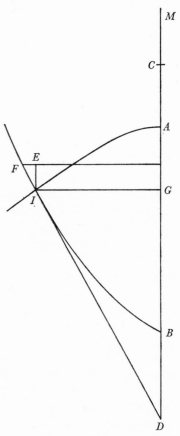

Suppose A series of curves of the same nature wch have all the same Axis *AB* and Vertex *A*. Let *AI* be one of them and *BI* the Curve wch cuts it at right angles, Because the Curve *AI* is given its subnormal *GD* will be given, But this subnormal is the subtangent of the curve *BI*. therefore $FE:EI::IG:GD$. that is calling AG x and IG y; $\dot{y}:-\dot{x}::y:GD$. (here $\dot{x}$ must be negative for when y increases x decreases, [)] by this Analogy I get an equation and from the equation I obtain a value of the Latus Rectum. This value of the latus Rectum I put in its Room in the equation wch expresses the nature of the curve proposed and I have a new exquation [*sic*] expressed by x y and their fluxions wch gives the nature of the curve required. For example suppose the curve *AI* a Parabola whose equation is $2lx = yy$ then

$$GD = l \text{ and } \dot{y}:-\dot{x}::y-\frac{\dot{x}y}{y} = l^{[4]}$$ this value of l being put in the equation for the parabola gives $\frac{-2x\dot{x}y}{\dot{y}} = yy$ and $-2x\dot{x} = y\dot{y}$ hence $a^2-x^2 = \frac{1}{2}y^2$ or $2a^2 - 2x^2 = y^2$ Hence the curve *BI* is an Ellipse whose Greater Axis is double in Power to its lesser.

If the curves Proposed are Hyperbolas, whose center is C transverse Axis $AM = 2a$, their equation is $\frac{2lax+lx^2}{2a} = y^2$ and the subnormal is $\frac{la+lx}{2a}$ and $\dot{y}:-\dot{x}::y:\frac{la+lx}{2a}$ hence $l = -\frac{2ay\dot{x}}{a+x\times\dot{y}}$ wch value of l being put in the equation

282

of the Hyperbola gives $\dfrac{-4a^2yx\dot{x}-2ayx^2\dot{x}}{2ax\times\overline{a+x}\times\dot{y}}=y^2$ and $\dfrac{-x^2\dot{x}-2ax\dot{x}}{x+a}=y\dot{y}$ that is

$-x\dot{x}-a\dot{x}+\dfrac{a^2\dot{x}}{a+x}=y\dot{y}$ or $\dfrac{2a^2\dot{x}}{a+x}=2x\dot{x}+2a\dot{x}+2y\dot{y}$ and takeing the Fluents we

have Fluent $\dfrac{2a^2\dot{x}}{a+x}+$ a Given quantity $=x^2+2ax+y^2$. From wch we draw the

followin[g]

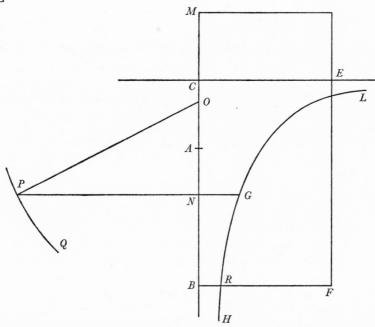

construction.[4a] at the center C Assymptotes CA CE describe the hyperbola LGH
whose rectangle inscribed is $=2a^2$. make BF perpendiculare to BA and $=$ to it.
and make up the rectang. $MF[.]$ in AB take any point N, and let NO be a mean
proportional [between] NA and NM, at the center O with the Radius OP whose
square is equal to the rectangle MF minus the Hyperbolick space $BNGR$ describe
the Arch PQ and let NG produced meet with it in P the point P will be in the
curve required The curve will be an oval Figure whose greatest Ordinate is at A.
Mr Sterling [5] an undergraduat here has likewise solved this Problem, I have also
received the solution of this and several other Problems from Mr Pemberton [6]
If you please you may put mine in the Transactions [7] I am Sr

Yours most obliged Humble servant

JOHN KEILL

For
The Honoured Sr Isaack Newton
at his house in St Martins Street
near Leicester fields
            Westminster

NOTES

(1) Printed in Rigaud, *Correspondence*, II, pp. 421–4. The postmark is 24FE, and the letter must have been written the previous day in Oxford. The year is given by the content.

(2) That is, the problem Leibniz had sent to Conti for Newton; see Letter 1170, note (6), p. 254. The problem, originally raised publicly in 1697 (see Letter 1201, notes (7) and (8)) aroused a great deal of interest amongst both continental and English mathematicians.

Those who attempted to solve the problem included James Stirling, Henry Pemberton (see notes (5) and (6) below) and John Machin (see Rigaud, *Correspondence*, I, p. 267). Most of these solutions were of limited value; either they applied to special cases only, as a result of confusion over the original statement of the problem (see Letter 1170, note (3)) or they were general, but incomplete. Newton's own attempt at a solution (see Number 1187*a*) was extremely feeble.

(3) Keill here effectively only outlines a general approach, and then gives solutions for two simple cases, in one of which he has to resort to geometrical construction.

Hence suppose we write the first set of curves (exemplified by *AI*) as $u = u(x, l)$, where u is the ordinate and l a parameter, and the set to be determined (exemplified by *BI*) as $v = v(x, a)$ where a is a parameter. If the two sets intersect at right angles we may write, at the point of intersection

$$-v' = \frac{1}{u'} \quad \dots (1) \qquad \qquad \text{(i.e. } \frac{\dot{y}}{-\dot{x}} = \frac{y}{GD} \text{ in Keill's figure)}$$

$$\text{and } u = v \quad \dots (2)$$

Using equation (1) we can write l in terms of x, u and u'; and hence in terms of x, v, and v'. Substituting this expression for l into u in equation (2), we obtain a differential equation in x, v, and v' which can, in principle, be solved.

Henry Pemberton had in fact written to Keill on 11 February 1716 giving an outline of the same method as an algorithm. He wrote again on 24 April 1716, with further details (see U.L.C. Lusasian Papers, Res. 1893(*a*), Packet 2).

(4) Keill should have written $\dot{y} : -\dot{x} :: y : l$, hence $-\dfrac{\dot{x}y}{\dot{y}} = l$.

(4a) We follow Keill's MS, figure, although this does not accurately reflect the construction. Here $AN = x$ and $PN = y$.

(5) James Stirling (1692–1770) of Balliol College, Oxford. His solution (essentially the same as Keill's) appeared on pp. 15–19 of the Appendix to his *Lineæ tertii ordinis Newtonianæ sive illustratio tractatus Newtoni de enumeratione linearum tertii ordinis* (Oxford, 1717). See Charles Tweedie, *James Stirling: a sketch of his life and work* (Oxford, 1922), and Letter 1325, note (1), vol. VII.

(6) Henry Pemberton (1694–1771), who was to become the third editor of the *Principia*, had not yet taken his M.D. degree at Leiden (1719) but had entered himself as a medical student at that university in October 1712 and again in August 1714 (compare his biographer's statements quoted by I. Bernard Cohen in *Isis*, **54** (1963), 320). He was well known to Keill already, and by implication his name at least was known to Newton also. His solutions to Leibniz's problem may be found in letters addressed to Keill dated 11 February 1716 and 24 April 1716 (U.L.C. Lucasian Papers, Res. 1893(*a*), Packet 2).

(7) Keill's solution did not appear in the *Philosophical Transactions*.

1187 NEWTON TO CONTI
26 FEBRUARY 1716
From a holograph draft in the University Library, Cambridge.[1]
Reply to Letter 1170; for the answer see Letter 1197

Leicester Fields, London. 26 *Feb.* 171$\frac{5}{6}$

Sr

You know that the Commercium Epistolicum conteins the ancient Letters & Papers preserved in the Archives & Letter Books of the Royal Society & Library of Mr Collins relating to the dispute between Mr Leibnitz & Dr Keill & that they were collected & published by a numerous Committee of Gentlemen of severall nations appointed by the R. Society for that purpose. Mr Leibnitz has hitherto avoided returning an Answer to the same; for the Book is matter of fact & uncapable of an Answer. To avoid answering it he pretended the first year that he had not seen this Book nor had leasure to examin it, but had desired an eminent Mathematician to examin it.[2] And the Answer of the Mathematician (or pretended Mathematician) dated 7 June 1713 [N.S.], was inserted into a defamatory Letter dated 29 July [N.S.] following, & published in Germany without the name of the Author or Printer or City where it was printed.[3] And the whole has been since translated into French & inserted into another abusive Letter (of the same Author, as I suspect)[4] & answered by Dr Keill in July 1714,[5] & no answer is yet given to the Doctor [6]

Hitherto Mr Leibnitz avoided returning an Answer to the Commercium Epistolicum by pretending that he had not seen it. And now he avoids it by telling you that the English shall not have the pleasure to see him return an Answer to their slender reasonings (as he calls them) & by endeavouring to engage me in dispute about Philosophy & about solving of Problems, both which are nothing to the Question.[7]

[I have left off Mathematicks 20 years ago & look upon solving of Problemes as a very unfit argument to decide who was the best Mathematician or invented any thing above 50 years ago. And] [8] As to Philosophy [it is as little to the purpose] [8] He colludes in the significations of words, calling those things miracles wch create no wonder & those things occult qualities whos causes are occult tho the qualities themselves be manifest, & those things the souls of men wch do not animate their bodies, His Harmonia præstabilita is miraculous & contradicts the daily experience of all mankind, every man finding in himse[l]f a power of seeing with his eyes & moving his body by his will. He preferrs Hypotheses to Arguments of Induction drawn from experiments,

285

accuses me of opinions wch are not mine, & instead of proposing Questions to be examined by Experiments before they are admitted into Philosophy he proposes Hypotheses to be admitted & beleived before they are examined. But all this is nothing to the Commercium Epistolicum.

He complains of the Committee of the Royall Society as if they had acted partially in omitting what made against me. But he fails in proving the accusation. For he instances in a Paragraph concerning my ignorance, pretending that they omitted it, & yet you will find it in the Commercium Epistolicum pag 74 lin. 10, 11, & I am not ashamed of it. (9) He saith that he saw this Paragraph in the hands of Mr Collins when he was in London the second time, that is in October 1676. It is in my Letter of 24 October 1676, & therefore he then saw that Letter. (10) And in that & some other Letters writ before that time I described my method of fluxions. And in the same Letter I described also two generall methods of series, one of wch is now claimed from me by Mr Leibnitz.

I beleive you will think it reasonable that Mr Leibnitz be constant to himself & still acknowledge what he acknowledged above 15 years ago, & still forbeare to contradict what he forbore to contradict in those days.

In his Letter of 20 May 1675 [N.S.] (11) he acknowledged the Receipt of a Letter from Mr Oldenburg dated 15 Apr. 1675 with several converging series conteined therein. (12) And I expect from him that he still acknowledge the receipt thereof. Many Gentlemen of Italy France & Germany (you your self being one of them) have seen the original Letters & the entries thereof in the old Letter books of the Royal Society, & the Series of Gregory is in the Letter of 15 Apr 1675, & in Gregories original Letter dated 15 Feb. 1671.

In a Letter dated 12 May 1676 [N.S.] (13) seen by the same Gentlemen he acknowledged that he then wanted the method for finding a series for the Arc whose sine was given, & by consequence that he wanted it when he wrote his Letter of 24 October 1674 (14) And I expect that he still acknowledge it.

In the Acta Eruditorum for May 1700, in answer to Mr Fatio who had said that I was the oldest inventor by many years, Mr Leibnitz acknowledged that no body so far as he knew, had the method of fluxions or differences before me & him, & that no body before me had proved by a specimen made publick that he had it. (15) Here he allowed that I had the method before it was published or communicated by him to any Body in Germany that the Principia Philosophiæ were a proof that I had it, & the first specimen made publick of applying it to the difficulter Problemes, And I expect that he still continue to make the same acknowledgement, At that time he did not deny what Mr Fatio affirmed, & nothing but want of candor can make him unconstant to himself.

In a Letter to me dated 7 March 1693 [N.S.][16] & now in the custody of the R.S. he wrote, Mirifice ampliaveras Geometriam tuis seriebus, sed edito Principiorum opere ostendisti patere tibi quæ Analysi receptæ non subsunt. Conatus sum Ego quoque notis commodis adhibitis quæ differentias & summas exhibent, Geometriam illam quam transcend[ent]em appello, Analysi quodammodo subjicere nec res male processit. And what he then acknowledged he ought still to acknowledge.

In his Letter of 21 June 1677 [N.S.] [17] writ in Answer to mine of 24 Octob 1676 wherein I had described my Method partly in plain words & partly in cyphers, he said that he agreed with me that the method of tangents of Slusius was not yet made perfect, & then set down a differential method of Tangents published by Dr Barrow in the year 1670, & disguised it by a new notation, pretending that it was his own & shewed how it might be improved so as to perform those things wch I had ascribed to my method, & concluded from thence that mine differed not much from his, especially since it facili[t]ated Quadratures. And in the Acta Eruditorum for October 1684, in publishing the Elements of his method he added that it extended to the difficulter Problemes which without this Method or another like it, could not be managed so easily.[18] He understood therefore in those days that in the year 1676 when I wrote my said Letter I had a method which did the same things with the method wch he calls differentiall, & he ought still to acknowledge it, especially now the sentences in cyphers are decyphered & other things in that Letter relating to the method are fully explained, & the compendium mentioned therein is made publick.

In his Letter of 27 Aug. 1676 [N.S.] [19] he represented that he did not beleive that my Methods were so generall as I had described them in my Letter of 13 June preceding,[20] & affirmed that there were many Problemes so difficult that they did not depend upon Equations nor Quadratures, such as (amongst many others) were the inverse Problemes of tangents, And by these words he acknowledged that he had not yet found the reduction of Problems to Differential Equations. And what he then acknowledged, he acknowledged again in the Acta Eruditorum for April 1691 pag. 178, & ought in candor to acknowledge still.[21]

Dr Wallis in the Preface to the two first Volumes of his works published in April 1695, wrote that I in my two Letters written in the year 1676 had explained to Mr Leibnitz the Method (called by me the method of fluxions & by him the differential method) invented by me ten years before or above (that is, in the year 1666 or before) & in the Letters which followed between them, Mr Leibnitz had notice of this Paragraph & did not then contradict it nor found any fault with it.[22] And I expect that he still forbeare to contradict it.

But as he has lately attaqued me with an accusation wch amounts to plagiary: if he goes on to accuse me, it lies upon him by the laws of all nations to prove his accusation on pain of being accounted guilty of calumny. He hath hitherto written Letters to his correspondents full of affirmations complaints & reflexions without proving any thing. But he is the agressor & it lies upon him to prove his charge.

I forbear to descend further into particulars. You have them in the Commercium Epistolicum & the Abstract thereof, to both of which I refer you.

I am

Sr

Your most humble

and most obedient Servant

Is. NEWTON

NOTES

(1) Add. 3968(38), fos. 564–5. There are seven further holograph drafts, all apparently of Newton's reply to Conti, in Add. 3968(38), fos. 558–73, but only the one we print here is dated and signed. There are great differences between the drafts, two of which refer to Newton in the third person. Two others begin (fos. 571 and 573)

'Sr

The more I consider the Postscript of Mr Leibnitz the less I think it deserves an answer. For it is nothing but a piece of railery from the beginning to the end.'
However the content of all the drafts is essentially contained in the version we print here.

The letter was printed by Newton, without Conti's knowledge and after Leibniz's death, in Raphson, *History of Fluxions*, pp. 100–3 (see also Letter 1170, note (2)), word-for-word as in the draft we follow here, but with the usual differences of style introduced by the printer. Des Maizeaux printed a French translation in his *Recueil* (II, pp. 20–9) and it has often been reprinted since.

The letter took some time to reach its destination if Conti's account (see his letter to Taylor, dated 11 May 1721, printed in Brewster, *Memoirs*, II, pp. 432–3, and in vol. VII of this *Correspondence*) is to be believed. According to Conti he arranged a meeting of foreign ministers at Newton's instigation, to discuss the Leibniz affair and to examine the *Commercium Epistolicum* papers in the Royal Society archives. Baron von Kilmansegge, speaking for all the ministers, suggested that Newton should himself write a letter to Leibniz, and the King approved. But Newton had in fact already written his reply, and sent it to Conti for transmission to Leibniz. The letter, however, remained in London for over a month, and the Baroness von Kilmansegge (see Letter 1203) arranged for Pierre Coste to translate it into French; Conti wrote his own covering letter (see Letter 1190), and showed it to De Moivre for correction.

(2) Johann Bernoulli, though Newton was still unaware of the fact.

(3) The *Charta Volans* (Number 1009); compare Arnold's Letter 1181.

(4) The *Remarques*, printed in Letter 1018.

(5) Keill's 'Answer' in the *Journal Litteraire de la Haye*; see Letter 1053a and notes.

(6) See Letter 1138 and others subsequently where Leibniz declines to answer Keill.

(7) As regards 'philosophy' Newton means the Leibniz–Clarke correspondence (see Letter 1173, note (3)); for the 'Problems' see the postscript to Letter 1170, and notes.

(8) The words in square brackets have been struck out and do not appear in the printed text.

(9) 'Sed in simplicioribus vulgoque celebratis Figuris, vix aliquid relatu dignum reperi quod evasit aliorum conatus; nisi forte *Longitudo Cissoidis* eiusmodi censeatur.' ('But in the simpler and commonly known figures I have found scarcely anything worth recording, that has escaped the efforts of others, unless perhaps the length of the cissoid may be so considered.') See Letter 1170, p. 251.

(10) The *Epistola Posterior*; see Letter 187, vol. II.

(11) To Oldenburg; see *Commercium Epistolicum*, p. 42 and Gerhardt, *Briefwechsel*, pp. 122–4.

(12) Printed in *Commercium Epistolicum*, p. 39, and in Gerhardt, *Briefwechsel*, pp. 113–22, where the letter is dated 12 April 1675 N.S.

(13) Letter 158, vol. II, p. 3.

(14) Letter 126, vol. I; the date should be 16 October N.S.; Newton made an obvious slip.

(15) The passage in question reads (see Gerhardt, *Leibniz: Mathematische Schriften*, v, p. 347): 'Interim considerandum relinquo . . . de Methodo summi momenti valdeque diffusa circa maxima et minima fuisse actum, quam ante Dn. Newtonum et me nullus quod sciam Geometra habuit, uti ante hunc maxime nominis Geometram nemo specimine publice dato se habere probavit.' ('Meanwhile I leave the consideration . . . of [what] was done concerning the method (or the highest importance and remarkably general) touching maxima and minima, a method which no mathematician known to me had before Mr. Newton and myself, just as no one before this mathematician of great renown had proved that he possessed it by giving an example to the public.')

(16) Letter 407, vol. III, p. 257. ('You have brought a wonderful fulfillment to geometry by your series and in your published book of the *Principles* you have shown that certain things lie open to you which are not embraced within the conventional analysis. I too have tried to annex to analysis in a special way that geometry which I call *transcendent*, employing a convenient notation for handling differences and summations; and the thing has not gone badly.') This short passage shows that Leibniz was more impressed by Newton's work on series than anything else, and that he was far from recognizing Newton's possession of any particular method or system of analysis.

(17) Letter 209, vol. II; the passage mentioned is on p. 213.

(18) *Acta Eruditorum* for October 1684, pp. 467–73; see Gerhardt, *Leibniz: Mathematische Schriften*, v, pp. 200–26; in particular pp. 225–6. It is hard to see how these words can be taken as admitting the prior existence of Newton's 'other method.'

(19) Letter 172, vol. II, see p. 64; however, Leibniz referred not to 'Newton's methods', but to the method by which problems are 'ad series infinitas reduci'—which somewhat alters the significance of the passage. Leibniz never denied Newton's pre-eminence in this.

(20) Letter 165, vol. II; the *Epistola Prior*.

(21) 'Nec dissentit conclusio circa relationem inter tempora et velocitates in gravi per medium descendente. Hanc enim ad sectorem hyperbolicum reduxit Newtonus, ad seriem infinitam Hugenius, quam invenit pendere a quadratura hyperbolæ, nos ad logarithmos . . . tanquam perfectissimum talia exprimendi modum præbentes.' ('Nor is the conclusion inconsistent with the relationship between the times and the velocities of a heavy body descend-

ing through a medium. For Newton reduced this to a hyperbolic sector, Huygens to an infinite series which he found depended on the quadrature of the hyperbola, [and] ourselves to logarithms . . . as offering the most perfect method of expressing such things.') See Gerhardt, *Leibniz: Mathematische Schriften*, VI, p. 144.

(22) See John Wallis, *Opera mathematica*, III, (Oxford, 1699) pp. 653–9, where Wallis's letter to Leibniz dated 1 December 1696 is printed.

1187a NEWTON'S 'SOLUTION' OF LEIBNIZ'S PROBLEM
[c. FEBRUARY 1716]
From drafts in the University Library, Cambridge [1]

Sr

Mr John Bernoulli in the Acta Eruditorum for October 1698 pag. 471 wrote in this manner. [2] Methodum quam optaveram generalem secandi [curvas] ordinatim positione datas sive algebraicas sive transcendentes, in angulo recto, sive obliquo, invariabili, sive data lege variabili, tandem ex voto erui, cui Leibnitio approbatore, ne γρύ addi posset ad ulteriorem perfectionem, et vel ideo tantum quod perpetuo ad æquationem deducat, in quo si interdum indeterminatæ sunt inseparabiles, methodus non ideo imperfectior est, non enim hujus sed alius est methodi indeterminatas separare. Rogamus itaque fratrem ut velit suas quoque vires exercere in re tanti momenti. Suscepti laboris non penitebit si felix successus fructu jucundo compensaverit. Scio relicturum suum quem nunc fovet modum qui in paucissimis tantum exemplis adhibere potest. These Gentlemen had been four or five years about Problemes of this kind, & to give the very same solution with that here mentioned might require a spirit of divination. But the Probleme may be generally solved after the following manner.

The Probleme

A series of Curves being given of one & the same kind, succeeding one another (in forme & position) in an uniform manner according to any general Rule find another Curve wch shall cut all the Curves in the said series, in any angle right, or oblique, invariable, or variable according to any Rule assigned.

The method of Solution

Let *BD* by any one of the Curves in the Series, *CD* the Curve wch is to cut it, *D* the point of intersection *AE* the common Abscissa of the two Curves & *ED* the common ordinate; & the two Rules will give [the angle of intersection &] the perpendicular to the Curve *CD* & the radius of its curvity at the point

D; & the position of the perpendicular will give the first fluxion & the curvity the second fluxion of the ordinate of the same curve CD. And so the Probleme will be reduced to equations involving fluxions & by separating or extracting the fluents will be resolved

Mr Leibnits [3] in the Acta Eruditorum for May 1700 pag 204, challenged Mr Fatio to solve this Probleme, [4] Invenire Curvam aut saltem proprietatem tangentium Curvæ quæ Curvas etiam transcendentes ordinatim datas secet ad angulos rectos. Let ED the Ordinate of the Curve CD be represented by the area of another Curve upon the same Abscissa AE & the first fluxion of this Ordinate will be represented by the Ordinate of the other Curve & the second fluxion by the proportion of this last Ordinate to ye subtangent of the other curve. And so the property of the Tangent of the other Curve is given.

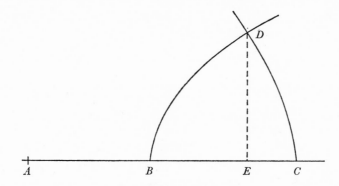

Let the Ordinate of the Curve desired be represented by the area of another curve upon the same Abscissa & the first fluxion will be represented by the Ordinate of this other Curve, & the second fluxion by the proportion of the Ordinate to the subtangent. And so the Probleme is reduced to the property of a Tangent

and [5] the first Rule will give the tangent of the Curve BD at the point D, & the second Rule will give the angle of intersection & tangent of the other Curve CD at ye same point D, & both the Rules together will give the Radius of the curvity of the other Curve CD at the same point D. Let the Abscissa AE flow uniformly & its fluxion be called 1, & the position of the Tangent of ye curve CD will give the second fluxion of the same Ordinate And so the Problem will be reduced to equations involving the first & second fluxions of the Ordinate of the Curve desired, & by reducing the Equations & extracting or separating the fluent (wch is not the business of this method) will be resolved.

There may be some Art in chusing the Abscissa & Ordinate or other

Fluents to which the invention of ye Curve *CD* may be best referred & in reducing the Equations after the best manner. But nothing more is here desired then a general method of resolving the Probleme without entering into particular cases.

The curvity of the intersecting Curve *CD* at ye point *D* is found by taking in the tangent of the Curve *CD* another point *d* infinitely near to ye point *D*, & finding the tangent at ye point *d* of the Curve in the series wch passes through that point *d*, & also the tangent of the Curve intersecting it at ye same point *d*; & upon the two intersecting curves at ye points of intersection *D* & *d* erecting perpendiculars. For these perpendiculars shall intersect one another at ye Center of the curvity of the intersecting Curves.

[*We omit the first part of the Latin version which is essentially the same, but the solution is different. It goes as follows:*] [6]

Solutio

Natura Curvarum Secandarum dat tangentes earundem ad intersectionum puncta quacumque, et anguli intersectionum dant perpendicula Curvarum secantium, et perpendicula duo coeuntia, per concursum suum ultimum, dant centrum curvaminis Curvæ secantis ad punctum intersectionis cujuscumque. Ducatur Abscissa in situ quocumque commodo, et sit ejus fluxio unitas, et positio perpendiculi dabit fluxionem primam Ordinatæ ad Curvam quæsitam pertinentis, et curvamen hujus Curvæ dabit fluxionem secundam ejusdem Ordinatæ. Et sic Problema semper deducetur ad æquationes. Quod erat faciendum

Scholium

Non hujus, sed alius est methodi æquationes reducere, et in series convergentes ubi opus est convertere. Problema hocce, cum Nullius fere sit usus, in Actis Eruditorum annos plures Neglectum, et insolutum Mansit. Et eadem de causa solutionem ejus non ulterius prosequor.

NOTES

(1) We have here used Add. 3968(25); fo. 371 is a holograph English version, fo. 369 a clerical copy of a Latin text, largely the same, but differing in the 'solution' which we give as the last two paragraphs here. The paper was published in *Phil. Trans.*, **29**, no. 347 (1716), 400, under the title 'Problematis olim in Actis Eruditorum Lipsiæ propositi Solutio Generalis.' —see note (6) below. The general history of the problem, and later variants of it, are discussed in M. Cantor, *Vorlesungen über Geschichte der Mathematik*, III (Leipzig, 1898), pp. 443–55. See also Letter 1170, note (6), p. 254.

Leibniz's letter with the challenge problem was communicated by Conti to Newton in February 1716 (see Letter 1170) and we suppose that these drafts were composed soon afterwards. Newton made several other, equally poor and inclusive, attempts to attack the problem (see for example U.L.C. Add. 3960(12), fo. 197 and Add. 3964(6), fos. 5–6 and other MSS. in private possession).

(2) 'I have at length worked out to my satisfaction a general method such as I desired for cutting curves given in position, whether algebraic or transcendental, ordinately, either in a right angle or obliquely (whether in an invariable way or according to some variable rule), to which, in Leibniz's opinion, no scruple can be added to bring it to ultimate perfection for the reason that it always leads to an equation; if in this some indeterminate quantities remain sometimes inseparable the method is not the less perfect on that account; for the way of separating out indeterminates belongs not to this but to some other method. Accordingly I beg my brother [Jakob I] to please to devote his talents also to a matter of such importance. If a happy success shall reward him with an agreeable prize he will not repent the task he has undertaken. I know he will relinquish the method he now cherishes which can be applied in only a few instances.' For the history of the problem among the continental mathematicians, see Letter 1201, note (7), p. 320.

(3) These two paragraphs have apparently been struck out.

(4) 'To find the curve or at any rate the property of the tangents of a curve which will cut given curves, including transcendental ones, the sequence of which is given, at right angles.' See also Letter 1201, note (8), p. 321.

(5) This version is written on the reverse of fo. 371.

(6) These two paragraphs are essentially as printed in the *Philosophical Transactions*:

Solution

The nature of the curves to be cut gives their tangents at any points of intersection whatever, and the angles of intersection give the perpendiculars of the intersecting curves, and when two perpendiculars coincide they give by their last coincidence the centre of the intersecting curve at the point of any intersection. Let the abscissa be drawn in any convenient position and let its fluxion be unity, and the position of the perpendicular will give the first fluxion of the ordinate to the curve which is appropriate to that point; and the curvature of this curve will give the second fluxion of this same ordinate. And so the problem may always be reduced to equations. Which is what was to be done.

Scholium

The reduction of equations and their conversion into converging series when the need arises belongs not to this method but to another. This problem here, as it is almost useless, has remained neglected and unsolved in the *Acta Eruditorum* for many years. And for the same reason I shall not pursue its solution any further.

1188 LOWNDES TO THE REFEREES
27 FEBRUARY 1716
From the copy in the Public Record Office[1]

Gentlemen

The Lords Commissioners of his Ma[jes]tys Treasury have Commanded me to signifye to You their Pleasure, that you give them an Acc[oun]t of the Number of Mr Flamsteeds Historia Cœlestis that have been printed, by whose directions, and at whose charge; Their Lordps likewise desire that you will inform them what Method of distribution was intended or agreed upon, as also how many have been disposed of in presents to any persons at any time and how many Remaine I am &c 27th Feb 1715/6

WM LOWNDES

NOTE

(1) T/27, 22, p. 51; compare Letter 1177 above. The present instruction was presumably written as a result of the receipt of a letter from the stationer Awnsham Churchill dated 22 February seeking instructions for the disposal of the volumes of *Historia Cœlestis* remaining on his hands.

Churchill's action prompted Flamsteed to write to the Lords of the Treasury, possibly also on the 27th or the day before, to add his complaint that no copies of the *Historia Cœlestis* had been delivered to him 'as a Present from his Majesty' despite the Warrant (Number 1171) issued to that effect. Understanding that Churchill was seeking instruction for the disposal of the remaining stock, Flamsteed now 'hopes and desires that Your Honors [the Treasury Lords] will determine nothing contrary to the good Intentions of his Majesty as well as of his late Royall Highness Prince George of Denmark, but order the said 300 Copies to be delivered immediately, according to his Majesty's Pleasure signifyd by the said Warrant.' (T/1, 198, no. 56).

After reading this plea on 29 February the Lords directed that the Referees be written to, to hasten their report and accordingly Lowndes wrote the following letter.

1189 LOWNDES TO THE REFEREES
29 FEBRUARY 1716
From the copy in the Public Record Office [1]

Gentlemen

The Lords Commissioners of his Matys Treasury are pleased to direct you to hasten your Report concerning Mr. Flamsted's Historia Cœlestis referred to You on the 27th Instant with all Convenient speed I am &c 29th Febry 1715/6

WM LOWNDES

NOTE

(1) T/27, 22, p. 52. Unfortunately, we have not traced the Referees' reply; however, on 23 March 1716 a Royal Warrant was issued to the Lords empowering them to cause Churchill to deliver 300 copies of the *Historia Cœlestis* to Flamsteed (*Cal. Treas. Books*, xxx (Part II), 1716, p. 146).

According to a note in Flamsteed's papers (see Baily, *Flamsteed*, pp. 318–19) 54 copies of the edition had been distributed already, 30 to the Treasury (bound), 10 to France for the Académie Royale des Sciences and the Observatoire, 10 'to those concerned in the impression', one each to Newton and Halley and two to Flamsteed himself. When the 300 copies were delivered to Flamsteed on 28 March (see Letter 1199) 46 copies would have remained in Churchill's hands—or as Flamsteed made it 39—if a full 400 sets of sheets were perfected.

1190 CONTI TO LEIBNIZ
[?] MARCH 1716

Extract from Des Maizeaux, *Recueil*.[1]
Reply to Letter 1170; for the answer see Letter 1197

Monsieur,

J'ai differé jusqu'à cette heure de repondre à votre Lettre, parce que j'ai voulu accompagner ma Réponse de celle que M. *Newton*, vient de faire à l'Apostille que vous y avez ajoutée.[2] Je n'entrerai dans aucun detail à l'égard de la dispute que vous avez avec M. *Keill*, ou plutôt avec M. *Newton*. Je ne puis dire qu'historiquement ce que j'ai vû, & ce que j'ai lû, & ce qu'il me manque encore de voir & de lire pour en juger comme il faut.

J'ai lû avec beaucoup d'attention & sans la moindre prevention le *Commercium Epistolicum*, & le petit Livre qui en contient l'*Extrait*. J'ai vû à la Societé Royale les Papiers Originaux des Lettres du *Commercium*:[3] une petite Lettre écrite de votre main à M. *Newton*; l'ancien Manuscrit que M. *Newton* envoya au Docteur Barrow & que M. *Jones* a publié depuis peu.

De tout cela j'en infere, que si on ôte à la dispute toutes les digressions étrangeres, il ne s'agit que de chercher si Mr. *Newton* avoit le Calcul des Fluxions ou infinitésimal, avant vous, ou si vous l'avez eu avant lui. Vous l'avez publié le premier, il est vrai; mais vous avez avoué aussi que Mr. *Newton* en avoit laissé entrevoir beaucoup dans les Lettres qu'il a écrites à Mr. Oldenbourg & aux autres. On prouve cela fort au long dans le *Commercium* & dans son Extrait. Quelles sont vos Réponses? Voila ce qui manque encor au Public, pour juger exactement de l'affaire.

Vos Amis attendent votre réponse avec beaucoup d'impatience, & il leur semble que vous ne sauriez vous dispenser de répondre, si non à M. *Keil*, du moins à M. *Newton* lui-même, qui vous fait un deffi en termes exprès, comme vous verrez dans sa Lettre.

Je voudrois vous voir en bonne intelligence. Le Public ne profite guere des Disputes, & il perd sans ressource pour bien des siecles toutes les lumieres que ces mêmes Disputes lui dérobent.

Sa Majesté a voulu que je l'informasse de tout ce qui s'est passé entre M. *Newton* & vous. Je l'ai fait de mon mieux, & je voudrois que ce fut avec succès pour l'un & pour l'autre.

Vôtre Probleme[4] a été resolu fort aisément en peu de tems. Plusieurs Geometres à Londres & à Oxford en ont donné la solution. Elle est générale; car elle s'etend à toutes sortes de Courbes soit Geometriques soit Mécaniques. Le Probleme est un peu équivoquement proposé: mais je croi que M. de Moivre[5] ne se trompe pas, en disant: qu'il faudroit fixer l'idée d'une suite de Courbes; par Exemple supposer qu'elles ayent la même soûtangeante pour la même Abscisse; ce qui conviendra non seulement aux Sections Coniques, mais à une infinité d'autres tant Geometriques que Mécaniques; on pourroit encore faire d'autres suppositions pour fixer cette idée.

Je vous parlerai une autre fois de la Philsophie de Mr. *Newton*. Il faut convenir auparavant de la Methode de Philosopher, & distinguer avec beaucoup de soin la Philosophie de Mr. *Newton*, des consequences que plusieurs en tirent fort mal à propos. On attribue à ce grand homme bien des choses qu'il n'admet pas, comme il l'a fait voir à ces Messieurs François qui vinrent à Londres à l'occasion de la grande Eclipse . . .[6]

A Londres le [7] *de Mars* 1716.

NOTES

(1) The letter was first printed in Des Maizeaux's *Recueil*; first edition (Amsterdam, 1720), II, pp. 12–15; it appears in *Phil. Trans.* 30, no. 359 (1719), 389, with the additional comment, 'M. l'Abbé Conti also spent some hours in looking over the old letters and letter-books, kept in the archives of the Royal Society, in order to see if he could find any thing, which made either for M. Leibnitz, or against Sir Isaac Newton, and had been omitted in the Commercium epistolicum Collinii et aliorum: but he could find nothing of that kind.'

The date has apparently never been completed. The letter was sent as a cover for Letter 1187.

Further light is thrown on Conti's actions with regard to the calculus dispute by his letter to Taylor dated 11 May 1721 (printed in Brewster, *Memoirs*, II, pp. 432–3 and vol. VII of this *Correspondence*). There he states that he showed the letter to both Newton and De Moivre before sending it, and that De Moivre introduced certain corrections (see also note (5) below). He mentions, too, that he himself was privy to many letters which Leibniz 'wrote for his justification' and that he showed these to Newton. Perhaps Conti means here not only Leibniz's letters to himself but also those addressed to Arnold and others.

(2) Letter 1187.

(3) The two phrases following confused Leibniz; see Letter 1197, note (14), p. 314. The first refers to Leibniz's letter to Newton, dated 9 March 1692/3, and the second to the *De analysi*, published by William Jones in 1711 as *Analysis per Quantitatum Series*. Leibniz thought both phrases referred to the same manuscript.

(4) See Letter 1170, last paragraph, p. 253, and notes.

(5) In his letter to Taylor (see note (1) above), Conti tells us that De Moivre explicitly suggested this addition.

(6) See Letter 1139 and notes; Monmort was amongst those who came to England to view the eclipse.

(7) Left blank by Conti; see note (1) above.

1191 NICOLAS RÉMOND TO LEIBNIZ
4 MARCH 1716
Extract from Gerhardt, *Leibniz: Philosophische Schriften*, III, p. 671

J'ai donné à mon frere [1] la lettre que vous lui avez fait l'honneur de lui ecrire, et je ne doute point qu'il ne m'apporte sa reponse au plustost. Au reste il me paroist que l'addition que vous m'avez envoiée au probleme, estoit necessaire. [2] Les Analystes Anglois et François n'en paroissent pas fort embarrassez et on devoit meme en mettre la solution dans les Transactions philosophiques. [3] J'envoie cette addition à Mr l'abbé Conti, et je ne doute point qu'elle ne rebatte l'orgueil des Anglois. J'aurai l'honneur de vous en instruire. M. l'Abbé Conti est tous les jours plus charmé de l'Angleterre et plus amoureux de Mr Newton. Il a eu l'honneur de souper avec le Roy d'Angleterre et aux propos de table il paroit bien que ce grand Prince a vecu avec Monsieur de Leibniz. Sa Majesté Britannique voulut savoir de lui l'historique de vostre dispute avec Mr Newton. Je lui ecris sur tout cela, common je dois, c'est-à-dire suivant ce que je dois à la verité et à mon attachement declaré pour vous, car vous devez compter, Monsieur, d'avoir en moi un admirateur tres sincere et un ami tres fidele.

NOTES
(1) Pierre Rémond de Monmort; see Letter 1139, note (5), p. 216.
(2) See Letter 1170, note (6), pp. 254–5.
(3) See Letter 1186 and notes.

1192 WOLF TO LEIBNIZ
4 MARCH 1716
Extract from Gerhardt, *Briefwechsel zwischen Leibniz und Wolf*, pp. 182–3.
Reply to Letter 1173

Quæ Keilius in Actis Anglicanis [1] contra Philosophica E.T. objecit, nullius sunt ponderis, immo ne nomine objectionis digna: recenset enim tantum nonnulla, in quibus E.T. dissidet a Newtono, quasi vero Newtoniana adeo sint manifesta, ut erronea censenda sunt, quæ cum iis non conveniunt. Miror autem, quod homo insulsus asserere non erubescat, [2] The editors of the Acta Eruditorum [3] have told the World, that Mr Newton denies, that the cause of gravity is mechanical . . . and Mr. Leibniz hath accused him [4] of making Gravity a natural or essential property of bodies, and an occult quality and miracle. And by this sort of railery they are perswading the Germans, that Mr. Newton wants judgment, and was not able to invent the infinitesimal method. Diserte enim Newtonus ait, causam gravitatis non agere [5] pro quantitate superficierum particularum, in quas agit, ut solent causæ mechanicæ, et vi spiritus cujusdam subtilissimi corpora crassa pervadente et in iisdem latente particulas corporum ad minimas distantias se mutuo attrahere etc. Immo ipsimet Angli (forsan ipse Keilius) in Diario Hagiensi, p. 217 scribunt de Newtono, il demontre, que la gravité n'est pas purement mechanique.

Translation

The points which Keill has raised against Your Excellency's philosophy in the English [*Philosophical*] *Transactions* [1] are of no weight, indeed are not worthy to be called objections; for he only repeats some things concerning which Your Excellency disagrees with Newton as though the Newtonian version were really so obvious, that anything in disagreement with it is to be taken as false. I am astonished, however, that the foolish man does not blush to claim that [2] . . . [Wolf here quotes in English; see page 297] . . . For Newton clearly says that the cause of gravity does not act [5] 'according to the quantity of the surfaces of the particles on which it acts, as mechanical forces do,' and that, by the 'force of a certain very subtle spirit pervading dense bodies and lying hidden in them, the particles of bodies mutually attract one another at the least distances,' etc. Indeed the English themselves (perhaps Keill himself) in the Hague *Journal* [*Literaire*] p. 217 write of Newton, 'il demontre, que la gravité n'est pas purement mechanique.'

NOTES

(1) That is, the *Recensio*, in the last pages of which Newton (but Wolf thought the author was Keill) had discussed Leibniz's philosophical views.

(2) Wolf quotes from the *Phil. Trans.* **29**, 1715, p. 223.

(3) An anonymous review of the second edition of Newton's *Principia* appeared in the *Acta Eruditorum* for March 1714, pp. 131–42; the reviewer wrote of Newton (p. 142) 'Vim gravitatis a causa aliqua ultro concedit, ast eam mechanicam esse regat, quia non agit pro quantitate superficierum sed materiæ solidæ' ('He submits that gravity arises from some further cause, and moreover denies this to be mechanical, because it acts not according to the quantity of the surfaces, but of the solid matter.')

(4) In his correspondence with Hartsoeker, published in 1712 (see Letter 918, vol. v).

(5) Wolf here quotes from the General Scholium of the second edition of the *Principia* (1713) p. 484. In the original '*superficierum*' is italicized to contrast with 'materia *solida*,' and '*vi* [spiritus]' is strengthened by 'et actio'.

1193 LEIBNIZ TO ARNOLD
6 MARCH 1716
From the copy in the British Museum [1]

Hanover ce 17 *Mars* 1716 [N.S.]

Le Problême des perpendiculaires a une suite de Courbes est de M. Bernoulli qui la resolu generalement dans toute son etendue. [2] Son fils la resolu dans un cas particulier des coniques, [3] et cest ainsi que quelques uns lauront resolu sur la communication de M l'Abbé Conti, mais on les attend au general— Je vous supplie Monsieur de dire a M. l'Abbè Conti, que les solutions des Cas particuliers faciles du Problême ne sont rien, et quil sagit de la solution generale. [4] Ainsi s'il ne menvoye que cela, il juge bien quon ny doit avoir aucun egard.

NOTES

(1) MS. Birch 4381, fos. 12–14. The extract, in Arnold's hand, was sent by Arnold to Chamberlayne on 16 May 1719 (see vol. VII).

(2) See Letter 1201, notes (7) and (8).

(3) See Letter 1201, note (10).

(4) This was the second time Leibniz made the request; see Letter 1179, note (2), p. 271.

1194 PIERRE RÉMOND DE MONMORT TO TAYLOR
20 MARCH 1716
Extracts from Taylor, *Contemplatio Philosophica* [1]

Le plus grand nombre s'est opposé a faire imprimer le morceau de Mr. Keil dans les Memoires de l'Academie [2] par la raison que Mr. Keil est etranger a l'Academie et que cela est contre les statuts. Je prise le parole, representai 1mo que le morceau est excellent, 2° que Mons. Newton est attaqué dans les memoires par Mr. N[icol]as Bernoulli qui non plus que Mons. Keil n'est pas membre de l'Academie. 3° que s'il etoit jamais possible de faire exception a une regle generale, c'etait au faveur d'un aussi grand homme que Mons. Newton. Je compte qu'il sera imprime s'il est avoué et reconnu de Mons. Newton ou de Mons. Halley au nom de la Societé Royale

... J'ai lu il y a quelques jours la copie d'une lettre [3] très curieuse que Mons. Leibniz a ecrit a l'Abbé de Conty homme de qualité et de merite qui est parmy nous et qui vous connoissez sans doutte. Si vous ne l'avez pas vu, demandez a la voir.

NOTES

(1) Pp. 85–6; on 22 April 1716 Taylor sent Newton a copy of the first of these two extracts (see Letter 1206).

(2) Keill's 'Defense du Chevalier Newton', eventually published in the *Journal Literaire de la Haye*; see Letter 1165, note (2), p. 246.

(3) Letter 1170.

1195 WARRANT TO THE MINT
26 MARCH 1716
From a copy in the Public Record Office.[1]
Reply to Letter 1185

George R.

Whereas Our late Royal Sister Queen Anne for the better Exercise of the Art of Graving, and that the Gravers of Her Mint might not lose the Skill they had Acquired, the Improvement of which would be the best Security of the Coine of Her Kingdome Did by Warrant under Her Royal Sign Mannual bearing date the Second day of November 1706 Impower the Master and Worker of Her Mint for the time being to give leave to the Said Gravers or any of them to make and Sell Such Medalls of Fine Gold and fine Silver as did not relate to State Affairs And that upon Such Medals as shoud be made for rewarding persons for their good Services to Impress the Service for which the Said Medalls are given with the Date, And did thereby also Impower, the Warden, Master and Worker, and Comptroller of Her Mint for the time being to direct and Require the Said Gravers and every of them to make such other Medals of fine Gold, fine Silver and fine Copper with plaine Historical Designs and Inscriptions in memory of great Actions as shoud be approved by the Said Warden, Master and Worker and Comptroller and to make Embossements puncheons Dyes and other Instruments requisite for Coining the Same according to Such Orders as they or any of them respectively should from time to time receive from the Master & Worker of the Said Mint and not otherwise, which Said Warrant being by the Demise of Our Said late Royal Sister become void, the Comm[issione]rs of Our Trea[su]ry have laid before us Your Memorial desiring the Same may be renewed, That the Gravers may improve their hands in the manner aforesaid when they are not employed about our Coine, We taking the Premisses into our Royal Consideration are Graciously pleased to Approve thereof, And Accordingly Our Will and pleasure is And Wee do hereby Authorise and Impower You the Master and Worker of Our Mint to give Liberty to the said Gravers to make such Medals from time to time of fine Gold and fine

Silver for the Improvement of the Art of Graving as do not relate to State
Affairs taking care that Such of them as shall be for the Rewarding persons
for good Services to Us have Impressed upon them the Service for wch such
Medals were given with the Date; And Our further pleasure is And Wee
do hereby Impower You likewise to direct and Require the Said Gravers
and every of them to make such other historical Medals in Memory of great
Actions as shall be approved of by the Warden Master & Worker & Comptroller of Our Mint aforesaid, And in order thereunto You the Master and
Worker of Our Mint are to direct the Said Gravers to make Such Embossements puncheons Dyes and other Instruments as shall be requisite for Coining
the Same. For all which this shall be Your Warrant. Given at Our Court at
St. James's the 26th day of March 1716 In the Second Year of Our Reigne

By His Mats: Command

R WALPOLLE [2]

WILL ST. QUINTIN

T: NEWPORT [3]

To Our Trusty and Wellbeloved the Warden
Master and Worker and Comptroller of Our
Mint now and for the time being

NOTES

(1) Mint/1, 8, p. 113. The antecedent documents are Letters 1134 and 1158. For a list of
historical medals struck by the Mint see Plate V (facing p. 233).

(2) Robert Walpole (1676–1745) was appointed First Lord of the Treasury and Chancellor
of the Exchequer on 10 and 12 October 1715, respectively.

(3) Thomas Newport (c. 1655–1719), a grandson of the first Earl of Bradford, and M.P. for
various constituencies since 1695, created Baron Torrington on 25 June 1716, served as a
Treasury Lord from 1715 to 1718.

1196 J. BERNOULLI TO WOLF
28 MARCH 1716

Extracts from the printed texts.[1]
For the answer see Letter 1208

EPISTOLA PRO EMINENTE MATHEMATICO, Dn.
JOHANNE BERNOULLIO, *contra quendam ex*
Anglia antogonistam scripta.

Qui Tibi [2] asseveravit, Vir celeberrime, quod inventio calculi integralis
proprio Marte obtenta sit a Dn. *Johanne Bernoullio*, nihil a veritate alienum
dicit; si præsertim hunc calculum a calculo differentiali, quem utique totum

illustri *Leibnitio* deberi etiam apud ipsum *Bernoullium* extra controversiam est, distinguere velimus . . .

[*Bernoulli now details papers, chiefly those in the* Acta Eruditorum, *where he has used the integral calculus. He mentions the* Charta Volans, *still as though the letter printed in it were not his own, and discusses Newton's shortcomings, in particular his apparent misunderstanding of second and higher order differences, and the problem of motion in a resisting medium. Towards the end of the letter he angrily rebuts Keill's contention that his own solution of the inverse problem of centripetal forces was based on Newton's Proposition 41, and continues:*]

Examinent etiam considerentque, quam brevi via quamque diversa a *Newtoniana* incesserit *Bernoullius*, dicantque postea, an alius quispiam præter antagonistam [3] sibi persuadere possit, meam [4] formulam ex Newtoniana esse desumtam . . .

[*The letter ends with a private message for Wolf:*]

Sufficiant tandem ista, quæ omnia limatissimo tuo judicio, vir nobilissime, submittere volui, ut si luce digna deprehendas, in Actis Lipsiensibus publicare possis: neque me reluctantem habebis si totius hujus Epistolæ contentum typis mandare volueris, mutatis mutandis et omissis omittendis. Consentio ut ante publicationem cum illust. Leibnitio communicetur, quia nollem eo invito aliquid mea ex parte in lucem prodiret; spero autem fore, ut neutiquam improbet quas præsertim congessi rationes validissimas, quibus quicquid ogganiat Keilius aliive sectatores, firmissime adstruitur, Newtonum eo tempore quo scripsit sua Principia philos. mathematica, nondum perspectam habuisse methodum differentiandi differentialia. Quod vero attinet ad formam sub quo optarem ut contenta hæc prodirent, poterunt conservare formam epistolæ, sed ita si placet mutandæ, tanquam ab anonymo, vel ab alio sive vero, sive ficti nominis scripta fuisset; ut verbo dicam, rem totam ea qua polles prudentia dirigas, ne Keilius suspicetur me hujus epistolæ scriptorem esse. [5] Ingratum enim mihi valde foret a Keilio bile sua perfricari et contumeliose traduci, ut solent ejus antagonistæ, postquam ille me hactenus satis humaniter tractavit. Quod superest vale etc. Basil. VIII Apr. 1716 [N.S.]

Translation

A LETTER WRITTEN ON BEHALF OF THE EMINENT MATHEMATICIAN, MR JOHN BERNOULLI,
against a certain English adversary

He [2] who assures you, famous Sir, that Mr John Bernoulli gained the invention of the integral calculus by his own efforts, speaks nothing but the truth; especially if we mean to distinguish this calculus from the differential calculus which even according to Bernoulli himself is, beyond all controversy, owed entirely to the great Leibniz . . .

Let them [his readers] study and reflect, how direct and different from Newton's was the route [to a solution] taken by Bernoulli, and let them afterwards declare, whether anyone but the adversary [3] himself can persuade himself that my [4] formula was derived from Newton's . . .

Let this at last be enough, then, of the things all of which I wished to submit to your most refined judgement, noble Sir, so that if you consider them worthy of publication you can print [them] in the *Acta* [*Eruditorum*] of Leipzig: nor will you find me reluctant if you wish to commit the contents of this whole letter to the press, changing what ought to be changed and omitting what ought to be omitted. I agree that it should be communicated to Leibniz before publication, because I do not wish anything to appear in print against his will by my doing; however I hope it will turn out that he will in particular not reject the very powerful arguments which I have drawn together from which, whatever Keill or his supporters whine, it is firmly established that Newton at the time when he wrote his *Principia Philos. Mathematica*, still had not understood the method of differentiating differentials. As regards the actual form in which I would choose that the contents of this [letter] appear, they can keep the form of a letter, but, please, so altered as though it had been written by an anonymous writer, or indeed by someone of another either real or fictitious name; in a word, you should arrange the whole thing with as much discretion as you can, lest Keill suspect this letter to have been written by me. [5] It would be exceedingly unpleasant for me to have Keill vent his spleen upon me, and to be rudely exposed to ridicule as his opponents usually are, after he has hitherto treated me quite politely. For the rest, farewell, etc.

Basel, 8 April 1716 [N.S.]

NOTES

(1) The letter, after some editing by Wolf, was printed *in extenso* in the *Acta Eruditorum* for July 1716, pp. 296–314; the first and second extracts we print are taken from pages 296 and 314 of that journal respectively. The third extract, the conclusion to Bernoulli's letter, giving private instructions for editing the article, was not of course printed there; we take it from an article by Johann Bernoulli's grandson, 'Anecdotes pour servir à l'Histoire des Mathématiques par M. Jean [III] Bernoulli', *Histoire de l'Académie Royale des Sciences et Belles-Lettres* for 1799 and 1800 (Berlin, 1803), pp. 32–50, where a lengthy discussion of the writing, transmission and editing of the whole letter is found, together with several extracts from the ensuing correspondence between Wolf and Bernoulli. In a continuation of this article in *Histoire de l'Académie Royale des Sciences et Belles-Lettres* for 1802 (Berlin, 1804), pp. 60–5, Johann [III] lists the variants between the draft of his grandfather's original letter and the version as printed in the *Acta Eruditorum*.

The burden of Johann [III]'s argument—and perhaps he shows some bias towards his ancestor—is that Wolf and Leibniz (who also saw the letter before publication (see Letter 1208, note (3)) did not edit Bernoulli's paper as much as he had expected them to do. Verbs and pronouns were changed from first to third person as appropriate (with the exception of the famous 'meam'; see note (4) below), and Keill's name was suppressed. But there was no attempt to soften any derogatory turns of phrase; indeed in one place, where Bernoulli simply wrote 'Keilius', Wolf has substituted 'antagonista audax' ('rash opponent'). Also Wolf has added certain passages stressing *Leibniz's* claim to certain mathematical demonstrations, and phrases praising both Leibniz and Bernoulli. Johann [III] suggests (but does not adduce any strong proof) that Bernoulli never read the edited version of the paper properly before it went into print. In fact, with the exception of the mistake over 'meam', Wolf and Leibniz had done no more and no less than the final paragraph of Bernoulli's letter explicitly requested.

Eventually, in 1718, after suspicion concerning the authorship of the letter had been aroused, particularly in the mind of Hermann and, of course, amongst the English mathematicians, Bernoulli was forced to confess that he himself had written the paper (see *Acta Eruditorum* for June 1718, pp. 261–2, in an appendix to a paper by Nikolaus Bernoulli.)

(2) Bernoulli's draft reads 'Quando ille (Doctor medicinæ Anglus)' ('When that man (an English Doctor of Medicine)') for 'Qui Tibi.' The doctor, presumably, is John Arnold; see Letter 1014, note (1), p. 27.

(3) Bernoulli's draft reads 'Keilium' for 'antagonistam'.

(4) Wolf should, of course, have changed 'meam' to 'Bernoullianam'; later he unconvincingly argued that 'meam' was a printer's error for 'eam', 'my formula' becoming 'that formula' (see Letter 1258). The English were quick to note the mistake (see Letter 1239). For Keill's doubt about the independence of Bernoulli's solution see Letter 1023.

(5) Keill, when he eventually saw Bernoulli's letter, recognized it, of course, as being from his hand. He at once set about replying to it (see his Letter 1239 to Newton). The article he prepared was published in the *Journal Literaire de la Haye* for 1719 (La Haye, 1720), pp. 261–87.

1197　LEIBNIZ TO CONTI
29 MARCH 1716

From Raphson, *History of Fluxions*, pp. 103–11.[1]
Reply to Letters 1187 and 1190; continued in Letter 1202

Monsieur,

C'est sans doute pour l'amour de la verité que vous vous etes chargé d'une espece de cartel [2] de la part de M. *Newton.* Je n'ay point voulu entrer en lice avec des enfans perdus, qu'il avoit detachés contre moy; soit qu'on entende celuy qui a fait l'Accusateur sur le fondement du *Commercium Epistolicum*, soit qu'on regarde la Preface pleine d'aigreur qu'un autre a mise devant la nouvelle Edition de ses Principes. Mais puiqu'il veut bien paroitre luy même, je seray bien aise de luy donner satisfaction.

Je fus surpris au commencement de cette Dispute d'apprendre qu'on m'acusoit d'etre l'Aggresseur; car je ne me souvenois pas d'avoir parlé de M. *N.* que d'une manere fort obligeante. Mais je vis depuis qu'on abusoit pour cela d'un Passage des Actes de *Leipzig* du Janvier 1705. ou il y a ces mots: *Pro differentiis L . . . sianis, D. N . . . nus adhibet, semperque adhibuit Fluxiones*; ou l'Auteur des Remarques sur le *Commercium Epistolicum* dit, *pag.* 108. *Sensus verborum est, quòd N . . . nus Fluxiones differentiis L . . . tianis substituit*: Mais c'est une Interpretation maligne d'un homme qui cherchoit noise: Il semble que l'Auteur de paroles inseré dans les Actes de *Leipzig* a voulu y obvier tout expres, par ces mots, *adhibet semperque adhibuit*; pour insinuer, que ce n'est pas apres la veue de mes differences, mais deja auparavant, qu'il s'est servi de Fluxions. Et je defie qui que ce soit de donner un autre but raisonnable à ces Paroles, *semperque adhibuit*. Au lieu qu'on se sert du mot *substituit* en parlant, de ce que le Pere *Fabri* avoit fait apres Cavallieri. [3] D'ou il faut conclure ou que M. *N.* s'est laissé tromper par un homme qui a empoisonné ces Paroles des Actes, qu'on supposoit n'avoir pas eté publiées sans ma connoissance, & s'est imaginé qu'on l'accusoit d'etre Plagiaire; ou bien qu'il a eté bien aise de trouver un pretexte de s'attribuer ou faire attribuer privément l'Invention du nouveau Calcul (depuis qu'il en remarquoit le succés, & le bruit qu'il faisoit dans le Monde) contre ses Connoissances contraires avouées dans son Livre des Principes, *pag.* 253. de la premiere Edition. Si l'on avoit fait connoitre qu'on trouvoit quelque Difficulté, ou sujet de plainte dans les Paroles des Actes de *Leipzig*, je suis assuré, que ces Messieurs qui ont part a ces Actes, auroient donné un plein contentement; mais il semble qu'on cherchoit un pretexte de rupture.

Je n'ay pas eu connoissance du *numerous Committee of Gentlemen of several Nations relating to the Dispute*; car on ne m'en a donné aucune part, & je ne say pas encore presentement les Noms de tous ces Commissaires, & particuliere-ment de ceux qui ne sont pas des Isles Britaniques, je ne crois pas qu'ils approuvent tout ce qui a eté mis dans l'Ouvrage publié contre moi.

Il est aisé a croire que j'ay eté quelque temps à *Vienne*, avant que d'avoir vû le *Commercium Epistolicum* deja publié, quoique j'en eusse des Nouvelles. Ainsi un ami sachant cela, [4] aussi zelé pour moi que les seconds de M. *N.* le peuvent être pour luy, a publié une Papier, que M. *N.* appelle diffamatoire (*defamatory Letter.*) Mais cette Piece n'etant pas plus forte que ce qu'on a publié contre moi, M. *N.* n'a pas droit de s'en plaindre. Si l'on n'a pas marqué l'Auteur ni le Lieu de l'Impression du Papier; on connoit assez le Nom & le Lieu de l'Auteur de la Lettre y inserée d'un excellent Mathematicien que j'avois prié de dire son Sentiment sur le *Commercium*, & cela suffit. M. *N.* (dont les Partisans ont marqué qu'il ne leur etoit pas inconnu) l'appelle un Mathe-maticien ou pretendu Mathematicien; & apres avoir fait inutilement des efforts pour le gagner, il le méprise contre l'opinion publique, qui le met entre ceux du premier rang, & contre l'evidence des choses verifiées par ses découvertes.

Lors que j'eus enfin le *Commercium Epistolicum*, je vîs qu'on s'y écartoit entire-ment du but, & que les Lettres qu'on publioit ne contenoient pas un mot qui peut faire revoquer en doute mon Invention du Calcul des Differences dont il s'agissoit. Au lieu de cela je remarquay qu'on se jettoit sur les Series, ou l'on accorde l'avantage à M. *N.* & que les Remarques contenoient de Gloses mal tournées, pour tacher de me decrier par des soubçons sans fondement quelque fois ridicules, & quelquefois forgés contre la Conscience de quelques uns de ceux qui en êtoient les auteurs ou approbateurs.

Pour repondre donc de point en point a l'Ouvrage publié contre moi, il falloit un autre Ouvrage aussi grand pour le moins que celuy là, il falloit entrer dans un grand detail de quantité de Minuties passées il y a 30 ou 40 Ans dont je ne me souvenois gueres; il me falloit chercher mes vieilles Lettres, dont plusieurs se sont perdues, outre que le plus souvent, je n'ay pas gardé les Minutes des miennes; & les autres sont ensevelies dans un grand tas de Papiers, qui je ne pouvois debrouiller qu'avec du Temps & de la Patience. Mais je n'en avois gueres le loisir, êtant chargé presentement d'Occupations d'une toute autre Nature.

De plus je remarquay que dans la Publication du *Commercium Epistolicum* on a supprimé des endroits qui pouvoient être au desadvantage de M. *N.* au lieu qu'on n'y a rien omis de ce qu'on croyoit pouvoir tourner contre moi par des gloses forcées. Comme je n'ay pas daigné lire le *Commercium Epistolicum* avec beaucoup d'attention, je me suis trompé dans l'Exemple que j'ay cité, n'ayant pas pris garde, ou ayant oublié qu'il s'y trouvoit; mais j'en citéray un autre: M. *N.* avouoit dans un des ses Lettres à M. *Collins,* qu'il ne pouvoit point venir à bout des Sections secondes (ou Segments seconds) de Spheroides ou corps semblables:[5] mais on n'a point inseré ce Passage ou cette Lettre dans le *Commercium Epistolicum*; il auroit été plus sincere par rapport à la Dispute, & plus utile au public, de donner le Commerce litteraire de M. *Collins* tout entier, là ou il contenoit quelque chose qui meritoit d'être lû; & particulierement de ne pas tronquer les Lettres, car il y en a peu parmi mes Papiers, ou dont il me reste des Minutes.

Ainsi tout consideré, voyant tant de marques de malignité & de chicane, je crûs indigne de moi d'entrer en discussion avec des gens qui en usoient si mal. Je voyois qu'en les refutant on auroit de la peine à éviter des Reproches, & des Expressions fortes, telles que meritoit leur Proceedé; & je n'avois point envie de donner ce Spectacle au Public, ayant dessein de mieux employer mon temps, qui me doit être precieux, & meprisant assez le Jugement de ceux qui sur un tel Ouvrage voudroient prononcer contre moi, d'autant que la Societé Royale même ne la point voulu faire; comme je l'ay appris par un Extrait de ses Registres.

Je ne crois point d'avoir dit (comme M. *N.* me l'impute) que les Anglois n'auroient point le plaisir, de me voir repondre à des raisonnements si minces;

car je ne crois point que tous les Anglois fassent leur Cause de celle de M. *N.* il y en a de trop habiles & de trop honnêstes pour épouser les passions de quelquesuns de ses adherens.

Apres cela, il m'accuse d'avoir voulu faire diversion, en combattant sa Philosophie, & en voulant l'engager dans des Problemes; mais quant à la Philosophie, j'ay donné publiquement quelque chose de mes Principes sans attaquer les siens; si ce n'est que par Occasion j'en ay parlé dans des Lettres particulieres, depuis qu'on m'en a donné sujet; & pour ce qui est des Problemes, je n'ay garde d'en proposer à M. *N.* car je ne voudrois pas m'y engager quand on m'en proposeroit à moi, nous pouvons nous en dispencer à l'age ou nous sommes, mais nous avons des amis qui y peuvent suppleer à notre Defaut.

Je ne veux point entrer icy dans le detail de ce que M. *N.* dit un peu aigrement contre ma Philosophie, car pour la sienne, ce n'en est point le lieu. J'apelle *Miracle* tous Evenement qui ne peut être arrivé que par la Puissance du Createur, sa Raison n'étant pas dans la Nature des Creatures, & quand on veut neamoins l'attribuer aux qualités ou forces des Creatures, alors j'appelle cette Qualité *une Qualité occulte à la Scholastique*; c'est à dire, qu'il est impossible de rendre manifeste, telle que seroit une pesanteur primitive; car les Qualités occultes qui ne sont point Chimeriques, sont celles dont nous ignorons la cause, mais que nous n'excluons point; & j'appelle *l'Ame de l'homme* cette Substance simple qui s'apperçoit de se qui se passe dans le corps humain, & dont les Appetits ou Volontés sont suivis par les Efforts du Corps. Je ne prefére pas les Hypotheses aux Arguments tirez de l'Induction des Experiences, mais quelquefois on fait passer pour Inductions generales ce qui ne consiste qu'en Observations particulieres, & quelquefois on veut faire passer pour une Hypothese ce qui est demonstratif. L'Idée que M. *N.* donne icy de mon Harmonie préetablie n'est pas celle qu'en ont quantité d'habiles gens hors d'Angleterre, & quelques uns en Angleterre; & je ne crois pas que vous même Mons. en ayez eu une semblable, ou l'ayés maintenant, à moins que d'être bien changé.

Je n'ay jamais nié qu'a mon second Voyage en Angleterre j'aye vû quelques Lettres de M. *N.* chez Monsieur *Collins,* mais je n'en ay jamais vû ou M. *N.* ait expliqué sa Methode des Fluxions, & je n'en trouve point dans le *Commercium Epistolicum.*

Je ne pas vû non plus qu'il ait expliqué la Methode des Series que je m'attribue, je crois qu'il veut parler de celle ou je prends une Series arbitraire, je l'ay fait avant mon second retour en Angleterre. Je ne nie pourtant pas, que M. *N.* n'eut pû l'avoir aussi, & ce n'est pas même une Invention fort difficile.

M. *N.* veut que j'avoue, & que j'accorde ce que j'ay avoué ou accordé il y a 15 Ans, ou autrement on devroit en attendre de luy autant; car il y a maintenant deux fois quinze Ans, que dans la premiere Edition de ses Principes, *pag.* 253, 254. il m'accorde l'Invention du Calcul des Differences, independement de la sienne, & depuis il s'est avisé je ne scay comment de faire soutenir le contraire.

Il est bon de savoir qu'a mon premier Voyage d'Angleterre en 1673. je n'avois pas la moindre connoissance des Series infinies, telles que M. *Mercator* venoit de donner, ny d'autres matieres de la Geometrie avancée par les dernieres Methodes, je n'etois pas même assez versé dans l'Analyse de Des Cartes, je ne traitois les Mathematiques que comme un *Parergon*, & je ne savois guere que la Geometrie practique vulgaire quoy que j'eusse vû par hazard la Geometrie des indivisibiles de *Cavalleri*, & un Livre de Pere *Leotaud*,[6] ou il donnoit les Quadratures des Lunules & Figures semblables, ce qui m'avois donné quelque curiosité; mais je me divertissois plustôt aux proprietez des Nombres, à quoy le petit Traité que j'avois publié presque petit garçon de l'Art des Combinaisons en 1666 m'avoit donné occasion,[7] & ayant observé des lors l'usage des Differences pour les Sommes, je l'appliquay à des suites de Nombres. On voit bien par mes premieres Lettres échangées avec M. *Oldenbourg*, que je n'etois guere allé plus avant, aussi n'avois je point alors la connoissance de M. *Collins*, quoy qu'on ait feint malicieusement le contraire.[8]

Ce fût peu à peu que M. *Hugens* me fit entrer en ces matieres, quand je le pratiquois à *Paris*, & cela joint au Traité de M. *Mercator* (que j'avois rapporté avec moi d'Angleterre parce que M. *Pell* m'en avoit parlé)[9] me fit trouver environ vers la fin de l'An 1673 ma Quadrature Arithmetique du Cercle, qui fût fort approuvé par M. *Hugens*, & dont je parlay à M. *Oldenbourg* dans une Lettre de l'An 1674. alors ny M. *Hugens* ny moi, nous ne savions rien des Series de M. *N.* ny de M. *Gregory*. Ainsi je crus être le premier qui eut donné la valeur du Cercle par une suite de Nombres rationaux; & M. *Hugens* le crut aussi, j'en écrivîs sur ce ton là à M. *Oldenbourg* qui me repondit qu'on avoit deja de telles Series en Angleterre, & l'on voit par ma Letter du 15 Juillet de 1674. & par la reponse de M. *Oldenbourg* du 8 Decembre de la même Année que je n'en devois avoir aucune connoissance alors, autrement M. *Oldenbourg* n'auroit pas manqué de me le faire sentir, si luy ou M. *Collins* m'en eussent communiqué quelque chose; mais je ne savois pas alors les Extractions des Racines, des Equations par des Series, ny les Regressions ou l'Extraction d'une Equation infinie; j'etois encore un peu neuf en ces matieres, mais je trouvai pourtant bientôt ma Methode generale par des Series arbitraries; & j'entray enfin dans mon Calcul des differences, ou les observations que j'avois faites encore fort jeune sur les differences des suites des Nombres, contribuerent à

m'ouvrir les yeux; car ce n'est pas par les Fluxions des lignes, mais par les
differences des Nombres que j'y suis venu, en considerant enfin que ces differences appliquées aux grandeurs qui croissent continuellement, s'evanouissent
en comparaison des grandeurs differentes, au lieu qu'elles subsistent dans les
suites des Nombres. Et je crois que cette voye est la plus analytique, le Calcul
Geometrique des differences qui est le même que celuy des Fluxions, n'etant
qu'un cas special, devient plus commode par les evanouissements.

M. *N.* allegue par apres les passages, ou j'accorde qu'il y a un Calcul
approchant de mon Calcul des differences, mais il pourra bien se souvenir
qu'il m'en a accordé autant, & s'il luy est permis de se retracter, pourquoy
ne me sera t'il pas permis d'en faire autant? Sur tout apres les verisimilitudes
que M. *Bernoulli* a remarquées, j'ay une si grande opinion de la Candeur de
M. *N.* que je l'ay crû sur sa parole, mais le voyant conniver à des accusations
dont la fausseté luy est connue, il etoit naturel que je commençasse de douter.

Je ne puis avouer ny desavouer aujourdhuy d'avoir écrit, ou receu des
Lettres écrites il y a plus de 40 ans telles qu'on les a publiées, je suis obligé
de m'en rapporter à ce qui se trouve dans les Papiers qu'on cite, mais je ne
remarque rien contre moi dans celles que M. *N.* allegue du 15 Avril & 20
May 1675. & du 24 Octobre 1676. sinon dans les faussetez du Glosateur, je
crois que c'etoit purement par distraction dans un sejour comme celuy de
Paris, ou je m'occupois à bien d'autres choses encore qu'aux Mathematiques,
& par l'éloignement que j'avois des Calculs, dont je craignois la longueur,
que j'ay demandé quelquefois à M. *Oldenbourg* la Demonstration ou la Methode
d'arriver a certaines choses ou j'aurois bien pû arriver moi même. Par exemple,
je crois d'avoir deja eu au douze de May 1676. ma Methode d'une Series
Arbitraire, qui m'auroit pû mener a des Series dont j'y demande la raison.
Car ayant consulté mon vieux Traité de la Quadrature Arithmetique achevé
quelque temps avant ma sortie de *France*, je me sers de la Series Arbitraire,
cependant les Series marquées dans cette Lettre sont une chose dont je consens
dêtre redevable à d'autres, & je crois de ne les avoir pas même connues en
1674.

N'entendant pas bien ce que M. *N.* allegue des Actes de *Leipzig* de May
1700, j'y ay regardé, & je trouve qu'il n'en a pas bien pris le sens. Il n'y est
point parlé de l'Invention du nouveau Calcul des differences, mais d'un
artifice particulier des *Maximis & Minimis*, qui est independant, & dont je
m'etois avisé bien du temps avant que M. *Bernoulli* eut proposé son Probleme
de la plus courte descente, mais dont je jugeois que M. *N.* se devoit être avisé
aussi, lors qu'il avoit donné la figure de son Vaisseau dans ses Principes. Ainsi
j'ay voulu dire, qu'il a fait connoître publiquement avant moi, qu'il possedoit
cet Artifice, ce que je ne pouvois pas dire du Calcul des Differences & des

309

Fluxions puisque j'en avois fait voir l'utilité publiquement avant la publication de ce Livre. Cet Artifice particulier de *Maximis & Minimis* n'est point
necessaire, quand ils s'agit simplement d'une grandeur (car alors la Methode
de M. *Fermat* perfectionée par les nouveaux Calculs suffit) mais quand il
s'agit de toute une Figure qui doit faire le mieux un effect demandé, il faut
autre chose.

M. *N.* hazarde icy une accusation mais, qui va tomber sur luy même. Il
pretend que ce que j'ay écrit pour luy à M. *Oldenbourg* en 1677. est un deguise-
ment de la Methode de M. *Barrow*. Mais comme M. *N.* avoue dans la *Pag.*
253 & 254. de la premiere Edition de ses Principes, *Me ipsi (tunc Methodum
communicâsse à Methodo ipsius vix abludentem præterquam in verborum & notarum
formulis*, il s'ensuivra que sa Methode aussi n'est qu'un déguisement de celle de
M. *Barrow*.

Je croy que luy & moi nous serons aisement quites de cette accusation:
Car une infinité de gens liront le Livre de M. *Barrow*, sans y trouver nôtre
Calcul; il est vray que feu M. *Tschirnhaus* qui s'apperceut un peu tard de
l'avantage de ce Calcul pretendoit qu'on pouvoit arriver à tout cela par les
Methodes de M. *Barrow*. Comme *l'Abbé Catelan* [10] François pretendit que
même l'Analyse de Des Cartes suffisoit pour toute ces choses, mais il etoit
plus aisé de le dire que de le montrer.

Cependant si quelqu'un a profité de M. *Barrow*, ce sera plûs tôt M. *N.*
qui a étudié sous luy que moi qui (autant que je puis m'en souvenir,) n'ay
veu les Livres de M. *Barrow* qu'à mon second Voyage d'Angleterre, & ne les ay
jamais lûs avec attention, parce qu'en voyant le Livre je m'apperçus que par
la consideration du Triangle Characteristique (dont les Cotez son les Elements
de l'Abscisse, de l'Ordonnée & de la Courbe) semblable a quelque Triangle
assignable, j'etois venu comme en me jouant aux Quadratures, Surfaces &
Solides dont M. *Barrow* avoit remply un Chapitre des plus considerables de
ses Leçons, outre que je ne suis venu a mon Calcul des Differences dans la
Geometrie qu'apres en avoir vû l'usage (mais moins considerable) dans les
Nombres, comme mes premieres Lettres dans le *Commercium Epistolicum* le
peuvent insinuer. Il se peut que M. *Barrow* en ait plus sçu qu'il n'a pas dit
dans son Livre, & qu'il a donné des Lumieres à M. *N.* que nous ne savons pas,
& si j'etois semblable a certains temeraires, je pourrois asseurer sur de simples
soubçons sans autre fondement que le Calcul des Fluxions de M. *N.* qu'el
qu'il puisse être, luy a eté enseigné par M. *Barrow*.

On peut bien juger que lors que j'ay parlé en 1676. des Problemes qui ne
dependoient, ny des Equations, ny des Quadratures, j'ay voulu parler des
Equations telles qu'on connoissoit alors dans le monde; c'est à dire, des Equations de l'Analyse ordinaire. Et on le peut juger de ce que j'ajoute les Quadra-

tures comme quelque chose de plus que ces Equations. Mais les Equations Differentielles vont au de la même des Quadratures, & l'on voit bien que j'entendois même parler des Problemes qui vont a ces sortes d'Equations inconnues alors au Public; cette objection se trouvoit deja dans les remarques an *Commercium*, mais je n'avois point crû que M. *N.* etoit capable de l'employer.

Je juge par un endroit de ma Lettre du 27 d'Aoust 1676. (*pag.* 65. du *Commercium Epistolicum*) que je devois deja avoir alors l'ouverture du Calcul des Differences; car j'ay dit d'avoir resolu d'abord par une certaine Analyse (*certa Analysi solvi*) le Probleme de M. de *Beaune* proposé à M. Des Cartes; si cette Analyse n'etoit que cela, on le peut resoudre sans cela; & je crois que Monsieur *Hugens* & Monsieur *Barrow* l'auroient donné au besoin comme beaucoup d'autres choses, mais selon ma maniere de noter, ce n'est qu'un jeu; je trouve une petite faute dans cette Page, il y a *ludus naturæ* au lieu de *lusus naturæ*,[11] mais cette faute etoit ancienne, & se devoit deja trouver dans la copie de ma Lettre[12] du 24 d'Octobre 1676. *pag.* 86. du *Commercium*) *Hos casus vix numeraverim inter lusus naturæ.* Je n'avois point entendu ce qu'il vouloit dire, mais à present je vois l'origine de la méprise.

Je ne saurois dire aujourdhuy si j'ay remarqué le passage de M. *Wallis,* ou il dit que M. *N.* savoit deja la Methode des Fluxions en 1666. Mais quand je l'aurois remarqué, je l'aurois laissé passer apparemment, etant fort porté alors à croire M. *N.* sur sa parole. Mais son dernier procedé m'a forcé d'être plus circomspect à cet égard.

M. *N.* dit que je l'ay accusé d'être plagiaire, mais ou est ce que je l'ay fait? Ce sont ses adherens quî ont paru intenter cette accusation contre moi, & il y a connivé. Je ne say pas s'il adopte entierement ce qu'ils ont publié, mais je conviens avec luy, que la malice de celuy qui intente une telle accusation sans la prouver, le rend coupable de calomnie.

Il finit sa Lettre en m'accusant dêtre l'aggresseur, & j'ay commencé celle cy en prouvant le contraire. Il sera fort aisé de vuider ce point preliminaire. Il y a eu du mesentendu, mais ce n'est pas ma faute; au reste je suis avec Zele.

Monsieur,

Hannover,   Vôtre tres Humble & tres Obeissant Serviteur

ce 9 *d'Avril* 1716. [N.S.]   L<small>EIBNIZ</small>

Apostille [13]

Vous avez donné, Monsieur, la solution d'un Problême que les Partisans de M. *Newton* n'avoient point trouvée jusqu'ici: car vous avez trouvé le moyen de me faire répondre en m'envoiant une Lettre de M. *Newton* lui-même.

Après cela vous n'aviez pas besoin de me faire des exhortations là dessus. Si la Question avoit été seulement, lequel de nous deux, de M. *Newton* ou de moi, a trouvé le premier le Calcul en question, je ne m'en mettrois point en peine. Aussi est-il difficile de décider ce que l'un ou l'autre peut avoir gardé *in petto*, & combien long-tems. Mais un Adherent de M. *Newton* a prétendu que je l'avois appris de lui; & depuis il a paru plus probable à quelques autres & même à M. Bernoulli, que la maniere de calculer que M. *Newton* a publiée dans les Oeuvres de M. *Wallis* a été fabriquée a l'imitation de mon Calcul des Differences deja publié. Il n'y a pas la moindre trace, ni ombre du Calcul des Differences ou Fluxions dans toutes les anciennes Lettres de M. *Newton* que j'ai vûës, excepté dans celle qu'il a écrite le 24 d'Octobre 1676, ou il n'en a parlé que par enigme: & la solution de cette enigme qu'il n'a donnée que dix ans après, dit quelque chose, mais elle ne dit pas tout ce qu'on pourroit demander. Cependant, prevenu pour M. *Newton*, j'ai eu autre fois la condescendance d'en parler, comme si elle disoit presque tout: & c'est après moi que d'autres en ont parlé de même. Mon honneteté a été mal reconnuë.

Vous me dites, Monsieur, que M. *Jones*, a publié une de mes Lettres à M. *Newton*: aiez la bonté de m'apprendre où. [14]

C'est aller un peu vite, que de dire, que mon Probleme a été resolu fort aisément. Je croi qu'il n'a point été resolu du tout. Car de donner quelques cas faciles, comme dans les Coniques, & de se restraindre au cas de la soutangeante &c. ce n'est pas faire grand chose. M. *Bernoulli* l'a resolu par une methode générale. On fixe assez l'idée en disant, qu'il s'agit généralement de toutes sortes de lignes qui ne different entr'elles dans leurs constructions que par les changemens d'une seule droite constante dans la ligne, & changeant de ligne en ligne. Prenez telle ligne qu'il vous plaira, vous aurez d'abord par cette methode une suite d'infinité d'autres.

Je m'etonne, Monsieur, que vous dites qu'*avant que de parler de la Philosophie de M.* Newton, *il faut convenir de la methode de philosopher.* Est-ce qu'il y a une autre Logique à Londres qu'à Hanover? Quand on raisonne en bonne forme sur des faits bien averez, ou sur des Axiomes indubitables; on ne manque pas d'avoir raison. Si les sentimens de M. *Newton* sont meilleurs qu'on n'a dit; tant mieux; je serai toujours bien aise de lui rendre justice.

[*Leibniz goes on to complain that Conti sends him news of no one but Newton, and asks for further information about the English intellectual scene.*]

Vous voyez que l'*Apostille* est pour vous, Monsieur: & la *Lettre* est plûtôt pour M. *Newton*; à l'exemple de celle qu'il vous a écrite.

NOTES

(1) Apparently neither original nor draft of this letter survives. Leibniz sent the letter to Monmort for Conti; see Letter 1198 following. From this covering letter and the Postscript at the close of the present one, it is obvious that Leibniz intended the whole (except the Post-script) to be communicated to Newton.

Newton drafted several sets of observations upon this letter, of which the final version is Number 1211.

Meanwhile, feeling that his letter would be delayed in reaching Conti via Monmort, Leibniz composed a second letter to the former to be sent directly to Conti (see Letter 1202).

(2) 'a challenge'; because Leibniz had not replied to the *Commercium Epistolicum*.

(3) For Cavalieri's work on the theory of indivisibles see D. T. Whiteside, *Archive for History of Exact Sciences*, **1** (1961), 312–30.

(4) Christian Wolf, to whom Leibniz had entrusted the publishing of the *Charta Volans* (Number 1009).

(5) Leibniz refers here to a letter from Newton to Collins, dated 20 August 1672 (Letter 90, vol. I) where Newton stated a complicated infinite series by which the second segment of an ellipsoid could be calculated. (The second segment is the volume cut off by two planes through the solid perpendicular to its axes.) Leibniz saw this letter while on his second visit to London in October 1676, and made excerpts from it, particularly noting that Newton credited Gregory with the first discovery of the series (see Hofmann, p. 183). Leibniz's complaint presumably is that Newton failed to make this acknowledgement in the Epistola Prior (vol. II, pp. 28 and 38), where he repeats the same result, and that he could approach no further than Gregory towards an elegant solution.

(6) Vincentius Leotaudius, *Examen Circuli quadraturæ hactenus editarum celeberrimæ*, . . . (Lugduni, 1654).

(7) *Dissertatio de Arte Combinatoria* (Leipzig, 1666).

(8) Leibniz did not meet Collins until his second visit to England in 1676; see Letters 186 and 205, vol. II, and Hofmann, pp. 182 and 189–93.

(9) For John Pell see Letter 17, note (2) (vol. I, p. 38); he met Leibniz in London in the spring of 1673 (see Hofmann, p. 15), and introduced him to Mercator's *Logarithmotechnia*. Leibniz's letter to Oldenburg of 6 October 1674 is extracted in Letter 126, vol. II, and is given in full in Hall and Hall, *Oldenburg*, XI (in press).

(10) Little personal is known of the Abbé de Catelan; (see André Robinet, 'L'Abbé de Catelan, ou l'erreur au service de la vérité,' *Revue d'Histoire des Sciences*, **11** (1958), pp. 289–301). He opposed the use of infinitesimals in mathematics, publishing *Logistique pour la science generale des lignes courbes, ou maniere universelle & infinie d'exprimer & de comparer les puissances des grandeurs* (Paris, 1691), as also (in the same year) *Principes de la science generale des lignes courbes*. His name occurs very frequently in the *Oeuvres Complètes de Christiaan Huygens*, VIII and IX.

(11) The phrase should read 'il y a *ludus naturæ* au lieu de *hujus naturæ*'; again, three lines later, *lusus* is printed for *ludos*, Leibniz's reference to de Beaune's curves 'of this nature' (*hujus naturæ*; vol. II, p. 64, line 10) was incorrectly copied by Collins for Newton as 'sport of nature' (*ludus naturæ*; vol. II, p. 129, line 21); an expression he could not understand. The misprint in Raphson, *History of Fluxions* confused the issue further, *ludus* and *lusus* being almost synonymous. Gerhardt, *Briefwechsel*, p. 281, silently corrects the error. See Whiteshide, *Mathematical Papers*, IV, p. 633, note (42).

(12) This should read 'the copy of my letter [of 27 August 1676 N.S. which Newton quoted in his reply] of 24 October.'

(13) The postscript, intended for Conti alone, is not printed in Raphson, *History of Fluxions*. We use the version in Des Maizeaux, *Recueil*, 1st edition (Amsterdam, 1720) pp. 67–71.

(14) Des Maizeaux notes that 'M. Leibniz confond ici le Traité de M. Newton *de Analysi* &c, publié par M. *Jones*; avec la Lettre de M. *Leibniz* à M. *Newton*, écrite en 1693.' Leibniz's confusion is a result of Conti's loose phraseology in his previous letter; (see Letter 1190, note (3)).

1198 LEIBNIZ TO MONMORT
29 MARCH 1716
From the copy by Newton in the University Library, Cambridge.[1]
Cover for Letter 1197

Monsieur

Je prends la liberté de vous envoyer les pieces d'un procés nouveau ou renovellé, Puisque vous avéz la bonté de vous interesser pour moy M. l'Abbé Conti, qui avoit fait des demarches de mediateur, m'a envoyé maintenant un cartel de defy de la part de M. Newton.[2] Je réponds à la lettre de l'un et de l'autre par la Lettre & par P.S. a M. l'Abbé [3] c'est-à-dire à M. Newton dans la Lettre et à M. l'Abbé dans le Postscriptum, et je suis bien aise, Monsieur, que vous et vos amis et particulierement M. l'Abbé Varignon, et d'autres personnes de L'Academie Royale des Sciences, à qui il en voudra faire part en sojent informes. Je vous supplie de garder la copies des Lettres de M. l'Abbé [4] et de M. Newton et d'envoyer ma reponse a M. l'Abbé. Vous voyes bien Monsieur, pourquoy j'ay voulu me servir de la voje de la France, au lieu de repondre directement d'icy. Si vous croyés, Monsieur, que cette repond vaille la peine qu'on en garde aussi une copie, cela depend de votre jugement. Mais je ne voudrois pas qu'on en imprimât rien sans mon consentement. Je ne fais point d'autres reflexions sur ces Lettres; on en fera assez sans moy . . .

P.S. Je vous envoye la lettre à M. l'Abbé Conti sub *sigillo volante*,[5] et il n'est point necessaire que vous la firmies. Je veux bien qu'on sache que vous l'aves veue, Monsieur, & que je suis bien aise que vous en soyes informé.

Hanover ce 9 *d'Avril.* 1716 [N.S.]

NOTES
(1) Add. 3968(31), fo. 445r.
(2) Letter 1187.
(3) See Letter 1197.
(4) Letter 1190.
(5) Under a loose seal; that is, an open letter.

1199 FLAMSTEED TO SHARP
29 MARCH 1716
Extract from the original in the Royal Society[1]

God has in some measure Granted me what I desired. Sr I.N. had contrived to dispose of ye printed Volumes of my Observations in such manne[r]
yt they should have been spread all over Europe as his gift to librarys &
ingenious p[er]sons with Halleys Copy of my spoyled Catalogue of the fixed
stars & a malitious preface preface [*sic*] of Hallys yt was wrote without my
knowledge to it as also his abstracts of ye planetary Observations taken with
ye Murall Arch of which I trusted a Copy into his hand to be printed March 20
1707/8: I was fully informed of his intent & therefore makeing my Application
by proper p[er]sons got his Ma[jes]ties order to have 300 Copies [of] them
delivered unto me [2] & last night my man brought [them] down to ye Observatory tho Mr Churchill was by agreement to print but 400. Sr. I N has
[sent] 3 Copies into Italy some say to ye Pope. one to the king of france one to
Mons Torcy [3] & DesMarets [4] each one ten to ye Royal Academy of Paris &
about 40 to ye Exchequer of which I am told ye French Envoy has had 17 so
yt theire is 9 or ten left in ye Exchequer. & 39 in Mr Churchills hands which
I am endeavoring to get into my own hands that I may hinder any more of
ye false Catalogues from going abroad or his very sorry abstracts which I
intend to sacrifice to TRUTH as soone as I can get leasur saveing some few
that I intend to bestow on yu & such freinds as you that are hearty lovers of
truth that yu may keep them by yu as Evidence of ye malice of Godlesse
persons & of [ye] Candor & Sincerity of ye freind that writes to yu, & con-
veys them into [your] hands. for I will not say I make yu a present of that
which is odious of it selfe & will be detested by every Ingenious man.

Pray let me know by yr next how I may send yu this prec[i]ous parcell.

NOTES

(1) Sharp Letters, fo. 99; printed in Baily, *Flamsteed*, p. 321.

(2) See Letter 1189, note (1).

(3) Jean Baptiste Colbert (1665–1746), Marquis de Torcy, nephew of the great Colbert;
he was a great diplomatist, and holder of important posts under the French crown.

(4) Nicolas Desmarets (1650–1721), another nephew of the great Colbert, financial
administrator, who had been dismissed from his post as controller-general of finances, which
he had held since 1708, in September 1715.

1200 NEWTON TO TOWNSHEND
[APRIL 1716]
From the holograph draft in the Mint Papers [1]

To the Rt Honble my Lord Viscount Townshend Principal Secretary of State [2]

My Lord

The silver wch your Lordship gave me to be assayed, was produced out of a pound weight Averdupois of Ore & weighed not fifteen pence but fifteen penny weight & some grains when it first came out of the Ore. [3] It had some dirt sticking to ye bottom of it & a piece cut off & flatted with the dirt & sent by my Lord Mayor to the Mint to be assayed, proved only x dwt ob[olus] better then standard, because the Assay was spoiled by the dirt wch stuck fast to the assay piece. By two assays wch I caused to be made of clean pieces cut off from the silver, it proved xvii dwt better then standard. Now fifteen penny weight of such fine silver is worth four shillings & two pence. And therefore the Ore is exceeding rich, a pound weight averdupois holding 4s 2d in silver. This silver holds no gold

Two ounces Troy of the Ore wch your Lordp gave me to be assayed yeilded upon the first melting three penny weight of silver wch upon the Assay proved two penny weight worse then standard, & therefore was worth $9d\frac{1}{4}$, & after this rate a pound weight averdupois of the Ore produces 22 penny weight of silver wch is worth about 5s 7d.

An ounce Troy of the same Ore yeilded upon the first melting 1 dwt 12 gr & this being melted again wth a convenient flux pounder [sic] left 1 dwt & 10 gr of fine silver, & after a third melting there remained 1 dwt wanting 4 gr, some of the silver being lost among the scorias. This last silver upon the Assay proved xiij dwt better then standard.

The Ore holds little or no copper. It is silver Ore, but where it grows doth not yet appear to me.

All wch is submitted to your Lordps consideration

Is. NEWTON

NOTES

(1) III, fo. 268. In another hand [Conduitt's?] is written at the foot of the page 'An acc[oun]t of the ore taken out of Sr John Erskin's mine.' (See note (3) below.)

(2) Charles Townshend (1674–1738), second viscount, one of the principal politicians of his age, Robert Walpole's brother-in-law, and immortalized (misleadingly) as 'Turnip' Townshend from his agricultural innovations at Rainham, Norfolk. His political fortunes brightened by the Hanoverian succession, Townshend had been appointed Secretary of

State for the Northern Department on 17 September 1714; he was dismissed in December 1716.

(3) A full account of events connected with the silver mine opened at Alva in Clackmannanshire, Scotland (a little north of Alloa on the Forth) was written by Newton himself in a draft refutation of the claims of one James Hamilton to further reward for his services (see Mint Papers, III, fo. 246). A 'vein of Lead Ore holding some silver,' discovered in December 1714, was worked until February 1716; it was abandoned because the landowner, Sir John Erskine (probably the second son of Sir Charles Erskine of Alva, the baronetcy having been created in 1666; see also Letter 1216, note (1)), had joined the Jacobite rebellion led by the head of his family (and namesake) the Earl of Mar. James Hamilton, 'who had been imployed in smelting the Ore brought some of it to London in March or April (1716), & made the Lord Maior acquainted with the richness of the Mine, & Mr Haldane the brother in law of Sir John made the matter known at Court . . . It was at first proposed to send down Sr Isaac Newton [aged 73!] to examin the Mine, but he represented himself unacquainted with such matters, & declined recommending any body else in point of skill, saying that it would be better to send down somebody of rank from the Kings silver Mines in Germany. Whereupon Mr. Justus Brandshagen was proposed by others.'

Nothing is known of Brandshagen (previously mentioned in Letter 1174), save that he also claimed a medical qualification. A Warrant for the journey of Brandshagen and Hamilton to Scotland was signed by the Prince Regent on 20 August 1716.

1200A   HALDANE TO NEWTON
[c. APRIL 1716]
From the original in the University Library, Cambridge.[1]
For the answer see Letter 1200B

Sr

I was at My Lord Tounshends this morneing where I found him in Conversation with a German [2] who had bin recomended to his Lo[rdshi]p by Count Bothmar and others as on who had a great deal of knouledge and experience in what relats to Mines and Minerals his Lo[rdshi]p ordered the Gentleman to come to my Lodgeings att four a Clock this afternoon that I might bring him to uait upon you I send this to knou if you ar to be att home about that tym or if you can not that you will be pleased to Lett me knou when it will be easier and more convenient for you   I am with the profoundest rspect

Sr

Kings street near St James
Square Monday 2 a Clock

Your most humble Servant

HALDANE

To
The Honourable Sr Isack Newton

NOTES

(1) Add. 3970(3), fo. 608. For the context of this letter see Letter 1200, note (3).

(2) Presumably Justus von Brandshagen, who it is likely was acquainted with Count Bothmar (the chief Hanoverian agent in London and a friend of Leibniz) and also with Leibniz himself; see Letter 1174.

1200B NEWTON TO HALDANE
[*c*. APRIL 1716]
From the draft in the University Library, Cambridge.[1]
Reply to Letter 1200A

Sr

I should be heartily glad to discourse the gentleman recommend[ed] by Count Bothmar & others for his skill in & experience in things relating to Mines & Mineralls & for that end will be at home till after five a clock this afternoon

NOTE

(1) Add. 3965(18), fo. 689.

1201 LEIBNIZ TO J. BERNOULLI
2 APRIL 1716
From Gerhardt, *Leibniz: Mathematische Schriften*, III/2, pp. 959–60.
For the answer see Letter 1210

Opportune literas Tuas accipio,[1] renovata jam lite Anglicana. Newtonus ipse, cum videret mihi Keilium indignum responsione haberi, in arenam descendit, literis ad Dnum. Abbatem Contium [2] scriptis, qui ad me misit. Ego respondi,[3] et versionem Anglicæ Newtoni Epistolæ cum responsione me ad Dn. Remondum misi Parisios, Abbati Contio transmittendam, et amicis Parisinis ostendendam. Ex Gallia Tibi omnia communicabuntur, miraberis tam levibus argumentis actum. Potissimum est, me aliquoties ipsi inventum concessisse, ergo nunc salvo candore negare non posse. Respondeo, me tantam de ipsius candore tunc opinionem habuisse, ut quidvis affirmanti facile crediderim, nunc dum accusationi contra me connivet, imo accedit, quam falsam novit, dubitare de ejus sinceritate coactum. Epistolam quam Tuam esse scit,[4] ait a Mathematico vel Mathematicum affectante scriptam (*par un Mathématicien ou prétendu Mathématicien*) quasi merita Tua ignoret. Totam chartam, cui Epistola Tua inserta est, vocat diffamatoriam, quasi magis famam lædat, quam addita Commercio Epistolico.

Dominus Arnoldus mihi scripsit, Keilium in novo quodam Transactionum loco [5] contendere, Te quoque ignorare Calculum differentialem, sed homo indignus est cui respondeatur.

Cæterum Contius, qui ad partes novorum amicorum nonnihil accedere videtur, scribit Anglos facile solvisse Problema duarum Serierum curvarum invicem perpendicularium; Moivræum enim, præter alios, scilicet ut figeret ideas, rem reduxisse ad subtangentem; verba Contii sunt: *il faut supposer la même soutangente pour la même abscisse*, quæ non satis intelligo. [6] Quidquid vero sit, hoc non est solvere problema, sed ejus casum, Problema ipsum jam in Actis Lipsiensibus proposuimus Majo anni 1697, pag. 211, [7] et cum Fatius insurrexisset, Maji 1700, pag. 204. [8] Commodum autem evenit, ut exemplum a Te acceperim, quod non ita facile solutum iri judicas; nihil potuit fieri accommodatius. Id nunc Contio mittam, [9] ut habeant, in quo ideas figant, tantisper dum solutionem generalem inveniant. Interim Domini Filii Tui solutionem elegantem casus Hyperbolici vel Elliptici ad Dn. Wolfium misi, ut Actis Eruditorum Lipsiensibus [10] inseri curet.

Translation

I receive your letter [1] opportunely, the dispute with the English having now been renewed. Newton himself, since he saw that I regarded Keill as unworthy of an answer, has entered the ring, having written a letter to the Abbé Conti, [2] who has sent [it] to me. I have replied, [3] and I have sent a translation of Newton's English letter, with my answer, to Mr [Nicolas] Rémond in Paris, for sending on to the Abbé Conti, and for showing to his Paris friends. It will all be communicated to you from France, and you will be surprised at the presentation of such trifling arguments. The chief [of these] is that I have several times allowed him the discovery and therefore cannot deny it without loss of reputation. I reply that I had such a [high] opinion of his honesty at the time, that I would readily have believed whatever he affirmed, but now that he winks at accusations against me, or rather agrees with what he knows is false, I am forced to doubt his sincerity. The letter, which he knows is yours, [4] he says is written by a mathematician or a pretended mathematician (par un Mathématicien ou pretendu Mathématicien) as if he were ignorant of your work. He calls the whole paper [the *Charta Volans*] in which your letter is inserted defamatory, as if it were more damaging to a reputation, than what was added to the *Commercium Epistolicum*.

Mr Arnold has written to me that Keill in some fresh place in the [*Philosophical*] *Transactions* [5] has argued that you also are ignorant of the differential calculus, but the man does not deserve to be answered.

For the rest, Conti, who seems inclined to take sides with his new friends, writes that the English have easily solved the problem of the two series of curves [cutting] mutually at right-angles, for [De] Moivre, among others (to be more precise) has reduced the problem to the subtangent; Conti's words are, 'il faut supposer la même soutangente pour la mmeê abscisse', which I do not quite understand.[6] Whatever the truth of it, this is not to solve the problem, but a [special] case of it. We have already proposed the problem itself in the *Acta [Eruditorum]* of Leipzig for May 1697, p. 211,[7] and, when Fatio thrust himself forward, in May 1700, p. 204.[8] However, it happens conveniently that I should have received an example from you, which you judge is not to be solved so easily; nothing could have been more suitable. I will now send it to Conti,[9] so that they have it to build their ideas on while they discover a general solution. Meanwhile I have sent to Mr Wolf your son's elegant solution for the case of the hyperbola or ellipse, in order that he should see that it is inserted in the *Acta Eruditorum* of Leipzig[10]

NOTES

(1) Johann Bernoulli, in a letter of 29 February 1716 (see Gerhardt, *Leibniz: Mathematische Schriften*, III/2, pp. 957–9) had provided a new problem for Leibniz to pose to the English. See note (7) below, and Letter 1202, notes (1) and (4).

(2) See Letter 1187.

(3) See Letter 1197.

(4) Leibniz was mistaken, since Newton was not yet willing to allow Bernoulli's authorship of the excerpt of Letter 1004 quoted in the *Charta Volans* (Number 1009); Newton was unlikely to have seen Leibniz's acknowledgement of the fact in the *Nouvelles Literaires* (see Letter 1172). Indeed, much later Newton accepted Bernoulli's written disclaimer (see Letter 1321, vol. VII).

(5) That is, Keill's 'Defense'; see Letter 1165, note (2), p. 246; for Arnold's mention of the paper to Leibniz see Letter 1181, note (14), p. 276.

(6) For English solutions of the problem, see Letter 1186 and notes.

(7) *Acta Eruditorum* for May 1697, p. 211. This paper contained both a very general statement of the problem of curves cutting orthogonally, and the basis for the new, special problem which Leibniz, on Bernoulli's suggestion, was to pose to the English in Letter 1202.

Johann Bernoulli set the general problem as part of his long-continued mathematical warfare with his brother, Jakob Bernoulli. Jakob retaliated in 'Solutio Problematis Fraterni in Actis Mens. Mai 1697 p. 211 propositi de Curva infinitas Logarithmicas ad angulos rectos secante', *Acta Eruditorum* for May 1697, p. 231, where he solved the problem for a few very specialized classes of curves. Johann replied again in 'Johannis Bernoulli Annotata in Solutiones Fraternas Problematum quorundam suorum editas proximo Actorum Maio', *Acta Eruditorum* for October 1698, p. 466; and there he noted that he and Leibniz had discussed the general problem in 1694, in private correspondence. He mentions his own letter to Leibniz of 2 September 1694 [N.S.] where he first poses the problem, and Leibniz's reply of 6/16 December 1694, where an outline of an approach to a solution is given. (The letters are printed in Gerhardt, *Leibniz: Mathematische Schriften*, III/1, pp. 143–57). Leibniz writes (see Gerhardt, p. 157):

> Pene exciderat Problema inveniendi curvam quæ ordinatim positione datis occurrat ad angulos rectos. Cujus Methodus meo judicio consistit in duabus æquationibus, una continente relationem inter x, y et constantem quamdem in curva positione data, sed pro diversibus talibus ordinatim datis variabilem b; altera continente valorem ipsius $dy:dx$ in curva quæsita, expressam ex proprietate perpendicularium in curva positione data, cujus æquationis ope datus ipsius b valor per $dy:dx$, y, x, pro re nata, quarum duarum

æquationum ope tollendo b, habetur æquatio differentialis primi gradus pro relatione inter x et y.

(The problem of finding the curve which meets at right angles in succession [curves] given by position has almost been solved. In my opinion, the method for this relies on two equations, one containing the relationship between x, y and some constant [b] in the curve given by position, but with b different for each of the different curves [in the family]; the other containing the value of dy/dx in the required curve found from the property of perpendicularity to the curve given by position, by the help of which equation the value of b itself is given in terms of dy/dx, y, x, as they now are; eliminating b by the help of the two equations, a differential equation of the first degree is obtained for the relationship of x and y.)

The problem remained one of those under continual discussion both publicly and privately amongst the continental mathematicians. The course of the argument may be followed in *Johannis Bernoulli Opera Omnia* (Lausanne and Geneva, 1742) where not only Johann's own papers are published, but also extracts of related work; see also Letter 1170, note (6), p. 254.

The special problem which Bernoulli here constructs is 'Invenire curvam quæ Omnes Cycloides communis initii normaliter secat'. ('To find the curve which cuts normally all Cycloids having a common starting point'.) Inspired by Huygens' concept in his *Traité de la Lumière* of wave-fronts at all points normal to the direction of propagation of the rays of light, Bernoulli postulates that if in a refractive medium a constant *vis refractiva* acts vertically downwards at every point, the family of cycloids he describes, because they are brachistochrones, are also the refractive paths of rays from a common starting point. The synchrone—the curve on which a number of particles, setting off down the family of cycloids at the same time, will lie after a given period of time—will thus be analogous to the wave-front formed by the rays of light.

Bernoulli seems to have silently passed over his work on this problem when he sent Leibniz the new problem to pose to the English—it is Leibniz who here emphasizes that the new problem, giving for the case $n = \frac{1}{2}$ the differential equation for a cycloid, was related to this earlier problem; see Letter 1202, note (4).

(8) 'G.G. L[eibnitii] Responsio ad Dn. Nic. Fatii Duillerii imputationes', *Acta Eruditorum* for May 1700, p. 204.

(9) See Letter 1202.

(10) Nikolaus [II] Bernoulli, 'Problema: data serie linearum per rectæ in eadem Linea constantis variationem prodeunte invenire aliam seriem linearum, quarum quævis priores omnes ad angulos rectos secet.' *Acta Eruditorum* for May 1716, pp. 226–30.

I202 LEIBNIZ TO CONTI
3 APRIL 1716
From Des Maizeaux, *Recueil*.[1]
Continuation of Letter 1197

Hanover ce 14. *d'Avril* 1716 [N.S.]

MONSIEUR,

Pour ne vous point faire attendre, je vous dirai par avance que j'ai répondu d'abord à l'honneur de votre Lettre,[2] & en même tems à celle que Mr. *Newton* vous a écrite; & j'ai envoyé le tout à Mr. *Remond* à Paris, qui ne manquera pas de vous le faire tenir. Je me suis servi de cette voie, pour avoir des temoins neutres & intelligens de notre Dispute: & M. *Remond* en fera encore part à d'autres. Je lui ai envoyé en même tems une copie de votre Lettre, & de celle de Mr. *Newton*. Après cela vous pourrez juger, si la mauvaise chicane de quelques-uns de vos nouveaux Amis m'embarrasse beaucoup.

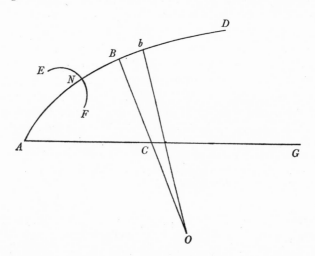

Quant au Probleme dont quelques-uns parmi eux ont voulu resoudre des cas particuliers pour en fixer, disent ils, les idées, il y a de l'apparence qu'ils se seront jettez sur des cas faciles, car il y en a dans les Courbes transcendantes aussi bien que dans les ordinaires; mais il s'agit d'une solution générale. Ce Probleme n'est point nouveau. M. Jean *Bernoulli*[3] l'a deja proposé dans le mois de May des Actes de Leipsic 1697. p. 211. Et comme M. *Fatio* méprisoit ce que nous avions fait; on en repeta la proposition pour lui & pour ses semblables dans les Actes de May 1700. p. 204. Il peut servir encore aujourd'hui à faire connoitre à quelques-uns, s'ils sont allez aussi avant que nous en Methodes; & en attendant qu'ils trouvent le moyen de parvenir à la solution

322

générale, ils pourront essayer ce qu'ils peuvent en fixant les idées sur un cas particulier, qu'on leur propose dans le papier ci-joint. Sa solution vient encore du même M. *Bernoulli*. Ainsi vous aurez la bonté de ne pas vous rendre trop tôt aux insinuations de ceux qui nous sont contraires; comme lorsqu'ils vous ont fait accroire que notre Probleme leur étoit aisé. Je suis avec zele, Monsieur, Votre &c

PROBLEMA [4] *continens casum specialem Problematis generalis de invenienda serie Curvarum quarum quælibet sit ad aliam seriem Curvarum perpendicularis.*

   Super recta AG *tanquam axe ex puncto* A *constructis Curvis quotcunque, qualis est* ABD, *ejus naturæ ut radius osculi ex singulis singularum Curvarum punctis* B *eductus* BO *secetur ab axe* AG *in* C *in data semper constante ratione, ut nempe sit* BO: [ad] BC [::] [ut] M: [5][ad] N. *Construendæ jam sunt trajectoriæ qualis est* ENF, *priores Curvas* ABD *secantes ad angulos rectos.*

NOTES

   (1) (Amsterdam, 1720), II, pp. 26–8; the letter had been published anonymously by Newton in *Phil. Trans.* **30**, no. 359 (1719), 925–7, together with a criticism of the new problem. A holograph draft of this criticism is in U.L.C. Add 3967(4) fo. 38v. Newton wrote

Thus far this Letter. The words in wch M. Leibnitz first proposed the general Probleme to M. l'Abbe Conti were: Trouver une line, *BCD* qui coupe à angles droits toutes les courbes d'une suite determinée d'une même gendre; par example, toute les Hyperboles *AB, AC, AD*, qui ont le meme summet & le même centre: & cela par une voye generale. And in the Acta Eruditorum Anno 1698 p. 470, 471 he calls the curves in this determinate series Curvæ ordinatim datæ & positione datæ. And by all this the series of curves to be cut is given & nothing more is to be found then the other series which is to cut it at right angles. But Mr Leibnitz being told that his Probleme was solved, he changed it into a new one of finding both the series to be cut & the other series wch is to cut it. And the particular Probleme proposed in this Letter is a special case, not of the general Probleme first proposed, but of this new double general Problem. And the first part of this double Probleme (vizt by any given property of a series of Curves to find the Curves) is a general Problem harder then the former & of wch Mr Leibniz had a general solution. Mr Leibnitz In a letter to Mr John Bernoulli dated $\frac{6}{16}$ Decem 1694 & published in the Acta Eruditorum for Octob. 1698, pag 471 set down his solution of the 1st Probleme when the given series of Curves is defined by a finite equation expressing the relation between the Absciss & ordinate. The same solution holds when the equation is a converging series, or the property of the Curves to be cut can be reduced to such an equation by ye *Analysis per series numero terminorum infinitas.* But Mr Leibnitz was for doing it without converging series.

The new problem here posed (see note (4) below and Letter 1201, note (7)) presented far more difficulty than the earlier one; Newton himself published no attempts at a solution. In England and on the continent, however, many mathematicians took up the challenge (see M. Cantor, *Vorlesungen über Geschichte der Mathematik,* III (Leipzig, 1898), pp. 450–5). There is no doubt that Johann Bernoulli fully recognized the subtlety of the problem he had suggested; there was no one to match him in mathematical ability at this time—the argument

323

between Newton and Leibniz, now old men and well past the prime of their intellectual capabilities, merely served as a medium through which Bernoulli could display his skills.

(2) Letter 1190.

(3) In a later letter to Monmort of 28 March 1717 (Letter 1237), Bernoulli was indignant that Leibniz had disclosed his authorship of the problem.

(4) 'A problem containing a special case of the general problem of finding a series of curves each of which is normal to another series of curves.

Let any number of curves, such as ABC, be constructed from the point A upon the straight line AG as axis, their nature being such that for each curve the radius of curvature BO, drawn from any point B of the curve, is cut by the axis AG at C, in a constant, given ratio—that is to say $BO/BC = M/N$. Now the trajectories such as ENF are to be constructed, cutting the former curves ABD at right angles.' The problem had been sent by Bernoulli to Leibniz in his letter dated 29 February 1716, couched in almost identical terms; see Gerhardt, *Leibniz: Mathematische Schriften*, iii/2, p. 958. He there discusses a faulty and incomplete solution.

Bernoulli correctly specified the solution of the problem in the case $n = \frac{1}{2}$ in the *Acta Eruditorum* for May 1697, p. 210, but without proof; he must, however, have possessed a method of solution in order to contrive the problem. Jakob Hermann published a solution in the *Acta Eruditorum* for August 1717, pp. 348–52, and Brook Taylor, in *Phil. Trans.* **30**, no. 354 (October–December 1717), 695–701 gave the same solution, but in fluxional form.

(5) In the text the expression is misprinted as $BO :^{ad} BC^{ut} N :^{ad} N$. Bernoulli had simply written $BO : BC = 1 : n$.

1203   LEIBNIZ TO THE BARONESS VON KILMANSEGGE

7 APRIL 1716

From Des Maizeaux's *Recueil* (second edition, 1740), ii, pp. 33–46 [1]

Hanover ce 18. *d'Avril* 1716. [N.S.]

MADAME,

Je suis bien aise que des Dames aussi éclairées que vous l'êtes, prennent connoissance de ma controverse avec Mr. *Newton*. Si vous ne voulez pas prendre la peine de pénétrer dans l'embarras des figures des Calculs, dont vous seriez capable de venir à bout, autant que qui que ce soit, vous pénétrerez assez dans l'Historique pour n'être point surprise: & vous autres Dames Hanoveriennes à Londres, ou ici, vous ne devez point être fâchées, ce semble, qu'il y ait quelque chose, en quoi Hanover, tout petit qu'il est, ne céde point au grand Londres. Il céde en grandeur, en richesses, en tout ce qu'il vous plaira; mais non pas en affection pour le Roi, ni par rapport au mérite des Dames, ni en Géométrie. Voici le fait.

Etant venu en France l'an 1672, jeune garçon, comme il est aisé de croire, j'apportai de nos Universitez, toute autre connoissance, que celle de la profonde Géométrie. Le Droit & l'Histoire étoient mon fait. Je me plaisois pourtant à la Mathématique pratique, & je m'étois un peu exercé aux propriétez des Nombres, ayant publié un petit Livre sur l'Art des Combinaisons dès l'an 1666.[2] & je fis même une remarque considérable sur les différences des suites (ou *series*) des Nombres, où d'autres n'avoient pas assez pris garde. A Paris je me fourrois dans les grandes Bibliothéques, & je cherchois des Pièces rares, sur-tout en Histoire; mais je ne laissois pas de donner encore quelque tems aux curiositez de Mathématique. Je fis un tour à Londres, & m'y trouvant au commencement de l'année 1673. quoique je n'y fisse point un long séjour, je ne laissai pas de faire connoissance avec Mr. *Oldenbourg*,[3] Secrétaire de la Societé des Sciences, que le Roi Charles II. avoit érigée: & comme j'aimois un peu la Chymie, je pratiquai aussi Mr. *Boyle*, chez qui je rencontrai un jour un Mathématicien, nommé Mr. *Pell*;[4] & lui ayant conté une certaine observation, que j'avois faite sur les Nombres, il m'apprit qu'un Holsteinois, qui se trouvoit à Londres, nommé Mr. *Mercator*, l'avoit faite aussi dans un Livre publié depuis peu sur la figure qui s'apelle Hyperbole. Je cherchai ce Livre, & je l'apportai avec moi en France.[5]

Comme j'y pratiquai Mr. *Huygens* de Zulichem, inventeur du Système de Saturne & des Pendules, & grand Géometre, je commençai à prendre goût aux méditations Géométriques. J'y avançai en peu de tems, & trouvai une suite de Nombres (ou *series*) qui faisoit pour le Cercle ce que celle de Mercator avoit fait pour l'Hyperbole. La découverte fit du bruit à Paris. Mr. *Huygens* la fit valoir; & cela joint à d'autres raisons, fit qu'on me destina une place dans l'Académie Royale des Sciences.[6] Nous crumes que j'étois le premier, qui avois fait quelque chose de tel sur le Cercle; & j'en écrivis sur ce ton-là à *Mr. Oldenbourg* en 1674, avec qui auparavant je ne m'étois point entretenu de telles choses, quoique nous eussions déja échangé plusieurs Lettres. Mr. *Oldenbourg* m'écrivit qu'un Mr. *Newton* à Cambridge avoit déja donné des choses semblables, non-seulement sur le Cercle; mais encore sur toutes sortes d'autres figures, & m'en envoya des essais. Cependant le mien fut assez applaudi par Mr. *Newton* même. Il s'est trouvé par après, qu'un nommé Mr. *Gregory*, avoit trouvé justement la même *series* que moi. Mais c'est ce que j'appris tard.

Mais ce n'est pas de quoi il s'agit: car j'allai plus avant & joignant mes anciennes observations sur les différences des Nombres à mes nouvelles méditations de Géométrie, je trouvai environ en 1676. (autant qu'il m'en peut souvenir) un nouveau Calcul, que j'appellai *le Calcul des Différences*, dont l'application à la Géométrie produisoit des merveilles. Mais devant retourner en Allemagne, où feu Monseigneur le Duc Jean Frideric,[7] Oncle de notre Roi, m'avoit appellé la même année, & voulant profiter du peu de séjour qui me restoit à Paris, on peut bien juger que je n'eus point le tems de demeurer

325

long-tems dans mon Cabinet, & de méditer beaucoup, pour faire valoir d'abord
ma nouvelle découverte. Je passai par l'Angleterre & par la Hollande. Etant à
Londres, mais très-peu de jours, je fis connoissance avec Mr. *Collins*, qui me
montra plusieurs Lettres de Mr. *Newton*, de Mr. *Gregory* & d'autres, qui rouloient
principalement sur les *series*. Etant arrivé à Hanover je reçus de Mr. *Oldenbourg*
en 1677. une Lettre que Mr. *Newton* lui avoit écrite pour m'être communiquée,[8]
où il disoit pouvoir mener les Tangentes d'une figure donnée sans ôter les irra-
tionelles, & aussi réciproquement, qu'il avoit deux Méthodes pour trouver la
figure propre aux Tangentes d'une nature donnée; & il cacha l'une & l'autre
sous des lettres transposées. Je répondis à Mr. *Oldenbourg* par une Lettre
donnée à Hanover le 21. de Juin 1677. & je lui envoyai ma Méthode, que je
jugeois fournir tout ce que Mr. *Newton* promettoit des siennes en enigme.

Les choses en demeurérent-là, & j'eus quelque loisir de pousser mes méditations tant sur ces matiéres, que sur d'autres. Quelques années après des Amis
à *Leipsic* de concert avec moi, commencérent un Journal des Savans en Latin,
qui se devoit donner tous les Mois, & qui a toujours été continué depuis. Je
m'engageai d'y fournir quelque chose de tems en tems. Cela commença en
1682. Je publiai alors ma *series* pour le Cercle dont j'ai parlé ci-dessus. En 1684,
je publiai le nouveau Calcul des Différences, que j'avois inventé & gardé
presque neuf ans sans me presser de le publier. Cette invention dont on
reconnut l'usage par l'application à des questions difficiles, fit du bruit. Le
Marquis de l'Hospital, Vice-Président de l'Académie des Sciences à Paris,
fit un Livre exprès là-dessus. On s'en servit en France, en Italie, & même
en Angleterre. Mais personne ne s'y signala davantage que Messieurs *Bernoulli* en Suisse.

En 1687. Mr. *Newton* publia son Livre intitulé: *Principes Mathématiques de
la Nature.* Il dit en Latin page 253. 254. ce qui donne ce sens en François:
Dans le Commerce des Lettres que j'ai eu il y a dix ans (par l'entremise de Mr. Oldenbourg) *avec Mr.* Leibniz, *très-habile Géometre, lorsque je lui fis savoir que j'avois
une Méthode de determiner les quantitez les plus grandes ou les plus petites, de mener
des Tangentes, & d'effectuer d'autres choses semblables en termes sourds aussi-bien
qu'en termes rationaux, que je cachai sous des lettres transposées, qui renfermoient ce
sens*: 'une équation donnée, qui contient des quantitez fluents, trouver les
Fluxions, & réciproquement:' *ce célèbre personnage me répondit, qu'il étoit tombé
sur une Méthode, qui faisoit aussi cet effet, & la communiqua; qui ne différoit guère de
la mienne, que dans les termes, & dans les caractères.* Ainsi, Mr. *Newton* ne me con-
testa point d'avoir trouvé la chose de mon chef. J'eus aussi l'honnêteté de
dire publiquement, & de faire dire à mes Amis, que je croyois que Mr.
Newton avoit eu de son chef quelque chose de semblable à mon invention.

Mais en 1711. que j'étois depuis environ 27. ans en possession de l'invention, il y eut des gens en Angleterre qui poussez, ce semble, par des mouvemens d'envie, s'avisérent de me le contester. On prit pour prétexte certaines paroles du Journal de Leipsic [9] de l'an 1705. qu'on expliquoit malignement, comme si elles disoient que Mr. *Newton* l'avoit prise de moi, quoiqu'il n'y ait pas un mot qui le dise. Là-dessus on m'accusa par une espèce de retorsion prétendue, d'avoir plutôt appris la chose de Mr. *Newton.* On porta la Societé Royale de Londres à donner commission à certaines personnes d'examiner les vieux Papiers sans m'en donner aucune part, & sans savoir si je ne recuserois point quelques Commissaires, comme partiaux. Et sous prétexte du rapport de cette Commission, on publia un Livre contre moi en 1711. sous le titre de *Commerce Epistolique,* où l'on inséra des vieux Papiers, & des anciennes Lettres, mais en partie tronquées; & on supprima celles qui pouvoient faire contre Mr. *Newton.* [10] Et ce qui est le pis, on y ajouta des Remarques pleines de faussetez malignes, pour donner un mauvais sens à ce qui n'en avoit point. Mais la Societé Royale n'a point voulu prononcer là-dessus, comme j'ai appris par un Extrait de ses Registres. [11] & plusieurs personnes de mérite en Angleterre (même des Membres de la Societé Royale) n'ont point voulu prendre aucune part à ce qui s'est fait contre moi.

On me manda la nouvelle de la publication de ce Livre, avant que le Livre me fût rendu; & ayant su qu'on en avoit envoyé un Exemplaire au célèbre Mr. *Jean Bernoulli,* qui connoissoit à fond l'invention dont il s'agissoit, & l'avoit fait valoir mieux que personne par de belles découvertes, & qui étoit tout-à-fait impartial, je le priai de m'en dire son sentiment. Il répondit par une Lettre datée de Bâle le 7. de Juin 1713. dont voici l'extrait traduit en François. [12]

"Il paroît que Mr. *Newton* a fort avancé par occasion le Doctrine des *Series,* en se servant de l'extraction des Racines, qu'il y a employée le premier. Et il semble qu'il y a mis toute son étude au commencement, sans avoir songé à son *Calcul des Fluxions,* ou *des Fluants,* ou à la réduction de ce calcul à des opérations analytiques générales en forme d'Algorithme ou de règles Arithmétiques ou Algébraïques. Ma conjecture est appuyée sur un indice très-fort. C'est que dans toutes les Lettres du *Commerce Epistolique,* on ne trouve point la moindre trace, ni ombre des Lettres comme x, ou y, pointeés d'un, deux, trois ou plusieurs points mis dessus, qu'il employe maintenant à la place de dx, ddx, $dddx$; dy, ddy, $dddy$ &c. Et même dans l'Ouvrage des Principes Mathématiques de la Nature, où il avoit si souvent occasion d'employer son Calcul des Fluxions, il n'en dit pas un mot, & on ne voit aucune de ces marques; & tout s'y fait par les lignes des figures, sans aucune certaine Analyse déterminée, mais seulement d'une maniére qui a été employée non-seulement par lui, mais encore par Mr. *Huygens,* & même en quelque façon par Torricelli, Roberval, Cavallieri, & autres. Ces lettres pointées n'ont paru que dans le 3. Volume des Oeuvres de Mr. *Wallis*, plusieurs années après que le Calcul des Différences fut déja reçu partout. Un autre indice qui fait conjecturer que le Calcul des Fluxions n'est point né avant celui des Différences, est que la véritable maniére de prendre les Fluxions, c'est-à-dire, de différencier les différences, n'a pas été connue à Mr. *Newton*. C'est ce qui est manifeste par ses *Principes Mathématiques*, où non-seulement l'accroissement constant de la grandeur x, qu'il marqueroit à présent par un point, est marqué par un o; mais même une fausse règle est donnée pour les degrez ultérieurs des différences. Par où l'on peut juger, qu'au moins la véritable maniére de différencier les différences ne lui a point été connue, quand elle étoit déja fort en usage auprès d'autres.''

On publia cette Lettre : & je crus avec raison pouvoir opposer ce jugement à celui de tous ceux qui pourroient approuver la maniére, dont on en avoit usé contre moi dans la publication du Livre intitulé *Commerce Epistolique*. Et ne pouvant répondre à ce Livre de point en point, & en même tems d'une maniére digne de moi, sans faire un autre Livre plus gros, & y employer bien du tems, parce que j'aurois tâché de faire sentir aux Adversaires, qu'il leur manque encore quelque chose, je m'en dispensai pour lors, ayant des occupations plus nécessaires & presque indispensables; outre que j'avois appris d'ailleurs que leur Livre ne faisoit guère d'impression dans le Monde. Mais il est arrivé par hazard, qu'on leur donne un Os à ronger. Mr. *Bernoulli* avoit proposé un Problême dans un Journal de Leipsic; & on y insista dans un autre Journal de Leipsic, pour ceux qui méprisent nos Méthodes & s'en font accroire. On l'a répété mainténant pour ceux qui prétendent que la Méthode des Fluxions de Mr. *Newton* leur suffit. Quelques-uns s'y sont appliquez; mais on ne croit pas jusqu'ici qu'ils en viennent aisément à bout.

Enfin depuis quelques semaines Mr. *Newton* a paru lui-même contre moi par une Lettre écrite à un Ami.[13] Elle m'a été communiquée; & j'ai fait passer ma Réponse par la voye de France, afin que ce qui se fait entre nous soit connu des personnes neutres & intelligentes.

Voilà, Madame, l'Histoire de notre controverse, qui ne peut manquer de vous ennuyer. Mais on ne pouvoit pas vous en donner une pleine information sans s'étendre; & on ne sauroit éviter que les Juges ne bâillent quelquefois, quand ils ont à prendre connoissance de Procès aussi longs & aussi grands que le nôtre. Mais si les Spectateurs ne s'ennuyoient pas, ils prendroient trop de plaisir : ils se divertiroient à nos dépens. Pour moi, je ne veux pas me mettre en colére pour vous faire rire, vous & vos Amies. Vous pouvez faire (sans comparaison) comme un Cordonnier à Leyde, dont j'ai mis autrefois l'Histoire dans une Epigramme Latine. Quand on disputoit des Thèses à l'Université, il ne manquoit jamais de se trouver à la Dispute publique. Enfin quelqu'un qui le connoissoit lui demanda s'il entendoit le Latin? *Non*, dit-il, *& je ne veux pas même me donner la peine de l'entendre.* "Pourquoi venez-vous donc si souvent dans cet Auditoire, où l'on ne parle que Latin? *C'est que je prends plaisir à juger des coups.*" Et comment en jugez-vous sans savoir ce qu'on dit? *C'est que j'ai un autre moyen de juger, qui a raison.* "Et comment?" *C'est que quand je vois à la mine de quelqu'un qu'il se fâche, & qu'il se met en colére, je juge que les raisons lui manquent.*"

Il vous seroit aisé, Madame, à vous & à votre Correspondante [14] de savoir la Géométrie, aussi-bien que nous; mais vous ne voulez pas en prendre la peine, & vous en voulez cependant juger en vous divertissant. Quoi qu'il en soit, je crois avoir fourré ici de quoi vous donner le moyen de prendre connoissance du point historique, sans que vous ayez besoin d'entrer dans la Géométrie; & du moins vous me trouverez assez résolu, & assez gai pour un plaideur. Les Lettres que nous échangeons Mr. *Newton* & moi, vous instruiront du reste. Je suis avec respect,

MADAME,

Votre, &c.

NOTES

(1) Sophia Charlotte (*c.* 1673–1725), Countess von Platen and Hallermund, married Johann Adolph, later Baron von Kilmansegge. He was Master of the Horse to George I and was also marginally involved in the calculus dispute; see Letter 1187, note (1), p. 288. Her mother had been mistress to Ernst August, father of George I, whose mistress she herself was. She came to England with the King in 1714 and was created Countess of Darlington in 1722. From her bulk she was known as the 'Elephant and Castle'.

In Des Maizeaux, *Receuil* the present letter is followed (pp. 47–52) by the undated 'Apostille d'une Lettre de Mr. Leibniz a Mr. Le Comte de Bothmer' which traverses in a more peevish tone much the same ground as the present letter, but dwells more at length on the iniquity of Newton's supposition that the celebrated passage in the *Acta Eruditorum* for January 1705: 'Pro differentiis igitur Leibnitianis D. Newtonus adhibet, semperque adhibuit Fluxiones ...' implied that fluxions were *ex post facto* modelled on differentials.

(2) *De arte combinatoria* (Leipzig, 1666).

(3) Leibniz had been in correspondence with Oldenburg since 13 July 1670; Leibniz's visit was the occasion of their first meeting and development of personal liking.

(4) Compare Leibniz to Oldenburg, 3 February 1672/3 (Hall & Hall, *Oldenburg*, IX, pp. 438–43), with reference to Boyle, Pell and Mouton.

(5) Nicolaus Mercator, *Logarithmotechnia* (London, 1668).

(6) Leibniz was elected Academician in 1675, and *Associé Étranger* on 28 January 1699 N.S. after the reorganization.

(7) Duke of Brunswick-Lüneberg (1625–79). Leibniz moved to Hanover, the ducal seat, in 1676. The duchy became an electorate in 1692.

(8) See Letter 1187, note (10), p. 289.

(9) The *Acta Eruditorum*.

(10) This is, perhaps, the only assertion in this letter which is wholly unworthy of Leibniz.

(11) See Letter 1092.

(12) For the second time, Leibniz admits Bernoulli's authorship of this statement, which he had promised to conceal.

(13) Letter 1187.

(14) 'Madame de Pölnitz, à Hanover [Des Maizeaux's note].'

1204 LEIBNIZ TO ARNOLD
17 APRIL 1716
From the copy in the British Museum[1]

Hanover ce 28 dAvril 1716 [N.S.]

J'ay receu la lettre de M. Newton a M. l'Abbè Conti et j'y ai deja rèpondu.[2] Mais jai envoie ma reponce en France a M. [Nicolas] Remond a Paris, Ami et correspondent de M. l'Abbè Conti pour la faire tenir à cet Abbè, dont il m'avoit envoiè autre fois une lettre. Et je le fais afin que toute notre dispute soit communiquèe a des Connoisseurs a Paris. Monsieur l'Abbè Conti ne m'avoit point envoie de solutions, apparemment parce quil aura appris par ma lettre quelles ne sont point ce quon demande. Je luy ai envoiè un exemple special du probleme general, qu'il pourra proposer a resoudre a ses nouveau amis.

NOTES

(1) MS. Birch 4281, fos. 12–14. The extract, in Arnold's hand, was sent by Arnold to Chamberlayne on 16 May 1719 (see vol. VII), as was a similar extract from Leibniz's letter to him of 25 May 1716.

(2) That is, Newton's Letter 1187 and Leibniz's reply, Letter 1197.

1205 CUNNINGHAM TO NEWTON
20 APRIL 1716
From the original in King's College Library, Cambridge[1]

Sr Isac Newton

Venice ye 1st May 1716

Hon[oure]d Sr

In Febry last I wrot to you,[2] yt I had shewn the Abstract of the Commercium to the Marqs Polleni and Dr Micolotti,[3] who are soe good judges of your merit, as to have a singular esteem for you, Polenni is an Ingenious sinceer worthy honest Gent, is very sensible that ye commercium is unanswerable, The frivolous reply [4] yt was made to it 3 years agoe by some of Mr Libnitz friends, confirms him more and more in yt opinion. And he had the same observations on yt sheet, yt truelie you and I had, Monsigr Bianchini [5] owns to me the receipt of a Coppy of the Abstract of ye Commercium yt I sent to him by My Ld Harrold [6] He gives his most humble service to you, and owns the obligations he is under to you, but had not then read the Abstract, but when he has, I shall ask him his sentimt of it, I think to goe to Padua for a few days next week, where I shall see Pollennus, he tells me yt your book of principles has inflamed about 20 or 25 of his acquaintance into the studie of Nature and Mathematicks, and yt they altogether follow your way. I find all yt speak of you, wch are many, have a true sentiment not only of your sublime learning, but alsoe of your solid judgement and candure, and these they say the[y] draw from your way of writing. I have been talking to some of the scatches, [7] you have been pleased to communicat to me in conversation, of chronologie, Every body says tis a pity you doe not put them in writing and I am Sr Isac, of their mind. I wish it wer possible for you sr to get the better of your inflexible Modestie,

Sr In my former to you I told you that I intended to ease you of the trouble of keeping for me the 400 pd Exchqr Notes and desired you might sell them in March Last, and yt I would Order Mr Thomas Williams Marcht in London to call at your house for them or the money and take his receipt for your acquitance. I have not heard from him on that head, wch makes me think the l[ette]r has been miscarryed, soe perhaps has yours, please to let me know if it [is] soe, or if he called at your house, I shal write to him the next week to mind it If you write to me give it in to Mr Pringle Undersecritary at Mr Secritary Stanhops office, [8] And I am wth great respect

Sr   Your most obedt hble servt and friend

ALEXR CUNNINGHAM

NOTES

(1) Keynes MS. 141(B); microfilm 1011.25.
(2) Letter 1183.
(3) Presumably Pietro Antonio Michelotti (1673–1740) who practised medicine in Venice, and was a member of the Medical College there. He was elected Fellow of the Royal Society in 1718. For Giovanni Poleni see Letter 1183, note (3), p. 280.
(4) Presumably the *Charta Volans*, Number 1009.
(5) See Letter 1183, note (2).

(6) Anthony Grey (1695–1723), eldest son of the Duke of Kent, who predeceased his father. The Duke of Kent had been created Earl of Harold in 1706.

(7) Sketches. About this time Newton gave Conti permission to make a copy of his draft essay on ancient chronology for the private use of the Princess of Wales. Conti, on his return to France broke trust with Newton, and showed the manuscript to a number of friends, including Fréret, an antiquary, who had it translated into French, and added his own critical commentary. The manuscript was eventually printed in 1725 as *Abrégé de Chronologie de M. le Chevalier Newton, fait par lui-même, et traduit sur le manuscrit Anglais.* Newton was annoyed, (although the printer, Cavelier, had given him warning of their impending publication, which Newton chose to ignore) and wrote a reply, 'Remarks on the Observations made on a Chronological Index of Sir Isaac Newton', *Phil. Trans.* **33**, no. 389 (1725), 315. See Brewster, *Memoirs,* II, pp. 301–12, and Letters of 30 April 1724, 9 March 1725 and 27 May 1725, vol. VII.

(8) Robert Pringle was Undersecretary of State for Scotland; for James Stanhope, Secretary of State, see Letter 1249, note (1), p. 401.

1206 BROOK TAYLOR TO NEWTON
22 APRIL 1716
From the original in the Library of the Royal Society of London[1]

Sir,

The great loss to our Family of my good Dear Mother has made it necessary for me to make hast home, and I find the circumstances of our Family will not suffer me to be in Town before the rising of the Royal Society; wherefore I am under a necessity to beg the favour of You, Sir, to excuse me for not attending you in Crane Court, and that you will be pleased to get Mr Desaguliers,[2] or some other Person to do the Secretaries business at the Meetings of the Society; and I hope I shall another time have an opportunity of making the Society some amends for my present absence.

Upon my coming to London on Tuesday night I found a letter from Mr Monmort dated the 31 March N:S: wherein he gives me the following account of what pass'd at the French Academy relating to Dr Keils Paper, which it seems they don't care to print.[3]

Le plus grand nombre . . . au nom de la Societé Royalle.[4]

These are Mr Monmorts own words which I thought it my duty to communicate to you, not knowing what sort of an account Mr Fontenelle may have given in his letter to Dr Halley.[5] Mr Monmort in all his letters to me seems to take a particular pleasure in expressing the great respect he has for you, Sir, and in one of his last he tells me he has sent to me a hamper of Champagne wine, and begs your acceptance of 50 bottles of it.[6] I can send it from hence either by Land carriage or by Water, If you will be pleased to let me know whither I shall direct it I will send it assoon as it comes to my hands. Pray, Sir, do me the favor to make my most humble service acceptable to Mrs Barton

I am

Sir

Your most faithful

and most obedient servant

BROOK TAYLOR

Bifrons near Canterbury [7]
22 *April* 1716
To
    The Honble Sir Isaac Newton
at his house in St. Martin's Street near Leister Fields

London

NOTES

(1) MS. MM 5.49, printed in Brewster, *Memoirs*, II, pp. 509–10.

(2) Jean Théophile Desaguliers (1683–1744), the 'Curator and Operator of Experiments' for the Royal Society, took the minutes of the meetings from 12 April to 17 May as a result of the absence of the two secretaries.

(3) The paper referred to here is Keill's 'Defense', eventually published in the *Journal Literaire de la Haye*, **8** (Part II, 1716), 418–33; see Letter 1165, note (2), p. 246.

(4) Taylor here quotes the passage we have printed in Letter 1194.

(5) A copy of this letter of 8 March [N.S. presumably], in Keill's hand, was seen by Edleston (*Correspondence*, p. 187, note ∗) but cannot now be found. He quotes the sentences: 'Nous ne cedons point ici aux Anglois meme en estime et en veneration pour Mr Newton. Et l'Academie voudroit fort qu'il fust possible' for it to print Keill's paper, but it was not.

(6) This gift, intended in fact for Catherine Barton, is mentioned in Monmort's letter to Taylor of 12 April 1716 N.S. printed in Taylor's *Contemplatio Philosophica*, pp. 93–5 and partially in Brewster, *Memoirs*, II, p. 491. It is in this letter that Monmort, with the height of gallantry, expatiates extravagantly on that lady's charms and capacities.

(7) The home of Taylor's parents.

1207 FLAMSTEED TO NEWTON
23 APRIL 1716
From the holograph original in the University Library, Cambridge [1]

The Observatory Aprill 23. [*Monday*] 1716

Sr

    Pray return me by ye Bearer my servant Joseph Crosthwait my 4[t]o MS of night notes from Nov: 1678 to to [*sic*] Feb 1684: which it seems was not at hand when you returnd those of ye preceedeing & following yeares

333

With ye same I desire you to returne also the 175 MS sheets of Observations made with my Mural Arch which were trusted into your hands March 20. 1707/8

Togeather with so much of ye Copy of my Catalogue as was delivered into yr hands sealed up at your own request with my *Originall Copys* of what is printed, which have not been yet returned.

And if with them you send me the Copper plates belonging to what is printed you will oblige me & prevent farther trouble from Sr

Your humble Servant

JOHN FLAMSTEED   MR

My man has your receipt for ye 4to book of Night which on ye delivery of ye book to him he will return to you & give you Receipts for what other things you send by him

JF

NOTE

(1) Add. 4006, no. 28; there is an amanuensis copy in Flamsteed MSS. vol. 53, fo. 108. The manuscripts Flamsteed mentions here are those discussed in Letters 1148 and 1151.

1208   WOLF TO J. BERNOULLI
29 APRIL 1716

Extract from an article by Johann [III] Bernoulli in the Berlin *Mémoires*.[1]
Reply to Letter 1196

Quod me ea docere volueris quæ circa historiam calculi integralis imo et differentialis hactenus ignota mihi fuere, grata mente agnosco, dataque occasione tam publice quam privatim cum aliis communicabo. . . . Quæ [2] de nugatore Keilio prolixe scribis, omnia perplacent; cumque in illorum gratiam, quibus de stricturis ejus judicare non datum est opportunum judicem ut fastus ejus retundatur, Epistolam integram cum Leibnitio [3] te non invito communicabo, et ipsius consilio decernam, qua forma et quo habitu in publicum prodire debeat. Sane homo insulsus est perfrictæ adeo frontis &c.

Translation

What you were pleased to teach me about the history of the integral calculus and indeed the differential calculus also, was until then unknown to me, as I gratefully admit; given opportunity, I shall impart it to others both publicly and privately . . . All that [2] you write in detail about that worthless fellow Keill pleases me; and since I judge it opportune that Keill's arrogant pride should be deflated, for the satisfaction

334

of those who are not [themselves] able to judge his criticisms, I shall (as you are agreeable) send the whole letter to Leibniz [3] and reveal [to you] his opinion, about the form and dress in which it should appear before the public. Surely that bungling man is so shameless . . .

NOTES

(1) 'Anecdotes pour servir à l'Histoire des Mathématiques par M. Jean [III] Bernoulli', *Histoire de l'Académie Royale des Sciences et Belles-Lettres* for 1799 and 1800 (Berlin, 1803), pp. 44–5. See also Letter 1196, note (1), p. 303.

(2) That is, the contents of the 'Epistola pro Eminente Mathematico'; see Letter 1196.

(3) Bernoulli had asked Wolf to ensure that Leibniz had approved the 'Epistola pro Eminente Mathematico' before publication.

1209 COTES TO JONES
5 MAY 1716
Extract from Rigaud, *Correspondence*, I, pp. 272–4 [1]

I must now beg your assistance and management in an affair, which I cannot so properly undertake myself, especially by letters. You know Sir Isaac has left his sixth form imperfect, and under a limitation.[2] This, though it does not lessen my opinion of the author, yet it appears as an eyesore to me in so beautiful a work. The very great respect and honour, which is due to him upon all accounts, makes me wish it were removed by himself. Pray therefore let him know that I can take off that limitation, and make this form as perfect as the others. And use all the address you have to make him set upon the same thing, and let me know that he has done it. My design is to mention it in my treatise;[3] that he hearing from you what I had done, did himself, at your request, reconsider his sixth form, and very easily made it perfect.

I here send you a transcript of what I have set down for my memory about it. You see I happened to alter the appearance of the form a little putting $\frac{d\dot{z}z^{\theta\eta-1}}{e+fz^{2\eta}+gz^{4\eta}}$ instead of $\frac{d\dot{z}z^{\theta\eta+\frac{1}{2}\eta-1}}{e+fz^{\eta}+gz^{2\eta}}$, but this makes no real difference as to the matter in hand. There are three ways (and I think there can be no more) of resolving the fluxion proposed into fluxions of other forms. The first way resolves it into two fluxions of Sir Isaac's second form; this way is coincident with that which he has taken. The other two ways do each resolve it into four fluxions of Sir Isaac's fifth form. By collecting the several parts, and substituting the values of p, q, r, s, t, you will very easily be satisfied of the truth of these rules.

335

Positis $p = \frac{1}{2}f + \sqrt{\frac{1}{4}ff - eg}$ $q = \frac{1}{2}f - \sqrt{\frac{1}{4}ff - eg}$

$r = \sqrt{eg}$ $s = \sqrt{-fg + 2\sqrt{eg^3}}$ $t = \sqrt{-fg - 2\sqrt{eg^3}}$

I. Erit $\dfrac{d\dot{z}z^{\theta\eta-1}}{e+fz^{2\eta}+gz^{4\eta}} = \dfrac{g}{q-p} \times \dfrac{d\dot{z}z^{\theta\eta-1}}{p+gz^{2\eta}} + \dfrac{g}{p-q} \times \dfrac{d\dot{z}z^{\theta\eta-1}}{q+gz^{2\eta}}$

II. Vel $= \dfrac{g}{2r} \times \dfrac{d\dot{z}z^{\theta\eta-1}}{r+sz^{\eta}+gz^{2\eta}} + \dfrac{gg}{2rs} \times \dfrac{d\dot{z}z^{\theta\eta+\eta-1}}{r+sz^{\eta}+gz^{2\eta}} + \dfrac{g}{2r} \times \dfrac{d\dot{z}z^{\theta\eta-1}}{r-sz^{\eta}+gz^{2\eta}}$

$\qquad\qquad - \dfrac{gg}{2rs} \times \dfrac{d\dot{z}z^{\theta\eta+\eta-1}}{r-sz^{\eta}+gz^{2\eta}};$

III. Vel $= \dfrac{g}{2r} \times \dfrac{d\dot{z}z^{\theta\eta-1}}{r+tz^{\eta}-gz^{2\eta}} - \dfrac{gg}{2rt} \times \dfrac{d\dot{z}z^{\theta\eta+\eta-1}}{r+tz^{\eta}-gz^{2\eta}} + \dfrac{g}{2r} \times \dfrac{d\dot{z}z^{\theta\eta-1}}{r-tz^{\eta}-gz^{2\eta}}$

$\qquad\qquad + \dfrac{gg}{2rt} \times \dfrac{d\dot{z}z^{\theta\eta+\eta-1}}{r-tz^{\eta}-gz^{2\eta}}.$

Si e, g, sunt signis dissimilibus, solvitur per I.

Si e, g, sunt signis similibus, et f dissimili, solvitur per II.

Si e, f, g, sunt signis similibus, et $ff > 4eg$, solvitur per I.

Si e, f, g, sunt signis similibus, et $ff < 4eg$, solvitur per II.

Si e, g, sunt signis similibus, f dissimili, et $ff > 4eg$, solvitur per I, II, III.

Your's, &c.

R. Cotes.

Be pleased to communicate the contents of this letter to Sr. Is. only. I have some thoughts of touching Mr. L., and would not have his agents inform him upon what points; which might produce something to hinder my design.

NOTES

(1) This is probably the last extant letter from Roger Cotes, who died on 5 June. In the earlier part of the letter, dealing with novel methods of integration, Cotes claims ability to integrate 'by measures of ratios and angles', the expression

$$\frac{z^{\theta\eta + \frac{\delta}{\lambda}\eta - 1}}{e + fz^{\eta} + gz^{2\eta} + hz^{3\eta}}$$

where θ and δ are integers, and λ a whole power of 2. The integration of the expression was later proposed to the continental mathematicians for solution, and this first part of Cotes' letter was published in *Phil. Trans.*, **32**, no. 372 (1722), 139–50; (see also Edleston, *Correspondence*, pp. 230–5). The whole letter is also discussed by Robert Smith in his preface to Cotes' tables of integrals; see Roger Cotes, *Harmonia Mensurarum* (Cambridge, 1722), pp. 113–19.

(2) Cotes here refers to the sixth form in Newton's revised table of integrals first published in his *Tractatus de quadratura curvarum*, in appendix to his *Opticks* of 1704. This had been the fourth form in Newton's 1671 tract on series and fluxions; see Whiteside, *Mathematical Papers*, III, p. 256, note (549). There Newton integrated expressions of the form

$$\frac{z^{\theta\eta+\frac{1}{2}\eta-1}}{e+fz^\eta+gz^{2\eta}}$$

for the cases $\theta = 0$ and $\theta = 1$, by first reducing them to terms corresponding to the second form, that is, $\dfrac{z^{\theta\eta+\frac{1}{2}\eta-1}}{a+bz^\eta}$.

However, as he wrote to Leibniz in the *Epistola Posterior* (see vol. III, p. 119), the reduction was valid only for $f^2 \geqslant 4eg$ (since in other cases a and b involve imaginary quantities). As Cotes says, there is no difficulty in removing the limitation by performing the reduction in a different way.

(3) Cotes refers to his *Harmonia Mensurarum*, eventually published posthumously by Robert Smith in 1722. Very extensive tables of integrals are given on pp. 121–247. Cotes' Form 30, on pp. 160–1, is equivalent to a generalized version of Newton's Form 6, where θ can be any integer.

1210 J. BERNOULLI TO LEIBNIZ
9 MAY 1716

Extract from Gerhardt, *Leibniz: Mathematische Schriften*, III/2, pp. 960–2.
Reply to Letter 1201; for the answer see Letter 1213

Bene se habet, quod Newtonus ipse tandem in arenam descenderit, pugnaturus sub proprio suo nomine, et seposita larva.[1] . . . Quidquid sit, spero nunc veritatem historicam melius detectum iri, siquidem Newtonus pro suo, quem habere suppono et confido, candore res gestas fideliter enarrabit, eorumquæ que a Te producentur, veritatem publice agnoscet.

Ex Gallia nihil aliud intellexi, nisi quod a Domino Monmortio (Frater ni fallor Remondi, qui Tecum in Commercio est) Agnatus meus literas accepit,[2] in quibus aliquid de ea re commemorat confirmatque idem, quod de Contio habes, qui nimirum perscripsit, Anglos exemplum Hyperbolarum ad angulos rectos secandarum solvisse; quod equidem non miror, quia exemplum facillimum et a Filio quoque meo solutum.[3] Addit vero, Anglos jactare generalem solutionem problematis,[4] sed dubito solutionem fore talem, qualem desideramus. Nam si per *solvere* intelligant exhibere utcunque aliquam æquationem differentialem ex indeterminatis invicem permixtis complicatam, eorum solutio non erit perfecta, quia nulla constructio inde deduci potest, concessis licet figurarum quadraturis.[5] Ideoque cum forte pro solutione exempli postremo a me transmissi dederint aliquam æquationem, in qua indeterminatas cum suis differentialibus sibi mutuo intricatas observaveris, nullo modo acquiescendum erit, sed insistendum et urgendum, ut quod superest absolvant, separentque a se invicem indeterminatas, sine quo gloriari non queunt: se solvisse exemplum, hoc est enim ex eorum numero, ubi separatio ita succedit, si rite tractentur, sed hic singulari quadam arte opus est. Taylorum, subjungit Montmortius,[6] etiam solvisse prius exemplum Hyperbolarum, tum etiam

jactasse, se quoque habere exempla quædam, quæ Tibi vicissim propositurus
sit, refertque unum a Tayloro perscriptum, quod hoc est: *Determinare Tra-*
jectorias normaliter secantes curvas, quibus respondet hæc æquatio zzddx = 2xdz²,
in qua supponitur dz *constans; et definire naturam harum curvarum, quibus ista æquatio*
respondet. Post brevem applicationem solvi hoc problema, et postea etiam solvit
Agnatus meus, amboque invenimus, quod triplex curvarum species satisfaciat
æquationi isti zzddx = 2xdz², nempe Parabolas, Hyperbolas et Curvas quasdam, quæ sunt trium dimensionum. Quod si nunc Parabolæ illæ normaliter
sunt secandæ, erit Trajectoria aliqua Ellipsis: si vero Hyperbolæ illæ, habebi-
mus pro Trajectoriis alias Hyperbolas prioribus similes et concentricas;
si Curvæ illæ trium dimensionum sint secandæ, erit etiam Trajectoria altioris
ordinis: sic itaque pars posterior problematis triplicem solutionem admittit:
prior vero etiam triplicem, nisi per conditionem aliquam, qualem Taylorus
addit (quam vero non satis intelligo) restringatur ad unam aliquam simpli-
cem.[7]

Miror quomodo Newtonus scire potuerit, me auctorem esse Epistolæ
illius,[8] quam inseri curasti chartæ illi contra Newtonum publicatæ, cum
tamen nemo mortalium sciverit, me illam scripsisse, nisi Tu, ad quem scripta
est, et ego, a quo scripta est. Fortassis autem expressio ista: *par un Mathéma-*
ticien ou prétendu Mathématicien, alium habet sensum, quam putas: potest enim
etiam ita sumi, quasi Newtonus crediderit, Epistolam istam esse suppositiam
et tanquam a Mathematico quodam conficto exaratam, revera tamen ab
ipso Auctore chartæ inventam et intrusam; quod si rem ita sumas, videbis
per *un prétendu Mathématicien* intelligendum esse Mathematicum confictum et
nunquam existentem. Vellem Arnoldus locum indicasset in Transactionibus,
ubi Keilius dicet, me quoque ignorare Calculum differentialem:[9] interim
parum me moveret, quod Keilius ex ira furiosus contra me deblaterat; etsi
crederem, verum esse quod Arnoldus retulit, sed cum nec ex Gallia, nec
aliunde simile quid audiverim, Arnoldus forte deceptus est, eo quod intellexerit Keilium alicubi dicere, me usum Serierum convergentium Newtoni
non satis intelligere. Alias enim Keilius,[10] si me dicere vellet Calculi differentialis ignarum, sibimet ipsi turpiter contradiceret, quippe qui in Diario
Gallico Hagiensi,[11] ubi contra Chartam illam, de qua supra, disputans ad
me provocat tanquam ad Judicem idoneum et Calculi differentialis callentissi-
mum, quique adeo quam optime decidere possim, annon problema illud:
Data area curvæ, invenire ejus applicatam (quod Newtonus jam dudum solverit)
idem sit, quam hoc: *Datæ quantitatis invenire differentialem.* Præterea alia in me
cumulat elogia, quæ omnia ejus sunt naturæ, ut me necessario Calculi
differentialis peritissimum crediderit, adeo ut vel calumniator, vel mente cap-
tus censendus esset, si nunc contrarium diceret.

Translation

It is a good thing that Newton has at last entered the ring himself, in order to fight under his own name,[1] and laid aside his mask . . . Whatever it may be, I hope now the historical truth will be more clearly discovered, if only Newton will, with that candour which I suppose and trust him to possess, tell faithfully the things which have happened, and will publicly acknowledge the truth of what you have put forward.

I have learnt nothing from France, except that my nephew has received a letter from Mr Monmort[2] (the brother, if I am not mistaken, of [Nicolas] Rémond, who is in correspondence with you) in which he mentions something about the matter and confirms what you had from Conti, who doubtless communicated it to him, that the English have solved the case of the Hyperbolas cutting at right angles; which does not of course surprise me, because it is the easiest case, and has also been solved by my son.[3] Indeed, he adds that the English boast of a general solution of the problem,[4] but I doubt it will be such a solution as we wish for. For if they understand by *solving* the presenting of some differential equation or other, complicated by mutually intermixed indeterminates, their solution will not be perfect, since no construction can be deduced from it, even if the quadrature of figures be granted.[5] Therefore since they might perhaps give as a solution of the example lately sent [them] by me, some equation in which you will observe the indeterminates muddled up with their own differentials, that is by no means to be accepted but it must be insisted and pressed upon them that they are to finish what remains to be done, and separate the indeterminates from each other, without which [step] they are in no position to boast that they have solved the case; for this [case] is one of those where the separation succeeds if they carry it out rightly. But this needs some special method. Monmort adds[6] that Taylor has also solved the first case of the hyperbolas, and even boasted that he also has some problems which he will propound to you in exchange, and he [Monmort] reports one [problem] written out by Taylor, which is this: *To determine the trajectories cutting normally curves which correspond to the equation* $z^2 . d^2x = 2x . dz^2$, *in which* dz *is considered constant; and to define the nature of the curves to which this equation itself corresponds.* After a brief consideration I have solved this problem, and later, indeed, my nephew solved it, and we both discovered that a triple series of curves satisfies the equation itself, $z^2 . d^2x = 2x . dz^2$, namely, parabolas, hyperbolas, and certain curves of third degree. But if now these [curves] to be cut normally are parabolas, the trajectory will be a certain ellipse: if indeed they are hyperbolas, we will have other hyperbolas, similar to and concentric with the first, as trajectories; if the curves to be cut normally are those of third degree, the trajectories too will be of a higher order: thus the latter part of the problem allows a triple solution; indeed even the first [part] is triple, unless by some condition such as Taylor adds (which I do not properly understand) it is restricted to some simple [case].[7]

I wonder how Newton could have known that I was author of that letter,[8] which you took care was included in that paper published against Newton, when actually no living soul knew that I had written it except you to whom it was written, and I, by

339

whom it was written. Perhaps, however, that very expression: *par un Mathématicien ou prétendu Mathématicien* has a meaning other than you think: for it could even be read as if Newton had believed that the letter itself was not genuine, but written as the work of some fake mathematician, when really invented and inserted by the author of the paper himself, because if you look at the matter in this way, you will see that by *un prétendu Mathématicien* is to be understood a fake mathematician, [one who] never existed. I wish Arnold had indicated the place in the [*Philosophical*] *Transactions*, where Keill says that I also am ignorant of the differential calculus:[9] meanwhile, it would disturb me very little that Keill, raging with anger, should babble furiously against me; even if I believed what Arnold reported to be true, but since neither from France nor from any other place have I heard anything similar, perhaps Arnold is deceived because he has understood Keill to say somewhere [10] that I do not sufficiently understand Newton's use of converging series. For otherwise Keill, if he means to say that I am ignorant of the differential calculus, disgracefully contradicts himself, for it is certainly he who, in the French *Journal* [*Literaire*] [11] of the Hague, when disputing against the paper mentioned above, challenges me (as though I were a capable judge, and one highly experienced in the differential calculus) as one well qualified to be able to decide whether that problem: *Given the area of a curve, to determine its ordinate* (which Newton had solved long ago) is not the same as this: *To determine the differential of a given quantity*. Besides, he heaps other praises on me, all of them of such a stamp that he must necessarily have believed me most skilled in the differential calculus, so that he must be judged either a slanderer or fickle-minded if he should say otherwise.

NOTES

(1) Bernoulli refers to Newton's correspondence with Conti.

(2) Nikolaus Bernoulli had been in correspondence with Pierre Rémond de Monmort for some years; some of their early letters, on probability theory, were already published as an appendix to the second edition of Monmort's *Essay d'analyse sur les jeux de hazard* (Paris, 1713). Nikolaus also wrote directly to Leibniz concerning the letter from Monmort discussed here (see Gerhardt, *Leibniz: Mathematische Schriften*, III/2, pp. 992–4).

(3) See Letter 1201, note (10), p. 321, for Nikolaus [II] Bernoulli's solution.

(4) Presumably Bernoulli refers to Newton's 'solution'; see Number 1187a.

(5) That is, allowing that the quadratures may be effected once the differential equations are put into directly quadrable form.

(6) Nikolaus quoted directly from Monmort's letter in his own letter to Leibniz, dated 24 October 1716 N.S.: (see note (2) above). Taylor's problem was there worded slightly less generally, thus: 'Invenire curvam quæ per datum punctum transeat, et ad angulos rectos secet curvas omnes, per aliud punctum transeuntes, et expressas per æquationem $zzddx = 2xdz^2$, nempe z fluente uniformiter.' ('To find the curve which passes through a given point, and cuts at right angles all curves passing through a different point, and expressed by the equation $z^2 . d^2x = 2xdz^2$, where z is uniformly flowing.')

Taylor clearly saw this as a problem similar in style to the one Bernoulli had set—the curves to be cut are not given completely, but in terms of a differential equation.

(7) Bernoulli's conclusions are of course correct; the general solution of the equation is

$x = az^2 + b/z$. Unless we impose some further condition on the parameters a and b, there is no simple general solution to the problem of finding the curves orthogonal to these. The condition Taylor adds (see note (6) above), that all the curves to be cut should pass through one point, imposes a relationship on the parameters a and b, so that $x = az^2 + (p-aq)/z$ where p, q are constant and a is the varying parameter. (The 'triple solution' Bernoulli refers to is this general conic, with the two degenerate cases, $x = az^2$ (parabolas) and $x = k/z$ (hyperbolas intersecting at infinity).) Bernoulli clearly used this condition imposed by Taylor in the solution he describes. However, his computation of the corresponding normal trajectories is in error.

(8) The excerpt of Letter 1004, printed in the *Charta Volans* (Number 1009).

(9) See Letters 1181, note (14), and 1201, note (5).

(10) In his 'Answer' in the *Journal Literaire de la Haye* for July and August 1714, p. 345 (see Letter 1053a, note (1)).

(11) *Ibid.*, p. 331.

1211 NEWTON'S 'OBSERVATIONS' ON LETTER 1197
[MAY ?1716]
From Raphson, *History of Fluxions*, pp. 111–19 [1]

OBSERVATIONS upon the preceding EPISTLE [2]

Mr. *Leibnitz* by his Letter of the 29th of *December* 1711 [N.S.]. [3] justified the Passage in the *Acta Eruditorum* for *January* 1705. *pag.* 34, and 35. and thereby made it his own, and now endeavours in vain to excuse it, pretending that the Words *adhibet semperque adhibuit* are maliciously interpreted by the Word *substituit*. But in the Interpretation which he would put upon the Place, he omits the Words *igitur* and *quemadmodum*, the first of which makes the Words, *semperque adhibuit*, a Consequence of what went before, and the latter makes them equipolent to *substituit*; neither of which can be true in the Sense which Mr. *Leibnitz* endeavours now to put upon the Words. [4] He has therefore accused me. In both his Letters to Dr. *Sloan*, (that dated the 4th of *March* [N.S.], and that dated the 29th of *December* 1711 [N.S.].) [5] he pressed the Royal Society to condemn Dr. *Keill*; and before I meddled in this matter challenged me to declare my Opinion. His Words in his second Letter are; *Itaque vestræ æquitati committo, annon coercendæ sint vanæ & injustæ* [Keillii] *vociferationes, quas ipsi* Newtonio *viro insigni, & gestorum optimè conscio, improbari arbitror; ejusque sententiæ suæ libenter daturum indicia mihi persuadeo.* The Words are civil, but the Sense is, That I must either condemn Dr. *Keill*, or enter into a Quarrel with Mr. *Leibnitz*, as has happen'd; and therefore he is the Aggressor. For it is very well known here, that I constantly endeavoured to avoid these Disputes, till they were pressed upon the Royal Society and me.

In his Letter of the 4th of *March st. n.* 1711. he pressed the Royal Society to condemn Dr. *Keill* without hearing both Parties; and when the Doctor put in his Answer, Mr. *Leibnitz* refused to give his Reasons against the Doctor, and call'd it *Injustice* to expect it from him, and yet persisted in pressing them against him, and thereby put them upon a Necessity of appointing a Committee to search out old Papers, and give their Opinion upon them. If they did it without him, it was his own Fault: He was for over ruling them, and called it *Injustice* to expect that he should defend his Candor, and plead before them. If they gave him no Opportunity to except against any of the Committee, it was because he refused to be heard, and they had a sufficient Authority to appoint a Committee without him, and he had no Right to except against what they did for their own Satisfaction. If they have not yet given Judgment against him, it is because the Committee did not act as a Jury, nor the Royal Society as a formal Court of Justice. The Committee examined old Letters and Papers, and gave their Opinion upon them alone, and left room for Mr. *Leibnitz* to produce further Evidence for himself. And it is sufficient that the Society ordered their Report, with the Papers upon which it was grounded, to be published; and that Mr. *Leibnitz* in all the three Years and four Months which are since elapsed, has not been able to produce any further Proof against Dr. *Keill* than what was then before them.

Mr. *Leibnitz* saith, That the Letter which I call *Defamatory*,[6] being no sharper than that which has been published against him, I have no Reason to complain. But the sharpness of the Letter lies in Accusations and Reflections, without any Proof; which way of Writing is unlawful and infamous, and never used but in a bad Cause. The sharpness of the *Commercium* lies in Facts which are lawful and fit to be produced. The Letter was published in a clandestine, back-biting manner (as defamatory Papers use to be) without the Name of the Author, or Mathematician, or Printer, or City where it was printed, and was dispersed above two Years before we were told that the Mathematician was *John Bernoulli*;[7] the *Commercium* was printed openly at *London* by Order of the Royal Society.

The Mathematician to whom Mr. *Leibnitz* appealed from the Royal Society, I called a Mathematician, or pretended Mathematician, not to disparage the Skill of Mr. *Bernoulli*, but because the Mathematician in his Letter of the 7th of *June* 1673,[8] cited Mr. *Bernoulli* as a Person distinct from himself; and Mr. *Leibnitz* lately caused that Letter to be reprinted [9] without the Citation, and tells us, that the Mathematician was Mr. *Bernoulli* himself: And whether the Mathematician, or Mr. *Leibnitz* is to be believed, I do not know.[10] Mr. *Bernoulli* had the differential Method from M. *Leibnitz*, and was the chief of his Disciples, and gave his Opinion for his Master in the *Acta Leipsica* before he saw the *Commercium Epistolicum*; at which Time he was *homo novus, & rerum anteactarum parum peritus*,[11] as Mr. *Leibnitz* objected against Dr. *Keill*; and what he wrote after he saw the *Commercium* was in his own Defence, and his Skill in Mathematicks will not mend the matter.

He complains that the Committee have gone out of the way, in falling upon the Method of *Series*: But he should consider that both Methods are but two Branches of one general Method of *Analysis*:[12] I joyn'd them together in my Analysis[.] I interwove them in the Tract [13] which I wrote in the Year 1671, as I said in my Letters [14] of the 10th of *December* 1672, and the 24th of *October* 1676. In my Letter of the 13th of *June* 1676,[15] I said, that my Method of *Series* extended to almost all Problems, but became not general without some other Methods, meaning (as I said in my next Letter) the Method of Fluxions, and the Method of Arbitrary *Series*; and now to take those other Methods from me, is to restrain and stint the Method of *Series*, and make it cease to be general. In my Letter of the 24th of *October* 1676, I called all these Methods together, my general Method. See the *Commercium Epistolicum, pag.* 86. *lin.* 16. And if Mr. *Leibnitz* has been tearing this general Method in pieces, and taking from me first one Part, and then another Part, whereby the rest is maimed, he has given a just Occasion to the Committee to consider the Whole. It is also to be observed, That he is perpetually giving Testimony for himself, and it's allowed in all Courts of Justice to speak to the Credit of the Witness.

He represents, That the Committee of the Royal Society have omitted Things which made against me, and printed every Thing which could be turned against him by strained Glosses; and to make this appear, he produces in his last Letter but one, an Instance of my Ignorance omitted by them, but confesses now that he was mistaken in saying that it was omitted, and saith that he will cite another Instance. He saith, That in one of my Letters to Mr. *Collins*, I owned that I could not find the second Segments of Sphæroids, and that the Committee have omitted this.[16] If they had omitted such a passage, I think they would have done right, it being nothing to the purpose. But on the contrary, Mr. *Collins* in a Letter to Mr. *James Gregory* the 24th of *Decemb.* 1670,[17] and in another to Mr. *Bertet* the 21st of *Feb.* 1671, both printed in the *Commercium Epistolicum, pag.* 24, 26. wrote, That my Method extended to second Segments of round Solids. And Mr. *Oldenbourg* wrote the same thing to Mr. *Leibnitz* himself the 8th of *December* 1674.[18] See the *Commercium Epistolicum, pag.* 39. So you see that Mr. *Leibnitz* hath accused the Committee of the Royal Society, without knowing the Truth of his Accusation, and therefore is guilty of a Misdemeanour. The Committee were so far from acting corruptly against Mr. *Leibnitz*, that they took no Notice of his Ignorance of Geometry in those Days, and omitted several other things which made strongly against him, such as were the two Letters in my Custody, and the Paragraph in the Preface to the two first Volumes of Dr. *Wallis*'s Works relating to this matter, and that a Copy of *Gregory's* Letter of the 5th of *Septemb.* 1670.[19] was sent to Mr. *Leibnitz* in *June* 1676,[20] amongst the Extracts of *Gregory's* Letters.

[21] The Committee in their Report,[22] affirmed, That they had extracted from the ancient Letters, Letter-Books, and Papers, what related to the Matter referred to them: All which Extracts delivered by them to the Society, they believed to be genuine and authentick. Mr. *Leibnitz* accuses them for not printing the Letters entire (including as well what did not relate to the matter referred to them, as what did relate to it,) as if it were not lawful to cite a Paragraph out of a Book, without citing the whole Book. Thus he complains, that the *Commercium Epistolicum* should have been much bigger. But when he is to answer it, he complains that it is too big, and would require an Answer as big as itself. And so the ancient Letters and Papers must be laid aside, and the Question must be run off into a Squabble about Philosophy and other matters: And the great Mathematician, who in his Letter to Mr. *Leibnitz*, dated the 7th of *June* 1713 [N.S.] [23] concealed his Name, that he might pass for an impartial Judge, must now pull off his Mask, and become a Party-man in this Squabble, and send a Challenge by Mr. *Leibnitz* to the Mathematicians in *England*, as if a Duel, or perhaps a Battel with his Army of Disciples, were a fitter way to decide the Truth, than an Appeal to ancient and authentick Writings; and Mathematicks must henceforward be filled with Atchievements in Knight-errantry, instead of Reasons and Demonstrations.

Mr. *Leibnitz* acknowledges, that when he was in *London* the second time,[24] he saw some of my Letters in the Hands of Mr. *Collins*, especially those relating to *Series*, and he has named two of them which he then saw, *viz.* that dated the 24th of *October* 1676, and that in which he pretends that I confessed my Ignorance of second Segments. And no doubt he would principally desire to see the Letter which contained the chief of my *Series*, and particularly that which contained those two for finding the Arc by the Sine, and the Sine by the Arc, with the Demonstration thereof, which a few Months before he had desired Mr. *Oldenbourg* to procure from Mr. *Collins*; that is, the *Analysis per æquationes numero terminorum infinitas*. But yet he tells us, that he never saw where I explained my Method of Fluxions, and that he finds nothing of it in the *Commercium Epistolicum*, where that Analysis and my Letters of the 10th of *December* 1672, 13th of *June* 1676, and the 24th of *Octob.* 1676, are published.[25] I suppose he means, because he finds no prick'd Letter there. And by the same way of arguing, he and Mr. *Bernoulli* may pretend that they find nothing of the Method of Fluxions in the Introduction to the Book of Quadratures.

He saith also, That he never saw where I explain the Method claimed by me, in which he assumes an arbitrary *Series*. If he pleases to look into the *Commercium Epistolicum, pag.* 56 and 86, he will there see that I had that Method in the Year 1676, and five Years before.[26] And Dr. *Wallis* in the second Volume of his Works, *pag.* 393. *lin.* 32. has told him, That this Method needs no further Explication than what I there gave of it. Mr. *Leibnitz* might find it himself, but not so early; and second Inventors have no Right.

He pretends, that in my Book of Principles, *pag.* 253, 254, I allowed him the Invention of the *Calculus Differentialis*, independently of my own; and that to attribute this Invention to my self, is contrary to my Knowledge there avowed. But in the Paragraph there refered unto, I do not find one Word to this purpose. On the contrary, I there represent, that I sent Notice of my Method to Mr. *Leibnitz* before he sent Notice of his Method to me; and left him to make it appear that he had found his Method before the Date of my Letter; that is, eight Months at the least before the Date of his own. And by referring to the Letters which passed between Mr. *Leibnitz* and me ten Years before, I left the Reader to consult those Letters, and interpret the Paragraph thereby. For by those Letters he would see that I wrote a Tract on that Method, and the Method of *Series* together, five Years before the writing of those Letters; that is, in the Year 1671. And these Hints were as much as was proper in that short Paragraph, it being besides the Design of that Book to enter into Disputes about these Matters.

He saith, That when he was in *London* the first Time, which was in *January* and *February* 1673, he knew nothing of infinite *Series*, nor of the advanced Geometry, nor was then acquainted with Mr. *Collins*, as some have maliciously feigned. But who hath feigned this, or what need there was of feigning it, I do not know. At that Time Dr. *Pell* gave him Notice of *Mercator's Series* for the *Hyperbola*, and he carried *Mercator's* Book with him to *Paris*, tho' he did not yet understand the higher Geometry. And any of those to whom Mr. *Collins* had communicated mine and *Gregory's Series*, might give him Notice of them,[27] without his being acquainted with Mr. *Collins*.

He saith, That after his coming from *London* to *Paris*, his first Letters were of other matters than Geometrical, till Mr. *Huygens* had instructed him in these Things; and that he found the Arithmetical Quadratures of the Circle towards the End of the Year 1683, and began to write of it to Mr. *Oldenbourg* the next Year, and found the general Method by Arbitrary *Series* a little after, and the *Differential Calculus* in the Year 1676, deducing it from the *Series* of Numbers; and that in his Letter of the 27th of *August* 1676 [N.S.], by the Words, *certa Analysi*,[28] he meant the Differential Analysis.[29] And am not I as good a Witness that I invented the Methods of *Series* and Fluxions in the Year 1665, and improved them in the Year 1666, and that I still have in my Custody several Mathematical Papers [30] written in the Years 1664, 1665, and 1666, some of which happen to be dated; and that in one of them dated the 13th of *Novemb.* 1665, the direct Method of Fluxions is set down in these Words: [31]

PROB. *An Equation being given, expressing the Relation of two or more Lines, x, y, z, &c. described in the same time by two or more moving Bodies,* A, B, C, &c. *to find the Relation of their Velocities, p, q, r,* &c.

Resolution. *Set all the Terms on one side of the Equation, that they become equal to nothing. Multiply each Term by so many times* $\frac{p}{x}$ *as x hath Dimensions in that Term.*

Secondly, *Multiply each Term by so many Times* $\frac{q}{y}$ *as y hath Dimensions in it.*

Thirdly, *Multiply each Term by so many Times* $\frac{r}{z}$ *as z hath Dimensions in it,* &c.

The Sum of all these Products shall be equal to nothing. Which Equation gives the Relation of p, q, r, &c. And that this Resolution is there illustrated with Examples, and demonstrated, and applied to Problems about Tangents, and the Curvature of Curves. And that in another Paper dated the 16th of *May* 1666, [32] a general Method of resolving Problems by Motion, is set down in Seven Propositions, the last of which is the same with the Problem contained in the aforesaid Paper of the 13th of *Novemb.* 1665. And that in a small Tract written in *Novemb.* 1666. [33] the same Seven Propositions are set down again, and the Seventh is improved by shewing how to proceed without sticking at Fractions or Surds, or such Quantities as are now called *Transcendent.* And that an Eighth Proposition is here added, containing the Inverse Method of Fluxions so far as I had then attained it, namely, by Quadratures of Curvilinear Figures, and particularly by the three Rules upon which the *Analysis per Æquationes numero terminorum infinitas,* is founded, and by most of the Theorems set down in the *Scholium* to the Tenth Proposition of the Book of Quadratures. And that in this Tract, when the Area arising from any of the Terms in the Valor of the Ordinate cannot be expressed by vulgar Analysis, I represent it by prefixing the Symbol □ to the Term. As if the *Abscissa* be *x,* and the Ordinate $ax - b + \frac{bb}{a+x}$, the Area will be $\frac{1}{2} axx - bx + \square \frac{bb}{a+x}$. And that in the same Tract I sometimes used a Letter with one Prick for Quantities involving first Fluxions; [34] and the same Letter with two Pricks for Quantities involving second Fluxions. And that a larger Tract which I wrote in the Year 1671, [35] and mentioned in my Letter of the 24th of *Octob.* 1676, was founded upon this smaller Tract, and began with the Reduction of finite Quantities to converging *Series*; and with the Solution of these two Problems: 1. *Relatione Quantitatum fluentium inter se data, Fluxionum relationem determinare.* 2. *Exposita æquatione Fluxiones Quantitatum involvente, invenire relationem Quantitatum inter se.* And that when I wrote this Tract, I had made my Analysis composed of the Methods of *Series* and Fluxions together, so universal, as to reach to almost all Sorts of Problems, as I mentioned in my Letter of the 13th of *June* 1676.[36] and that this is the Method described in my Letter of the 10th of *Decemb.* 1672.[37]

In the Year 1684.[38] Mr. *Leibnitz* published only the Elements of the *Calculus Differentialis*, and applied them to Questions about Tangents, and *Maxima & Minima,* as *Fermat* and *Gregory* had done before; and shewed how to proceed in these Questions, without taking away Surds, but proceeded not to the higher Problems. The *Principia Mathematica*[39] gave the first Instances made publick of applying this *Calculus* to the higher Problems; and I understood Mr. *Leibnitz* in this Sense in what I said concerning the *Acta Eruditorum* for *May* 1700. pag. 206.[40] But Mr. *Leibnitz* observes, that what was there said by him, relates only to a particular Artifice *de Maximis & Minimis,* with which he there allowed that I was acquainted, when I gave the Figure of my Vessel in my Principles. But this Artifice depending upon the Differential Method as an Improvement thereof, and being the Artifice by which they solved the Problems which they value themselves most upon (those of the *Linea celerrimi Descensus,* and the *Linea Catenaria* and *Velaria,*) and which Mr. *Leibnitz* there calls *a Method of the Highest Moment, and greatest Extent*; I content my self with his Acknowledgment, that I was the first who proved by a Specimen made publick, that I had this Artifice.

In the Year 1689, Mr. *Leibnitz* published the principal Propositions of this Book as his own, in three Papers,[41] called, *Epistola de Lineis opticis, Schediasma de resistentiâ Medii & Motu projectilium gravium in medio resistente, & Tentamen de Motuum Cælestium Causis*; pretending that he had found them all before that Book came abroad. And to make the principal Proposition his own, he adapted to it an erroneous Demonstration,[42] and thereby discovered, that he did not yet understand how to work in second Differences. And this was the second Specimen made publick, of applying the Method to the higher Problems. Hitherto this Method made no Noise, but within a Year or two began to be celebrated.

Dr. *Barrow* printed his Differential Method of Tangents in the Year 1670.[43] Mr. *Gregory* from this Method, compared with his own, deduced a general Method of Tangents without Calculation; and by his Letter of the 5th of *Septemb.* 1670, gave Notice thereof to Mr. *Collins.*[44] *Slusius,* in *Novemb.* 1672. gave Notice of the like Method to Mr. *Oldenbourg.*[45] In my Letter of the 10th of *Decemb.* 1672,[46] I sent the like Method to Mr. *Collins,* and added, That I mentioned it to Dr. *Barrow* when he was printing his Lectures; and that I took the Methods of *Gregory* and *Slusius* to be the same with mine, and that it was but a Branch or Corollary of a general Method, which without any troublesome Calculation extended not only to Tangents, but also to other abstruser Sorts of Problems concerning the Crookedness, Area's, Lengths, Centers of Gravity of Curves, &c. and did all this, even without freeing Equations from Surds; and I add, That I had interwoven this Method with that of Infinite *Series*, meaning, in the Tract which I wrote in the Year 1671. Copies of these Two Letters were sent to Mr. *Leibnitz* by Mr. *Oldenburg* in the Extracts of *Gregory*'s Letters, in *June* 1676;[47] and Mr. *Leibnitz* in his Letter of the 21st of *June* 1677 [N.S.],[48] sent nothing more back than what he had Notice of by these two Letters, namely, Dr. *Barrow*'s *Differential Method of Tangents* disguised by a new Notation, and extended to the Method of Tangents of *Gregory* and *Slusius*, and to Equations involving Surds, and to Quadratures. But this is not the Case between me and Dr. *Barrow*. He saw my Tract of Analysis in the Year 1669,[49] and was pleased with it. And before his Lectures came abroad, I had deduced the Method of Tangents of *Gregory* and *Slusius* from my general Method. But Mr. *Leibnitz* in those Days knew nothing of the higher Geometry, nor was yet acquainted with the vulgar Algebra.

In his Letter of the 27th of *August* 1676 [N.S.], he wrote thus:[50] *Quod dicere videmini plerasque difficultates (exceptis Problematibus Diophantæis) ad Series infinitas reduci, id mihi non videtur. Sunt enim multa usque adeo mira & implexa, ut neque ab æquationibus pendeant, neque ex Quadraturis. Qualia sunt ex multis aliis Problemata Methodi tangentium inversæ.* And when I answered, That such Problems were in my Power, he replied (in his Letter of the 21st of *June* 1676,) That he conceived that I meant by *Infinite Series*, but he meant by *Vulgar Equations*. See the Answer to this in the *Commercium Epistolicum, pag.* 92.[51]

He saith, That one may judge, that when he wrote his Letter of the 27th of *August* 1676 [N.S.], he had some Entrance into the Differential *Calculus*, because he said there, that he had solved the Problem of *Beaune certa Analysi*, by a certain Analysis.[52] But what if that Problem may be solved *certa Analysi* without the Differential Method? For no further Analysis is requisite than this; That the Ordinate of the Curve desired, increases or decreases in Geometrical Progression, when the *Abscissa* increases in Arithmetical, and therefore the *Abscissa* and Ordinate have the same Relation to one another, as the Logarithm and its Number. And to infer from this, that Mr. *Leibnitz* had Entrance into the Differential Method; is as if one should say, That *Archimedes* had Entrance into it, because he drew Tangents to the Spiral, squared the *Parabola*, and found the Proportion between the Sphere and the Cilynder, or that *Cavallerius*, *Fermat* and *Wallis* had Entrance into it, because they did many more things of this kind.

P. S.[53]

When the Committee of the Royal Society published the *Commercium Epistolicum*, the Letters and Papers in my Custody were not produced. Among them were the following Letter of Mr. *Leibnitz*, dated $\frac{7}{17}$ of *March* 1693, and a Letter of Dr. *Wallis*'s, dated the 10th of *April* 1695; both which upon a fresh Occasion two Years ago, were produced, examined, and left in the Archives of the Royal Society.[54] The first shews what Opinion Mr. *Leibnitz* had of this matter before he knew my Symbols, or any thing more of the Method of Fluxions than what he learnt from my Letters and Papers writ in or before the Year 1676, or from the *Principia Philosophiæ Mathematica*, and by Consequence before I could deceive him; and that he then gave me the Precedence. The second (compar'd with the Preface to the Doctor's Works) shews what Opinion the *English* Mathematicians, and some others abroad, had of this Matter, when they heard that the Differential Method began to be celebrated in *Holland* as invented by Mr. *Leibnitz*. The first of these two Letters, and Part of the second, are hereunto subjoyn'd.[55]

NOTES

(1) These 'Observations' were composed by Newton for the Appendix to Raphson, *History of Fluxions* (see Letter 1170, note (2)) pp. 111–18 with the heading '*Cum* D. Leibnitius *adduci non posset, ut vel Commercio Epistolico responderet, vel probaret quæ pro habitu affirmabat, cumque præcedentes Epistolas in Galliam prius mitteret quam earum tertia in Angliam veniret, & prætenderet se hoc facere, ut testes haberet, & alias etiam adhiberet contumelias:* Newtonus *minime rescripsit, sed Observationes sequentes in Epistolam illam tertiam scriptas, cum amicis solummodo communicavit.*' ('As Mr Leibniz could not be persuaded either to answer the *Commercium Epistolicum* or to prove what (as is his way) he [merely] asserts, and as he sent the preceding letters [Letters 1170, 1187, 1197; Leibniz sent copies of these to Monmort, see Letter 1198,] to France before the third of them had reached England, and claimed that he was doing this in order that he might have witnesses, and also behaved insolently in other ways, Newton has made no reply but has only imparted to his friends the following Observations upon that third letter').

In Des Maizeaux's *Recueil* the postscript is dated '18/29 de May 1716', but it is not clear on what evidence he assigned this date. Drafts in Newton's hand exist in the University Library, Cambridge, Add. 3968; one partial draft was printed by S. P. Rigaud in *An Historical Essay on the First Publication of Sir I. Newton's Principia* (Oxford, 1838), Appendix, pp. 20–4. We have compared the text in Raphson with that in Add. 3968(31), fos. 447–8; apart from a few major additions which are inserted in the printed version and which we note below, differences between the two are only the usual ones of spelling and orthography.

Leibniz remained unaware of Newton's 'Observations' at least until July 1716 and clearly expected a personal reply: he wrote to Arnold on 24 July,

'Je Nai plus de Nouvelles de l'Abbé Conti depuis que je lui ai escrit une lettre [Letter 1202] qui n'a rien d'offensant, mais qui lui a fait sentir que m'appercevois de la Partialitè, mais qui je ne m'en soucious point. Je ne scai si M. Newton repliquera a Ma Reponce [Letter 1197]

elle pourroit bien avoir servi a le desabuser.' (British Museum, Sloane MS. 4281, fos. 12–14; the extract, in Arnold's hand, was sent by Arnold to Chamberlayne on 16 May 1719 (see vol. VII).)

(2) The 'preceding Epistle' ('foregoing' in Newton's draft) is, of course, Leibniz's letter to Conti, Letter 1197.

(3) See Letter 884, vol. v.

(4) The remainder of the paragraph is omitted in the draft.

(5) See Letters 822 and 884, vol. v.

(6) That is, the excerpt of Letter 1004 printed in the *Charta Volans* (Number 1009).

(7) Bernoulli was later explicitly to deny to Newton the authorship of Letter 1004; see Letter 1201, note (4), p. 320.

(8) Newton should have written 1713, referring to Letter 1004; Raphson, Des Maizeaux and Gerhardt all repeat his error.

(9) See Letter 1172. Bernoulli's reference to himself in his letter of 7 June 1713 [N.S.], as printed in the *Charta Volans* had, of course, been inserted by Leibniz; see Number 1009, note (8), p. 21.

(10) The remainder of this paragraph is omitted in the draft.

(11) 'An upstart, with little deep knowledge of what has gone before'; thus had Leibniz referred to Keill in Letter 884.

(12) The draft omits the words 'of Analysis'.

(13) *Tractatus de Methodis Serierum et Fluxionum*; see Whiteside, *Mathematical Papers*, III, pp. 32–372.

(14) Letter 98, vol. I, and Letter 188, vol. II.

(15) Letter 165, vol. II.

(16) See Letter 1197, note (5), p. 313.

(17) Letter 22, vol. I, p. 54; Turnbull, *Gregory*, pp. 153–9.

(18) Letter 130, vol. I, p. 330.

(19) Letter 18, vol. I.

(20) See Letter 164, vol. II.

(21) This paragraph is omitted from the draft.

(22) For the events leading up to the formation of the *Commercium Epistolicum* Committee, see the Introduction to vol. v, pp. xxi–xxvii.

(23) Letter 1004. It seems clear from this paragraph that Newton knew perfectly well that the problem which Leibniz had sent to Conti for the English mathematicians had originated with Bernoulli; see Letter 1170, and Letter 1201, note (7), p. 320.

(24) See Letter 1197, note (8), p. 313.

(25) The rest of the paragraph is omitted from the draft. See also note (29) below.

(26) The sentence following is omitted from the draft.

(27) The draft reads 'some of them'.

(28) See vol. II, p. 64; '1683' in this sentence is an obvious misprint for 1673.

(29) The whole of the rest of this paragraph, and the 'Problem', are omitted from the draft; instead we find the much shorter passage:

And am I not as good a Witness that I found the method of fluxions in the year 1665 & improved it in the year 1666 & that before the end of the year 1666 I wrote a small Tract on this subject which was the grownd of that larger Tract wch I wrote in the year 1671 (both which are still in my custody,) & that in this smaller Tract tho I generally put letters for fluxions as Dr Barrow in his Method of Tangents put Letters for differences, yet in giving a general Rule for finding the Curvature of Curves I put the letter $\mathfrak{X}$ with one prick for first fluxions drawn into their fluents & with two pricks for second fluxions drawn into the square of their fluents, & that when I wrote the larger of those two Tracts I had made my Analysis composed of those two methods so universal as to reach to almost all sorts of Problemes as I mentioned in my Letter of 13 June 1676.

Possibly Newton set down this account from memory, and later looked through his manuscripts thoroughly; the tracts mentioned here are discussed in more detail in the final printed version (see notes (30)–(37) below). There is evidence that Newton had genuinely forgotten, in his draft document, that his early use of 'pricked' letters differed from his final fluxional calculus; the phrases 'drawn into their fluents' and 'drawn into the square of their fluents' are interlineated additions. It is hard to tell from the manuscripts whether the letter $\mathfrak{X}$ he uses is the specialized symbol of the October 1666 tract, or the x he used in later work. Whatever the case, it is clear that Newton in the printed version of the passage purposely revised the phraseology in an attempt to conceal the differences between his early and late use of 'pricked' letters (see note (34) below).

(30) See Whiteside, *Mathematical Papers*, I, pp. 145–448.

(31) *Ibid.*, p. 383.

(32) *Ibid.*, pp. 392–9.

(33) Presumably Newton means the October 1666 tract, *ibid.*, pp. 400–48.

(34) Newton's claim is true, but misleading. In the tract he did use a form of dot notation (*ibid.* p. 421); thus the symbol

$$\mathfrak{X} = x\,\frac{\partial f}{\partial x} \text{ where } \mathfrak{X} = f(x,y); \quad \mathfrak{X} = y\,\frac{\partial f}{\partial y}, \text{ etc.}$$

Hence the symbols involved fluxional quantities, but did not *directly represent* fluxions. See also note (29) above, and Letter 1053a, note (10), p. 91.

(35) *De Methodis Serierum et Fluxionum*. See Whiteside, *Mathematical Papers*, III, p. 32. The problems Newton refers to begin on p. 74 and p. 82 respectively.

(36) Letter 165, vol. II.

(37) Letter 98, vol. I.

(38) That is, in 'Nova Methodus pro Maximis et Minimis, . . .' *Acta Eruditorum* for October 1684, pp. 467–73. See Gerhardt, *Leibniz: Mathematische Schriften*, V, pp. 220–6.

(39) *Principia*, Book II, Lemma 2.

(40) 'G.G.L. Responsio ad Dn. Nic. Fatii Duillerii Imputationes . . .' *Acta Eruditorum* for May 1700, pp. 198–208. See Gerhardt, *Leibniz: Mathematische Schriften*, V, pp. 340–9. The relevant excerpt was printed in the *Commercium Epistolicum*, p. 107: 'Quam [*methodum*] ante Dominum *Newtonum* & me nullus quod sciam Geometra habuit; uti ante hunc maximi nominis Geometram, NEMO specimine publice dato se habere probavit; ante Dominos *Bernoullius* & Me nullus Communicavit.'

(41) The three papers were published in the *Acta Eruditorum* for January 1689, pp. 36–8 and pp. 38–47, and for February 1689, pp. 82–96, respectively. For the 'Tentamen' and Newton's critique of it, see Number 1069a.

(42) The rest of this sentence is omitted from the draft. See also Number 1069a, note (10), p. 122.

(43) Isaac Barrow, *Lectiones Geometricæ* (London, 1670).

(44) Letter 18, vol. I.

(45) See Hall & Hall, *Oldenburg*, IX, Letter 2100 and note (10); the letter was not printed in the *Commercium Epistolicum*.

(46) See Letter 98, vol. I; in the letter, which Newton here paraphrases, the names of Gregory and Sluse were not explicitly mentioned, and the date of the 1671 tract (see note (35) above) was not given.

(47) See note (15) above.

(48) Letter 209, vol. II.

(49) Newton in Raphson, *History of Fluxions*, and Gerhardt, *Briefwechsel* following him, wrongly puts 1699.

(50) Letter 172, vol. II, p. 64 and (translation) p. 71.

(51) That is, Leibniz's letter to Oldenburg, 12 July 1677; see Letter 210, vol. II.

(52) Descartes discussed what has become known as 'De Beaune's Equation' in a letter of 20 February 1639 [N.S.] (*Oeuvres*, eds. Ch. Adam and Paul Tannery: *Correspondance*, II, pp. 514–17). Florimond de Beaune's formulation of this first authentic instance of the 'inverse problem of tangents' is lost but it may be expressed in modern form as finding the curve whose differential equation is $dy/dx = a/(y-x)$. Public interest in the De Beaune equation was revived by Johann Bernoulli in 1692 (see Whiteside, *Mathematical Papers*, III, pp. 96–7, note (140)). For Newton's familiarity with this equation from about 1672, and Leibniz's attack on it in 1676, see *ibid.*, III, pp. 84–5, note (109) and IV, p. 633, note (42). It seems that Leibniz's claims for his success with this equation, in his notes of 1676 and his letter to Newton of 17/27 August 1676, were not justified by his real analysis of it.

(53) The postscript does not appear in the draft.

(54) Newton produced these two letters at a meeting of the Royal Society on 5 May 1715, and it was ordered that they be placed in the archives with the original of the *Commercium Epistolicum* correspondence. His statement in the text suggests that Des Maizeaux's date for this paper, which we here follow, may be one year too early.

(55) That is, printed in the Appendix to Raphson, *History of Fluxions*, pp. 119–20; both letters were addressed to Newton (see vol. III, Letter 407 and vol. IV, Letter 498). Newton concludes the appendix with two short extracts from Wallis' *Opera mathematica* (Oxford, 1693) and a section from Fatio's *Lineæ brevissimi descensus* (London, 1699), (see vol. V, p. 98, note (3)), adding a few brief comments of his own.

1212 THE MINT TO THE TREASURY
26 MAY 1716
From the original in the Public Record Office[1]

To the Right Honorable the Lords Commiss[ione]rs of his Majes. Treasury

May it please your Lordshipps

In Obedience to your Lordshipp's Order Signified to us by Mr Lowndes the 24th instant,[2] we humbly lay before your Lordshipps the following Account shewing the several Quantitys of Tin unsold, and what remains due to the respective persons in whose hands the said Tin lyes for their security, viz.

	Tuns	due to him
In Mr. Moses Beranger hands in Hollande about	1130	£39 000
Sr. Theod. Janssen about	380	22 000
Sr John Lambert et al	900	11 000
In Cornwall & Devon exclusive of Christmass & Ladyday's Tin	170	
In the Tower of London	1500	
	4080	

Mint Office the 26th May   RICH. SANDFORD

1716   ISAAC NEWTON

NOTES

(1) T/1, 199, no. 24.

(2) We have not found a record of this. However, on 1 May 1716 the Treasury Lords ordered that next day the Principal Officers of the Mint must present themselves with an account of the tin in hand, whether in London or in the country, and of what revenue might be expected from weekly sales of tin. There was anxiety because the miners had not been paid recently (*Cal. Treas. Books*, xxx (Part II), 1716, p. 19).

1213   LEIBNIZ TO J. BERNOULLI
27 MAY 1716

Extracts from Gerhardt, *Leibniz: Mathematische Schriften*, III/2, pp. 962–3.
Reply to Letter 1210; for the answer see Letter 1217

Solutionem Trajectoriarum perpendicularium ad Hyperbolas Domini Filii Tui reperies in Actis Lipsiensibus.[1] Adjectum est, inservire ad intellectum problematis generalis, quod ad explicandos progressus in Calculo infinitesimali inservire possit.[2] Equidem dictum non est, per se tamen intelligitur, in generali illo saltem efficiendum esse, ut res reducatur ad æquationem differentialem primi gradus, et in specialibus reducendam rem ad quadraturas, quoties per notas hactenus artes licet.

Misi Contio problema Tuum speciale, et videbimus, quid Taylorus vel alii in eo sint præstituri.[3]

Accepi Taylori *Methodum*, quam vocat, *incrementorum.*[4] Est applicatio Calculi differentialis et integralis ad numeros, vel potius ad magnitudines generales. Ita Angli equos, ut in Proverbio est, adjungunt post currum. Ego incepi calculum differentialem a numerorum seriebus, eoque utiliter usus sum ad

353

summas serierum numericarum, et postea animadvertens in Geometria differentias et summas dare quadraturas, et multa ob incomparabilitatem evanescere in lineis, via naturali perveni a Calculo generali ad specialem geometricum seu infinitesimalem. Isti contra procedunt, nempe quod veram inveniendi methodum non habuerunt. In toto suo Libello neminem citat, nisi Newtonum.

. . . Serram etiam Philosophicam nunc cum Newtono, vel quod eodem redit, cum ejus Hyperaspita Clarkio, Regis Eleemosynario,[5] me reciprocare fortasse jam intellexeris. Scis, Keilium[6] et Præfatorem[7] novæ Editionis Principiorum Newtoni etiam Philosophiam meam pungere voluisse. Itaque scripseram ego forte Serenissimæ Principi Regiæ Walliæ, pro excellenti ingenio suo harum rerum non incuriosæ, degenerare nonnihil apud Anglos Philosophiam vel potius Theologiam Naturalem; Lockium et similes dubitare de immaterialitate animæ, Newtonum Deo tribuere sensorium, quasi spatio tanquam organo sensationis opus habeat; inde alicui in mentem venire posse, quasi non sit nisi anima mundi secundum veteres Stoicos. Eundem Auctorem Dei Sapientiæ et perfectionibus derogare, dum velit Mundum esse Machinam non minus imperfectam, quam horologia nostrorum artificum, quæ sæpe retendi debent aut alias corrigi; ita Machinam Mundi, secundum Newtonum et asseclas, correctione quadam extraordinaria subinde indigere, quod parum sit dignum Deo Auctore. Mea sententia Deum omnia tam sapienter ab initio constituisse, ut correctione non sit opus, quæ imprudentiam arguat. Serenissima Princeps Walliæ excerpta hujus Epistolæ Clarkio communicavit. Is scriptum contra Anglico sermone ipsi dedit, quod illa ad me misit; respondi, replicavit; duplicavi, triplicavit; ego novissime quadruplicavi, seu ad tertium ejus scriptum respondi. Inter alia improbat formulam a me in Theodicæa usurpatam, quod Deus sit Intelligentia supramundana, tanquam a me a mundi gubernatione excludatur. Ego quæsivi, an ergo velit Deum nihil aliud esse, quam Intelligentiam mundanam, seu animam mundi? Male excusat doctrinam Newtonianam de spontanea virium activarum diminutione et tandem cessatione in mundo, nisi a Deo reparentur. Ex quo intelligitur, Newtonum ejusque asseclas veram scientiam rei dynamicæ nondum habere. Ex nostris enim principiis semper servatur eadem quantitas virium. Male etiam excusat phrasin Newtonianam de spatio sensorio Dei. Et quia spatium hodie est Idolum Anglorum, ego ipsi ostendo, spatium non esse aliquid reale absolutum, non magis quam tempus, sed ordinem quemdam generalem coexistendi, uti tempus est ordo existendi successive. Itaque esse aliquid ideale, quod si creaturæ tollerentur, non futurum esset, nisi in ideis Dei. Ostendi etiam, secundum Newtonum crebris miraculis ad sustentandum naturæ censum opus esse, et ex Clarkii excusationibus deprehendo, ipsum non habere bonam notionem miraculi. Ipsi enim miracula tantum secundum nos a naturalibus differre videntur, tanquam minus usitata: sed secundum Theologos et veritatem, miracula (saltem ea, quæ sunt superioris ordinis, velut creare, annihilare) transcendunt omnes naturæ creatæ vires. Itaque quidquid ex naturis rerum inexplicabile est, quemadmodum attractio generalis materiæ Newtoniana aliaque ejusmodi, vel miraculorum est, vel absurdum. Fortasse nonnihil adhuc continuabitur nostra collatio, in qua absunt quæ offendere possint, et videbo quo res sit evasura. Hujusmodi enim collationes mihi ludus jocusque sunt, quia in Philosophia.

Translation

You will find your son's solution of the [problem of the] trajectories normal to hyperbolas in the *Acta* [*Eruditorum*] of Leipzig.[1] Something has been added which may serve to explain the progress of the infinitesimal calculus,[2] in order to facilitate understanding of the general problem. It has not actually been said, although it is readily understood, that in that general case at least so much must be done that the matter is reduced to a differential equation of the first degree, and that in special cases it is to be reduced to quadratures, as often as the hitherto known procedures permit.

I have sent Conti your particular problem,[3] and we will see what Taylor, or others, will accomplish in that respect.

I have received Taylor's *Method of Increments*[4], as he calls it. It is the application of the differential and integral calculus to numbers, or rather to general magnitudes. Thus the English 'put the cart before the horse' as the saying goes. I began the differential calculus from series of numbers, and I made good use of it for the sums of series of numbers, and, afterwards, observing that differences and sums give quadratures in geometry, and that because of [their] incomparability many [quantities] vanish into lines, I arrived by a natural route from the general calculus at the special geometrical or infinitesimal [calculus]. They [the Newtonians] proceed in a contrary fashion, because in fact they did not have the true method of discovering. In the whole of his book he cites no one but Newton

. . . Perhaps you will already know that I am now grinding a philosophical axe with Newton, or, what amounts to the same thing, with his champion, Clarke, the royal almoner.[5] You know that Keill,[6] and the writer [7] of the Preface to the new edition of Newton's *Principia*, wished to attack even my philosophy. Therefore I had casually written to Her Royal Highness the Princess of Wales who on account of her fine intellect is not uninterested in these matters, that philosophy, or rather natural theology, declines somewhat in England; that Locke and similar men doubt the immateriality of the soul, that Newton attributes a sensorium to God, as though He has need of space to be, as it were, an organ of sensation; whence anyone might suppose that He is nothing else but the spirit of the world according to the old Stoics. [I have also written] that the same author diminishes the wisdom and perfections of God, so long as he wishes the world to be a machine no less imperfect than our craftsmen's clocks, which often must be rewound or otherwise corrected; that thus the world-machine, according to Newton and his followers, on this account repeatedly requires some extraordinary correction, which is hardly worthy of God its Architect. In my opinion God has made everything so wisely from the beginning, that the correction (implying His lack of foresight) is not necessary. Her Royal Highness the Princess of Wales has communicated to Clarke excerpts of this letter. He gave her a paper written against it in the English language, which she sent to me. I have answered and he has replied; I have replied a second time, and he a third; I have written most recently a fourth time, that is, I have answered his third paper. Amongst other things, he disapproves of the form of words used by me in the *Theodicæ*, that God is a supra-mundane Intelligence, as though He is excluded by me from the government of the world. I have asked whether, therefore, he wishes God to be nothing other than a world-based Intelligence, or soul of the world. He makes out a feeble case for the Newtonian doctrine of the spontaneous decrease of active forces and their final cessation in the world, unless God renews them. From this one infers that Newton and his followers have as yet no true knowledge of the science of dynamics. For by our principles the same quantity of forces is always conserved. He even makes a weak case for the Newtonian phrase concerning the sensory space of God. And because today space is the idol of the English, I show him that space is not something really absolute, any more than time is, but a certain general order of coexisting, just as time is the order of successively existing. Therefore it is something ideal, because if the creation were abolished it would not be, except in the ideas of God. I have also shown that, according to Newton, frequent miracles are necessary for maintaining the natural order, and from Clarke's excuses, I perceive that he himself does not have a good notion of a miracle. For in his view miracles seem distinct from natural occurrences only according to our apprehension, as being less ordinary; but according to theology and truth miracles (at least those which are of a higher order, such as to create or destroy) transcend all natural, created forces. Therefore whatever is unexplicable from the natures of things, such as the Newtonian universal attraction of matter and other things of that kind, is either miraculous or absurd. Perhaps our encounter (from which all occasions for giving offence are absent) will be further protracted; I shall see how it turns out. For encounters of this sort, because they are in philosophy, are a game and amusement to me.

NOTES

(1) See Letter 1201, note (10), p. 321.

(2) It is not clear what Leibniz means here; the paper by Nikolaus [II] ends with a brief discussion of a number of special cases, but little is said about the general problem. Nikolaus later published a more general account of the problem in *Acta Eruditorum* for June 1718, p. 248. There he brings to light further correspondence about the matter, including a letter dated 10 July 1716 from Johann Bernoulli to Monmort where he discusses the impossibility of a complete, general solution.

(3) See Letter 1202, note (4), p. 324.

(4) Brook Taylor's *Methodus Incrementorum Directa & Inversa* (London, 1715) by no means

deserved all of Leibniz's scathing comments; true it used fluxional notation (in a form peculiar to Taylor) and cited only Newton among contemporaries; but it included Taylor's theorem (see p. 23) and a number of other useful devices.

(5) See Letters 1173, note (3), p. 259, and 1138, note (1), p. 214, for the Leibniz–Clarke correspondence and Leibniz's theological ideas.

(6) Leibniz refers to the passage in Keill's 'Answer', p. 348 (see Letter 1053a, note (1)) where Keill refers to Leibniz's 'Tentamen' as 'le morceau de Philosophie le plus incomprehensible qui ait jamais paru.'

(7) Roger Cotes, of course; see Letter 1173, note (3), p. 259, and Letter 980, note (3) (vol. v, p. 387).

1214 ROBERT BALLE TO NEWTON
28 MAY 1716
From the orighal in Trinity College Library, Cambridge [1]

Sr:

Having yesterday bin with my Lord Parker, [2] I acquainted him how you had resolved to waite on his Majestie, at which he was much pleased, saying, he would willingly accompany us injoyning me to acquaint you therewith. I shall attend to heare when the time is fixt, & where to meete, & with pleasure take this oportunity to assure you that I am

Sr: Your most devoted &

most obleidged humb: servt:

ROBERT BALLE

Camdn: House Kens:
28: *May* 1716
For Sr: Isack Newton

NOTES

(1) R.16.38A³. Almost nothing is known of the writer of this letter, who was elected F.R.S. on 30 November 1708. He is mentioned in the Journal book of the Royal Society of London as a member of Committees, but be published nothing. He died, perhaps, in 1733. He was not, apparently, a son of William Ball, the astronomer, nor of Dr Peter Balle, the physician; nor a graduate, nor a lawyer. He may have been a merchant since he was reported in Italy not long before his presumed death.

(2) Presumably Sir Thomas Parker (1666?–1732), Lord Chief Justice. He was later appointed Lord Chancellor and created Earl of Macclesfield.

1215 ARNOLD TO LEIBNIZ
28 MAY 1716
Extract from the original in the Niedersächsische Landesbibliothek, Hanover [1]

Je loue fort le tour que vous avès fait prendre a Vos lettres pour Monsieur
labbè conti, Jai peur quon nagira pas trop honnêtement envers vous cest
pourquoi vous ne pouvès pas estre trop sur vos gardes, Mr lAbbè Conti est
persuadè de la verite de tout ce quils lui disent, et il n y a pas beaucoup de
difficulte den imposer a un qui entend si peu du calcul il ma voulu prover le
dernier fois que jai estre ches lui que dans le page 76 du *commercium epistoli-*
cum [2] estè contenu des regles generales du calcul differ[en]tiel et indegrall
[*sic*] et que vous et Monsieur Bernoulli naviès fait que nous donnes des
Exemples Jugès de la de son savoir. Je crois quil a estè trompè par le *dz*
elevè a quelques puissance quil aura pris pour un differentiel.

NOTES

(1) See Bodemann's catalogue, no. 17.

(2) Page 76 of the *Commercium Epistolicum* is part of Newton's letter to Oldenburg of 24
October 1676 (see Letter 188, vol. II) where Newton quotes from the integral tables of his
1671 tract (vol. II, p. 138; Whiteside, *Mathematical Papers*, III, pp. 246–55), since printed in
De quadratura curvarum. The integrals take the form (for example) $dz^{\eta-1}/(e+fz^{\eta}+gz^{2\eta})$. Here
d, e, f, g are given constants; Arnold claims that Conti mistook dz for a Leibnizian differential.

1216 LEIBNIZ TO ROBERT ERSKINE
22 JUNE 1716
Extract from the original in the Niedersächsische Landesbibliothek, Hanover [1]

Ma dispute avec M. Clarke defenseur de M. Newton dure encore mais
j'éspere qu'elle sera bientost finie, car je luy envoye maintenant une response
assés ample à son dernier ecrit, la quelle éclaircit les choses à fonds; ainsi je
crois qu'apres cela, je n'auray plus grand chose à dire sans repetition, et s'il
ne se rend point à la raison, je le laisseray là comme invincible. Aussi tost que
cela sera fait j'envoyeray au *Acta Eruditorum* de Leipzig [une] petite relation [2]
de nostre controverse, en forme de lettre addressée à vous Monsieur.

NOTES

(1) See Bodemann's catalogue, no. 15; the letter is printed in W. Gurrier, *Leibniz in*
seinem Beziehung zu Russland und Peter dem Grossen (Leipzig, 1873), p. 361.

Robert Erskine (or as he and Leibniz spelled the name, Areskin), a Scot, was a son of Sir

Charles Erskine, first baronet, of Alva (see Letter 1200). He was entered at Edinburgh University without graduating (1691), and after apprenticeship to a surgeon in that city travelled to Paris (1697) and then Utrecht where he proceeded M.D. in 1700. Returning to London be became friendly with Cheyne and was elected F.R.S. (as 'Areskine', 30 November 1703) but left for Russia the next year where he became physician to Prince Menshikov (Letters 1100 and 1111). In 1713 he became personal physician to Peter the Great and in the following year 'Archiater'. A great favourite with the Czar, he accompanied Peter on his European tour of 1716 when, in Germany, he met Leibniz. He returned to Russia and died there in 1719, leaving a library of over 4000 volumes.

(2) Leibniz did not carry out this plan.

1217 J. BERNOULLI TO LEIBNIZ
3 JULY 1716
Extract from Gerhardt, *Leibniz: Mathematische Schriften*, iii/2, pp. 965-6.[1]
Reply to Letter 1213

Taylori librum pariter nondum vidi: Montmortius ab aliquo jam tempore promisit, se duo ejus exemplaria nobis transmissurum, sed forsan occasionem mittendi nondum habuit. Facile credo Taylorum neminem citare, nisi Newtonum: hoc enim multis Anglis in more est, ut omnia invideant cæteris, vel omnia ad se vel ad suos derivent. Sic Taylorus Theoriam meam de Centro oscillationis pene totam, ut scribit Cl. Hermannus, ex Actis Lipsiensibus [2] desumsit, eamque, ne plagium esset nimis manifestum, obscuritatis quodam peplo ita involvit, ut fere inintelligibilis sit, quam tamen clarissime exposui. Ais, quod suarum artium specimen exhibere volens vix habeat nisi jam dicta, quodque totus Liber scriptus sit satis obscure; hoc ego non miror, quomodo enim quæ aliena sunt sua facere posset, nisi obscuritatem studio affectaret ad furtum celandum, sed hoc maxime miror, quod dum Angli illi sunt omnium impudentissimi plagiarii, eo temeritatis procedant, ut aliis hoc vitium exprobrare audeant. *Quis tulerit Gracchos* [3] etc.

Placet quod Cl. Wolfius epistolam [4] meam Tibi miserit; hoc enim ab eo petii, quia nolebam aliquid circa materiam, quam continet, a me publicari Te inscio, nedum invito: nunc mihi gratissimum contingit, quod pro æquitate Tua agnoscis mihique tribuis, me sine alterius ope ad artem summandi vel integrandi pervenisse, atque mea potissimum opera calculum infinitesimalem celebrem redditum fuisse: ego vero vicissim lubenter patior, quod addidisti, Tibi quoque aliquam summandi integrandive atque etiam Exponentiales adhibendi artem non defuisse; si bene memini, simile quid in epistola illa jam dixi, imo fortius rem eandem in Tui honorem expressi. Tecum quoque sentio, præstare, ut Keilius non nominetur, quam ut nominando ansam ei demus sibi applaudendi.

De disceptatione philosophica, quæ Tibi est cum Clarkio, nihil ante intellexeram. Ex iis, quæ refers, video nihil tam absurdum proferri posse a Newtono, quod inter Anglos non inveniat Patronos ac Defensores suos. Hi non disputant, ut veritatem tueantur, sed quia de Nationis gloria agi putant, quando vident, Magistrum suum, in cujus verba jurarunt, in discrimine causæ suæ sive bonæ sive malæ (hoc non attendunt) versari. Hinc dubito, utrum hoc tantum ab ipsis sis consecuturus, ut agnoscant Newtonum errare posse, aut omnino aliqua in re errasse. Mihi quoque dudum absona visa est ejus doctrina de spontanea virium diminutione et tandem cessatione in mundo: siquidem per se clarissimum mihi apparet nullam vim destrui, quæ non simul effectum edat sibi æquivalentem, quia nihil tendit ad sui annihilationem: effectus autem nihil est aliud, quam vis ipsa efficienti substituta, ita ut eandem virium quantitatem servare necesse sit. Dicit Newtonus alicubi in *Principiis Philosophiæ Naturalis*,[5] Vortices Cælestes Cartesii ideo admitti non posse, quia ob partium suarum attritionem et frictionem tandem a motu cessarent: sed jam sibi ipsi contradicit. Si enim, secundum ipsum, jactura virium in Mundo reparari a Deo, et tota Machina mundana subinde quasi retendi debet, annon et idem Cartesius in Vorticum suorum defensionem reponere posset, quod nempe, si vel maxime per attritionem partium in motu retardarentur, Deus tamen decrementum motus resarcire possit, eos quandoque per novam impulsionem in pristinam celeritatem incitando.

Translation

I also have not yet seen Taylor's book: Monmort promised some time ago that he would send us two copies of it, but perhaps he has not yet had an opportunity to send it. I can easily believe that Taylor cites no one except Newton: for it is a characteristic of the English that they begrudge everything to other [nations] and attribute all things to themselves or to their nation. Thus Taylor has taken my theory of the Centre of Oscillation almost complete from the *Acta [Eruditorum]* of Leipzig,[2] as Mr Hermann writes, and, in order that the plagiary should not be excessively obvious, he so wraps up the theory in some cloak of obscurity that what I have most clearly expounded is almost unintelligible. You say that while eager to present a sample of his skills he has hardly said anything that was not said before, and that the whole book is written obscurely enough; I myself do not wonder at this, for how can he pass off as his own things which belong to others, unless he affects obscurity on purpose in order to hide the theft; but it surprises me extremely that while the English themselves are the most impudent plagiarists of all, they carry on with such temerity that they dare to charge others with this crime. *Who could bear [to hear] the Gracchi*[3] etc.

I am glad that Mr Wolf has sent you my letter;[4] for I asked this of him because I did not wish anything about the subject it deals with to be published by me without your knowledge, still less against your wish: now it gives me great pleasure that with your usual fairness you acknowledge and allow it to me that I arrived at the art of summation and integration without the help of anyone, and that I have very markedly increased the fame of the infinitesimal calculus by my work. In return I very gladly admit the additional point you made, that you also were not deficient in the art of summation and integration, and even in the art of using exponentials; if I remember well, I have said something similar already in that letter, indeed, I have expressed the same thing more strongly in your honour. Also, with you, I feel it is better that Keill should not be named, as by naming him we would give him an excuse for flattering himself

I knew nothing previously of that philosophical discussion, which you are carrying on with Clarke. From what you report I see nothing so absurd can be advanced by Newton, but that it finds patrons and defenders among the English. They do not maintain an argument in order to defend the truth, but because they think the glory of their nation is at stake when they see their master, on whose word they swear, to be in danger of losing an argument, whether [his case be] good or bad (this they do not consider). Hence I doubt whether you can expect [even] this much from them, that they will acknowledge Newton to be capable of error, or at any rate to have been mistaken in any one particular. To me, too, his theory of the spontaneous decrease of forces and their eventual cessation in the world has long seemed incongruous: since it seems to me most clearly self-evident that no force is destroyed without at the same time giving rise to an equivalent effect, because nothing tends towards its own annihilation: however, the effect is nothing other than the force itself transformed into accomplishment so that it is necessary that the same quantity of force be conserved. Newton says somewhere in the *Principia Philosophiæ Naturalis* [5] that for this reason the Cartesian celestial vortices cannot be admitted, because on account of the attrition and friction of their particles they will finally cease to move: but now he contradicts himself in [saying] this. For if according to him, the loss of forces in the world has to be made up by God, and the whole world-machine repeatedly rewound, as it were, why may not Descartes himself reply, in defence of his vortices, that if their motion is very markedly retarded by the attrition of their particles God can nevertheless restore the loss of motion, by accelerating them from time to time to their original speed by means of a new impulse?

NOTES

(1) This is the last excerpt of the Leibniz–Bernoulli correspondence which we print. Four more letters were exchanged between the two men before Leibniz's death on 3 November 1716; (see Gerhardt, *Leibniz: Mathematische Schriften*, III/2, pp. 967–73). None of these has very direct bearing on Newton's affairs, although all discuss mathematical matters: Leibniz's letter of 15 July 1716 reports Hermann's comments on the problem of curves cutting at right angles; Bernoulli's reply of 11 August criticizes Brook Taylor's *Methodus Incrementorum* (see note (2) below); Leibniz, in his final letter of 12 October, again discusses his correspondence with Clarke, and criticizes Taylor's work. Leibniz's death probably prevented him from seeing Bernoulli's reply of 31 October, where he briefly discusses the English solutions of his problem.

(2) 'Johannis Bernoulli Meditatio de natura centri oscillationis', *Acta Eruditorum* for June 1714, p. 257. Brook Taylor had indeed dealt with the same material in his *Methodus Incrementorum*, pp. 95–100 and Bernoulli publicly, but mildly, accused Taylor of plagiary, in the 'Epistola pro Eminente Mathematico', published in the *Acta Eruditorum* for July 1716 (see Letter 1196, note (1)). Taylor, in his solution of the problem of intersecting curves in the *Philosophical Transactions* for 1717 (see Letter 1202, note (4)) included a few equally mild comments about Bernoulli's incompetence. Bernoulli was not impressed by Taylor's paper, but the argument between the two men would probably have died down, had not Monmort fanned the flames by repeatedly suggesting to Taylor that Bernoulli's comment in the 'Epistola' was strong enough to require an answer. Taylor eventually, in the spring of 1719, conceived the plan of writing a paper conjointly with Keill against Bernoulli, but nothing came of it. By the middle of May he had prepared his own answer to Bernoulli, but was still doubtful about publishing it. However, largely as a result of further pressure from Monmort, his paper eventually appeared in *Phil. Trans.* **30**, no. 360 (1719), 955–63, as 'Apologia D. Brook Taylor, J.V.D. & R.S. Sec, contra V.C. J. Bernoullium Math. Prof. Basiliæ.' In 1719 and 20, Monmort acted as intermediary for increasingly heated correspondence between Bernoulli and Taylor. (For further details, see vol. VII of this *Correspondence* in preparation; and consult Taylor's letter to Monmort dated 5 February 1720, which is printed in Taylor's *Contemplatio Philosophica*, pp. 109–18.)

(3) 'Quis tulerit Gracchos de seditione querentes?' ('Who could bear [to hear] the Gracchi complaining of sedition?' Juvenal, *Satires*, Sat. ii, line 24). The point of Juvenal's sarcasm is: who can tolerate hypocrisy?

(4) That is, Bernoulli's draft of the 'Epistola pro Eminente Mathematico'; see Letter 1196, note (1), p. 303.

(5) Presumably Bernoulli recalls either the general discussion in Cotes' Preface, or the final paragraphs of the Scholium to Prop. 40, Book II, both passages appearing for the first time in the second edition. Bernoulli's wit is somewhat cynical: he attempts to save a particular hypothesis by appropriating what Newton had intended as a general truth about the universe—the inevitable degradation of motion that renders the laws of mechanics assymmetrical with respect to time.

1218 CHRISTOPHER TILSON TO THE MINT
17 JULY 1716
From the copy in the Public Record Office.[1]
For the answer see Letter 1225

Whitehall Trea[su]ry Chambers 17 July 1716.

The Rt. Honble: the Lords Comm[issione]rs of His Mats. Trea[su]ry are pleas'd to Refer this Bill for Silver and Engraving of Several publique Seals [2] to the Principal Officers of His Mats. Mint who are to peruse His Mats. Warrants directing the Several publique Seals to be prepared and the Respective Certificates of their being delivered pursuant thereto and Examine into the Reasonableness of the prizes Set down for the Same And thereupon to make their Report to their Lordships with their Opinion what is fit to be done therein which in the Absence of the Secretarys is Signifyed by their Lordships Command by

CHRIS TILSON

NOTES

(1) Mint/1, 7, fo. 96. The account for the seals engraved, given on the verso of fo. 95, amounts to £788. 15s. 8d. There is another copy in T/4, 9, p. 288.

Christopher Tilson (1670–1742), Clerk to the Treasury from 1685 to his death, held in addition several other minor official posts and entered the House of Commons in 1727 (*History of Parliament*, II, p. 469).

(2) These seals had been engraved by John Roos, chief engraver of seals to Anne and George I, who had died on 15 June 1716 (*Cal. Treas. Books*, XXX (Part II), 1716, p. 560). He was succeeded by James Girard.

1219 J. BERNOULLI TO ARNOLD
4 AUGUST 1716
From the copy in the British Museum [1]

Bâle le 15 *Aoust* 1716 [N.S.]

Le Probleme dont M. Leibniz vous a ecrit que jai donne une solution generale consiste a determiner la courbe qui coupe a angle droit une infinitè de Courbes donnèes par un même loi; [2] si je me souviens bien je vous ai fait part autrefois de cette solution; ce qu'il y a de plus difficile la dedans, cet que si les courbes donnèes sont transcendentes elles rendent souvent impossible la solution generale; souvent aussi dans les courbes donnèes Algebraiques on arrivè a une equation dont les indeterminèes ne se laissent pas separer. M. Leibniz m'a aussi mandè [3] qu'il avoit receu une lettre de M. Newton mais quelle estoit ecrite a M l'abbè Conti et que celui ci la lui avoit communiquèe sur la priere de M Newton; [4] que M Newton avoit fait cette demarche après avoir remarquè que M Leibnitz ne daignoit pas de repondre a M Keil, ainsi que cette lettre de M Newton n'avoit estè escrite que pour entrer lui même en lice avec M. Leibnitz bien loin de s'accommoder sur l'affaire des calculs.

NOTES

(1) M.S. Birch 4281, fos. 12–14. The extract, in Arnold's hand, was sent by Arnold to Chamberlayne on 16 May 1719 (see vol. VII).

(2) Possibly this news was contained in the missing portion of Leibniz's letter to Arnold of 17 April 1716, from which we have printed his extract in Letter 1204.

(3) See Letter 1201.

(4) See Letter 1187 and Number 1187a.

363

1220 WILLIAM NEWTON TO NEWTON
7 AUGUST 1716
From the original in the University Library, Cambridge[1]

Honored Sir

In obedience to my fathers Commands I am oblig'd to continue here at london for sometime longer Sr Wm St Quintin[2] haveing promis'd my friends[3] in the Countrey that I shall be prefer'd in the Customes, now seeing that I have Qualified my self for that purpose, & that I am in expectation dayly to come into Mr John Selby's place at Whitby, he being to be promoted to a Collection in the South, & that now Sir the Welfare of our Family seems much to depend on my good success. Yr kindnes has been extraordinary to me, for otherwise I might have suffer'd very much. I apply dayly to the Treasury & hope that Mr John Selbys Warrant & mine will be granted us & that how glad I shall be to be in a Condition to support my Dear Father, that has liv'd well in ye World, Sr you must not think much with me for being so free with you, I must Confess I've been a great trespasser upon you but I hope yr Goodness will pardon me, & that my Dear father prays dayly you may continue my friend, & not let us sink now, when there is so fair a probality [*sic*] of my getting now into business, I've taken abundance of pains & the great fatigue I've undergone & hardships as are almost inexpressible. I hope I shall demonstrate when please god I am in business my gratitude to you & how much I am

Honored Sir

Yr most humble & most obliged

servant   WM NEWTON

Augt ye 7th
1716
For Sr Isaac Newton
at his house in St Martins Street[4]
nigh Leister Fields
    These

NOTES

(1) Add 3968(41), fos. 6 bis–7 v. Newton has used the paper for anti-Leibniz drafts about fluxions, mentioning the 'fluxions scholium' in Book II of the *Principia*. It is not possible to say positively whether the writer was or was not a relation of Sir Isaac's. He wrote again to Newton for help on 26 March 1717 (see Letter 1236).

(2) See Letter 1135, note (5), p. 210. Presumably Sir Isaac had approached him on the writer's behalf.

(3) Relations.

(4) 'Lane', originally written, had been struck out and 'Street' substituted in a different hand.

1221 LOWNDES TO NEWTON
20 AUGUST 1716
From the copy in the Public Record Office.[1]
For the answer see Letter 1224

Sr

Two persons who have been recommended to his Ma[jes]ty as well skilled in Working Mines the One named Justus Brandshagen & the other James hamilton being Ordered forthwith to go downe to Scotland to Work or try the Working of the Silver Mine lately discovered there The Lords Comm[issione]rs of his Mats. Trea[su]ry are pleased to direct You to consider what Authority is necessary for them to be Invested with, and what Instructions may properly be given them [2] And to make Your Report thereupon to their Lordps with all convenient Speed I am &c

20 *Aug* 1716   WM LOWNDES

NOTES

(1) T/27, 22, p. 102. Compare Letter 1200, note (3), p. 317.

(2) On this same day the Prince of Wales as Regent (the King being at Hanover) had issued a Royal Warrant to the Treasury for the payment of £60 to Brandshagen and £30 to James Hamilton, plus daily allowances during their service in Scotland (see *Cal. Treas. Books*, xxx (Part II), 1716, p. 410).

1222 NEWTON TO THE TREASURY
21 AUGUST 1716
From the holograph original in the Public Record Office.[1]
For the answer see Letter 1223

To the Rt Honble the Lords Comm[issione]rs
of his Majts. Treasury

May it please your Lordps

I humbly pray your Lordps Order for issuing out of the Coynage Duty any summ not exceeding thirty & six pounds for defraying the charges of a dinner for the Jury at the Tryall of the Pix [2] & allowing the same in my Accounts.

All wch is most humbly submitted to your Lordps great wisdome

ISAAC NEWTON

Mint Office
Aug. 21. 1716

NOTES

(1) T/1, 205, no. 19.
(2) On 29 August 1716.

1223   THE TREASURY TO NEWTON
24 AUGUST 1716
From the copy in the Public Record Office.[1]
Reply to Letter 1222

After Our Hearty Commendations Wee do hereby Authorize and Impower You to Defray the Charges of a Dinner for the Jury at the Approaching Triall of the Pix [2] Out of any the Moneys remaining in Your hands for the Use and Service of His Majestys Mint Taking care that the Charges thereof in all particulars do nott exceed the Sum of Thirty Six pounds And this shall be as well to you as to the Auditors for allowing thereof upon Your Acc[oun]t a Sufficient Warr[an]t

Whitehall Treasury Chambers 24th: *day of August* 1716.

R WALPOLE

WM. ST. QUINTIN

R: EDGCUMBE [3]

To Our Very Loving Freind Sr. Isaac
Newton Knt. Master & Worker of
His Mats. Mint.

NOTES

(1) Mint/1, 8, p. 114. There is another copy in T/53, 24, p. 532.
(2) The trial was to take place on 29 August 1716.
(3) Richard Edgcumbe (1680–1758), M.P. for various Cornish boroughs from 1701 to 1742; he was raised to the peerage in 1742. Edgcumbe was only briefly a Treasury Lord in 1716–17 (and again from 1720 to 1724), though he was a close friend and supporter of Walpole and 'managed' Cornwall politically on his behalf.

1224 NEWTON TO THE TREASURY
25 AUGUST 1716
From the copy in the Mint Papers.[1]
Reply to Letter 1221

To the Rt. Honble. the Lords Commissioners of his Majesties Treasury

May it please your Lordships

The design of sending down two Persons well-skilled in assaying and working of minerals, Mr. Brandshagen and Mr. Hamilton, to Sr. John Ereskins Mines in the Parish of Alva five miles from Sterling East and by North;[2] being as I presume, that they should in the first place survey the Mine with the Mountain about it and Assay the Ore and make a Report upon the whole that it may certainly appear whether the Same be a silver mine and of what value before they begin to dig & work the Ore, and it being intended that in doing this they should act under Mr. Haldane, Senr. the Brother in Law of the Said Sr. John Ereskine, and Mr. Haldane living near the mine and having expressed himself willing to encourage and Supervise the Said two men and forward their design: I humbly propose pursuant to your Lordships Directions the following Instructions as proper for this purpose.

1. That in the presence of Substantial Witnesses, the Said two Persons cause to be broken off from each of the two Veins of Ore which are in the said Mine, about Six or Eight pounds of Ore, and Seal up the same in two papers with Inscription upon them denoting what Vein each parcell is taken from: the Inscriptions to be signed by the Witnesses, and the two Parcells to be packed up together and sent to London to the Lords Commissrs. of his Majestys Treasury to be Assayed in London by their Order. the Witnesses may be Mr. Haldane and one or two of his Sons who are Parliament Men and Mr. Drummond Warden of the Mint at Edinburgh or any other Gentleman of Credit whom my Lord Lauderdale General of the said Mint shall please to send thither for his own Satisfaction, and the Satisfaction of the Government, and any other Person or Persons whom your Lordships shall order to be present.

2. That in the presence of the same Witnesses other pieces of Ore be broken off from each of the two Veins and Assayed and the Assays reported by the said two persons and repeated once or twice if need be for the Satisfaction of the Witness and Signed by the said Witnesses in Testimony that the Assays were made before them and sealed up and sent also to the Lords Commissioners of his Majestys Treasury. And that a Description of the said

two Veins in Breadth and depth and distance from one another be also sent and which way they run and what sort of Earth or Stone the two Veins are lodged in, and what is the depth of the Mine and the distance of each Vein from the Surface of the Mountain, and whether in that Mine there be any Bedds of Silver or Copper Ore besides the two Veins.

3. That the Casks (or old Hogsheads and Barrells) which were filled with about 40 Tunns of Ore dugg out of the said Mines by Order of the Lady Ereskine in the time of the late Rebellion and buried on the North west Side of her house just by the Gate thereof be enquired after, and a Report made thereof and of what that Ore holds by the Assay.

4. That the Burn or Channel made in the south side of the Mountain by floods running down about three or four Furlongs westward from the Said Mine within Sr. John Ereskins part of the Mountain, be well viewed to see what Sparrs and other Signs of Minerals or Metalls be found there, to Assay them and Report the Produce.

5. About two miles westward from this Silver Mine there is a Copper Mine said to be very rich in Copper and Silver so as in a pound weight of Ore to hold about half a pound of Copper and twelve penny weight of Silver, and to belong to one who went into the Rebellion, If two or three pieces of that Ore can be procured, let it be Assayed to See how much Copper & how much Silver it holds. And if there be any other Mines within three or four miles of the Silver Mine, let them examin what Silver the Ore may contain.

6. Let them view all Sr. John Ereskin's part of the Mountain and observe what Signs of Minerals may appear any where above ground and report what they find. And let them give the best Account they can of any other Mines of Copper or Lead which they can hear of in that Mountain.

I humbly propose also that Mr. Haldane Senr. be desired by a Letter from your Lordships to see that the two Gentlemen sent down do put these things in Execution, and to give them Directions from time to time to do what he may think further proper for giving his Majesty and his Royal Highness and your Lordships Satisfaction in this Matter. And that the two Gentlemen sent down be ordered to observe the directions of Mr. Haldane in making these and such like Enquiries and Observations. And that the Earl of Lauderdale General of the Mint at Edenburgh, be also desired by a Letter from your Lordships to give the two Gentlemen sent down his Protection and Encouragement, and to send either Mr. Drummond Warden of that Mint or some other intelligent and Credible Person to the Mine, who may there see the Silver Ore dug out of the Rock and Assayed and packed up to be sent to London, and may give his Lordship and the Government an Acc[oun]t thereof: And that Mr. Hamilton do assist Mr Brandshagen.[3]

If this Mine shall prove a true Silver Mine, the Kings Council learned in the Law may be consulted about the Right which either the King may have to it as a Royall Mine, or the Commissioners for the forfeited Estates; and about the Right which either of them may have to the 40 Tunn of Silver Ore already dug up; and about the Authority requisite to secure them in order to work the Mine and Smelt the Ore, and also about the Kings Right in the Copper Mine two miles westward as a Royall Mine rich in Silver, tho not belonging to Sr. John Ereskine. And when the Fact and the Law is known, the Silver Mine and the Ore may be Seized as the Law shall direct, and the Kings Right in the other Mine may be also asserted.

All which is most humbly submitted to
your Lordships great Widsome

Mint Office
25. *Aug.* 1716.

NOTES

(1) III, fos. 231–2; there is a holograph draft at fo. 253, but this is a fair copy in a clerical hand.

(2) On the Alva mine see Letter 1200, note (3), p. 317. Three days before writing the present letter Newton had taken a lengthy deposition from James Hamilton about the mine (Mint Papers, III, fo. 233). After detailing the local topography, this describes the mine as penetrating about two or two and a half fathoms from the surface, and as containing 'two veins of ore running horizontally the one almost three foot above the other, the upper vein was about 22 inches broad from top to bottom & about 18 inches wide, the other about 14 inches broad or deep & about the same wideness with the former.' It claims that Sir John Erskine had 134 ounces of fine silver from the mine, and that after he joined Mar's rebellion 40 tons of ore were stored in barrels hidden near his house by Lady Erskine. Shortly after Hamilton ran away and revealed the secret in London. The same deposition draws attention to the silver-rich copper lode about two miles west of the mine at Alva.

(3) The eight words after the colon have been added in Newton's hand.

1225 THE MINT TO THE TREASURY
27 AUGUST 1716
From the copy in the Public Record Office.[1]
Reply to Letter 1218

To the Rt: Honble: the Lords &c.

May it Please Your Lordships.

In Obedience to Your Lordships Order of Reference of the 17 of the last Month upon the Bills hereunto annexed of Mr. John Roos His Mats. late Engraver of Publick Seals, Wee humbly Report to Your Lordships, that wee have Considered & Examined the Same, and do find by the Perusal of the Several Warrants He has Produced to us, That His Majesty has directed the making of the Several Seals in the Said Bills mentiond, and by Several Certificates and Receipts it appears that they were delivered to the Respective Offices or Officers of State they were Ordered for, and by other Certificates that they were of the Weight Expressed in the Bills.

Wee further humbly Certifye to Your Lordships that wee have Examined the Prices and Rates of the Said Seals and find them to be the Same with those allowed to the Said Mr. Roos for the like Seals Engraved in Her late Majestys Reign, and with those paid both to his Predecessor Mr. Harris and to Mr. East Ingraver to His Majesty King James.[2] And as it Appears to us upon Examining the Impressions of the Said Seals, and upon comparing them with the former that the worke now performed by Mr. Roos is good, Wee are humbly of Opinion He may deserve the Price sett down in his Bill.

All which is most humbly Submitted to Your Lordships great Wisdom

Mint Office the 27 *August* 1716

R. SANDFORD

Is. NEWTON

NOTES

(1) Mint/1, 7, fo. 96.

(2) Henry Harris was chief engraver from 1690 to 1704, when he died. His predecessor, Thomas East, was the uncle of John Roos.

1226 'D.S.' TO NEWTON
18 SEPTEMBER 1716
From the original in the Bodleian Library, Oxford[1]

London Sep: 18: 1716

Sr

I have undertaken to assert ye following Position which some Gentlemen Contradict Your Decision will Determine what each of us are to pay a Considerable Wager depending

Position

Admitting a Shipp in Harbour & there be a Bullet Let Drop by Anyone from ye Top of ye Main Mast it will Certainly fall perpendicular to some exact Poynt beneath as if ye Shipp were in its swiftest Motion   I am Your very Humb. Sert.

D: S

Pray Direct for Mr. D. S. to
    be Left at Toms Coffee House
            Cornhill

    For
Sr Isaac Newton
    In London
[By Lesterfeilds]

NOTE

(1) New College MS. 361, II, fo. 70. Perhaps this was just a stupid joke at Newton's expense. Note that the 'problem' is inverted so as to be ridiculous.

It is well known that the question, whether a stone would fall from the top of the mast of a ship in swift, smooth motion to a point perpendicularly beneath, was discussed by Galileo in his *Dialogo sopra i due Massimi Sistemi del Mondo* (Florence, 1632), by Gassendi in *De motu impresso a motore translato* (Paris, 1642) and by others. For recent articles on the origins of this problem see P. Ariotti in *Annals of Science*, **28** (1972), 191–203 and D. Massa, *ibid.*, **30** (1973), 201–11.

1227 LOWNDES TO NEWTON
12 NOVEMBER 1716
From the copy in the Public Record Office[1]

Sr

I am commanded by the Lords Commissioners of his Mats. Treasury to Send You the inclosed Letter from Mr Drummond of Edinburgh,[2] and to desire you to attend their L[ordshi]ps thereupon tomorrow morning at 11 of the Clock.[3] 12 Novr 1716

W LOWNDES

NOTES

(1) T/27, 22, p. 134.

(2) See Letter 1227a.

(3) There is no minute of a meeting of the Treasury Lords on 13 November. On the fourteenth Newton was instructed to write to Brandshagen and the Hamiltons in the terms employed by him in Letter 1229, and also to prepare further instructions for them, to be laid before the Lords. (*Cal. Treas. Books*, xxx (Part ii), 1716, p. 44.)

1227a WILLIAM DRUMMOND
TO THE TREASURY
3 NOVEMBER 1716

From the original in the Public Record Office.[1]
For the answer see Letter 1229

Sir

I received the favour of your Letter some time agoe, With the honour of a Copie of His Highness Instructions to Doctor Brandshagen and his Assistant James Hamilton, about the Survey and tryall of Sr John Erskine's Mine.

The Doctor arrived here about a fourtnight agoe he has since been frequently with My Lord Lauderdale and me, and has been preparing necessary materialls for executting his instructions, wherein I have given him all the assistance I could, and every thing is in readines. And now within these few dayes Sir John Erskine is arrived, and offers immediately to goe and show where the Mine is, And My Lord Lauderdale and I are likewise ready to sett out, but the Doctor complains that he wants money, not onlie for defraying the charge of the tryalls but for his own and the Hamiltouns subsistence. Mr Haldane of Gleemeges [2] by whom he expected to be supplied with every thing (as the Doctor sayes) telling him that he had no manner of Orders about furnishing him with money, And without money he sayes he cannot proceed. Thus the matter att present stands, And on this account the execution of the Instructions came to suffer a delay, which the Season of the year makes not soe convenient, I perceiving indeed that the Doctor absolutely wanted, and he being a stranger, Out of my duety to the Government, your Recommendation, and regaird to the mans necessities, for his and the Hamiltons present subsistence, have advanced him £16: ster [3]

I humbly judged it propper to give you this account that you might know, how the matter is, Mr Haldane has been in the country for severall dayes, I shall alwayes be verry ambitious of anie opportunity to show with how much respect I am

Sir

Your most Humble and

most obedient servant

W. DRUMMOND

Edinburgh. Novr: 3d
          1716

NOTES

(1) T/1, 201, no. 1. The letter is endorsed: '12 Nov 1716 Sir Isaac Newton to [be] here tomorrow morn'. The background is as follows: after the Prince of Wales' Warrant had authorized the journey of Brandshagen and James Hamilton to Scotland (see Letter 1221, note (2)), Lowndes wrote to Lauderdale (as General of the Edinburgh Mint) explaining their mission and enclosing a copy of their instructions. He further explained how Drummond (Warden of the Edinburgh Mint) and Haldane were to co-operate in the search for the Alva silver (T/17, 3, pp. 528–30; see *Cal. Treas. Books*, xxx (Part ii), 1716, p. 442). Meanwhile, Newton himself had advanced £30 to James Hamilton's brother Thomas in order that he might join the party; this sum was repaid to him on 5 September (*ibid.*). No advance, apparently, had yet been made to Brandshagen and James; see Letter 1229.

(2) See Letter 1224, *ad init.*

(3) In accordance with Newton's Letter 1229, a money warrant for £100 was issued to Fauquier on 20 November, for transfer to Brandshagen and James Hamilton (*ibid.*, p. 554). For the settlement with Brandshagen see Letter 1244.

1228 NEWTON TO THE TREASURY
14 NOVEMBER 1716

From the holograph draft in the Mint Papers [1]

To the Rt Honble the Lords Commissioners
of his Majties Treasury

May it please your Lordps

For diminishing the expences about Sr John Areskine's Mine & making the greater dispatch in that affair,[2] I most humbly offer to your Lordships consideration whether it may not be advisable that Dr Justus Brandshagen & the two Hamilton's who are sent down to Scotland to view the Mine be ordered to smelt the Ore which lyes buried in Casks by Sr John Areskines house so soon as they have despatched the Report which by the Warrant of his Royall Highness they are already ordered to make; provided they find

373

the silver produced out of that Ore to be more then sufficient to pay all the charges of smelting it. All three understand the smelting of ores, & can consult & advise one another & therefore may each of them work apart in several furnaces for making the greater dispatch, unless they find it more convenient to work together at one or two great furnaces. I make this Proposal upon presumption that the Ore is worth smelting because it was buried in Casks for that reason, & that they can find a convenient place for setting up one or more furnaces. As fast as the Cakes of silver come from the Test, they may be marked with the Roman numbers I. II. III. IV. V. VI. etc stamped on them with a chissel; & a list of the number & weight of every Cake may from time to time be sent up to your Lordships or your Order, that the weight & value of all the silver extracted may be ascertained for preventing imbusselments. [3] And when these Cakes are melted into large Ingots, the Ingots may be numbered & weighed in the same manner. And the number & weight of every Regulus or lump of coarse silver got out of a pot of ore may be entred in books. And while they are preparing furnaces & smelting the Ore aforesaid, the Report which they are going to make may be considered here in relation to the Mine.

All which is most humbly submitted to your Lordships [great] wisdom

ISAAC NE[WTON]

Mint Office.
14 *November* 1716.

NOTES

(1) III, fo. 260. The paper is torn, removing part of Newton's signature.
(2) See Letter 1227, note (3).
(3) Embezzlements.

1229 NEWTON TO DRUMMOND
15 NOVEMBER 1716
From the copy in the Scottish Record Office.[1]
Reply to Letter 1227a

St. Martins Street near Leicester-
fields, London 15th *Novemr.* 1716

Sir

Upon the reading your letter in the Treasury The Lords Commissioners of the Treasury were pleased to order One hundred pounds for Dr. Brandshagen to bear the charges of himself and the two Hamiltons and the expenses of making the assays and finishing the Report which they are ordered to make, and to pay the Sixteen pounds which you have advanced to him, An order is drawing for paying the money out of the Exchequer, to Dr. John Francis Fauquier my deputy in the Mint.[2] And if either you or Dr. Brandshagen please to get a return for it, it shall be payd by Dr. Fauquier to the order of Dr. Brandshagen, This order may be sent either to Dr. Fauquier immediatly, or inclosed in a letter to me. I beg the favour that you would press Dr. Brandshagen and the two Hamiltons to lose no more time, but make all the convenient hast they can to finish the Enquiry and report which they are ordered to make   I am

Sir

Your most humble &

most obedient servant

Signed Is. NEWTON

NOTES

(1) Lauderdale Muniments: Mint Papers. We are most grateful to Captain the Hon. G. E. I. Maitland-Carew, of Thirlestane Castle, for permission to print this letter, which is a contemporary copy of the original (hence the word 'Signed' beside Newton's name).

(2) This action was taken as a consequence of a decision by the Treasury Commissioners on the previous day (see *Cal. Treas. Books*, xxx (Part II), 1716, pp. 44 and 554).

1230 LOWNDES TO THE MINT
24 NOVEMBER 1716
From the copy in the Public Record Office[1]
For the answer see Letter 1250

Whitehall Treasury Ch[ambe]rs 24 Nov: 1716

The Rt. Honble the Lords Comm[issioner]s of His Mats. Treasury are Pleased to Refer this Petition and the Ac[coun]t [2] Annexed unto the Warden, Master & Worker & Comptr[oller] of His Mats. Mint who are to Consider the Same and to Report to their Lordships their Opinion as to the Reasonableness of the Said Acc[oun]t and how the same should properly be paid the Petitioner

WM LOWNDES

NOTES

(1) Mint/1, 8, p. 121; there is another copy in T/4, 9, p. 299.
(2) See Letter 1230*a*.

375

1230a THE ENCLOSED PETITION AND ACCOUNT
From the copy in the Public Record Office [1]

To the Rt: Honble: the Lords Comm[issione]rs. of His Mats: Trea[su]ry
The Humble Petition of Mr. Richard Barrow

Sheweth

That your Pet[itione]r. was about Six years past, on the Death of Mr. Weddale Appointed by Craven Peyton Esqr. when Warden of the Mint to Prosecute Clippers and Coiners and Utterers of false Mony.

That your Pet[itione]r. was paid his Bill of Charges for Such Services to Mich[aelm]as 1713. [2] And that there is now due to him for two Years Services ending at Mich[aelm]as 1715 (being then dismist from such Employmt.) the sum of [£]295 : 15 : 06 As by the Bills hereto Annexed may Appear. [3]

Your Pet[itione]r. therefore humbly Prays Your Lordships to Refer his Bills to the Principal Officers of His Mats. Mint, and that they may Report the Same to your Lordships
And Your Pet[itione]r. Shall Ever Pray

NOTES

(1) Mint/1, 8, p. 114.

(2) See Letters 1132 and 1133.

(3) Presumably Barrow was dismissed (on 30 September 1715) by the new Warden, Sir Richard Sandford; compare Letter 1147. The detailed account occupies Mint/1, 8, pp. 115–20, and includes Barrow's salary for two years of £120; see also Letter 1157.

1231 CONTI TO NEWTON
29 NOVEMBER 1716
From the original in the Library of King's College, Cambridge [1]

Hanover, 10 *xbre* 1716 [N.S.]

Monsieur

Je vous demande pardon si ie n'ay pas pû vous ecrire iusque a cette heure. Je suis tombé malade depuis que ie suis icy, et ie ne suis pas encore revenu de ma maladie. [2] Je n'ay vû ni Le Roy, ni La Cour, et ie suis obligé de garder la chambre depuis vingt iours.

M. Leibniz est mort; et la dispute est finie. [3] Il y a laissé plusieurs lettres, et plusieurs manuscrits [4] qu'on imprimera, avec des manuscrits d'autres scavants, une quelque Traité de M. Des-Cartes qui n'est point paru jusque icy. Il y a des Dialogues sur les articles de la Teodicer; une i[n]struction au Prince

376

Eugene sur les exercices militaires; une instruction au Czar pour faire fleurir les arts et les sciences dans son Pais; beaucoup des remarques sur la Langue Universelle, et sur l'etimologie des mots; Comme ie esperre que le Roy me donnera la permission de voir les papiers, ie remarquerai s'il y a quelque chose touchant votre dispute, mais peut-etre qu'on cachera ce qui ne fait point d'honneur a la memoire de M. Leibniz.[5] On a comencé de travailler sur sa vie. M. Wolfius aura le soin d'ecrire tout ce qui appartient aux Mathematiques:

M. Leibniz a travaillé pendant toute sa vie a inventer des machines[6] qui n'ont point réussi. Il a voulu faire une espece de moulin a vent pour les mines, un Carosse, qui tire sans cheveaux: un Carosse, qui se change, en chaise a porteurs, et un charette; iusque des Souliers a ressort. Il y a deux modelles de sa machine arithmetique, mais elle est tres composee, et on en dit qu'elle n'est alafin que la machine de Pascal multiplié.

Vous aurez vû l'insolente dissertation, qu'on a imprimé dans les Actes de Lipsic au mois de Juin. M. Bernoulli pretend a cette heure d'être l'inventeur de calcul integral;[7] Je suis seur que la dissertation vous fera rire.

Je ne scay pas si l'Ambassadeur de Venise[8] vous a prié de proposer ala Société Royal M. le Marquis Orsi[9] Senateur de Boulogne, et un de plus grand scavants que nous avons en Italie. Il est celebre en France par plusieurs livres, qu'il a ecrit et c'est un Segneur, qui a beaucoup de merite et de talent. On dit, qu'il a refusé autrefois d'être Cardinal. Il s'est addressé a moy pour vous prier de cette grace, et ie le fais volontier, car ie connois le meure et le scavoir de M. le Marquis Orsi.

Si il y aura quelque chose de nouveau touchant l'affaire de M. Leibniz je vous en informerai avec toute l'exactitude.[10] Il n'y a peut-etre une persone plus interessé pour votre gloire, que moy. J'en ay l'obligation, et meme l'inclination. Je suis avec tout le zele, et en vous priant de faire mes compliments a Madame votre Niece.[11]

Monsieur

Votre très-semble, et tres obeissant

serviteur

CONTI.

NOTES

(1) Keynes MS. 140; microfilm 1011.24; printed in Brewster, *Memoirs*, II, pp. 434–5. Conti, a great favourite of George I, had agreed to accompany him on a visit to Germany, partly in the hope of meeting Leibniz. Leibniz had died, however, before his arrival. Conti returned to London in 1717, and finally to France in 1718.

(2) Conti suffered from asthma.

(3) Leibniz died on 4 November 1716; but the dispute was not thereby finished. Ahead was the publication of the Leibniz–Clarke correspondence, of the Appendix to Raphson, *History of Fluxions*, another edition of the *Commercium Epistolicum*, and Des Maizeaux's *Recueil*.

(4) It is impossible to identify with certainty the manuscripts Conti refers to, but there is a wealth of material on all the subjects mentioned in the Leibniz Archiv at the Niedersächsische Landesbibliothek in Hanover. See E. Bodemann, *Die Leibniz-Handschriften der Königlichen Öffentlichen Bibliothek zu Hannover* (Hanover and Leipzig, 1895).

(5) It was from Conti that Des Maizeaux later obtained numerous MS. letters and papers of Leibniz for inclusion in his *Recueil*, first published in 1720. See the Preface to the *Recueil*, and Letter 1295, note (1), p. 457. Conti himself never published any significant portions of Leibniz's work.

(6) There are numerous manuscripts relating to these in the Leibniz Archiv at Hanover (see note (4) above).

(7) Conti refers to Bernoulli's statements in the 'Epistola pro Eminente Mathematico', published in the *Acta Eruditorum* for July, not June (see Letter 1196, note (1)).

(8) An ambassador from Venice attended a meeting of the Royal Society on 23 February 1716.

(9) Giovanni-Giuseppe Orsi (1652–1733) had studied mathematics as a young man, but later his interest turned towards literary matters. He was elected a Fellow of the Royal Society on 30 November 1716.

(10) In fact no further correspondence took place between Conti and Newton.

(11) Catherine Barton; see introduction to vol. v, p. xliv.

1232 NEWTON TO BRANDSHAGEN
[Late 1716]
From the holograph draft in the Mint Papers [1]

Sr

I thank you for the description you gave me of your proceedings & of the form of the vein of silver ore. By the description which I have lately met with it is not a round vein like the body of a tree but a broad flat vein like the leafe of a Table. It is about four five or six inches thick for the thickness varies. It is covered over on either side with a crust of spar about six inches thick. The spar is mixed with some Ore, but the Ore in the middle between the two crusts of spar is the richest & the whole thickness of the Ore & spar together is about 17 or 18 inches. This broad vein runs both downwards from the bottom of the levell & also northwards from the further end of the levell proceeding both downwards & northwards into the mountain like a wall rising up from the foundations of the mountain almost to the top of it & running cross the mountain from north to south. I send you this description that you may examin it. And if it be true, you will find the vein of Ore not only at the bottom of the levell under the shaft but also at the further end of the Levell, rising up from the bottom of the Levell to a considerable height at the northern side of the Mine. This account I had from one who has seen the place. And I send it to you that if all the rubbish be not carried out of the Levell, you may cause it to be carried out till you come to the firm rock at the further end of it & there observe if you can find any signs of the vein running northwards into the Mountain For its possible that James Hamilton might there see two pieces of this vein & take them for two veins running upon a levell north & south. I am told that they began to dig the Ore at the bottom of the Levell & so dug it upwards letting it drop down into the Level as fast as they broke it off from the rock. And therefore its probable that some part of the vein may be found above the bottom of the Levell at that end thereof next the mountain. If upon sea[r]ching, you & Mr Hamilton make any new discovery, you need not give my Ld Lauderdale & Mr Drummond any trouble about this particular but only acquaint Mr Haddon of Geaneagles [2] therwith if he is at his house during these Holydays, & send me an account thereof as soon as you can.

NOTES

(1) III, fo. 269. The letter may have been written in the late summer of 1716 or about Christmas 1716 (compare Letters 1228 and 1229).

(2) *Sic*; possibly for Gleneagles, on the other side of the hills north of Alva.

1233 HORATIO WALPOLE TO NEWTON
11 JANUARY 1717
From the copy in the Public Record Office [1]

Sir

Mr Haywood the Lt. Gov[erno]r of Jamaica having by his Letter (of wch I inclose you a Copy) [2] acquainted my Lords of the Trea[su]ry, that the Lord Archibald Hamilton upon his going off from that Island deposited with him 1481 Ounces of Spanish Silver Coyn, and thereby desired (in as much as the said Money was a seizure & supposed to belong to pirates) their Lordships Directions what he should do therewith; Their Lordships have Ordered the said Lieut Gov[erno]r to consigne the said Silver to you by the first & safest Conveyance as you will find by the inclosed Copy of the Letter to him on that Occasion; [3] I am therefore by their Lordps Command to desire you (as soon as the said Silver comes to hand) to cause the same to be Coyned, & pay the proceed thereof into the Rec[eip]t of his Mats. Exchequer taking a Tally for the same in the name of the said Mr Haywood, as the proceed of the said Silver paid in by your hands I am &c   11th January 1716

H WALPOLE

27-2

NOTES

(1) T/27, 22, p. 153. Horatio Walpole (1678–1757) was a younger brother of Robert (later Earl of Orford) and uncle of the celebrated writer and dilettante of the same name. He long occupied a seat in Parliament (for Norfolk boroughs), was eventually ambassador at the Hague and in Paris, and was created Baron Walpole in the year before his death. From 1715 to 1717 he was one of the Secretaries of the Treasury, an office he was to occupy again from 1721 to 1730.

(2) This copy and another mentioned below are not, of course, entered with the present letter. Peter Heywood had replaced Hamilton as Governor of Jamaica when the latter was dismissed in May 1716, accused of 'encouraging and being concerned in fishing upon the Spanish Wrecks and robbing them' (see *C.S.P. Colonial (America and West Indies), 1717–18*, p. 47). Hamilton had been appointed Governor on 15 July 1710.

(3) William Lowndes' letter to Heywood, dated 20 December 1716, is calendared in *Cal. Treas. Books*, xxx (Part II), 1716, p. 601.

1234 MONMORT TO NEWTON
14 FEBRUARY 1717
From the original in Kings College Library, Cambridge[1]

Paris the 25th of febr. 17$\frac{16}{17}$

Sir

I have received by the hands of the chevalier de Truier the Magnificent and Gentile present [2] which you have sent me. I regard it as a pretious monument of your goodness & of the honour you have pleased to do me. How glorious is it for me! and What subject of vanity for my wife to wear ornaments given by Mr Newton and chosen by Mrs Barton, whose wit and tast are equal to her beauty?[3] I leave my spouse to revenge herself. for my part, penetrated with a sincere gratitude, I confine myself at present to assure you Sir that I have for you not only all the respect that your great name Inspires, but that nothing can equal the tender and perfect attachement with wich I have the honour to be

Sir

Your most humble

& most obedient Servant

REMOND DE MONMORT

NOTES

(1) Keynes MS. 101(B); microfilm 931.11.

(2) Presumably Newton's answer to Monmort's gift of fifty bottles of champagne; see Letter 1206.

(3) Monmort had earlier sung the praises of Newton's niece Catherine Barton in a Letter to Brook Taylor; see Brewster, *Memoirs*, II, p. 491.

1235 AMBROSE WARREN TO NEWTON
15 MARCH 1717
From the original in the University Library, Cambridge[1]

March ye 15th. 171$\frac{6}{7}$

Sir

According to an Order of the Trust[2] on the 19th: of November last for quarterly Meetings of the Trustees &c You are Desired to meet the Rest of the Trustees, Governors and Directors at the Chapel Vestry Room, On Monday next being the 18th. Instand, by 5 of the Clock Afternoon Precisely, upon matters of Moment relating to the said Trust.

I am

Sr

Your most Dutyfull

Servant

AMB WARREN.[3]

To
Sr Isaac Newton
Knight
    Humbly

NOTES

(1) Add. 3965 (18), fo. 678; the letter is in the hand of a clerk, except for the signature and address, which are in Warren's hand.

(2) Newton was one of the nine trustees of Archbishop Tenison's Chapel (now St Thomas's), Regent Street. See Number 642, note (2) (vol. IV, p. 380).

(3) Ambrose Warren, Agent for the Trustees.

1236 WILLIAM NEWTON TO NEWTON
26 MARCH 1717
From the original in the University Library, Cambridge[1]

Sr Isaack   *Marshalsea in South[wark]*
                                             March the 26th: 1717

Haveing applyd to all my friends and acquaintance in london, now seeing your Goodness contributes so freely to my Enlargmt, & that all the difference is about the Fees, my two Creditors not willing to pay them, so wth. much ado one Mr Charles Dent my Acquaintance, tho a poor man hath raised for me twenty shillings & I cannot doe any more, if I were to perish here, to compleat the Fees, God knows it is a very dismal thing to perish for hunger. Dear Sr pray pardon my Importunities, my life being in danger such a weakness I have upon me, that my health is much impair'd. If I dye must end my life miserably here God knows, Dear Sr pray let not ye good intention be now diverted, my liberty may be now had, and there is but this difference in ye paying the Fees wch. Dear Sir must be owing much to yr goodness. Therefore I humbly beg it may be done. I may live to make you a grateful return. I am affraid that if I get not out in a day or two I may be continued longer upon ye accompt that ye Keeper of ye Prison is to carry down to Reigate above 60 Fellons to be tryd there Therefore Dear Sr for Christ Jesus sake deliver me out of this sad place. And I shall have reason to pray for you as long as I live who at this time craves leave to subscribe himself

Dear Sir

Yr most Humble & most

Obedient Servant

WM NEWTON

Sr

	lb	s	
Mr Simpson one of my Plaintiffs			
insists on ———————— one pound ten shillings	01	10	
The other Mr Lewen ——————— 40 shillings	02	0	
	03	10	
Yr kindness	03	04	6
Mr Dent	01	00	0

Over and above paying ym Rests 0 – 14 – 6 towards ye paymt of my Fees I could not live on the air & my bedroom tho God knows very bad paymt for 6 weeks twenty shillings

So Dear Sr be pleasd to add twenty shillings more to wt you first designd, then I may got out honourably

If I outlive my father
& there be anything left
To me you shall be reimburst

For Sr Isaack Newton at
his House in St Martin-street
near Leister fields
        These

NOTE

(1) Add. 3965(17), fos. 638v–639r. Nothing is known about the writer, who had once before applied to Newton for relief; see Letter 1220. The Marshalsea was, of course, primarily a debtors' prison.

1237 J. BERNOULLI TO MONMORT
28 MARCH 1717
From a copy in the Burndy Library [1]

Je vous proteste, Mons[ieu]r que Je n'ai jamais eu la pensee de me commettre avec Messieurs les Anglois, ni d'entrer en lice quand même quelqu'un d'eux m'attaqueroit, bien loin de les defier le premier. le temp & le repos me sont trop precieux pour les consumer en vaines desputes mais voicy ce que c'est. Mons[ieu]r Leibnitz m'ayant demandé si Je ne pouvois pas luy fournir quelque probleme [2] pour le proposer a Messieurs les Anglois et en particulier a M. Keil pour la solution duquel seroit requise une adresse particuliere dont on ne pust s'aviser aisement sans la connoissance de quelques unes des methodes que nous avions trouvées dans le temps que J'etois encore en Hollande, & que M. Leibnitz ne trouvoit pas apropos d'en faire part encore au public, me priant pour cela de menager le secret affin de s'en servir un jour utilement contre ceux qui voudroient nous braver, comme il arrive aujourdhuy. Pour faire donc plaisir a Monsr. de Leibnitz, Jay imaginé un probleme qui me paroissoit avoir les qualites telles qu'il pouvoit souhaitter. Je luy en fis part avec une double solution, affin qu'il pût, s'il le jugeoit apropos, le proposer aux Anglois mais sous son nom. Jay sujet d'etre etonné de voir que M. Leibnitz m'ait produit comme auteur et proposant de ce probleme, et cela malgre moy et même a mon inscû. [3] Vous aura donc la bonté de desabuser Mr. Newton de l'opinion ou il est a cet egard; et de l'assurer de ma part que Je n'ai jamais eu le dessein de tenter Messieurs les Anglois par ces sortes de deffis, et que Je ne desire rien tant que de vivre en bonne amitié avec luy, et de trouver l'occasion de luy faire voir combien J'estime son rare merite, en effect Je ne parle jamais de luy qu'avec beaucoup d'eloges. Il seroit pourtant a souhaitter qu'il voulût bien prendre la peine d'inspirer a son ami Mr. Kiel des sentiments de douceur et d'equité envers les etrangers, pour laisser chacun en possession de ce que luy appartient de droit, et a juste titre, car de vouloir nous exclure de toute pretention ce seroit une injustice criante. Voicy cependant le probleme dans les propres termes que Je l'ay communique a Mons[ieu]r Leibnitz. Puisque vous temoignes le desirer.

383

[Then in Newton's hand]

Super [4] recta *AG* tanquam Axe ex puncto *A*, educere infinitas Curvas, qualis est *ABD*, ejus naturæ ut radii osculi in singulis punctis *B* et ubique ducti *BO*, secentur ab Axe *AG* in *C* in data ratione. ut nempe sit $BO.BC::1.n$. Deinde construendæ sunt Trajectoriæ *EBF* primas curvas *ABD* normaliter secantes.

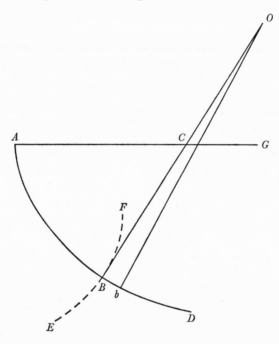

Thus far John Bernoulli in a Letter to Mons[ieu]r Remond de Monmort dated 8 April 1717

NOTES

(1) Printed in Brewster, *Memoirs*, II, pp. 437–8. This extract was sent by Monmort to Taylor for Newton, and Taylor passed it on almost immediately (see Letter 1334i, vol. VII), but without letting Monmort know that he had done so (see Letter 1280). The letter is clearly a reply to one from Monmort to Bernoulli, now lost.

(2) Bernoulli now outlines the history of the problem (see Letter 1170, note (6)) sent by Leibniz to Conti for Newton, and reveals that it originally came from him.

(3) Bernoulli's righteous indignation is hardly justified, nor the comments which follow where he tries to divorce himself entirely from the calculus dispute. Although he had wanted to keep his part in the matter secret, his anonymous involvement—in his comments in the *Charta Volans* (Number 1009), in the problem posed to the English mathematicians, and in the 'Epistola pro Eminente Mathematico' (see Letter 1196)—was considerable, and his protestations of innocence ring false.

(4) See Letter 1202, note (4), p. 324. Bernoulli had sent the problem to Leibniz on 29 February 1716.

1238 J. BERNOULLI TO ARNOLD
28 MARCH 1717
From a copy in the British Museum [1]

Bale ce 8 *Avril* 1717 [N.S.]

Je vous rends graces de la communication que vous m'avès faite de ce qui a
parû l'annèe passèe dans les transactions de Londres et surtout de la description entiere que vous m'avès envoièe de la piece [2] dont le titre estoit *Problematis olim in actis eruditorum Lipsiæ propositi Solutio generalis &c* jai lû la pretendue solution que l'auteur donne de ce Problême mais je voi quil s'y donne
un air de Maitre, sans en meriter le Nom, il traite la Chose Cavalierement, et
conclud enfin selon la formule ordinaire de *quod erat faciendum,* cela est bon
pour des Ignorants qui se laissent eblouir par la mine de confiance que notre
soluteur [*illegible*] il ne faut pas estre trop clairvoyant pour decouvrir que ce
qu'il dit ne signifie rien au fond, ce qui fait quil na garde dappliquer a un
seul exemple de cette belle solution; un exemple sur des courbes transcendentes comme celui que M Leibnitz a proposè a quelques Mathematiciens
anglois auroit trop montrè la foiblesse de la methode de notre Soluteur

NOTES

(1) MS. Birch 4281, fos. 12–14. The extract, in Arnold's hand, was sent by Arnold to
Chamberlayne on 16 May 1719 (see vol. VII).

(2) That is, Newton's 'solution' of the general problem posed by Leibniz to Conti, of
determining the set of curves cutting a given set at right angles; see Number 1187*a.*

1239 KEILL TO NEWTON
17 MAY 1717
From the original in the University Library, Cambridge [1]

Oxford May 17*th* 1717

Honoured Sr

Dr Halley informed me that there was an Answer [2] in the Acta Lipsiæ to
what I had wrote about Bernouli and that you had ordered it to be sent to
me, but I never received it, it has been some way miscaried I desire you would
inquire by whom it was sent, A freind of mine brought me the Acta the other
day and I was amazed at the impudence of Bernoulli I believe there was
never such apeice for falshood malice envie and ill nature published by a
Mathematician before, It is certainly wrote by himself, for tho he speaks of

Bernoulli always in the third person yet towards the latter end of his paper he foregott himself, and says that no body but the Antagonist can persuade himself that *My* formula was taken from Newtons this is within ten lines of the end.[3] I shall be in town about a fourthnight hence and then I will set about the answering him as he deserves [4] mean time I remain

Sr

Your

most faithful and obedient

Humble Servant JOHN KEILL

NOTES

(1) Add. 3985, no. 18.

(2) The 'Epistola pro Eminente Mathematico'; see Letter 1196.

(3) See Letter 1196, note (4), p. 304.

(4) Keill's answer was 'Lettre de Monsieur Jean Keyll, Docteur en Medicine, & Professeur en Astronomie de l'Université d'Oxford, à Monsieur Jean Bernoulli Professeur en Mathematiques de l'Université de Basle: Ecrite en Avril 1617 [*sic*]', published in the *Journal Literaire de la Haye* for 1719, Part II (La Haye, 1720), pp. 261–87. 1617 is clearly an error for 1717, but the month too, seems wrong—the present letter indicates that Keill did not see Bernoulli's article until May.

Two drafts of Keill's paper are in U.L.C. Res. 1893(*a*), (Lucasian Professorship, Box I, Packet 5); the second is more complete and contains considerable portions not included in the printed version. There is a fair copy of this extended draft in U.C.L. Add. 3968(23), fos. 339–62, and this is followed by Newton's suggested additions and modifications, in English (fos. 363–7), which were not all in the end adopted.

Most of Newton's proposed changes relate to details, but he did seek to improve on Keill's draft by rewriting some longer passages. Thus he would have preferred the following opening:

Sr

> I was not a little surprized to find in the Acts of Leipsic of the Month of July 1716 an Anonymous Letter written in your defense. It is in a style which I will not accuse of rudeness & insolence towards the English: but I shall beg leave to take some notice of the Authors extraordinary behaviour in point of candour & sincerity in passing by every thing which makes against you & acknowledging none of the errors into wch you are fallen & of wch I have accused you. Ces méprises sont si palpables &c.

(The printed text resumes: 'Ces erreurs pourtant sont palpables . . .') Newton proposed at one point to add the following paragraph:

> But this is not the first instance of this kind. When the Commercium Epistolicum came abroad, & Mr Leibnitz who had complained of me to the R. Society appealed from them to you & desired you to examin the matter, & you (according to your usual modesty) took upon you to act as a Judge between him & the Committee of the R. Society; you gave judgement in your Letter of 7th June 1713 without setting your name to it & cited your self as a Witness for your self in these words (quemadmodum ab eminente quodam Mathematico

dudum notatum est,) & Mr Leibnitz inserting this Letter into another of his own dated 29 July 1713 recommended the anonymous author to the world as an impartial Judge. And this Anonymous Letter being afterwards translated into French was published in that language as yours, & the citation was omitted, it being indecent for you to appeal to your self & to call your self an Eminent Mathematician. And now it appears by your last Paper that your designe was to claime the inverse method of fluxions as your own by the name of the integral method, & by consequence that when you set up for a Judge between Mr Leibnitz & the Committee of the R. Society, & cited your self as a Witness for your self & for him: you were in a conspiracy with him to share the method between you, & for that end you agreed with one another that you should be both judge & witness in your own cause. And this is your modesty, your candour, your integrity & your justice. But since you make a practice of writing controversial abusive papers without setting your name to them & of applauding your self in them: whenever I meet with such anonymous papers wherein you are applauded or cited as a witness or your enemies abused: I shall for the future look upon them as written by your self or at least by your procurement, unless the contrary appears to me.

The first person is, of course 'Keill' and the content of Newton's draft was printed on pp. 263–4 of the *Journal Literaire*.

It is obvious that Newton's thoughts did not run strongly towards moderation when he took pen in hand. However, for unknown reasons (perhaps because he was influenced by such foreign moderates as Conti and Monmort) Newton held on to Keill's paper after its author had submitted it to Newton (see Letter 1269) so that its publication was delayed two years; it was probably the renewal of the quarrel by Bernoulli and his pupil Kruse in the *Acta Eruditorium* for 1718 that persuaded Newton to allow the printing both of this 'Lettre' and Keill's *Epistola ad Bernoulli*. In a draft, probably of early 1720 (see Letter 1334 I, vol. VII), Newton wrote: 'I have prevailed with Dr. Keill during the two last years to suspend publishing what he has written against Mr Bernoulli, tho I cannot in justice hinder him perpetually from defending hims[elf] from the usage he has met wth from persons imployed by Mr Bernoulli.'

There is very little that is new in Keill's paper. He points out the editorial slips in the 'Epistola pro Eminente Mathematico' which give away Bernoulli's authorship of it, and claims that Bernoulli is probably author of the *Charta Volans* (Number 1009) and of the excerpt from a letter to Leibniz printed in it (Letter 1004). He makes the point that Bernoulli's solution to the inverse problem of central forces using calculus is exactly that of *Principia*, I, Proposition 41, written differently (see Whiteside, *Mathematical Papers*, VI, p. 350, note), adding (p. 280):

La difference que j'y trouve est que celle de Monsieur Newton est beaucoup plus claire, plus courte, & a beaucoup plus l'air de celle d'un Géométre que la vôtre; car il est plus d'un Géometre d'exprimer une quantité par l'aire d'une Courbe que par $ab - \int\phi x$. Il est à observer que pour déguiser votre formule, & afin qu'elle ne ressemblât pas si fort à celle de Monsieur Newton, vous avez pris la quantité x^2 qui est hors du vinculum dans la forme de Monsieur Newton, & par multiplication vous l'avez portée dans le vinculum.

He expresses his rancour at being referred to as 'that Englishman', and not by name and title. Then he repeats all the arguments already brought forward in the *Recensio*, with particular reference to Bernoulli's own part in the dispute.

1240 NIKOLAUS BERNOULLI TO NEWTON
20 MAY 1717
From the original in King's College Library, Cambridge [1]

Viro Summo
Isaaco Newtono Eq. Aur.
S.P.D.
Nicolaus Bernoulli.

Accepi ante aliquot menses nitidissimum exemplar operis Tui nunquam satis laudati de Principiis Philosophiæ Naturalis, [2] quod cum salute plurima mihi dono misisti. Pro tanto munere quamvis dignas Tibi referre gratias nequeam, in id tamen omnibus, quibus possum, modis incumbam, ut Tibi constet, Te tam insigne beneficium non in indignum aut ingratum collocasse. Summa cum delectatione nec sine fructu, ut spero, lectioni Excellentissimi operis Tui nova ista editione insigniter aucti et emendati operam dabo, quoties Academici labores (sustineo enim hic Cathedram Mathematicam ad quam superiore anno a Seren. Republ. Veneta vocatus fui) [3] aliave negotia et studia non impedient. Cæterum ut benignum in me animum, cujus novum testem pretiosissimum Tuum donum esse voluisti, porro conserves, enixe rogo. Vehementer etiam opto, ut Tibi persuadeas, me de observantia Tibi debita, deque beneficiorum a Te præcipue et ab Insigni vestro Hallejo in Anglia [4] acceptorum memoria nihil unquam remissurum. Vale.

Patavii d. 31. *Maij* 1717 [N.S.]

Translation

Nikolaus Bernoulli
presents a grand salute
to the great Sir Isaac Newton

I received some months ago a most elegant copy of your work on the *Principia Philosophiæ Naturalis*, [2] which can never be praised enough, sent by you as a gift to me with a grand salutation. Although I am unable to return you the thanks you deserve for so great a gift, nevertheless I will undertake to do everything in my power to convince you that you have made so notable a present to one who is neither ungrateful nor unworthy of it. I shall proceed to the perusal of your most excellent work (notably corrected and enlarged in this new edition) with the greatest delight and not without profit (I hope) whenever I am not prevented from so doing by my university work (for I occupy the Chair of Mathematics here to which I was summoned last year by the Venetian Republic) [3] and by other business and researches. Moreover I earnestly beg you continually to preserve this spirit of goodwill towards myself, of which you meant this most precious gift of yours to be a testimony. I am extremely eager, too, that you should be quite sure that I shall never renounce the deference due to yourself nor the memory of the kindnesses that I received in England [4] from yourself particularly and your distinguished countryman Halley. Farewell.

Padua 31 May 1717 [N.S.]

NOTES

(1) Keynes MS., no. 92; microfilm 931. 1.

(2) The second edition.

(3) In 1716 Nikolaus [I] Bernoulli succeeded Hermann as Professor of Mathematics in Padua. In 1722 he returned to Basel University.

(4) Nikolaus visited England in 1712, and while there brought to Newton's notice the error in Book II, Prop. 10 of the *Principia*; see Letter 951*a*, vol. v, pp. 348–9.

1241   H. SMITHSON TO NEWTON
26 MAY 1717
From the original in the Bodleian Library, Oxford [1]

May ye 26th 1717

Honour'd Sr

Here is committed to this Goile [2] one [*illegible*] agoe and a coman strumpet about the City and to wit an old offender that goes for his wife taken with divers Instruments and False myn'd [3] Monney in their Custody I believe I can obtaine a large discovery Off ruin [to] them if I had money to bear them Company and to humor them which I will and Sr diligently present if you please to lay Yr Summe, on me soe to do; I am Sr your poore destress'd Servt to command

HEN; SMITHSON

from ye Kings Ward in
the Marshalsea [4]
pray Sr favouring
with Your Answer
        for
Sr Isaac Newton att his
    Home in St Martins Street
                London

NOTES

(1) Bodleian New College MS. 361, II, fo. 53. The letter has been much written over by Newton and is now difficult to decipher. For Henry Smithson see Letters 1060 and 1065.
(2) Gaol.
(3) minted.
(4) See Letter 1236, note (1), p. 383.

1242 NEWTON TO THE TREASURY
MAY 1717
From the holograph draft in the Mint Papers[1]

To the Rt Honble the Lords Commissioners of his Mats Trea[su]ry.

May it please your Lordps

In obedience to your Lordps verbal Order that I should lay before your Lordps a Proposal or Memorial about coyning copper-moneys:[2] I humbly represent that the Copper be imported into a Mint by weight in clean barrs nealed & of a due fineness & size for cutting out of them blanks of such a weight as his Ma[jes]ty shall appoint; that the fineness be such that the Barrs when heated red hot will spread thin under the hammer without cracking; that the scissel be delivered back to the Importer by weight, & the Importer be paid for the excess of the Copper imported above the scissel returned back, after the rate of [*blank*] per pound weight averdupois; that it be in the power of the Mint-master to refuse such copper as doth not beare the assay or is not well sized nealed & cleaned; that when a parcel of copper-money, suppose a Ton or two, is coyned, the same be well mixed by shovelling it forwards & backwards in a heap before sufficient witnesses & then assayed before them in four or five distant places & the assays entred in books, & the tale of the heap estimated by taking a medium of all the assays, & the money then put into baggs by weight to be delivered to the people, & the weight & price of the baggs entred in books, & three or four or perhaps five pence allowed in every quarter of an hundred weight for preventing complaints about the tale; & out of every heap assayed, four or five pieces be put into a box & kept to be examined at the end of the year before whom your Lordps shall appoint; & a Remedy of about an half penny in the pound weight allowed for accidentall errors. And the Mint-master, out of the produce of the coynage, as fast as it shall arise, to pay for the Copper imported after the rates aforesaid, & be discharged upon taking back his Notes, & to pay also for putting the buildings & coyning Tools in repair at the first setting up of this coinage & for such new Tools & other things as shall be wanting; & account annually to the king. And that the King may at any time stop this coinage during pleasure.

The Officers requisite in this service are, A Mint-master with a Deputy. A Smith to forge the Dyes & Puncheons. A Graver for graving & polishing them. A Moneyer or body of Moneyers for cutting out the Blanks & coyning them & taking care of the coining Tools & keeping them in repayr. A Clerk for seeing the moneys assayed & weighed & entring the proceedings in books. Another Clerk (who may be called the Kings Clerk) for doing the like in be-half of the King & his people & for making a Controllment Roll upon oath. And an Auditor for examining the Account. The assays may be made by the Moneyer or Smith or any Labourer, & the barrs & scissel weighed by the Moneyer & the Agent of the Importer together.

All which is most humbly submitted to your Lordps great Wisdome

ISAAC N[EWTON]

Mint Office
May 1717

NOTES

(1) II, fo. 365.

(2) At the Treasury on 29 April 1717: 'My Lords having received his Majesty's pleasure that new farthings and half pence should be coined, do, in order thereto, order the following advertisement to be inserted in the [London] Gazette:

His Majesty thinking it necessary that new farthings and half pence should be coined and made of the finest British copper, the Treasury Lords give this notice thereof so as all persons who are minded to supply his Majesty's Mint with the finest British copper for that purpose may give in proposals to their Lordships in writing sealed and left with their Secretaries or one of them at any time before the 25th of May next.' (*Cal. Treas. Books*, XXXI (Part II), 1717, pp. 13 and 285.)

This notice was published on the following day. It seems likely that the verbal instructions were given to Newton at this time (compare Letter 1249).

1243 FATIO TO NEWTON
15 JUNE 1717
From the original in King's College Library, Cambridge[1]

Worcester June the 15th 1717
In Foregate-Street, at ye Sign of the
Cabinet

Honoured Sir,

This is to acquaint You that I have agreed with Mr Benjamin Steele[2] the Watchmaker at £15. for him to make the Watch for Dr Bentley. It will

391

be with four pierced Rubies and four Diamonds; and, I hope, will be worth the Money.[3] I beg leave to insert here this Note for Dr Halley, that he may correct by it the little Manuscript[4] of mine which is in his hands. I had at first supposed the Sun's Excentricity to be of 17 parts of the Radius 1000; which I had from Monsieur Cassini. Then thinking I had been mistaken, I supposed it to be of 34 parts. And having now seen your Book,[5] I suppose it of $16\frac{15}{16}$ parts. I am, with much respect, and as I have always been,

Honoured Sir

Your most humble and

most obedient Servant

N. Facio

Corrigenda in Paradoxo Astronomico

Paragrapho	pro	lege
sed tota &c	ne vel tribus Horæ Minutis superat	ne vel duobus Horæ Min. sup
Etenim &c	$2\frac{1}{3}$	$1\frac{1}{5}$
Describat &c	2′ 42″	1′.24″ 8‴
Jam de &c	2′.42″	1′.24″ 8‴

To Dr Halley Secretary
  of the Royal Society

To the Honble Sir Isaac Newton
President of the Royal Society &c
at his House
near the lower end of Leicester Fields
                London

NOTES

(1) Keynes MS. 96(D); microfilm 931.5.

(2) No maker of this name is recorded. However, Mrs Josephine Fellows of Hanley Castle, Worcestershire, kindly reports that a pair-cased watch and an eight-day brass clock movement bearing his name, at least, are still in existence.

(3) Fatio was a keen advocate of the use of jewelled movements in watches; in 1704 he had attempted, with Jacob Debaufree [Jacob de Beaufré], to take out a Patent for a 'New Invention' of using 'Precious and more Common Stones' in clocks and watches. This attempt to obtain a monopoly of the technique had angered the English (see *The English Clock and Watchmakers Reasons against the Frenchmans Bill* [n.d.] and also Bernard Gagnebin in *Notes and Records of the Royal Society*, **6** (1949), 111–12).

392

(4) The manuscript was apparently never published. However, much later, Fatio produced a little book entitled *Navigation improv'd: being chiefly the Method for finding the Latitude at Sea as well as by Land by taking any proper Altitude, Together with the Time between the Observations* . . . (London, 1728). There he gives a 'new trigonometrical rule' for finding latitude at sea from two measurements of the Sun's altitude, and the time between the measurements, but confesses that in fact Henry Sherwin (presumably in *Mathematical Tables*, 1705) has already published the Rule, although in a very cumbersome way. Fatio then says that this 'Rule makes part of the *Latin* Manuscript written by me in 1716, for the Improvement of navigation.'

At the time of writing this letter, Fatio had returned to England and was living in a state of semi-retreat in Worcestershire. From there he continued to write letters to Halley, to Whiston, to the Royal Society and to others concerning astronomical problems, but also became increasingly interested in mystical matters.

(5) In his discussions of Lunar theory in Book III of the second edition of the *Principia*, Newton gave the value $16\frac{15}{16}/1000$ for the solar eccentricity, in both Prop. 29 and in the Scholium to Prop. 35. The quantity does not occur in the first edition; in the third edition the value is changed to $16\frac{11}{12}/1000$ in the Scholium to Prop. 35, (see Koyré and Cohen, *Principia*, II, pp. 629 and 659). Cassini gives the value 17/1000 in his *Elemens de l'Astronomie* (Paris, 1694) p. 29, as an improvement of the value of 18/1000 given in the Rudolphine tables.

1244 TILSON TO NEWTON
17 JUNE 1717
From the copy in the Public Record Office.[1]
For the answer see Letter 1247

Sir

Inclosed you receive a Petition [2] Exhibited to My Lords of the Trea[su]ry by Mr Brandshagen for reward & charges in pursuing the business of inspecting & trying the Mine in Scotland belonging to Sr John Erskine which was adjudged to contain great quantity of silver; Their Lordps desire you to consider the case of the said Brandshagen, as also of the two Mr Hamiltons that were Encouraged to go on the same affair, and to Report a true State thereof to their Lordps together with your Opinion what is fit to be done therein so as the Warr[an]t signed by his Ma[jes]ty in favour of the said Mr Brandshagen and One of the Hamiltons may be discharged, as also the further demands of any of them upon the aforesaid Acco[un]t,[3] This in the Absence of the Secretarys &c. 17th June 1717

CHR TILSON

NOTES

(1) T/27, 22, p. 198.

(2) This is not, of course, with the minute; something of its content may be gathered from Newton's reply. Compare also Letter 1229.

(3) Compare Letter 1221.

1245 JOHN CORKER TO NEWTON
21 JUNE 1717
From the original in the Bodleian Library, Oxford [1]

Sr

My misfortune is so great which makes me trouble you at this time is that I have been out of Bisness so long and all my mony spent by Resonn that my famaly fell ill when they Came to Town and then my Wife Dying, my Daughter falling ill of the Small Pox and not fitt for servis yeat wherfore I humbly Crave your pardon in Takeing this freedom with you Sr as Letting my Case be known to you and Dow humbly Crave your Asisdance in this my afares which is all at Present from who was and is and Remeanies your Most Dewtyfull Servant to Command

JOHN CORKER

London
June:21:1717

Sr Tould you three Month agoe that my Wife was Dead and I wood a been willing to aworne your Leuevrey [2] if you had tould me that you wanted a footman when your man went away and I humbly Pray your one ansar for my Case is very hard

NOTES

(1) Bodleian New College MS. 361, II, fo. 42v.
(2) 'to have worn your livery'.

1246 FONTENELLE TO CHAMBERLAYNE
24 JUNE 1717
Extract from Brewster, *Memoirs*, II, pp. 289–90 [1]

You complain of me, (in your book of the lives of the french philosophers) after so civil a manner, that I think myself obliged to return you an answer. [2] I confess to you sincerely that till we had seen here the *Commercium Epistolicum*, it was commonly believed here that Mr Leibnitz was the first inventor of the Differential Calculation, or at least the first publisher therof, tho it was well known that *Sir Isaac Newton* was master of the secret at the same time; but as he did not challenge it, we could not be undeceived, and what I said concerning it was upon the credit of the common belief, which I did not find contradicted. But since it is so now, I promise you I will change my language whenever there is an opportunity, for I do assure you that it has been my study all my lifetime, to keep myself free from any partiality, whether national or personal, nothing being my concern but truth.

NOTES

(1) We have used the copy of Brewster's *Memoirs* in the University Library, Cambridge, which has been collated by H. R. Luard with the original manuscripts.

The manuscript is described in the *Sotheby Catalogue*, Lot 176, and by Brewster, as being in the hand of Catherine Barton. The Catalogue gives its date as 5 July 1717 [N.S.], Brewster as 5 February 1717. The former appears to be correct.

(2) Fontenelle refers to Chamberlayne's *The Lives of the French, Italian and German Philosophers* (London, 1717). Following the dedication, which is dated 25 March 1717, is a postscript, in which Chamberlayne writes:

> My Lord, [Lord Parker] I think my self bound to mention, by way of Postscript, a kind of a Negative Injustice and Affront which the Ingenious Historian of the Transactions of the *Academy* [that is, Fontenelle] has put upon our Nation, and more immediately upon the *Royal Society*. Your Lordship will observe in the Lives of the Marquis *de L'Hopital*, and Messieurs *Viviani*, *Guglielmini*, and [Jakob] *Bernoulli*, Four of the greatest *Mathematicians of Europe*, and all Members of the said *Academy*, that there is an Account given of the *Differential Calculation*, and the Invention thereof, which I think is every where ascribed to the late Mons. *Leibnitz*, . . .; at least it is no where attributed, as it ought, to our own Countryman Sir *Isaac Newton*, the First Mathematician in the World. Now it is notorious that the Writers of the *Acta Leipsiensia* make Mons. *Leibnitz* the Author of the said Differential Calculation; but it is not less known to your Lordship, that Dr. *Keill*, in the *Commercium Epistolicum*, has done our *British Philosopher* justice; and has fully proved that which Mr. *L.* did in some manner acknowledge to me (when I attempted to reconcile these two Great men) that Sir *I.N.* might be the first Inventor, but that he himself had luckily fallen about the same Time upon the same Notions.

The four Eloges mentioned by Chamberlayne are to be found in Fontenelle's *Eloges des Academiciens avec l'Histoire de l'Académie Royale des Sciences*, i (La Haye, 1740; facsimile edition, Brussels 1969).

1247 NEWTON TO THE TREASURY
27 JUNE 1717
From the holograph draft in the Mint Papers.[1]
Reply to Letter 1244

To the Rt Honble the Lords Commissioners of his
Majties Trea[su]ry

May it please your Lordps

In obedience to your Lordps Order of June 17th instant, upon the Petition of Dr Justus Brandshagen, directing me to state the accounts of the Doctor & those of Thomas & James Hamilton in relation to Sr John Ereskin's Mine: I humbly represent that by the Warrant of his Royal Highness [2] the Doctor

395

was allowed 60£ & James Hamilton 30£ to enable the Doctor to provide some things in London & both of them to go down into Scotland to view & examin the said Mine; & after their arrival at Edinburgh the Dr was to be allowed 20s per day & James Hamilton 10s per day by the said Warrant during their stay in Scotland upon the said work, or so long as your Lordps should see cause to pay the said allowances. And that Thomas Hamilton was afterwards by a warrant of the Lord Commissioners of the Treasury allowed also 30£ to go to Scotland wth his brother, & had a verbal promise, as I was then informed, to have the same allowance with his brother of 10s per day. And that the two Hamiltons arrived at Edinburgh Sept 14th & the Doctor Octob. 13th last, & the Doctor & Thomas Hamilton left Edinburgh Feb. 19th following to return into England, but James Hamilton went to Alloua to watch the Mine & see that no Ore be dug & carried away.

By the said Warrant therefore the Doctor is to receive 129£ for 129 days beside 44£ 13s 6d paid by him to the two Hamiltons, & besides the charges he hath been at in clearing the Mine of rubbish, & providing iron tools for the labourers & timber to prop & secure the Mine, amounting to 8£ 12s 0d, & besides his further charge of fitting up a Laboratory wth Furnaces, an Assay Table, Mortars, Sives, Charcoal, Pitcoal, Nitre & other things necessary for making the Assays, amounting to about 10£; & the charge of seven fair copies of the Report on large paper for yr Lordps & others concerned;[3] & of providing baggs & boxes for the Ore, amounting to 6£. 8s. 0d. All which summs amount unto 199£. 3s. 6d. whereof he hath received 150£. And so there remains due to him 49£. 3s. 6d.[4]

By the same Warrant James Hamilton is to receive 79£ for 158 days, whereof he hath received of the Doctor 21£. 13s. 3d. And there remains due to him 57£. 6s. 9d.

And if Thomas Hamilton be allowed equally with his brother (for he went down upon that supposition, & was the most serviceable man of the three) there will be 79£ due to him on that account, whereof he hath received of the Doctor 23£. 3s. 3d. & there remains due to him 55£. 16s. 9d.

The total due to all three is 162£. 10s. 0d, besides what your Lordps shall see fit to allow to James Hamilton for his attendance at the Mine ever since Feb. 19th to see that no Ore be dug & carried away untill he be discharged. For wch service I conceive that 20£ may be a sufficient recompence if he be dismist immediately.[5]

All wch is most humbly submitted to your Lordps great Wisdome

Is. NEWTON,

Mint Office
27 *June*. 1717

NOTES

(1) III, fo. 263.

(2) Of 20 August 1716 (see Letter 1221 and Mint Papers, III, fos. 239 and 255).

(3) Brandshagen's report (of 29 April 1717) may be found in Mint Papers, III, fos. 221–6; it includes a very crude map of the mine, a verbal description of the veins and ores contained in them, and the tabulated results of quantitative assay of a number of samples of ore.

(4) Obviously Newton derived these expenses from Brandshagen's petition. Advances from their *per diem* allowances had been made to the delegates (compare Letter 1229) on 20 November 1716 (£100) and 7 February 1717 (£50); see *Cal. Treas. Books*, XXXI (Part II), 1717, p. 134.

(5) The Treasury accepted Newton's recommendations for the payment of the three delegates to 19 February 1717, and the award to James Hamilton thereafter. Subsequently the three claimed payment for the period between 19 February and the submission of their report on 29 April. Their claim was refuted by Newton in his draft narrative of the whole Alva mine affair (probably written in 1724) which contains allusions to correspondence now missing (see Mint Papers, III, fo. 246):

(i) On 8 January 1717 and again on 24 January the commissioners in Scotland for the mine (the Earl of Lauderdale, Mr Haldane and Mr Drummond) wrote to Newton describing 'the form, bigness & richness of the vein of Ore, & moved that he should acquaint the Lords of the Trea[su]ry that this [mine] survey was over & that the Report was preparing: & in the second that everything was then in readiness & that they intended to have sent up the Report with assays & Ore by express that night (Jan 24), but were stopt by a sudden resolution of Brandshagen & the two Hamiltons[,] directly contrary to the Princes Order[,] that the Report should not come up before them but they would bring it up themselves . . .'

(ii) 'On the said second Letter a Question was insinuated whether Brandshagen & the two Hamiltons should all of them leave Edinburgh without order, & Sr I. Newton Jan. 31 by Order wrote back that Tho. Hamilton should stay to watch the mine & see that no Ore be carried away till the Kings pleasure should be known.'

(iii) William Drummond replied to this letter from Newton on 16 February 1717 explaining that Thomas was much the most capable of the trio, and that therefore the commissioners in Scotland proposed to send him to London and detain James Hamilton to look after the mine. They confessed 'themselves ashamed of the delays' that had occurred.

Brandshagen and Thomas Hamilton left Edinburgh for London on 19 February, but still delayed two months before handing over their report. Newton argued that the whole case for payment after their departure was fraudulent and the delays deliberate: 'By the Princes directions they were to make hast, & yet they were above five months in Scotland in executing their Commission tho they might have done it in less than half the time.' Even so, the work had been finished by the end of January, and all after that was mere 'loitering' for which Newton was not responsible and no payment was deserved. And it was not he who had ordered *James* Hamilton to stay at the mine!

1248 LOUIS-JEAN LEVESQUE DE POUILLY
TO NEWTON
3 JULY 1717
From the holograph original in the University Library, Cambridge[1]

Monsieur

Voicy deux nouvelles methodes pour prendre la difference des meridiens. je n'en fais pas asses de cas pour les regarder comme un present qui merite de vous estre offert. je ne pourray cependant m'empescher de les estimer beaucoup si vous ne desaprouves point le pretexte que je prens de vous les presenter pour vous ecrire et vous assurer de mes respects. mais pourries vous trouver mauvais Monsieur que vous ayant l'obligation d'un bien aussy grand que la connoissance du vray, je cherche a vous en temoigner ma reconnoissance, car enfin, graces a vous, il est permis presentement d'estre initié aux mysteres de la nature et admis a la connoissance de ses secrets les plus caches. mais il n'en est point de vos decouvertes comme de ces systemes fameux qui ne sont appuyes que sur des conjectures, et qu'il est facile de renverser par d'autres conjectures. c'est sur des fondemens inebranlables que vous aves elevé ce magnifique édifice de votre philosophie. les yeux des ignorans n'en appercevront jamais toutes les beautes. mais ceux qui s'y connoistront le regarderont toujours avec etonnement comme le chef d'oeu[v]re et la merveille de l'esprit humain. de ce grand nombre d'hommes qui sont penetrés d'admiration pour vous j'ose me flatter que personne ne l'est plus que moy. et s'il y a sur la terre, quelque chose d'aussy grand que votre esprit c'est sans doute le respect avec lequel je suis

Monsieur

Votre tres humble et

tres obeissant serviteur

LEVESQUE DE POUILLY

A Paris ce 14*eme*
de juillet 1717

je scais Monsieur que si la Superiorité de votre esprit vous eleve infinement au dessus des autres hommes, la bonté de votre cœur vous rabaisse jusqua eux. C'est ce qui me fait esperer que vous voudres bien me faire scavoir si vous ne trouveres pas mauvais que je vous assure de temps en temps de mes tres humbles respects mon adresse Monsieur est rue des postes pres de l'ancienne estrapade

[*The enclosed methods*]

Methodes
pour prendre la difference des Meridiens

1^0 les etoiles fixes, Saturne, et jupiter n'ont point de parallaxe sensible aussy bien que Mars, Venus et Mercure pour peu qu'ils soient eleves au dessus de l'horizon, et la terre peut se regarder comme un point par rapport a la distance qui les separe de nous

2^0 Soit un grand Cercle qui passe par le centre de plusieurs etoiles fixes, supposons un observateur au centre de la terre qui voye quelqu'une des planetes entrer dans ce grand cercle. tous les autres observateurs qui seront sur le globe de la terre et a qui la parallaxe de ces planetes sera insensible, les verront aussy au meme instant entrer dans le meme plan et par la difference des heures qu'ils conteront pour lors ils pourrons juger de la difference des meridiens

3^0 il suit de la 1ere qu'il n'est necessaire que ce plan passe precisement par le centre de la terre

4^0 des Ephemerides exactes dans lesquelles on calculeroit a quelle heure une planete entreroit dans un plan qui couperoit la terre et passeroit par le centre de certaines etoiles fixes tiendroient lieu d'un observateur

autre methode

1^0 la lune met environ 49′ 12″ de plus que le soleil a faire sa revolution journaliere

2^0 si la lune decrivoit d'un mouvement sensiblement uniforme dans l'espace de 24 heures un cercle parallele a celuy du Soleil, la difference des ascensions droites du soleil et de la lune resteroit toujours la meme [.] que la lune parcoure une minute de son cercle sensiblement parallele a l'equateur soit un cercle de declinaison qui passe par le centre de la lune, ce cercle determinera l'ascention droite de la lune qui sera d'une minute plus grande qu'elle n'estoit par la 10eme du 2eme de Theodose [2] la difference des ascensions droites sera donc diminuée d'une minute par le mouvement de la lune, mais par la supposition le soleil dans le meme temps parcourt une minute de son cercle, donc son ascension droite sera aussy augmentée d'un minute et par consequent la difference restera la meme

3^0 il y auroit donc toujours une egale portion d'equateur interceptée entre les cercles de declinaison qui passeroient par son centre et celuy du soleil et par consequent on conteroit la meme heure dans tous les meridiens a l'instant qu'elle y passeroit

4^0 qu'on concoive 24 meridiens eloignes l'un de l'autre de 15 degres, soit imaginée une lune se mouvoir comme on la supposé dans les propositions precedentes, partir d'un certain point a une certaine heure et arriver a la meme heure dans tous les meridiens que la vraye lune parte en meme temps du meme point elle reviendra au meridien dont elle est partie 49′ 12″ plus tard que cette lune fictice, elle arrivera donc au premier de ces 24 meridiens 2′ 3″ plus tard que l'autre, c'est a dire qu'il sera pour ce second meridien 2′ 3″ de plus qu'il n'etoit pour le premier meridien quand la lune y passoit il sera de meme 4′ 6″ pour le second 12′ 18″ pour le sixieme 24′ 36″ pour le douxiesme 36′ 54″ pour le dixhuitieme et ainsy de suite

5⁰   un observateur n'aura qua examiner a quelle heure precise la lune passe deux fois de suite par le meme meridien, il distribuera ensuite la quantité de temps dont le jour lunaire surpasse le jour solaire dans tous les meridiens a proportion de leurs eloignement si quelque autre personne observe dans un autre pays l'heure du passage de la lune par le meridien, il sera facile par la comparaison de les observations d'en deduire le difference des meridiens

6⁰   des ephemerides exactes pourront tenir lieu du premier observateur

NOTES

(1) Add. 3972, fos. 13–16. The content of the letter makes it virtually certain that Newton was the addressee, though the envelope has been lost. Louis-Jean Levesque de Pouilly (1691–1750), according to the *Nouvelle Biographie Générale*, early studied mathematics in Paris, where 'un des premiers en France, il s'efforca d'expliquer l'admirable ouvrage des *Principes*'. By 1718 he was one of the editors of *L'Europe savante*, published at the Hague, and made his reputation as a man of letters; notably a work called *Théorie des sentiments agréables* (Geneva, 1744) was highly popular.

(2) Theodosius of Bithynia, *De sphæra libri duo*, first printed at Venice in 1518. The work was written in the first century B.C.

1249   LORD STANHOPE TO THE MINT
[20] JULY 1717
From the original in the Mint Papers.[1]
For the answer see Letter 1253

Gentlemen.

Enclosed I send You by the Commands of the Lords Commissioners of his Majesties Treasury the several Proposals for supplying his Majesties Mint with British Copper for the Coyning of Farthings and halfe pence delivered in to their Lo[rdshi]ps pursuant to his Mats. pleasure [2] signified by their Lor[dshi]ps in that behalfe in the Gazette the thirtieth of April last,[3] Their

Lordps direct You to consider the same and Report Your Opinion upon the said Propositions and upon all other matters relating to the Furnishing the Copper and the making of the Farthings and halfe pence intended I am

Gentlemen

Your most humble Servt

STANHOPE

Trea[su]ry Chambers
July 1717
Officers Mint

NOTES

(1) II, fo. 366. There is a minute in T/27, 22, p. 206, giving the date. James, first Viscount Stanhope (1673–1721), soldier and politician, had been appointed First Lord of the Treasury on 12 April and Chancellor of the Exchequer on 15 April 1717.

(2) This decision had been taken on 17 July; see *Cal. Treas. Books*, XXXI (Part II), 1717, p. 30.

(3) See Letter 1242, note (2), p. 391. Twelve offers of copper for minting had been received in response to the advertisement in the *Gazette*. The names are given in T/27, 22, p. 206 as follows:

1. Mr Neal	7. W.F.
2. Sir Isaac Newton	8. Richard Jones
3. Mr Chambers	9. Jonathan Holloway
4. Mr Chambers *et al.*	10. Samuel Green
5. John Parker & Partners	11. Mr Wood
6. Henry Robinson	12. John Applebee

No doubt Newton's name was included for his submitted ideas on the whole scheme for coining copper. The Treasury wrote again to the Mint on 2 August 1717 (Letter 1252) asking it to hasten its report.

1249A CHARLES STANHOPE TO THE MINT
26 JULY 1717

From the copy in the Public Record Office.[1]
For the answer see Letter 1278

The Right Honble. the Lords Comm[issione]rs of His Mats. Treasury are Pleas'd to Refer this Bill to the principal Officers of His Mats. Mint who are to Peruse his Mats. Warrants directing the several Publick Seals to be prepared and the Respective Certificates of their being delivered pursuant thereto & Examine into the Reasonableness of the Prices set downe for the same & thereupon make their Report to their Lordships

C. STANHOPE

Whitehall Treasury Chambers 26 *July* 1717

401

NOTE

(1) Mint/1, 7, p. 101. This is written on the bill for £1226. 5s. 7d submitted on behalf of the estate of the late John Roos (or Ross), engraver of seals; this was for his work in making nineteen seals for the new monarch, including those for the Courts of King's Bench and Common Pleas, the Commonwealth of Massachusetts, and the colonies of New York, Virginia, and the Bahamas etc.

Charles Stanhope (1673–1760), cousin of Lord Stanhope, served him as Undersecretary of State, 1714–17; then he was transferred as Secretary of the Treasury (with Lowndes), 1718–21, whence he became Treasurer of the Chamber, 1722–27. He was deeply involved in the South Sea Company scandal but was acquitted of crime in the subsequent parliamentary inquiry (*History of Parliament*, II, pp. 433–4).

1250 NEWTON AND SANDFORD TO THE TREASURY
31 JULY 1717
From the copy in the Public Record Office.[1]
Reply to Letter 1230

To the Rt. Honble: the Lords Com[missione]rs of His Ma[jes]ties Treasury.

May itt Please Your Lordpps

In Obedience to Your Lordpps Order of Reference of the 24th of Nov[em]ber last upon the Annexed Bill of Mr Barrow for Prosecuting Clippers and Coyners from Mich[aelm]as 1713 to Mich[aelm]as 1715.[2] Wee humbly Certifye Your Lordpps yt Wee have Exa[mi]ned the S[aid] Bill, & by the Vouchers do find yt ye Services therein mentioned were performed. And are humbly of Opinion they may deserve the Sum of 250 *lb* to be paid by the Ma[ste]r and Worker of His Ma[jes]ties Mint upon your Lordpps Order [3]

All wch: is most humbly Submitted to Your Lordpps great Wisdom

RICHD. SANDFORD

IS. NEWTON

Mint Office ye 31*th. July*
1717

NOTES

(1) T/1, 208, no. 11; in a clerical hand, including the signatures.

(2) Number 1230*a*.

(3) This recommendation was accepted by the Treasury Lords on 6 August (*Cal. Treas. Books*, XXXI (Part II), p. 41); the warrant for payment is Number 1254.

1251 NEWTON AND SANDFORD
TO THE TREASURY
31 JULY 1717
From the copy in the Public Record Office[1]

To the Right Honble: the Lords Commissioners of
His Majties. Treasury

May it Please your Lordships

The Tin Contract being determined the 1st. of June last, We are informed that the Tin Coyned[2] last Midsummer exceeds 500 Tuns.

A small quantity of it being already brought to Town has been sold at 3*lb*. 10*s*. per hundred to the Russia Merchants, some other quantitys are to be shipped off in Cornwall for Holland and the Streights at a Cheaper rate, and when two Ships more Loaden with Tin arrive at London to be sold We are credibly informed it will be offered at 3*lb*. 5*s*.[3]

As there is a prospect that a pretty good quantity will be transported both to Turkey and Italy before the winter comes on, We have thought it our duty humbly to represent to your Lordships that unless there be a discretionary power appointed to sell what remains in our hands at the market price, and to make such Contracts both with the Merchants and Pewterers as shall be judged to be most advantageous, the great stock in our hands is like to remaine unsold.

Which is humbly submitted to your

Lordships great Wisdom

RICH SANDFORD

IS. NEWTON

Mint Office
31*th. July*. 1717

NOTES

(1) T/1, 208, no. 12; in a clerical hand, bearing the signatures of Sandford and Newton.

(2) Produced; after the block tin had been smelted from the ore it was taken to one of the 'coinage towns' in the West, where the metal was assayed (by striking a piece off one corner of the block) and, if approved, the block was then 'coined' or stamped with the symbol of the King (or the Duke of Cornwall if he had been created). It was then merchantable.

(3) With free sale, the price was falling below the price of £76 per ton maintained during the years of the Crown monopoly which had just ended. As long ago as 1705 Newton had predicted, too pessimistically, that Cornish tin was over-produced and so its price must descend, perhaps as low as £45 per ton (vol. IV, pp. 466–7). Obviously the stock of tin held in the Tower was almost unsaleable at the monopoly price, when parcels coming fresh from the mines could be obtained for a lower one.

1252 HENRY KELSALL TO THE MINT
2 AUGUST 1717

From the copy in the Public Record Office.[1]
For the answer see Letter 1253

Gentl[emen]

The Lords Comm[issione]rs of his Ma[jes]tys Trea[su]ry desire you will hasten your Report [2] upon the several proposals for supplying his Ma[jes]tys Mint wth: British Copper for the Coining of Farthings & half pence wch: were referred to you the 20th of July last & transmit the same to their Lordps wth: all the speed you can I am &c 2d Augt. 1717

H: KELSALL [3]

NOTES

(1) T/27, 22, p. 213.

(2) The Treasury had previously requested a report; see Letter 1249.

(3) Henry Kelsall (?1692–1762), from 1714 to his death one of the four Chief Clerks of the Treasury, was a protégé of the Duke of Newcastle's, with whom he had been at school. He sat in Parliament from 1719 to 1734 (*History of Parliament*, II, p. 185).

1253 NEWTON TO THE TREASURY
3 AUGUST 1717

From the holograph draft in the Mint Papers.[1]
Reply to Letters 1249 and 1252

To the Rt Honble the Lords Comm[issione]rs of
his Mats Trea[su]ry.

May it please your Lordps

In obedience to your Lordps Order of Reference on the Proposals for importing barrs of fine copper into his Majts Mint for coyning copper money out of them, We humbly represent that upon giving the Proposers a meeting, We find that the Mint may be supplied with such barrs at nineteen pence per pound weight Averdupois or a little under, taking back the scissel at the same price but scarce under eighteen pence. The Master & Worker is willing to undertake the coynage out of such barrs at three pence farthing per pound weight & to defray all incident charges of assaying weighing coinage & putting off after a Mint is set up. And so a pound weight of Copper with the coinage will cost about $21\frac{1}{4}d$ or $22\frac{1}{4}$ & may be cut into 23 or 24 pence to answer all other charges. If there be a King's Clerk, he & the Auditor of this Mint may be paid what you Lordps shall order out of the profits of the coinage. The Master & Worker proposes to pay half the price of the Copper upon the importation thereof.

When by publick notice we gave the Proposers a Meeting some of them upon hearing how the copper was to be sized & assayed, withdraw. Mr Neale & Partners were in the country & are not yet returned to town. Mr Essington demanded 19*d* per pound weight for the best copper and 18*d* for a coarser sort. Mr Jones demanded 18½*d* for the best copper. Mr Hind demanded 18*d*. The copper he shewed us was very good & bore the Essay, & this morning he has delivered in some Barrs to be further examined. We are best satisfied with his Proposal, but have not yet had time to examin his barrs. Mr Parker demands the same price, but we have not yet seen his workmanship.[2]

If at any time it be thought fit to change the Reverse of the money it may be done by a Signe Manual, paying the Graver for a new Puncheon.

All wch &c

Ri. Sandford[3]

Is. Newton

Mint Office
Aug. 3, 1717

NOTES

(1) II, fo. 443; there is the beginning of a draft at II, fo. 326.

(2) On 10 August Newton was summoned to the Treasury, and there the Lords agreed that a contract should be made with Mr Hind (or, as he is sometimes called, Hines) to supply the Mint with copper at 18 pence the pound (*Cal. Treas. Books*, XXXI (Part II), 1717, p. 43). On the twelfth of the month £500 was granted to Newton by imprest so that he would have funds to begin the work of copper coinage (T/60, 9, p. 451).

(3) The name is written by Newton.

1254 WARRANT TO NEWTON
9 AUGUST 1717
From the copy in the Public Record Office[1]

After Our hearty Commendations Approving of the Report aforegoing These are pursuant to the Power to be given by Act of parliamt. in that behalfe to Authorise Direct & Require You Out of the Money in your Hands of the Coinage Duty for the Service of His Mats Mint to pay or Cause to be paid unto Rich Barrow Gent or his Assigns the Sum of two Hundred & fifty Pounds in full of his Bills to the Said Report Annexed & of all Charges Expences &

Demands for prosecuting of Offences in Counterfeiting or Diminishing the Coin of Great Britain from Mich[aelmas]s 1713 to Mich[aelmas]s 1715. And this shall be as well to you for Paymt. as to the Aud[itors] for allowing thereof upon your Acc[oun]t a Sufficient Warrt. Whitehall Trea[su]ry Chambers 9 Augt. 1717

> J WALLOP [2]
>
> GEO BAILIE [3]
>
> THO: MICKLETHWAITE [4]

To Our Very Loving Freind Sr Isaac Newton Knt.
Master & Worker of His Majestys Mint

NOTES

(1) Mint/1, 8, p. 121; compare Letter 1250. There is another copy with the account details (see Letter 1230a) repeated in T/53, 25, pp. 413–17.

(2) John Wallop (1690–1762), M.P. for Hampshire and (like the other two signatories) a Treasury Lord (until 1720). He was created Viscount Lymington in 1720 and Earl of Portsmouth in 1743. His son, also John, married in 1740 the younger Catherine Conduitt, Newton's great-niece, and it was through this marriage that possession of Newton's papers passed to the Portsmouth family.

(3) George Baillie (1664–1738) escaped to Holland after the Rye House Plot in the reign of Charles II, but returned with William III. He sat in the Scottish Parliament and later (1707) in that of the United Kingdom. A staunch Whig, he was a Lord of Trade (1710–12), then of the Admiralty (1714–17), and then of the Treasury (1717–25).

(4) Thomas Micklethwaite (1678–1718), heir of Sir John Cropley, M.P., was a protégé of James Viscount Stanhope, by whom he was appointed to the Treasury in 1717. A year later, when Stanhope left the Treasury, Micklethwaite became Lieutenant General of the Ordnance.

1255 NOTICE OF A MEETING OF THE COMMISSION FOR BUILDING FIFTY NEW CHURCHES
10 AUGUST 1717
From the original in the Mint Papers [1]

Sr

You are desired to Meet the Lords and others the Commissioners appointed by His Majesty for *Building* the **Fifty New Churches** *in and about* the *Cities* of London *and* Westminster, &c [2] *on Thursday the 15th Day* of *this augst* at *ten* of the Clock in the *fore* noon, at the *Pallace yard in Westmr* to proceed in the

Executing of the Authorities, and Transacting the Affairs of the said Commission. Dated the *tenth* Day of *August* 1717

WILLIAM WATERS, *Messenger.*

Sr Isaac: Newton Kt

NOTES

(1) II, fo. 336. This is a printed notice with the date, time and other variables written by hand. Newton has used the paper for (i) chronological notes (ii) draft orders relating to the copper coinage (iii) a note on the origins of Greek prose writing. Two similar printed notices, calling Newton to meetings of the Commission on 4 March 1717 and 2 July 1720 are to be found in the Yahuda Collection at the Jewish National and University Library, Jerusalem (Newton MSS. 7, packet 3). Another printed notice (Bodleian New College MSS. 361, II, fo. 77v) calls Newton to a meeting of the Commissioners for finishing St Paul's Cathedral upon 13 October [1719] (the year is not given in the notice). Newton attended a number of meetings of this latter Commission in the period 1715–21; see *The Wren Society*, **16** (1939), 116–18, 130, 132–6.

(2) In 1711 an Act had been passed imposing a duty on coals to pay for 'Building fifty new churches in or near the Cities of London and Westminster or the suburbs thereof.' Newton was one of the Commissioners appointed to implement the Act, who found themselves perpetually dogged by financial and other difficulties, requiring the passage of seven further Acts to elucidate the first. Only twelve churches were built between 1711 and 1730 (among them, St Mary-le-Strand and St Martin-in-the-Fields), while money intended for the new churches was diverted to the repair of Westminster Abbey and Greenwich Hospital, and even to the salary due to Wren for his work on St Paul's. The Act was finally repealed in late Victorian times. Architects employed by the Commissioners included Nicholas Hawksmoor, John James, Thomas Archer and James Gibbs.

As the original records of the commission have vanished it is not known what business (if any) was done on 15 August 1717. On the ninth the commission had resolved to ask for an additional £30000 (see H. Colvin, 'Fifty new churches', *Architectural Review*, **107** (January–June 1950), 189–96, and Sir J. Summerson, *Georgian London* (London, 1970), p. 84).

1256 LOWNDES TO NEWTON
12 AUGUST 1717

From a copy in the Public Record Office.[1]
For the answer see Letter 1264

Sir

The Lords Comm[issione]rs of his Ma[jes]tys: Trea[su]ry desire You to prepare & lay before their Lordps a State of the Gold & Silver Coins of this Kingdom both as to their Weight & Fineness with the Difference in Value between a Guinea & 21*sh*: & 6*d* in Silver together with Your Observations &

Opinion And their Lordps being Informed yt. great Quantities of Silver Coins are melted by sev[era]l persons for their private Lucre contrary to Law to direct You to propose what Method you think will be most Effectual for preventing & discouarging that evil practice   I am &c   12 Augt. 1717

W LOWNDES.

NOTE

(1) T/27, 22, p. 220*b*, a clerical minute. For the Treasury Lords' decision to thus invite Newton to prepare a paper see *Cal. Treas. Books*, XXXI (Part II), 1717, p. 44.

1257 NEWTON TO LORD STANHOPE
17 AUGUST 1717
From the holograph original in the Public Record Office [1]

Leicester Fields. Aug. 1717

May it please your Lordp

Mr Kelsal was with me this afternoon to forward the signing of the Warrant for setting on foot the copper coynage, [2] wch puts me upon acquainting your Lordship that I am not yet come to an agreement with the Moneyers & that some of those who propose to import copper at $17\frac{1}{2}d$ have been tampering with Hind the Brasier & his Partner to unsettle my agreement with him. [3] The moneyers insist stifly upon 2*d* per pound weight, & if I cannot get them to abate a farthing, I must intreate your Lordp & the other Comm[ission]ers that on their account a farthing be added to the $3\frac{1}{4}d$ allowed me. [4] And whilst those that propose to import copper at $17\frac{1}{2}d$ pr Lwt do not endeavour to satisfy me that they are able to manufacture the copper to my satisfaction & are willing to do it at that price, but rather avoid giving me that satisfaction & are underhand at work to undermine me by tampering with the men that I have recommended to your Lordps: I beg the favour of a few more days to see what I can do with the Moneyers & how the affair stands wth Mr Hind, who is at present out of town. I hope another week will be sufficient to settle every thing. And in the meane time the preparations will be making as fast as if the warrant was allready signed. I am

My Lord

Your Lordps most humble

& most obedient Servant

IS. NEWTON

NOTES

(1) T/1, 208, no. 28. The letter was enclosed with one from Kelsall to Charles Stanhope. In this Kelsall describes his meeting with Newton: 'Sir Isaac came Yesterday to the Treasury & desired the Warrant [for Coyning Copper Money] might be sent to Lord Stanhope for the Kings hand with a blank left for the Allowance to the Officers of the Mint for the charge of Coynage &c because of some difficulty he found to agree with the Moneyers about the price to be allowed them for their service, I waited upon him this afternnon & found he had several other Difficultys in this affair which I took the liberty to desire him to be put in writing to be sent to Lord Stanhope . . .' (P.R.O. T/1, 208, no. 28(a)).

(2) See Letters 1252 and 1253; the Warrant was issued on 13 September, see Number 1262 below.

(3) Compare Letter 1261 below.

(4) This would make the total cost per pound of coined money $21\frac{1}{2}d$, of which $2d$ would go to the moneyers and $1\frac{1}{2}d$ to Newton for fee and remaining Mint expenses; if the pound were cut into 23 pence, $1\frac{1}{2}d$ also would remain for the Crown's profit and any other outgoings.

1258 J. BERNOULLI TO WOLF
30 AUGUST 1717
From an article by Johann [III] Bernoulli in the Berlin *Memoires*.[1]
For the answer see Letter 1265

Cum nuper epistolam [2] a te transformatam in Actis mens. Jul. anni sup. obiter relegeram, visuris numquid ibi sit quod me tanquam auctorem ejus detegere posset, verbulum unicum inveni, unde attentus Lector a me scriptam esse suspicari possit. Etenim ultima demum epistolæ pagina quæ est 314 in illo Actorum Tomo lin. 10 a fine, legitur hoc, meam [3] formulam ex Newtoniana esse desumptam, ubi το meam mutari debuisset in το Bernoulli: Vereor autem ne Keilius, qui me proscindendi omnem occasionem arripiet, hoc contra me in usum suum sit versurus. [4] E re igitur esse credo, ut cogites de commoda hujus loci correctione in Erratis ponenda. Posses ex. gr. monere pro meam formulam legendum esse eam formulam, sed hoc tamen non satis quadrat: vellem itaque ut invenires modum commodiorem quo culpa in typothetam plausibiliter rejici posset; id prudentiæ tuæ relinquo.

Translation

When recently I casually re-read the letter [2] transformed by you in the *Acta* [*Eruditorum*] for July of last year, to see whether there might be something in it which could reveal me as its author, I found one single small word whence an attentive reader could suspect that it was written by me. For right on the final page of the letter, which is p. 314 in that volume of the *Acta*, 10 lines from the end, we read this, 'my formula taken from the Newtonian one,' where the word 'meam' [3] ought to have been changed into the word 'Bernoulli:' I fear, however, lest Keill, who snatches every opportunity for censuring me, will turn this against me to his own purpose.[4] Therefore I believe it to be advantageous that you should consider putting a suitable correction to this place in the errata. You could for example warn [the reader] that for 'my formula' is to be read 'that formula', but this still does not fit satisfactorily: I wish therefore that you would find a more convenient way by which the fault could plausibly be thrown back on the type-setting; this I leave to your prudence.

NOTES

(1) *Histoire de l'Academie Royale des Sciences et Belles-Lettres* for 1799 and 1800 (Berlin, 1803), p. 48; see also Letter 1196, note (1), p. 303. Edleston (*Correspondence*, p. 178) misdates the letter 18 September 1717 N.S.

(2) The 'Epistola pro Eminente Mathematico;' see Letter 1196. In March, 1717, Bernoulli had written to Wolf, in reply to Letter 1208, expressing satisfaction with the care Wolf had taken to conceal his authorship of the 'Epistola'. (An excerpt from this letter, and Wolf's reply, are given in the article cited in note (1) above.) Bernoulli now looks at the 'Epistola' again, and notices the slip which could give him away.

(3) See Letter 1196, note (4), p. 304.

(4) For Keill's projected reply to the 'Epistola', see Letter 1239, note (4), p. 386.

1259 NEWTON AND BLADEN TO THE TREASURY
5 SEPTEMBER 1717
From the original in the Mint Papers[1]

To the Right Honble. the Lords Commiss[ione]rs of His Majties Treasury

May it please your Lordships

Understanding that the Money arising by the coynage Act the last quarter ending at Michaelmas amounts to about 1400£. and that there is an Order from your Lordps. to the Auditor of the Exchequer to pay out of the same to the General of his Majties. Mint in North Britain the summe of 1200£.[2] We humbly represent to your Lordps that the charges of his Majties. Mint in the Tower of London in the said Quarter ending at Michaelmas amounts [*sic*] to about 1700£. whereof about 1100£. is the charges of coynage and that towards the payment of this summ there are only 310£. in our hands: And therefore humbly pray your Lordships that the said coynage Duty may be paid to the Master and Worker of the Mint as of course according to his Majties. General Warrant, or that else by the power lately given to your

Lordps. by Act of Parliamt, other monies may be paid to him in lieu of the same for enabling him to defray the charges of the said last Quarter, the Melter and Moneyers being out of pocket and the Clerks of the Mint not able to subsist without their Salarys

All which is most humbly submitted

to your Lordps. Great Wisdom

Is. NEWTON

MARTIN BLADEN

Mint Office
5th September 1717.

NOTES

(1) I, fo. 372. Since this letter, fairly written by a clerk and signed by the Master and Comptroller, remained with Newton, it was obviously not sent to the Treasury.

(2) A petition from Lauderdale for the payment of arrears of salary to the Edinburgh Mint was considered by the Treasury Lords on 30 July 1717, and a payment of £1200 ordered (*Cal. Treas. Books*, XXXI (Part II), 1717, p. 37).

1260 WARRANT TO NEWTON
11 SEPTEMBER 1717
From the original in the Mint Papers [1]

After Our hearty Commendations These are to Authorize and Direct you to deliver out of the Mint unto Henry Hines, who Contracts for the importing a certain quantity of Copper there for the Coyning of farthings & halfe pence, One Instrument or Tool, called a Cutter; You having informed Us, that the true Sizing of the Barrs of Copper that is to be imported by him, requires the use of the said Instrument or Tool, and that the Same may be safely delivered to him for that purpose, he giving to you a note under his hand, for the return thereof at any time when demanded, or as soon as the Service for which it is delivered to him Shall be performed And for so doing this Shall be Your Warrant. Whitehall Treasury Chambers 11th September 1717

STANHOPE

TORRINGTON [2]

J. WALLOP

To Our very loving Friend Sir Isaac
Newton Knt. Master and Worker of His
Mats. Mint

NOTES

(1) II, fo. 341, a clerical version, not signed. Presumably this is a formal permission to release from the Mint a special die, through which the bars would be drawn to size, of some potential value to illicit coiners.

(2) Thomas Newport (see Letter 1195 note (3)) was created Baron Torrington on 20 June 1716.

1261 LORD STANHOPE TO NEWTON
12 SEPTEMBER 1717
From the original in the Mint Papers.[1]
For the answer see Letter 1268

Whitehall Treasury Chambers 12th Septr: 1717

The Right Honble. the Lords Comm[issione]rs of his Mats. Treasury are pleased to referr this Petition to Sr Isaac Newton Knt Master & Worker of his Mats. Mint who is to consider the same and Report his opinion to their Lordps what is fit to be done therein.

STANHOPE

NOTE

(1) II, fo. 376. The instruction is written on the foot of a memorial from William Wood & Partners, without date.

The memorial states that though this firm offered fine copper near five pounds a Ton cheaper than others, another manufacturer was engaged to supply 30 tons to begin the coinage. But Wood has been told by Newton that the contract will be renegotiated after the 30 tons have been consumed, and that (on examination of a specimen of his metal) Wood is likely to receive the new contract. He now asks for a Warrant to supply copper at $17\frac{1}{2}d.$ the pound after the contract for 30 tons (at 18*d.* the pound) has run out, so that he may be well stocked with the metal.

This memorial should obviously be related to Letter 1257, Wood being one of the cheaper but disappointed vendors of copper there mentioned by Newton.

William Wood (1671–1730) was the Wolverhampton metal-merchant whose name was made odious later by Swift because of his Irish halfpence.

1262 WARRANT FOR THE COPPER COINAGE
[13 SEPTEMBER 1717]
Extracts from the copy in the Mint Papers[1]

OUR WILL and Pleasure is and We do hereby Authorize and Command You Sr. Isaac Newton Master and Worker of Our Mint in the Tower of London to receive into Our said Mint from time to time fine British Copper in Barrs or Filletts which when heated red hott will spread thin under the hammer without cracking which shall be of a due Size or thickness to be prescribed by you and out of the same to Coyne half Pence and farthings of such a bigness that forty and six halfpence or ninety and two farthings may make a pound weight averdupoise excepting such small errors as may happen in and by the unequall Sizing of the Barrs: Which errors you shall endeavour that they be not in excess or defect above the fortieth part of the whole weight and this not by designe but only by accident. [The price to be fixed by the Treasury, half to be paid to the importer on delivery of the copper at the Mint, half with the returned scissel. Imperfect metal is to be returned to the importer. Quantities to be fixed by the Treasury.] AND you shall cause our Effigies with the inscription GEORGIUS REX to be stamped on one side of each peice and the Effigies of a Britannia sitting upon a Globe with a Speare in her left hand and a Mirtle in her right and the inscription BRITANNIA stamped on the other side as in the late Copper Money, and under her the Date. [Random samples of the coin amounting to twentythree pence in value are to be checked for weight, and for quality by hammering when red hot. One penny in every seven pounds weight to be allowed as remedy in distribution. The moneyers not to distribute coin before the assay.

The process of coining to be controlled by a King's Clerk who is to keep proper books.] And our said Clerk shall also yearley make a Roll upon Oath of the weight and price of every parcell of new moneys coined and delivered from time to time by the Moneyers to the Master and Worker, And the said Master and Worker of our Mint shall account annually before the Auditor of our Mint for all the said Copper moneys coyned, AND be answerable to us for all the profits thereof above the charges, And Our said Auditor in auditing the said Accounts shall have all the same power as in auditing the Accounts for the coynage of Gold & Silver.

[The Master and Worker to have allowance in his accounts for expenses incurred in repairing buildings and tools] and the summe of three pence farthing by the pound weight for coining the said Copper Monies and for bearing and sustaining all manner of wast[e]s provisions necessaries and charges coming arising and growing in and about the coining Assaying weighing and delivering the said Copper and copper monies And the said Sr. Isaac Newton shall pay unto Our Clerk twenty shillings sterling by the Tun of all the Monies Coyned for his attendance on this Service. [Persons attending the Tower on this business to have free access.]

AND we do further command and require the Graver, Moneyers, Smith and all others attending on this Service to do their duty with diligence and Application and to observe the tasks and directions given them by our said

413

Master and Worker for coyning Our said monies well and with dispatch. AND for so doing this shall be your Warrant and the Warrant of all others concerned in this Coynage

To Our trusty and well beloved
Sr Isaac Newton Knight
Master and Worker of Our Mint

NOTE

(1) II, fo. 425. This is a fine clerical copy without date, signature or seal. There are drafts at fos. 409 and 439. The sentences within brackets are abbreviated from the manuscript.

The warrant is printed in *Cal. Treas. Books*, XXXI (Part III), 1717, pp. 575–7, from the King's Warrant Book, XXVIII, pp. 466–7.

The Mint's report on the coinage of copper (Letter 1253) was read at the Treasury on 6 August, when the Lords resolved 'to direct Sir Isaac Newton to prepare and lay before them the proper Warrants and Instructions relating to the [copper] coinage' and to allow him £500 for expenses (*Cal. Treas. Books*, XXXI (Part II), pp. 40 and 509). The present Warrant, drafted by Newton, adheres to the principles he had formulated in 1714 (Letter 1118).

1263 NEWTON TO SLOANE
13 SEPTEMBER [1717]
From the original in the British Museum[1]

Sr

I thank you for giving me timely notice of the Caveat. I think we should stick at no charge for defending the Legacy. What money shall be wanting for this purpose I'le advance till a Council shall be called. If you should see Dr Harwood[2] before me, pray desire him to have an eye upon this matter. I do not know the method of proceeding in these cases, but he can tell us. I will take the first opportunity to inform my self of what is to be done I am

Your most humble obedient servant

Is. NEWTON

Sept. 13.
For Dr Hans Sloane Bartt

NOTES

(1) MS. Sloane, 4060, fo. 74; printed in Edleston, *Correspondence*, p. lxxiv. Edleston dates the letter 13 September 1706, and presumes it refers to a legacy of £400 left to the Royal Society by John Wilkins in 1672 (*ibid.*, p. xxxviii). However, although there were some legal difficulties over Wilkins' legacy (see H. G. Lyons, 'The Society's first bequest', *Notes and*

Records of the Royal Society, **1** (1938), pp. 43–6) these were resolved by 1713; but the letter is directed to 'Dr Hans Sloane Bartt', and Sloane was not created baronet until 3 April 1716.

It seems likely that the legacy referred to is one out of several made to the Royal Society after this date; namely those of Francis Aston, Sir Godfrey Copley, Thomas Paget and Robert Keck (see *The Record of the Royal Society*, 1940, pp. 141–2). Of these, that of Thomas Paget best fits the facts. At a meeting of the Royal Society on 17 October 1717 'The Will of Dr. Thomas Paget was produced by which the Dr. hath Bequeathed to the Royal Society two Houses in Coleman Street London now lett for £97. 10s per annum as also three pictures Vizt. of Gassandus, T. Hobbs and Dr Henry Moore. The President was pleased to Undertake to Ask of the Executors the Deeds of the Estate.' At a Council meeting on 24 October 1717, the Society's solicitor, Frewin, was given a fee of two guineas for investigating the matter 'in order to make sure [of the Society's] Title to Dr Paget's Legacy'.

(2) John Harwood (1661–1731), LL.B. 1684, practised civil and canon law at Doctors Commons, he had been elected F.R.S. in 1686 and in 1716 was a member of the Council of the Royal Society.

1264 NEWTON TO THE TREASURY
21 SEPTEMBER 1717
From the holograph original in the Public Record Office.[1]
Reply to Letter 1256

To the Rt Honble the Lords Comm[ission]ers
of his Mats Trea[su]ry

May it please your Lordps

In obedience to your Lordps Order of Reference of Aug. 12th. that I should lay before your Lordps a State of the Gold & Silver Coyns of this Kingdom in weight & fineness, & the value of gold in proportion to silver with my observations & opinion, & what method may be best for preventing the melting down of the silver coyn: I humbly represent that a pound weight Troy of Gold eleven ounces fine & one ounce allay is cut into 44½ Guineas, & a pound weight of silver eleven ounces two penny weight fine & eighteen penny weight allay is cut into 62 shillings; & according to this rate, a pound weight of fine gold is worth fifteen pounds weight six ounces seventeen penny weight & five grains of fine silver, recconing a Guinea at 1£ 1s 6d in silver money. But silver in Bullion exportable is usually worth 2d or 3d per ounce more then in coyn. And if at a Medium such bullion of standard allay be valued at 5s 4d½ per ounce a pound weight of fine Gold will be worth but 14 lwt. 11oz 12 dwt 9 grs of fine silver in bullion. And at this rate a Guinea is worth but so much silver as would make 20s 8d. When ships are lading for the East Indies the demand of silver for exportation raises the price to 5s 6d or 5s 8d per ounce or above. But I consider not those extraordinary cases.

A Spanish Pistole was coyned for 32 Reaus [2] or four pieces of eight Reaus usually called pieces of eight, & is of equal allay & the sixteenth part of the weight thereof. And a Doppio Moeda [3] of Portugal was coyned for ten Crusados of silver & is of equal allay & the sixteenth part of the weight thereof. Gold is therefore in Spain & Portugal of sixteen times more value then silver of equal weight & allay according to the standard of those kingdoms. At wch rate a Guinea is worth 22s 1d. But this high price keeps their gold at home in good plenty & carries away the Spanish silver into all Europe, so that at home they make their payments in Gold & will not pay in Silver without a premium. Upon the coming in of a Plate fleet, the premium ceases or is but small: but as their silver goes away & becomes scarce, the premium increases, & is most commonly about six per cent. Which being abated a Guinea becomes worth about 20s & 9d in Spain & Portugal.

In France a pound weight of fine gold is recconed worth fifteen pounds weight of fine silver. In raising or falling their money, their kings' Edicts have sometimes varied a little from this proposition in excess or defect: but the variations have been so little that I do not here consider them. By the Edict of May 1709 a new Pistole was coyned for four new Lewises & is of equal allay & the fifteenth part of the weight thereof except the errors of their Mints. And by the same Edict fine Gold is valued at fifteen times its weight of fine silver. And at this rate a Guinea is worth 20s 8d½. I consider not here the confusion made in the monies in France by Frequent Edicts to send them to the Mint & give the king a Tax out of them. I consider only the value of Gold & Silver in proportion to one another.

The Ducats of Holland & Hungary & the Empire were lately current in Holland among the common people in their Markets & ordinary affairs at five Guilde[r]s in specie & five styvers, & commonly changed for so much silver moneys in three Guilder pieces, & Guilder pieces as Guineas are with us for 21s 6d sterling. At which rate a Guinea is worth 20s 7d½.

According to the rates of Gold to Silver in Italy, Germany, Poland, Denmark & Sueden a Guinea is worth about 20s & 7d, 6d, 5d, or 4d. For the proportion varies a little within the several governments in those countries. In Sueden Gold is lowest in proportion to silver, & this hath made that kingdom which formerly was content with copper money abound of late with silver sent thither (I suspect) for naval stores.

In the end of King Williams reign & the first year of the late Queen when foreign coyns abounded in England, I caused a great many of them to be assayed in the Mint & found by the assays that fine Gold was to fine silver in Spain, Portugal, France, Holland, Italy, Germany, & the northern kingdoms, in the proportions above mentioned, errors of the Mints excepted.

416

In China and Japan one pound weight of fine gold is worth but nine or ten pounds weight of fine silver; & in East India it may be worth twelve. And this low price of gold in proportion to silver, carries away the silver from all Europe.

So then by the course of trade & Exchange between nation & nation in all Europe, fine gold is to fine silver as $14\frac{4}{5}$ or 15 to one. And a Guinea at the same rate is worth between 20s 5d & 20s $8d\frac{1}{2}$, except in extraordinary cases, as when a Plate Fleet is just arrived in Spain, or ships are lading here for the East Indies, which cases I do not here consider. And it appears by experience as well as by reason that silver flows from those places where its value is lower in proportion to gold, as from Spain to all Europe & from all Europe to the East Indies, China & Japan, & that Gold is most plentifull in those places in which its value is highest in proportion to silver, as in Spain & England.

It is the demand for exportation which hath raised the price of exportable silver about 2d or 3d in the ounce above that of silver in coyn, & hath thereby created a temptation to export or melt down the silver coyn rather then give 2d or 3d for forreign silver[.] And the demand for exportation arises from the higher price of silver in other places then in England in proportion to gold, that is, from the higher price of gold in England then in other places in proportion to silver; & therefore may be diminished by lowering the value of gold in proportion to silver. If gold in England or Silver in East India could be brought down so low as to bear the same proportion to one another in both places, there would be here no greater demand for silver then for gold to be exported to India. And if Gold were lowered only so as to have the same proportion to the silver money in England wch it hath to silver in the rest of Europe, there would be no temptation to export silver rather then gold to any other part of Europe. And to compass this last there seems nothing more requisite then to take of about 10d or 12d from the Guinea, so that Gold may beare the same proportion to the silver money in England which it ought to do by the course of Trade & Exchange in Europe. But if only 6d were taken off at present, it would diminish the temptation to export or melt down the silver coyn, & by the effects would shew hereafter better then can appear at present, what further reduction would be most convenient for the publick. (4)

In the last year of K. William, the Dollars of Scotland worth about four shillings & six pence half penny, were put away in the north of England for 5s, & at this price began to flow in upon us. I gave notice thereof to the Lords Comm[issione]rs of the Treasury, & they ordered the collector of Taxes to forbear taking them & thereby put a stop to the mischief.

At the same time the Lewidors of France which were worth but seventeen shillings & three farthings a piece passed in England at 17s 6d. I gave notice thereof to ye Lds. Commissioners of the Treasury & his late Ma[jes]ty put out a Proclamation that they should go but at 17s, & thereupon they came to the mint & fourteen hundred thousand pounds were coyned out of them. And if the advantage of five pence farthing in a Lewidor sufficed at that time to bring into England so great a quantity of French money, & the advantage of three farthings in a Lewidor to bring it to the Mint: the advantage of $9\frac{1}{2}d$ in a Guinea or above may have been sufficient to bring in the great quantity of gold which hath been coined in these last fifteen years without any forreign silver.

Some years ago the Portugal Moedors were received in the West of England at 28s a piece. Upon notice from the Mint that they were worth only about 27s 7d, the Lords Commissioners of the Trea[su]ry ordered their Receivers of Taxes to take them at no more than 27s 6d. Afterwards many Gentlemen in the West sent up to the Treasury a Petition that the Receivers might take them again at 28s, & promised to get returns for this money at that rate, alledging that when they went at 28s their country was full of gold which they wanted very much.[5] But the Commissioners of the Treasury considering that at 28s the nation would lose five pence a piece, rejected the Petition. And if an advantage to the Merchant of 5d in 28s did pour that money in upon us: much more hath an advantage to the Merchant of $9\frac{1}{2}d$ in a Guinea or above, been able to bring into the Mint great quantities of gold without any forreign silver, & may be able to do it still till the cause be removed.

If things be let alone till silver money be a little scarcer, the Gold will fall of it self. For people are already backward to give silver for Gold, and will in a little time refuse to make payments in Silver without a premium as they do in Spain; & this premium will be an abatement in the value of the gold. And so the Question is whether Gold shall be lowered by the government, or let alone till it falls of it self by the want of silver money.[6]

It may be said that there are great quantities of silver in Plate, & if the Plate were coyned there would be no want of silver money. But I reccon that silver is safer from exportation in the form of Plate then in the form of money because of the greater value of the silver & fashion together. And therefore I am not for coyning the Plate till the temptation to export the silver money (wch is a profit of 2d or 3d an ounce) be diminished. For as often as men are necessitated to send away money for answering debts abroad, there will be a temptation to send away silver rather then Gold because of the profit which is almost 4 per cent. And for the same reason forreigners will chuse to send hither their gold rather then their silver.

All which is most humbly submitted to your Lordps great wisdome

ISAAC NEWTON.

Mint Office.
21 *Sept.* 1717

NOTES

(1) T/1, 208, no. 43, fos. 204–5; there are several near-contemporary manuscript copies, notably one in the Mint Papers, II, fos. 111–15. Related drafts are extremely numerous in the same volume of papers, for example at fos. 38, 65, 67, 69, 96–8, 100, 102, 107, 109, 236, and there are many other documents on the values of foreign coins. This letter was first printed in the *Daily Courant*, no. 5057, 30 December 1717 (see Letter 1271) and many times since (for example, Shaw, pp. 189–95).

(2) Reals.

(3) Double moidore. Newton here gives the nominal equivalences, not those established by commerce; as he goes on to explain, gold was in practice worth less in relation to silver.

(4) It was universally recognized that unworn silver coins had been removed from circulation, presumably for the reason given by Newton that they were a cheap source of bullion. Hence Newton at first proposed a reduction in the silver content of the silver coins (see Craig, *Newton*, pp. 106–9, relying here on Mint Papers, II, fo. 124). The alternative proposed, of increasing the exchange value of the silver coin in terms of gold, seems to have had no effect; the price of silver did not fall, and the supply of the metal to the Mint did not improve. England's was essentially a gold currency for the rest of the eighteenth century.

(5) See Letter 1095*a*.

(6) A Proclamation was issued on 22 December 1717 scheduling the reduced value of gold coins as recommended by Newton. See Letter 1270.

1265 WOLF TO J. BERNOULLI
30 SEPTEMBER 1717
From an article by Johann Bernoulli [III] in the Berlin *Mémoires*.[1]
Reply to Letter 1258

Quod si ad me mittere volueris notas tumultarias quas a te in schedam Keilianam[2] consignatas memoras, responsionem formalem sub nomine anonymi componam et Actis inseri curabo.[3] Sphalma typographicum[4] nuper admissum omnium commodissime tum una corrigeretur: quod quidem nunc mihi non consultum videtur, cum tuæ solutiones novæ problematis isoperimetrici in lucem prodeunt.[5]

Translation

If you wish to send me the hasty jottings which you mentioned having written on Keill's paper,[2] I would compose a formal answer under an anonymous name, and take care that it be placed in the *Acta* [*Eruditorum*].[3] The printing error[4] recently noticed would then be corrected with the greatest possible convenience along with it, which it does not seem prudent to me [to do] now, when your new solutions of the isoperimetric problem are being published.[5]

NOTES

(1) 'Anecdotes pour servir à l'Histoire des Mathématiques par M. Jean [III] Bernoulli,' *Histoire de l'Académie Royale des Sciences et Belles-Lettres* for 1779 and 1800 (Berlin, 1803), p. 49. See also Letter 1196, note (1), p. 303.

(2) Presumably Wolf refers to the *Recensio*, which he believed came from Keill's hand.

(3) Papers were later composed by Kruse, Mencke and Bernoulli's eldest son Nikolaus, under Johann [I] Bernoulli's direction, in his defence, against both Keill and Taylor. These were published in the *Acta Eruditorum* (see Letter 1321, vol. VII).

(4) See Letter 1196, note (4), p. 304. Since Bernoulli was eventually forced to admit his authorship of the letter, at the end of a paper by his son Nikolaus (see *Acta Eruditorum* for 1718, pp. 261–2), the claim that this was merely a printing error was never publicly made.

(5) Jakob and Johann [I] Bernoulli had been debating for about twenty years various problems concerning isoperimeters, and papers on the subject had already been published in the *Acta Eruditorum* and the *Mémoires de l'Académie Royale des Sciences* of Paris. Wolf refers here to Bernoulli's new solutions to be published in ' Remarques sur ce qu'on a donné jusqu'ici des problemes sur les isoperimetres', *Mémoires de l'Académie Royale des Sciences* for 1718, p. 100.

1266   NEWTON TO 'sGRAVESANDE
[OCTOBER 1717]
From a holograph draft in the Bodleian Library, Oxford [1]

Vir celeberrime

Epistolam tuam & chartas Italices scriptas quas una misisti communicavi cum Societate Regia quæ rem retulit ad quemdam ex Socijs Italica et mathematice doctum. Ipsa enim opinionem propriam de rebus dubijs nunquam profert. Socius autem ille lectis chartis observationes suas in Schediasmate composuit quod rem cum hac epistola accipies [2]

Your Letter I received together with the Papers [3] which accompanied it concerning the letting of the Rheno into the Po, & I communicated them to the R. Society. But I should acquaint you that ye Society make it a general Rule never to give their opinion in doubtful matters. They can give their testimony in matters of fact wch appear to them, but few of them are Mathematicians. They also avoid medling with civil affairs wch have no relation to Natural Philosophy. However, they desired one of their fellows [4] who is skilled in Mathematicks & understands the Italian tongue to peruse the same & upon considering them he drew up his observations upon them in a Paper wch you will receive from Mr Burnet. [5] I am

Sr

NOTES

(1) Bodleian New College MS. 361, II, fo. 42. The draft is a reply to a letter sent by 'sGravesande to Newton and shown to the meeting of the Royal Society on 17 October 1717. The following record is made in the Journal Book:

> The Case of Bononiæ and Ferrara in relation of the Desire of those of Bononia to prevent The Inundations of the River Rheno by letting the Waters of that River into the Great Po above Ferrara
>
> This these of Ferrara Oppose pretending that in Great Floods their own Lands which are now dry might be Over flowed by this Augmentation of the Po.
>
> The Bolognese mention that it cannot have that effect; and Signr. Manfredi for Bononia having Answered the Objections of Srs. Ceva and Moscatella Appeals to Forreign Mathematicians as Competent and Impartial Judges. On this Account the Book [see note (3) below] printed pro and con was sent by Mr. Gravsandt from holland to the President Desiring his Opinion upon the Matter, he was pleased to lay it before the Society And Dr. Halley was Ordered to look it Over and Report what he finds to be the State of the Case.

The problem of the flooding of the Po delta was a perennial one, and many drainage operations had already been carried out. Although the suggestions of Eustachio Manfredi, who was Professor of Mathematics at Bologna University, superintendent of waterworks at Bologna and wrote extensively on hydrostatics, were not implemented, later, in 1767 the Rheno was instead deflected westwards into the Po di Primaro, a distributary of the Great Po.

(2) 'Famous Sir,

I have communicated your letter, and the papers written in Italian which you sent with it, to the Royal Society, who referred it to one of the Fellows who is learned in Italian and in mathematics. For it never gives its own opinion on doubtful matters. However this Fellow, having read the papers, has drawn up his observations in the paper which you receive with this letter'.

(3) Presumably *Ragioni del Signor Giovanni Ceva commisario etc. e del Signor Doriciglio Moscatelli Battaglia prefetto etc. contra l'introduzione del Reno nel Pò Grande. Con la riposta alle medesime di Eustachio Manfredi matematico etc. Che contiene una piena informazione sopra i capi principali di questa materia* (Bologna, 1716).

(4) Halley; see note (1) above. At the meeting on 24 October 1717 the Journal book of the Royal Society reads 'Dr Halley Read a Short Extract of the Controversy between the Bolognese and Ferrarese about bringing the River Rheno which Overflows the Grounds of the Bolognese into the great Po. Those of Ferrara fear that this Addition of Water might make the Po pass its Banks to their Prejudice which the Bolognese endeavour to Refute by shewing that in no Circumstance the Addition of the Rheno, which is Agreed to Carry about a thirteenth part of the Water of the whole Po, can encrease it above 9 or 10 Inches Whereas the Banks of the Poe, are everywhere two or three Foot above the highest Floods. Upon the whole the Bolognesse seems to have the better of the Argument but no true Judgment can be made without a View of the Place, and an Exact Account of the Velocity of the Two Rivers.'

(5) William Burnet; see Letter 1028, note (1), p. 47.

1267 NEWTON TO THE TREASURY
23 NOVEMBER 1717
From the holograph draft in the Mint Papers[1]

To the Rt Honble the Lords Comm[ission]ers of his
Majts Treasury

May it please your Lordps

In obedience to your Lordps Order of Reference of the 19th Instant that an Account be laid before your Lordps of all the Gold & Silver coyned the last fifteen years & how much thereof has been Coined out of plate upon publick encouragements, & what copper money hath been newly coyned: it is humbly represented that since Christmas $170\frac{1}{2}$ to the 10th of this instant November there hath been coined in Gold $6630544\frac{1}{2}$ Guineas [2] recconing $44\frac{1}{2}$ Guineas to a pound weight Troy & in Silver 223380 pounds sterling recconing $3\pounds$ $2s$ to a pound weight Troy. And that part of this silver amounting to 143086 pound sterling was coyned out of English Plate imported by Acts of Parliament in the years 1709 & 1711, & another part amounting to $13342\pounds$ was coyned out of Vigo Plate in the years 1703 & 1704, & another part amounting to $45732\pounds$ was coined out of silver extracted from our own Lead Ore, & the rest amounting to $21220\pounds$ was coyned chiefly out of old Plate melted down & some of it out of Pieces of Eight.

The Graver of the Mint hath been hard at work ever since the last session of Parliament in making the Embossment & Puncheons for the halfpence & farthings & taking off a few Dyes from them. The making of a Embossment & Puncheon for a halfpenny takes up the time of about six weeks & there have been two embossments & two Puncheons made for the halfpence & one for the farthings. And now these are finished & some Dyes are made from the Puncheons, it will take up a little time to examin the copper & settle the best method of preparing sizeing nealing & cleaning it, & making it fit for the Mint: this being a manufacture different from that of coarse copper, & more difficult, & not yet practised in England. And as soon as this method is fixed, we shall begin to coine in quantity.

All wch is most humbly submitted to your Lord's great wisdome.

Mint Office
23 *November* 1717

(1) II, fo. 245r. The clerical copy submitted by Newton and returned to him by Lowndes on 21 December 1717 is at Mint Papers II, fo. 289.

(2) On *ibid.*, fo. 289 Newton has altered this figure to its equivalent in pounds sterling, reckoning the guinea at £1. 1s. 6d sterling—£7 127 835. Otherwise the two versions are identical.

1267a GOLD AND SILVER MONIES COINED, 1702-17

From the copy in the Mint Papers [1]

Gold and Silver Monies Coyned in
his Majesty's Mint within the Tower of London from
1st. January 1701/2 To the 20th. November 1717

	Gold Monies			Silver Monies			Silver Monies from Plate		
	Lwt	Oz	dwt	Lwt	Oz	dwt			
1702	3642	–	–	114	6	–			
1703	34	2	–	718	–	–			
1704	———————			4007	–	–			
1705	104	–	–	429	7	15			
1706	537	–	–	932	—	–			
1707	607	–	–	1174	–	–			
1708	1010	–	–	3751	–	–	Lwt	Oz	dwt
1709	2468	–	–	2223	7	15	23199	4	4
1710	3716	–	–	817	–	–			
1711	9324	–	–	1810	5	2	22957	6	18
1712	2855	–	–	1784	8	–			
1713	13137	–	–	2333	–	–			
1714	29526	–	–	1566	–	–			
1715	39090	–	–	1643	–	–			
1716	23765	–	–	1650	–	—			
1717	15186	–	–	948	–	–			
	145001	2	–	25901	10	12	46156	11	2

Memorandum that 3602 Lwt – 5 oz – 7 dwt, part of the 4007 Lwt of Silver Coyned in the year 1704, was silver melted from wrought plate brought in by the Comm[issione]rs for prizes after the Expedition from Vigo.

NOTE

(1) II, fo. 291. This table is obviously associated with the preceeding letter, and appears to have been submitted with it, though that is not specifically stated in the letter.

1268 NEWTON TO THE TREASURY
12 DECEMBER 1717
From the holograph original in the Public Record Office.[1]
Reply to Letter 1261

To the Rt Honble the Lords Comm[ission]ers of His Majties Treasury

May it please your Lordps

In Obedience to your Lordps Order of Reference that I should report my opinion upon the Petition of Mr William Wood for a Warrant to furnish the Mint with fine copper in barrs duly sized at $17\frac{1}{2}d$ per Lwt after the 30 Tunns agreed for at $18d$ per Lwt be coined,[2] I humbly represent that if Mr Wood will furnish the Mint with such copper, prepared by the battering Mills, as the specimen was which I had of him, I am ready to receive a Tonn or two every Month at that price till next Michaelmas, or longer if there be occasion. But if he intends to prepare it some other way, of which he has not yet given me a specimen I desire that a specimen of such work as he intends to furnish me with, may be sent into the Mint to be assayed before be begins to bring in a great quantity.

All which is most humbly submitted to your Lordps great wisdome

ISAAC NEWTON.

Mint Office
Decem. 12. 1717

NOTES

(1) T/1, 209, no. 19, fo. 222.

(2) That is, after the conditions of the contract made with Henry Hines and John Appleby on 22 August 1717 (Mint Papers, II, fos. 319 and 320) for the supply of 30 tons of copper for coinage had been fulfilled.

1269 JOSEPH INGLIS TO KEILL
19 DECEMBER 1717
Extract from the original in the University Library, Cambridge[1]

Your papers[2] have been In Sir Isaacs hands ever since they came into Mine, and as yet I have heard nothing about them, But assoon as I receive them I shall endeavour to forward them to Holland by the first sure hand.

NOTES

(1) Res. 1893(*a*); Lucasian Professorship, Box 1, Packet 3. Joseph Inglis was Keill's cousin.

(2) Keill's draft 'Lettre à Bernoulli', in reply to the article 'Epistola pro Eminente Mathematico' by Bernoulli in the *Acta Eruditorum* for 1716; see Letter 1239, note (4), p. 386.

1270 EDWARD NORTHEY TO NEWTON
21 DECEMBER 1717
From the holograph original in the Mint Papers[1]

Sr

it is by his majestys Commands I give you this trouble the Commons have addressed to his Ma[jes]ty to reduce Guyneas from one & twenty shillins & six [pence] to twenty one shillings & in proportion other Gold Coyn: there was a proclamation in King Charles the seconds time to raise the value of all gold Coyn as you will perceive by the inclosed[2] it is his ma[jes]tys desire yt you will be pleased to make the reduction of ye other Coyns mentioned in the paper as they are to bear a proportion to Guineas & yt you will be pleased to send it to my house in Essex Street this night or tomorrow morning for his Ma[jes]ty hath appointed a Council to sitt tomorrow to receive the Proclamation I am with the greatest respect

Sr

Yr most obedient humble

servt

EDW NORTHEY[3]

Satt night
21st *dec* 1717

For the Honourable Sr
Isaac Newton
   These

NOTES

(1) II, fo. 117. Compare Letter 1264.

(2) A copy of the draft proclamation; the other gold coins involved were the half-guinea piece, and pieces of two and five guineas, together with other coins previously passing at 23*s*. 6*d*. and 25*s*. 6*d* respectively.

(3) Sir Edward Northey (1652–1723), Attorney General from 1701 to 1707 and again from 1710 to 1717. He had been brought into the House of Commons by the Whigs in 1710 and sat until 1722.

1271 'M.M.' TO NEWTON
31 DECEMBER 1717
From the original in the Mint Papers[1]

Dec: 31*st*. 1717

Sr

Your Representation to the Treasury concerning the Coin of this Kingdom,[2] which was printed in the daily Courant of yesterday, is so judicious, instructive, and Satisfactory, that it deserves equal thanks, as admiration, and attention, and so far as I am able to judge of it, a standing Rule may be formed from it for regulating, and proportioning our Species at all times; For, as we are to watch what other Nations do in that respect and govern our selves accordingly, in the proportions between Gold, and Silver, and in putting a value upon each of them, you furnish with a method for adjusting that, with all ease, and plainness, and upon that acco[un]t (I presume) you proposed to lower Gold at present; Whether it will contribute to bring in Silver from abroad, or to bring out what may be hoarded at home, or cause our Gold to be exported is yet too early to determine; but if some good information may be credited, it is, that a great quantity has been carried away since the Proclamation, and particularly into Holland, where the profit was $1\frac{1}{2}$ per Cent by carrying Guineas thither before they were reduced here, and then one cannot wonder that they should go thither now that they are lower'd by Authority, which makes the profit near 4 per Cent upon them; But this being happen'd since the date of your paper, you cannot be said to be answerable for it, nor for others not considering the Rule you have laid down, so as to apply it to this raising of Gold in Holland; But instead of enlarging on this particular, I shall proceed to that part of your paper, which shews the watchful Eye you had on some private wags to raise both Gold, and Silver to the prejudice of the public, as receiving Scotch Dollars at 5 *sh*: and Moeders at 28 *sh*: both which were stopt on your representing it to the Treasury, and this seems to imply, how much it lys in the Treasury to regulate money-matters by timely interposing their Authority with Receivers, and other

426

Officers of Revenue, for as these are directed to take, or refuse money, so will other people thro the Nation take their measures, and be governed in all negotiations; which brings to mind the bad State our Coin was in by being clipt, and counterfeit to a great degree, and in a most notorious manner, which you cannot but remember, as well as the Millions it cost the Nation to retrieve the mischief:[3] Now Every one knows that the Scandalous pass which money was then brought to, did not happen all of a sudden and at once, but came to it by degrees, and in Several years, Suppose therefore, that the then Treasury (and you know who was at the head of it) had interposed when they perceived it, and as early as they might have done, and had given proper directions About it to the Receivers of the Revenue, One might imagine, it would have put a stop to it, as easily as there was once put to Dollars, and Moedors, and then if this had been done probably those Millions (at least a great deal thereof) might have been saved, If I am mistaken in this point, I hope you will set me right, I have often consider'd it, and could never be perswaded but the Treasury might have done exceeding much to have stopt (if not prevented) that pernicious practice of clipping and debasing Coin and thus I leave it with you; with adding one thing more which is about our paper-money (or Credit) the increasing or restraining which, seems equally to be in the power of the Treasury, as the Silver, and Gold Species, That it is increased, is not to be deny'd, But whether this increase is so good for the nation, is another consideration, and at this time seems to be well worth it, as it does, whether this increase has not been some cause of our being so much in want of Silver, and whether, if paper money be suffer'd so to multiply, and be not soon restrained it may not be some occasion of an equal want of Gold? You are an able Judge of these matters, and your Opinion of all of them, will oblige a great many people, but I being wholly unknown to you, I cannot desire it otherwise, than in the way (which was in print) that your Representation came to the hands of

Sr.

Your most humble Servant

M.M.

NOTES

(1) II, fo. 121. Newton either received few anonymous communications or retained few. This one illustrates the interest excited by his monetary opinions, and (perhaps) the extent of his fame.

(2) Letter 1264.

(3) The debased state of the currency in the 1690s had occasioned Charles Montagu's Great Recoinage of 1696, which possibly involved a loss of some five million pounds (in nominal values). Montagu's predecessor for several years as Lord High Treasurer was Godolphin.

1272 LOWNDES TO NEWTON
11 JANUARY 1718
From a clerical copy in the Public Record Office [1]

Sr

The Lords Com[missione]rs of his Mats. Trea[su]ry direct you to attend them here on Monday morning next about 10 a Clock to take their L[ordshi]ps Dirrections for preparing a New Indenture for the Mint [2] 11th Janry 1717 [/18]

WM LOWNDES

NOTES

(1) T/27, 22, p. 258.

(2) Strictly, a new Indenture between the Crown and the Master and Worker of the Mint was required as soon as the new monarch succeeded. It had become essential as it was now proposed to introduce a new gold coin, the quarter-guinea—an attempt to alleviate the shortage of silver coin. Newton did not act at once; see Letter 1277.

1273 CHARLES STANHOPE TO NEWTON
13 JANUARY 1718
From a clerical copy in the Public Record Office [1]

Gent

The Lords Comm[issione]rs of his Mats. Trea[su]ry direct You to Attend him there on Thursday next about 10 a Clock in the forenoon & in the mean time to send their Lordps an Acco[un]t how much Tinn remains in the Tower & p[ar]ticularly how much of the last Quantity Ordered to be sold. [2] I am &c 13 January 1717 [/18]

C: STANHOPE

NOTES

(1) T/27, 22, p. 258.

(2) The Treasury was anxious to wind up the 'tin affair' as rapidly as possible; a statement from the Crown's Cornish tin agents reporting on the great difficulty of preparing final accounts and disposing of their assets may be found at P.R.O. T/4, 9, pp. 361–2 (5 December 1717). Either in response to the present letter or a later one, accounts of the unsold tin and a statement of the salaries of those concerned at the Tower prepared by Newton were read at the Treasury on 25 February 1718 (P.R.O. T/4, 20, p. 94; T/1, 216, no. 37). A few days later the Treasury resolved that all the Tower tin salaries should cease from Christmas 1717 (*Cal. Treas. Books*, XXXII (Part II), 1718, p. 213). Thus Newton became £150 per annum poorer.

1274 INGLIS TO KEILL
14 JANUARY 1718
Extract from the original in the University Library, Cambridge[1]

Jan 14
1717/8

Dear Cousin

I acquainted Sir Is. Newton that you was fully satisfyd with his Corrections and Referr'd the whole to his judgement;[2] which he received very kindly, though he had been Impatient to hear from you. But you have forgott to send me back his paper, as we had done to take a copy of it And therefore, you must send it me to free Sir Is. of the Trouble of going over it again.

NOTES

(1) Res. 1893 (*a*); Lucasian Professorship, Box I, Packet 3.
(2) Compare Letter 1269. Inglis refers to Keill's drafted 'Lettre à Bernoulli', the reply to the 'Epistola pro Eminente Mathematico'; see Letter 1239, note (4), p. 386.

1275 LOWNDES TO NEWTON
17 JANUARY 1718
From a clerical copy in the Public Record Office.[1]
For the answer see Letter 1276

Sir

In Obedience to his Mats. Command upon an Address of the House of Lords You are forthwith to lay before that House An Acco[un]t attested by Your Selfe of all the Silver Money Coyned at the Recoynage of the Silver Money of England at the Tower of London & the several Country Mints & also of all the Silver Moneys & Gold Coyned at or in that time unto Michae-[lmas] last past.[2]

And the Lords Comm[issione]rs of his Mats. Trea[su]ry desire that You Will bring or send to this Office some time this day or early tomorrow Morning a Copy of the said Account, Which by their Lo[rdshi]pps Direction to You from &c 17 Jan 1717 [/18]

W LOWNDES

NOTES

(1) T/27, 22, p. 259.
(2) On 21 January 1718 Newton attended the House of Lords and in person delivered at the Bar of the House the two accounts required of the coinage, 1695–9 and 1699–1717. See Letter 1276.

1276 NEWTON TO THE TREASURY
20 JANUARY 1718
From the holograph draft in the Mint Papers.[1]
Reply to Letter 1275

An Account of the Silver moneys coyned
in the four years between Michaelmas 1695 and
Christmas 1696

	Lwt	oz	dwt	gr
At London	1642297.	2.	12.	14
York	99608.	4.	15.	9.
Bristol	147759.	6.	2.	22.
Exeter	147923.	8.	18.	20.
Chester	102360.	4.	16.	3.
Norwich	83849.	6.	0.	11.
	2223798.	9.	6.	7.

The whole weight after the rate of $3\pounds$. $2s$. $0d$ to the pound weight Troy, amounts unto 6,893,776$\pounds$. $4s$. $1\frac{1}{2}d$, in Tale And the Gold coined in the same time was by weight 19595 Lwt. 8oz., wch at $44\frac{1}{2}$ Guineas to the pound weight Troy amounts unto 872007 Guineas.

An Account [2] of the Gold & silver moneys coyned yearly from Christmas 1699 to Christmas 1717.

Years	Gold Moneys				Silver Moneys			
	Lwt	oz	dwt	gr	Lwt	oz	dwt	gr
1700	2701.	4.	14.	11.	4805.	10.	6.	16.
1701	26742.	0.	0.	0.	37477.	0.	0.	0.
1702	3642.	0.	0.	0.	114.	6.	0.	0.
1703	34.	2.	0.	0.	718.	0.	0.	0.
1704	00.	0.	0.	0.	4007.	0.	0.	0.
1705	104.	0.	0.	0.	429.	7.	15.	0.
1706	537.	0.	0.	0.	932.	0.	0.	0.
1707	607.	0.	0.	0.	1174.	0.	0.	0.
1708	1010.	0.	0.	0.	3751.	0.	0.	0.
1709	2468.	0.	0.	0.	25432.	0.	0.	0.
1710	3716.	0.	0.	0.	817.	0.	0.	0.

Years	Gold Moneys				Silver Moneys			
	Lwt	oz	dwt	gr	Lwt	oz	dwt	gr
1711	9324.	0.	0.	0.	24768.	0.	0.	0.
1712	2855.	0.	0.	0.	1784.	8.	0.	0.
1713	13137.	0.	0.	0.	2333.	0.	0.	0.
1714	29526.	0.	0.	0.	1566.	0.	0.	0.
1715	39090.	0.	0.	0.	1643.	0.	0.	0.
1716	23765.	0.	0.	0.	1650.	0.	0.	0.
1717	15186.	0.	0.	0.	948.	0.	0.	0.
	174444.	6.	14.	11.	114341.	8.	1.	16.

All the Gold moneys coined in these eighteen years, after the rate of $44\frac{1}{2}$ Guineas to ye pound weight amount unto 7762783 Guineas; & all the silver moneys after the rate of 3£ 2s to the pound weight Troy amount unto 354459£. 3s. 9d.

Mint Office   Is. NEWTON
Jan. 20. 171⅞.

NOTES

(1) II, fo. 263.

(2) This table is printed virtually without change in Rigaud, II, p. 434 and no doubt elsewhere.

1277 LOWNDES TO NEWTON
28 JANUARY 1718
From a clerical copy in the Public Record Office[1]

Sir

The Lords Com[missione]rs of his Mats Trea[su]ry having upon your attending them some time since given you directions to prepare the draught of an Indenture for the Mint, and the same not having yet been laid before their Lordps; This by their Lordps Commands is to remind you thereof and to direct you to transmit the said Draught to their Lordps forthwith[2] I am &c. 28th Janry 1717/18

W LOWNDES

NOTES

(1) T/27, 22, p. 262.

(2) Newton must have acted quickly after this reminder, for on 6 February Kelsall sent Newton's draft, including the schedules of salaries payable by himself and the Warden, to the Attorney and Solicitor General. A month later (6 March) the Law Officers were in-

structed to have the Mint Indenture pass the Great Seal. However, its becoming effective was further delayed (Number 1283) until 6 May 1718.

The complete Indenture is calendared in *Cal. Treas. Books*, xxxii (Part ii), 1718, pp. 223–5 (the Mint copy is in Mint/4, 47). We add here (Number 1277*a*) only Newton's draft for the new clauses concerning quarter-guineas.

1277*a* DRAFT CLAUSES FOR THE MINT INDENTURE
[1718]
From the holograph original in the Mint Papers[1]

—To make five sorts of money of Crown Gold,

One Piece which shall be called the quarter Guinea or five shillings & three pence piece running for five shillings & three pence sterling, & there shall be one hundred seventy & eight of these in the pound weight Troy.

One other piece wch shall be called the half Guinea or ten shillings & six pence piece running for ten shillings & six pence sterling & there shall be eighty & nine of these in the pound weight Troy.

—excepting only the Quarter Guineas. For because these pieces cannot be sized with the same exactness as the larger pieces of gold money may be, there shall be added to the Remedy in weight, one half grain for every four quarter Guineas in the pound weight Troy of the moneys tried. So always that the same default happen by Casualty, otherwise not.[2]

NOTES

(1) ii, fo. 418r; the draft was printed in the new Indenture, except for the final sentence.
(2) This sentence, not printed, is written in another hand—perhaps Northey's.

1278 NEWTON AND SANDFORD TO THE TREASURY
19 FEBRUARY 1718
From the copy in the Public Record Office.[1]
Reply to Letter 1249A

To the Rt Honble the Lords Comm[issione]rs of his Ma[jes]ties Treasury

May it please your Lordships

In obedience to your Lordships Order of Reference of the 26th. of July last past upon the Bill hereunto Annexed of the Administratrix of Mr John

Ross [2] deceased late Engraver of his Ma[jes]ties Seals, We humbly report to your Lordships that We have considered and Examined the same and do find by the perusal of the several Warrants she has produced to us, that his Ma[jes]ty had directed the making of the several Seals in the said Bill mentioned before the death of her husband and by several Certificates and Receipts it appears that they were delivered to the respective Offices, or Officers of State they were ordered for, and as they have been Weighed in the Mint, that they are of the Weight expressed in the said bill.

We further humbly certifie your Lordps that we have examined the prices of the said Seales, Which we find to be the same with those allowed to the said Mr Ross in his life time for the like Seals Engraved in her late Majties Reign, and with those paid to his Predecessors; And as it appears to us upon Examining the Impressions of the said Seals, and upon comparing them with the former remaining with us, That the work now performed by Mr Roos's administratrix is good, We are humbly of opinion She may deserve the prices sett down in the Bill amounting to [£1226. 5s. 7d]. [3]

All which is most humbly submitted to Your Lordships great Wisdom

RICH SANDFORD

ISAAC NEWTON

Mint Office the 19*th February*
1717[18]

NOTES

(1) Mint/1, 7, p. 102.
(2) John Roos or Ross; see Letter 1218, note (2), p. 363.
(3) Written in words in the copy.

1279 NEWTON AT THE TREASURY
28 FEBRUARY 1718
From the minute in the Public Record Office [1]

Whitehall Treasury Chambers 28th Febry 1717[18]
Present

[Lord Stanhope]	
Lord Torrington	Mr Wallop
Mr Baillie	Mr Micklethwaite

*   *   *

Sir Isaac Newton & Mr Essington & Partners are called in Mr Essingtons proposalls read [2]

Sir Isaac objects against the manner of assaying proposed by them and gives their Lordps an account of the method now used in the Mint.

Lord Stanhope acquaints them that Sir Isaac Newton shall deliver over to them a small Quantity of Farthings which shall be assayed in such manner as shall be agreed upon, And if the proposers will furnish the Mint with Copper which upon the assay shall prove to be as good as that now furnished and at a lower price than that now brought into the Mint, Their Lordps will give Sir I. Newton directions to Contract with them accordingly.

My Lords direct the Master of the Mint to Contract with the pet[itione]rs for 5 Tons of Copper at the same price now given, an assay of which is to be compared with a like assay of the Copper now coyning and the Difference between both to be stated and layd before my Lords

Mr Wood & Mr Jones may furnish the like Quantity (vizt 5 Tons each) at the same price & the same Conditions [3]

NOTES

(1) T/29, 23, pp. 202–3. Stanhope's name was presumably omitted by mistake from the list of Lords attending.

(2) At the previous meeting on the 24th, the Lords had repeated to several merchants and dealers in copper then present an offer already announced in the *Daily Courant*: anyone offering good copper at the current Mint price would receive 'encouragement'. The traders were directed to attend again on Thursday, when Newton and Mr Morris, the copper miller, would be summoned to meet them.

(3) The conditions were by no means acceptable to all the British copper industry. On 11 March the Lords received a letter (dated the eighth) from Messrs Clerke, Robinson and Chambers declining an offer to submit five tons, because 'we do not conceive that the conditions are such as will retrieve the Creddit wch. British Copper hath lost by the Coynadge being

hitherto Supplyed and manidged by those who where not artists in the refineing and working of Brittish Copper wch. we have attained by great expence and a long practis wherewith the Officers of the Mint are not acquainted.'

For Jones' proposal for coining copper see T/1, 213, no. 26. For Wood's negotiations see Letters 1261, 1286 and 1287.

1280 MONMORT TO NEWTON
16 MARCH 1718
From the original in the Burndy Library, Connecticut [1]

Mar. ye 27 1718

Sr

the person who wil have the honour to deliver you this letter is an ingenious man, and very understanding in Mechanicks, he is desirous of having the honour to wait upon you, and I am glad to take this opportunity to assure you of my respect, and at the same time to perform what both Mr. Bernoulli[s] [2] have desired of me very Pressingly in all their letters. they fear that their disputes with Mr. Keil may have made them loose the honour of your freind[s]hip and they seem to look upon that losse as a very unhappy accident. [3] I cannot but praise them in that for the care thay take of their own reputation but to execute the commission which they have given me several times I must have the honour to give you on accompt of all that they have writ to me relating to you.

Above a year ago Mr John Bernoulli having heard that you dit [*sic*] not approve the challenge which Mr Leibnitz had made in their name to the english Mathematicians of the Prob. of the trajectoryes, writ me the whole story of that affair, and desired me to send you an Extract of his letter. [4] Though it contained several curious things, and fit to discover the secret dispositions of M. Leibnitz, yet I did not dare send it you, for fear of interrupting you and giving you the trouble of an answer. therefore I directed it to my freind M. Taylor that he might shew it you. I know not whether my letter was lost, [5] but he made no answer, I writ again he answered me but sayd not a word relating to that subject. I see that Mr Bernoulli to excuse himself in the world has thought fit to relate this fact in the solution he has sent me of the problem of the trajectoryes which I received yesterday in writing, for it is not yet printed. [6] I here write you down the w[h]ole article The title of this Memoire is [7] Nic. Bernoulli John. f. de trajectoriis Curvas ordinatim positione datas ad angulos rectos vel alia data lege sectantibus qua occasione communicatur gemina constructio alicujus problematis a Leibnitio propositi de trajectoriis Orthogonalibus (it is the son that speaks for the father.[)]

Non equidem inficior problema ipsum a Patre fuisse suggestum sed nego cui aliqui ita interpretantur hoc ipsum fecisse ut provocaret ullum ex mortalibus[;] nedum eruditos Angliæ Mathematicos, quorum profundam sagacitatem precipue incomparabilis Newtoni data quavis occasione deprædicat et cum quibus pacem colere modo vellent esset id quod vehementissime cuperet. Prorsus enim adstipulatur Newtono existimanti illum imprudentiæ esse arguendum qui umbram captando h.e. lites serendo perdit quietem suam rem prorsus substantialem vid. comer. Epist. page 71.[8] sed ut intelligant quam sit a more parentis alienum alios ad certamen lacessere vel cum quoquam rixarum serram reciprocare, consultum duco indicare paucis rei historiolam. Ineunte nimirum anno 1715, in litteris Leibnitianis adse scriptis vidit problema quod vir inclytus transmiserat Illust. abbati C . . . eo fine ut ad pulsum Anglorum Analystarum tentandum, (sunt leibnitii verba) illus illis proponeret[.] Problema autem its sonabat. Invenire lineam *BCD* quæ ad angulos rectos secet omnes curvas determinati ordinis ejusd. generis. Ex gr. omnes hyperbolas ejusd. verticis et ejusd. centri *AB, AC, AD,* &c idque via generali.[9] Pater vero respondet quam difficile sit problema generaliter conceptum tam facile esse exemplum quod ille proposuerit[;] siquidem sit algebraicum et quale quidem ut illud vix mediocris ingenii vires eludere queat [10] etne dubitaret Leibnitius misit hinc solutionem hujus exempli evestigio inventam a me tum temporis satis juvene quam videre est in actis Leips. anno. 1716 page. 227[.] mirum itaque non fore addidit pater si excellentia Anglorum ingenia istius particularis exempli solutionem statim, sint datura. Rescripsit Leibnitius d. 31. januarii 1716 [N.S.] [11] se hyperbolas proposuisse non quasi problema in iis consisteret, sed ut intelligeretur, se enim diserte addidisse quæri methodum generalem, rogavit author ut novum sibi exemplum suppeditaret en verba ejus: quodsi mihi inquit Leibnitius, suppeditare exemplum voles, quod non particulari aliqua facilitate adjuvare putes sed ad generalem adigere rem gratam facies id enim pro specimine solutionis veræ domino Abbati nominare potero[.] vellem autem tale esse ut factis evolutionibus tandem ad quadraturas reducatur ne dicant ne [*read* non] a nobis quidem sufficientem solutionem dari posse[:] quanquam revera recurrendum sit ad differentias secundi gradus nostra autem Methodo inter primas consistatur &c[.] [12] rogatus pater non potuit non morem gerere tanto viro cujus merita in universam rem litterariam summopere venerabatur roganti[.] itaque in [*sic*] exemplum desumtum ex eadem materia quam selegerat Leibnitius de trajectoriis Orthogonalibus transmisit, suggessit problema de inveniendis et construendis lineis ad angulos rectos secantibus seriem curvarum quæ hanc habeant naturam ut cujuslibet in quolibet puncto radius convexitatis ad sui portionem ab axe resecto habeat datam rationem

hæc tum ita gesta sunt, num vero transilierit modestiæ limites exhibendo petenti problema quod proponeret tanquam suum non tanquam parentis mei qui hanc conditionem diserte stipulabat nunc æqui lectoris judicio relinquo[.] quis enim somniasset Bernoullium hujus problematis authorem existere nisi hoc ut conjecto ipse Leibnitius amico postea incaute propalanti privatim aperuisset [?] quo jure igitur imputabit quis Bernoullio ostentationis animum a quo si quisquam ipse semper abhorruit &c.

he writes to me what follows in a letter of March th[e] 17th. Oserois Je vous demander Mr si vous avez écrit a Mr. Newton comme vous vous en estes chargé, ce que je vous avois marqué touchant les sentimens que jay a son égard je vous serois extremement obligé si vous vouliez oster a Mrs les Anglois la fausse opinion ou ils sont a nostre égard comme si mon neveu et moy nous avions dessein dentrer en querelle avec eux et de diminuer le prix des decouvertes de Mr Newton je vous demande principalement cette grace par rapport a Mr Newton dont lestime et lamitié me sont tres pretieuses. Il seroit a souhaitter qu'il voulust inspirer a Mr Keil sa douceur et sa moderation. You may say Sr that these honourable and true testimonyes which he gives you and your nation do not agree with the memoire entitled *Epistola pro eminente* &c.[13] that is inserted in the Acts of Leipsic 1716. To that I know not what to answer and I think that Mr Bernoulli can do nothing better than to desown the memoire. a great freindship and an excessive zeal for his country has carryed Mr. Bernoullys freind to far, for I wil not believe for his honour that it had bin communicated to him. I took the freedom to tell franckly my opinion to Mr Bernoulli. he has answered me as follows [14]

. . . . je ne m'en suis meslé en aucune facon ni de la forme qu'il (mon ami) vouloit donner a la reponse ni des expressions dont il se serviroit et que je n'approuve pas touttes, il ma qualifie de tittres que je nai jamais eu la vanite dambitionner avec cela il a turlupiné Mr Keil dune maniere qui ne peut qu'aigrir son esprit cela ne me plaisoit pas jaurois souhaitte que mon apologiste eust dit les choses simplement et nettement sans toucher aux personalites cest ce que je luy aurois recommande avec emprestement sil mavoit communiqué son dessein, lorsqu'il moffrit par une lettre obligeante de vouloir defendre ma cause contre Mr Keil me priant seulement de luy envoyer les preuves authentiques lesquelles je ne pouvois pas luy refuser I doubt not but that he wil things [*sic*] fit to desown the memoire, for if it contains some think good and true there are several others which according to my opinion cannot bee sustained. If you think fit to honour me with any commission for Mr Bernoulli, I [15] wil do it with a great deal a pleasure and very faithfully. I have now performed what was desired of me being convinced that it would not displease you. My attachment to you is beyond all measure as well as my gratitude for all your favours and the respect with which I have the honour to be

Sr

yr most humble and

most obedt servant

RÉMOND DE MONMORT

I take the liberty to present my humble respects to your niece.[16] My wife presents her humble service to her also as wel as to you.

NOTES

(1) Newton forwarded the letter to Keill with Letter 1284. According to Brewster, a number of letters passed between Monmort and Bernoulli, and Monmort and Newton about this time (see Brewster, *Memoirs*, II, p. 71 and pp. 436–9); but this whole account is apparently based on this single extant document. He further confuses the issue by dating it once 7 March 1718, and once 27 March 1718, thereby implying a false multiplicity of letters.

(2) Johann [I] Bernoulli and his son Nikolaus [II] Bernoulli, who at this time was teaching at Basel and supporting his father in his arguments against the English mathematicians.

(3) Johann Bernoulli still denied authorship of his letter in the *Charta Volans*, and maintained that his argument was only with Keill; he later made the denial explicit in a letter to Newton (Letter 1320, vol. VII) and Newton was willing, at least at first, to believe him (see Letter 1323, vol. VII). Later, relationships between the two men deteriorated.

(4) We print this extract as Letter 1237.

(5) Clearly it did reach Newton safely in the spring of 1717; see Letter 1237, note (1), p. 384.

(6) The article was published in the *Acta Eruditorum* for June 1718, pp. 248–62; see J. Bernoulli, *Opera Omnia*, II, pp. 286–98 (Lausanne and Geneva, 1742). Monmort quotes the seventh section and part of the eighth section; his quotation is the same as the printed version except for orthographical detail—Monmort exhibits a sublime disregard for full-stops. The remainder of the article concerns the earlier history of the problem of curves cutting orthogonally, and also quotes the 'solution and construction' which Johann Bernoulli sent to Leibniz in his Letter of 11 March 1716 N.S.; see Letter 1202, note (4), p. 324.

(7) Nikolaus Bernoulli, son of Johann, on the trajectories cutting [other] curves, duly defined in position, either at right angles or according to some other given law, opportunity being taken to communicate a double construction of a certain problem concerning orthogonal trajectories proposed by Leibniz . . . [*after quoting this title Monmort proceeds directly to Section 7 of the paper as printed*] . . . I do not indeed deny that the problem itself was suggested by my father, but I give the lie to those who infer that he posed this same problem in order to provoke any living soul, emphatically not the learned English mathematicians, whose profound wisdom, especially that of the incomparable Newton, he proclaims publicly when any occasion arises, and with whom to be at peace is his most vehement desire,

438

if only they would allow it. In a word, he certainly agrees with Newton in thinking him to be considered most imprudent who, by taking offence, that is by sowing discord, 'destroys his peace, [which is] certainly a substantial good' (see *Commercium Epistolicum*, p. 71) [8] but so that they may understand how far it is from the character of my father to provoke others to strife or carry on a dispute with anyone, I think it right to disclose a short history of this affair in a few words. It was doubtless at the beginning of the year 1715 that he saw in a letter of Leibniz written to himself the problem which that celebrated man, in order 'to feel the pulse of the English analysts,' (the words are Leibniz's), had sent to the famous Abbé Conti to propose to them. The problem read like this: 'To find the line *BCD* which cuts at right angles all curves of a known order of the same type, (for example all hyperbolas *AB*, *AC*, *AD* with the same vertex and the same centre etc.) and this by some general method.' [9] My father indeed replied, that the problem taken generally is as difficult as the case which he had proposed is easy, since it is algebraic, and even of a sort which could hardly defeat the resources of a moderate talent; [10] and, so that Leibniz could not doubt it, he sent to him (from here) a solution of this case, instantly discovered by me (only a youth at that time) which is to be seen in the *Acta Eruditorum* for 1716, p. 227. 'It will not be surprising, therefore,' my father added, 'if the excellent talents of the English immediately discern a solution of this particular case.' Leibniz wrote back on 31 January 1716 [N.S.] [11] that he had proposed hyperbolas, not as if the problem lay in them, but so that it would be understood; for he had expressly added that a general method was required. The author [Nikolaus] then asked him to supply a fresh problem. These are his [Leibniz's] words: 'If you wish to supply me with a case,' says Leibniz, 'which you think is not made easy by some particular feature but compels [one] to generality, you will do something welcome to me for I could mention it as a specimen of a true solution to the Abbé [Conti]. I would wish further that it were such that, after transformations have been made, it may be reduced at last to quadratures, lest they should say that an adequate solution cannot be provided even by us. Although certainly differences of the second order are to be resorted to, yet our method consists of first [differences], etc.' [12] Being appealed to, my father could not refuse to obey when the request came from such a great man whose merit was venerated exceedingly in the whole literary world. Accordingly he sent a case taken from the same topic which Leibniz had chosen, concerning orthogonal trajectories; he suggested the problem of finding and constructing lines cutting at right angles a series of curves, which have this characteristic: that the radius of curvature of any one of them, at any point, has a given ratio to the part of it cut by the axis.

This is how things happened then. Whether he [Leibniz] trespassed beyond the limits of propriety in presenting the problem which he proposed for resolution as his own, rather than as my father's who had made this an express condition, I now leave to the candid reader's judgement. For who would have dreamt that Bernoulli was the author of this problem, unless, as I conjecture, Leibniz himself revealed it privately to a friend who afterwards carelessly made it public? By what right, therefore, will anyone attribute to Bernoulli that desire for ostentation which he always abhorred, if anyone did . . .

(8) The quotation is from the *Epistola Posterior*, 24 October 1676; vol. II, p. 114.

(9) See Letter 1170. The narrative here is scarcely accurate, in that it was Johann Bernoulli who had first proposed the issue of a challenge problem to Leibniz (Letter 1129) and the nature of the problem had then been discussed between them (e.g. Letter 1163). As usual, the Bernoullis are only too anxious to shift responsibility.

(10) Bernoulli had written to Leibniz in these terms in a letter of 15 January 1716 N.S.; see Gerhardt, *Leibniz: Mathematische Schriften*, III/2, pp. 953–5.

(11) See Gerhardt, *Leibniz: Mathematische Schriften*, III/2, pp. 956–7.

(12) *Ibid.*, p. 956.

(13) See Letter 1196.

(14) This extract is printed with modernized spelling in Brewster, *Memoirs*, II, pp. 439. This remarkable piece of fiction is worthy of an English rendering: 'I did not involve myself in any way either with the form that my friend proposed to give to the reply nor with the language that he used, which I do not wholly approve of. He has characterized me by titles which I have never been so vain as to aspire to, and besides he has maliciously teased Mr Keill in a way which could only embitter his spirit. That did not suit me. I could have wished that my apologist had said things simply and clearly without going on to personal remarks; that is what I would have strongly urged upon him if he had communicated his plan to me, when, in a courteous letter, he offered to undertake to defend my cause against Mr Keill, only begging me to send him authentic proofs, which I could not refuse to do.' In so far as this 'friend' had any real existence it was Wolf, whose effect on Bernoulli's text had been quite trivial (see Letter 1196 and notes).

(15) Newton did not take up Monmort's offer; he waited until Bernoulli himself sent a letter by Varignon (see Letter 1320, vol. VII) and replied by the same intermediary.

(16) Catherine Barton, for whom Monmort reveals a considerable predilection; see Letter 1234, and Brewster, *Memoirs*, II, p. 491.

1281 CONTI TO PIERRE DES MAIZEAUX
[MARCH 1718]
Extract from the original in the British Museum[1]

Rec'd April 1 1718 [2]

Monsieur

J'ay bien des belles choses a vous envoyer.[3] M. Remond[4] m'a donnè tout ce qu'il avoit de M. Leibniz; et vous en pourrez faire un petit volume. Je pense de vous envoyer les Lettres originales, mais avec la condition, que vous me L'envoyez;[5] ie ne doute pas de votre politesse, et ie n'ay[6] les marques. Ces Lettres ne vint pas a moy; je vous envoyerai tout cela par Mr L'Abbé Greco, qui partira apres Paques.

Je ne scais pas ce que M. Newton a cru àlà Lettre, que M. Monmort Luy a ecrit.[7] Il a preparé un grand manuscrit, ou il [a] attaqué la Philosophie, et les trois axiomes de ces Principes.[8]

NOTES

(1) Birch MS. 4282, fo. 258. The manuscript is very difficult to read. Pierre Des Maizeaux (1673–1745) fled with his family from France to Switzerland at the Revocation of the Edict of Nantes in 1685. He studied theology in Geneva, and also developed an interest in publish-

ing. In 1699, after various travels in Europe, he came to England where he remained for the rest of his life, was naturalized as Peter Des Maizeaux in 1708, and became absorbed into the expatriate community in London. Amongst his patrons was Halifax (possibly it was Des Maizeaux who was responsible for the *Memoirs of the Life of Charles Montague, late Earl of Halifax,* published anonymously in 1715) and it may be through him that he obtained an introduction to Newton. De Moivre, Chamberlayne and Conti were also amongst his friends. (See J. H. Broome, *An Agent in Anglo-French Relationships: Pierre des Maizeaux, 1673–1745* (London Ph.D. Thesis, 1949).)

His plan for a *Recueil de diverses Pièces sur la Philosophie, la Religion naturelle, l'Histoire, les mathematiques, etc. par Mrs Leibniz, Clarke, Newton et autres Autheurs célèbres* (Amsterdam, 1720) was typical of his publishing enterprises. It included in volume I the Leibniz–Clarke correspondence and in volume II, the correspondence between Newton, Conti, Chamberlayne and Leibniz and related letters. Des Maizeaux had begun preparations for the *Recueil* as early as 1716. In a letter from Leibniz to Des Maizeaux, dated 27 August 1716 N.S. (British Museum, Birch MS. 4284, fo. 210) we find that Des Maizeaux had then already set about collecting together the Leibniz–Clarke correspondence. Henri du Sauzet was to publish the *Recueil* and by 13 September 1717 we find him agitating for Des Maizeaux to send him a notice of it so that he may announce its forthcoming publication in his journal, the *Nouvelles Literaires* (British Museum, Birch, MS. 4288, fo. 5). Du Sauzet continually pressed Des Maizeaux for more copy, and was worried too by increasing financial difficulties; but the publication was further delayed in 1719 by alterations which Newton wished to introduce. Eventually, on 31 May 1720 N.S. Du Sauzet was able to write to Des Maizeaux, 'Notre Recueil est enfin achevé.' (British Museum, Birch MS. 4288, fo. 46).

Des Maizeaux gave a long account of the trials of the publication, and Newton's part in it, in a letter to Varignon of 30 March 1721 (see Letter 1359, vol. VII).

(2) The date is inserted by Des Maizeaux.

(3) Conti sent Des Maizeaux copies of the letters exchanged between Leibniz and Newton through himself—Des Maizeaux already knew of some of these from their publication in Raphson's *History of Fluxions*—and also additional letters, including those of Leibniz to the Baroness Kilmansegge and Count Bothmer, (see Letter 1359, vol. VII). Presumably they did *not* include the correspondence through Chamberlayne, which Des Maizeaux probably obtained later directly from Newton.

(4) Nicolas Rémond; much of his philosophical correspondence with Leibniz was published in volume II of Des Maizeaux's *Recueil.*

(5) *Sic,* apparently, but probably meaning 'me les renvoyé'.

(6) Meaning 'j'en ay les marques' ('I have evidence of it').

(7) Letter 1280.

(8) The title of the essay is given in another letter from Conti to Des Maizeaux, dated 5 July 1718 N.S. (British Museum, Birch MS. 4282, fo. 260) as *Critique sur la philosophie de M. Newton.* We have found no piece by Monmort bearing this title, but probably it became the 'Dissertation de M. de Montmor [*sic*], sur les Principes de Physique de M. Descartes, comparez à ceux des Philosophes Anglois', *L'Europe Savante* for October 1718, 5, Part 2 (La Haye, 1718), pp. 209–294. See also Letter 1366, vol. VII.

1282 LOWNDES TO NEWTON
19 MARCH 1718
From a clerical copy at the Public Record Office[1]

Sr

The Lords Commi[ssione]rs of his Mats. Trea[su]ry Command me to transmitt to You the enclosed Petition of John Mills with Mr Cracherodes Report thereupon,[2] which their Lordps desire You to consider & let them know if you have any Objection to the Paymt: of the Summ mentioned in the said Petition & Report Out of the 400*l* per Ann granted by Parliamt to the Mint for Apprehending & Convicting Clippers & Coyners I am &c 19th. March 1717/8

WM LOWNDES

NOTES

(1) T/27, 22, p. 282.

(2) John Mills, Constable of Holborn, Middlesex, sought £52.8*s*.3*d* due to him for apprehending and prosecuting false coiners. Anthony Cracherode, the Treasury Solicitor, duly certified that he had proceeded on a warrant signed by Lord Sunderland and Justice Ward. Mill's petition was read and approved by the Treasury Lords subject to Newton's approval on 27 March and Newton was authorized to pay him on 1 April (P.R.O. T/4, 20, p. 115; *Cal. Treas. Books*, XXXII (Part II), 1718, p. 292).

1283 NEWTON AT THE TREASURY
8 APRIL 1718
From the minute in the Public Record Office[1]

Whitehall Treasury Chambers 8th April 1718
Present

Mr Chancellour [of the Exchequer]	Mr Wallop
Mr Baillie	Mr Clayton [2]

Sir Isaac Newton attending according to Order[3] is called in & gives my Lords an Account of the progress made in Coyning Copper farthings & halfe pence, he says Six tons of Halfepence have already been issued out of the Mint.

My Lords ask him what preparations have been made in order to coin ¼ Guineas, pursuant to the Directions of the late Lords of the Trea[su]ry, he answers that the Dyes & puncheons are ready but that a power under the

Great Seal or by his Ma[jes]ties proclamation is wanting to Authorize him to coyn the same

My Lords admonish him to hasten the passing of the Indenture for Coyning Gold & Silver [4]

NOTES

(1) T/29, 24, p. 12.

(2) The old Treasury Commission headed by James Viscount Stanhope had been replaced on 20 March by a new one, led by Charles, Earl of Sunderland (1674–1722), as First Lord of the Treasury. John Aislabie (1670–1742), previously Treasurer of the Navy, replaced Lord Stanhope as Chancellor of the Exchequer. (For his active part in the South Sea Company's affairs Aislabie was dismissed from this office in 1721, declared guilty 'of most notorious, infamous, and dangerous corruption', expelled the House of Commons and for a time imprisoned in the Tower.) John Wallop and George Baillie continued from the previous Commission. William Clayton (1671–1752), created Baron Sandon in 1735, at this time M.P. for Woodstock, was a Treasury Lord until 1720 and again (after making his peace with Walpole) from 1727 to 1742 (*History of Parliament*, I, pp. 558–9).

(3) Newton's attendance on this day had been ordered by the Lords on 3 April.

(4) See Letter 1277, note (2), p. 432. It was not, in fact, until 30 October that Newton reported the completion of the dies for striking quarter-guineas; he was then ordered to coin all gold coming into the Mint into these tiny coins (*Cal. Treas. Books*, xxxii (Part II), 1718, p. 107). Compare also Letter 1309, vol. vii.

1284 NEWTON TO KEILL
2 MAY 1718

From the original in Trinity College, Cambridge.[1]
For the answer see Letter 1289

Dr Keill

I received about a month ago the inclosed Letter from Mr Monmort,[2] It conteins some extracts of Letters to him from Mr Bernoulli & his son. The chief point is that Mr Bernoulli denies that he is the author of ye Memoir entitled Epistola pro eminente &c that is inserted in the Acts of Lepisic 1716.[3] The Memoir it self lays it upon Mr Bernoulli by the words *meam solutionem*,[4] & if Mr Bernoulli is injured thereby it is not you but the author of the Memoir who has injured him. The injury is public & in justice requires a public satisfaction, not from you but from him that has done the injury. The question is therefore whether you will take notice of Mr Bernoulli's excusing himself in private or leave him to do it in publick. I have not yet returned any Answer to Mr Monmort, because I thought it best to stay till I had your

443

sense upon this matter. I think to discourse also [with] your friends [5] Dr
English & Dr Bower about it. I am

London. 2 *May* 1718

Your faithful friend &

humble servant

ISAAC NEWTON

I pray return Mr Monmorts
Letter by Dr Halley because
I am to answer it

For Dr John Keill, Professor of
Astronomy at
        Oxford

NOTES

(1) R.16.38, fo. 430; printed in Edleston, *Correspondence*, pp. 185–6.
(2) Letter 1280.
(3) Extracts of this are printed as Letter 1196.
(4) See Letter 1196, note (4), p. 304.
(5) Thomas Bower was elected F.R.S. on 23 October 1712. In 1703 he had become Professor of Mathematics at King's College, Aberdeen. In 1717 he resigned this post and came to London, where he lived for the rest of his life. 'Dr. English' might possibly be Joseph Inglis, Keill's cousin, who was acting as an intermediary between Keill and Newton (see Letter 1269). It is not clear why Newton should wish to consult either of these two about his letter from Monmort.

1285 NEWTON TO THE TREASURY
6 MAY 1718
From the draft in the University Library, Cambridge [1]

To the Rt Honble the Lords Comm[issione]rs of
his Majts Treasury

May it please your Lordps

Upon the passing of a new Indenture, I being to give security in 2000 *lb* besides my own bond, I humbly propose for my sureties Mr Hall the Comptroller of the salt Office in a Bond of one thousand pounds & Dr Fauquier my Deputy in a Bond of another thousand pounds. [2] [& pray that it may be referred to the proper Officers in the Exchequer Office to enquire into the sufficiency of these sureties.] [3]

All wch is most humbly submitted to your Lordps great widsome

ISAAC NEWTON

Mint Office
6 *May* 1718

NOTES

(1) Add. 3965(13), fo. 384.

(2) When a new Indenture of the Mint was passed at the beginning of Anne's reign Newton had named as his sureties for the same sums Fauquier and Thomas Hall. The latter had been appointed Assistant Master to Newton's predecessor, Thomas Neale, in 1696, and in 1702 was made Chief Clerk of the Mint (see vol. IV, p. 204, note (6), and pp. 375 and 392; the Letters Patent are in P.R.O. Mint/1, 5). He was replaced (confusingly enough) by Thomas Hill on 19 March 1717. He became a very rich man, being also a Comptroller of the Salt Duties, an office exercised by deputy. Hall named Newton as an executor of his will (see vol. VII, Letter 1306) and died on 25 February 1718. He was succeeded as Comptroller by his son Francis Hall, whom Newton here nominates as a surety (*Cal. Treas. Books*, XXXII (Part II), 1718, pp. 19 and 220).

(3) Presumably Newton meant to delete this sentence. On this same day the Treasury Lords referred Newton's proposal of sureties to the King's Remembrancer, whose Deputy, John Harding, reported on 7 May that 'the persons within named are of sufficient abilities to answer the several sums of Money for which they are respectively to be Security'. The Lords then demanded on the next day that bonds in the King's name be taken up from Newton's sureties (P.R.O. T/54, 25, p. 88).

A certificate dated 11 March 1720 to the effect that proper securities had been given is in Mint Papers, I, fo. 67.

1286 WILLIAM WOOD TO NEWTON
7 MAY 1718
From the original in the Mint Papers[1]

Wolverhampton
May 7th 1718

Sr

My son having acquainted me that, according to your order, he last weeke deliver'd you another Specimen of Copper, wch you were pleas'd to approve of, as you did ye former, for both ye goodness of ye metal & its workmanship: And yt you wo[ul]d have me send you my Proposals, in order to lay them before ye Lords of ye Treasury: Accordingly I have herewith sent 'em.[2]

I desire that they may be attended with your favour, wch if you please to vouchsafe me, I shall be carefull to preserve it, not only by sending such Copper as shall not leave any room for just reflections, but also endeavouring in ever[y] other respect to render my self agreeable.

I hope in a few months to be able to furnish you with 2 tons per weeke: But being yet uncertain whether quite so much, [I] thought best not to specify ye quantity in my Proposals otherwise than in generall terms.

When my attendance is requir'd, please to let me know, & I will immediately come up, interim am

Sr

Yr

Most Obedt H[umble] Sert

W WOOD

NOTES

(1) II, fo. 459r. The proposals mentioned were presumably sent to the Treasury by Newton, since they are now in T/1, 214, no. 14. Wood offered to submit the copper 'in plates roll'd to a due breadth & thickness, at ye price of 18*d* per lb' receiving back the scissel at the same price. He insisted on a minimum quantity of 100 tons, the weekly deliveries to be agreed with Newton.

With this proposal addressed to the Treasury Lords Newton probably also sent to them Letter 1287.

(2) In Letter 1268 Newton had already agreed to take some copper from Wood, and he had received renewed orders to do so on 28 February (Number 1279). However, Wood was dissatisfied with Newton's handling of the business and complained again to the Treasury in (probably) July; see Letter 1291.

1287 WOOD TO [NEWTON]
12 MAY 1718
From the original in the Public Record Office [1]

May 12*th* 1718

Sr

My son has acquainted me wth ye kind reception of my letter,[2] & favourable answer for wch I humbly return my hearty thanks.

In answer to what you were pleas'd to tell my son, Coll. Parker[3] shall not be concerned in ye Coinage, neither had I entertain'd any such design at first, but that I was led into it per mistake, thinking, (as he always told me) that he was yr old acquaintance; & favourite, & ye only person you intended for ye undertaking; & that otherwise I had no hopes of being concern'd. And this prevented my making application to you at first, before I gave in my Proposals.

As to ye price of ye Copper, tho $17\frac{3}{4}d$ is cheaper than can be afforded, &, I believe, than any one will sell, yet I will refer my self wholly to you in that matter: And (as I said in my former) shall be very carefull in every respect to be agreeable,[4] & approve myself

Sr

Yr Most Obedt H[umble] Servt

W. WOOD

Mr Wood jun in Dyers Court
in Alderman Berry [5]

446

(1) T/1, 214, no. 14(a), now with Wood's 'Proposals', see Letter 1286, note (2).

(2) Letter 1286. Newton became less favourable to Wood when he found that Wood wished to contract for as much as 100 tons.

(3) See Newton's memorandum, Number 1294.

(4) Is this repeated docility the covert offer of a future bribe?

(5) This address appears to be added in Newton's hand.

1288 NEWTON TO NICHOLAS TRON
17 MAY 1718

This draft letter, which we have not traced, is mentioned in the *Sotheby Catalogue*, Lot 169. It apparently concerns a book *De re Herbaria*. Nicholas Tron (or Nicolo Troni), a Venetian ambassador at the English Court, was elected a Fellow of the Royal Society on 10 November 1715, along with his compatriot Conti. Charles Tweedie, in *James Stirling* (Oxford, 1922), notes that it was Tron who invited James Stirling to visit Italy, and it was certainly to Tron that Stirling dedicated his *Lineæ Tertii Ordinis Newtonianæ* (Oxford, 1717), intended as a supplement to Newton's *Enumeratio Linearum Tertii Ordinis*.

1289 KEILL TO NEWTON
23 MAY 1718
From the original in the University Library, Cambridge.[1]
Reply to Letter 1284

Oxon May 23d 1718

Honoured Sr

I return you back the papers you sent me by Dr Halley,[2] with my most hearty thanks for communicating them to me, I find that Mr Bernoulli is sensible he has burnt his fingers, and would be glad to get off, but if he be in earnest, he ought to beg your pardon publickly for affirming you committed a mistake in your principles for not understanding 2d differences, for he now knows certainly that you made no mistake on that account, Mr Leibnits has told us that Mr Bernoulli was the Author of a scurrilous flying paper[3] wherein you [were] much abused he ought likewise to acquit himself of that or else beg pardon for writing of it, I will wait upon you assoon as I have finished my Astronomy[4] wch I hope will be in the space of a month. I am

Honoured Sr

your

Most obliged and Faithfull servant

JOHN KEILL

447

I am of your opinion that I ought to take no nottice of these letters and that it lyes on Bernoulli to clear himself [5] and produce the author of that scurrilous paper. There is another place in that paper where Mr Bernoulli speaks of himself in the 3d person in page 311 lin 6th Recordatus est verum esse quod dixi [6]

For
Sr Isaack Newton
    These

NOTES

(1) Add. 3985, no. 17.

(2) See Letter 1284, note (2).

(3) The *Charta Volans*, Number 1009. Keill is a little confused. Leibniz had revealed, in the *Nouvelles Litteraires* for 28 December 1715 [N.S.] (see Letter 1172), that Bernoulli was author of the excerpt of Letter 1004 printed in the *Charta Volans*. It was Leibniz himself who was responsible for the rest of the flying-paper.

(4) *Introductio ad veram astronomiam, seu lectiones astronomicæ* . . . (Oxford, 1718).

(5) Bernoulli was finally forced to admit authorship of the 'Epistola pro Eminente Mathematico' (to which he here refers) in an appendix to a paper by his son Nikolaus. See 'Nic Bernoulli Joh. F. De Trajectoriis curvas ordinatim positione datas ad Angulos rectos vel alia data lege secantibus; qua occasione communicatur germina constructio alicujus problematis a Leibnitio propositi de trajectoriis orthogonalibus: una cum Appendice de Epistola pro Eminente Mathematico Actis Lips Mens. Jul. A. 1716 inserta', *Acta Eruditorum* for June 1718, pp. 248–62. The appendix occupies pp. 261–2. Bernoulli, while admitting authorship, accused his editors of making the letter sound more insulting and belligerent than his original version.

(6) 'Epistola pro Eminente Mathematico', *Acta Eruditorum* for July 1716, p. 311, l. 6. Keill meant to write 'Mr Bernoulli speaks of himself in the first person.'

1290 CHARLES STANHOPE TO NEWTON
3 JULY 1718
From a clerical copy in the Public Record Office [1]

Sr

The Lords Comm[issione]rs: of his Mats. Trea[su]ry having received from the Lord Lieut. of Ireland [2] the Papers here inclosed which relate to the reducing the price of the Gold Coines in that Kingdome, their Lordps are pleased to direct You to consider the matter thereof As also how the Gold and Silvers Coins of this Kingdome may be Affected by such Reduction. And to Attend their Lordps with your Observations and Opinion thereupon in Writing as Soon as conveniently You can   I am &c. 3d. July 1718

C. STANHOPE

448

NOTES

(1) T/27, 22, p. 312.

(2) The letter from Charles, second Duke of Bolton, Lord Lieutenant of Ireland, enclosing one from the Irish Privy Council, had been read by the Treasury Lords on the previous day, who added (though Charles Stanhope did not) that Newton should particularly consider the repercussions of any change in Ireland on the English money (*Cal. Treas. Books*, XXXII (Part II), 1718, p. 73). What was desired was a reduction of the silver value of the gold guinea in Ireland from 23s to 22s.9d. The Lords Justices of Ireland made the same request (considered by the Treasury Lords on 4 August) and the Duke of Bolton wrote again on 25 November to inquire what progress had been made in this matter (P.R.O. T/1, 215, nos. 1 and 58).

We have not encountered any specific recommendations by Newton.

1291 LOWNDES TO NEWTON
15 JULY 1718
From the original in the Mint Papers[1]

Whitehall Trea[su]ry Chambers 15th July 1718

The Rt. Honble the Lords Comm[ission]ers of his Ma[jes]tys Trea[su]ry are pleased to Referr this Memorial to Sir Isaac Newton Knt Master & Worker of his Ma[jes]ties Mint who is to consider the same and Report to their Lordps a true State of the matter of fact together with his Opinion what is fit to be done therein [2]

W LOWNDES

Wm Wood Ref to Sr Isaac Newton

NOTES

(1) II, fo. 369v, written on the back of the (undated) memorial from William Wood, complaining that although he had offered to sell copper for coinage at $17\frac{1}{2}d$ per pound, and his son had attended on Newton at the Mint, Hines had been chosen instead to send in 20 or 30 tons at 18d the pound. Wood went on to to say that a Treasury Minute had been made that he should supply copper after that from Hines was used up and prays to be allowed to do so; Newton has approved Wood's copper and is willing to employ it for the Coinage.

(2) Wood's memorial was read at the Treasury on 9 July and ordered to be referred to Newton, who seems to have submitted no report on it until late in the year, for it was only on 13 December 1718 that 'My Lords agree to the Report' he had submitted on Wood, of which no record survives (*Cal. Treas. Books*, XXXII (Part II), 1718, pp. 60 and 77). Meanwhile Wood had reiterated his grievances (Mint Papers, II, fo. 367).

1292   THE TREASURY TO THE MINT
22 JULY 1718
From the minute in the Public Record Office [1]

Officers of the Mint to attend the Tryall of the Pix

Let the Warden, Master & Worker and Comptroller of his Mats: Mint within the Tower of London, and the rest of the Officers concerned take notice of his Mats pleasure signifyed in the within written Order of Council [2] Whitehall Treasury Chambers 22 July 1718

J Aislabie

Geo: Baillie

Wm Clayton

NOTES

(1) T/54, 25, p. 113.

(2) The Order of Council made at Kensington on 1 July appointed 4 August for the Trial of the Pyx, with the usual formalities.

Newton had submitted a memorial (now missing) requesting that such a trial take place, on 10 June 1718 (T/4, 20, p. 155).

1293   THE TREASURY TO NEWTON
31 JULY 1718
From the copy in the Public Record Office [1]

After Our hearty Commendations the aforegoing Bill [2] appearing to us by the Examination which you and the Comptroller of the Mint have passed thereupon to be Reasonable, Lett the same be satisfyed & paid out of the Moneys of the Coinage Duty which are Remaining, or shall be and Remain in your hands, Regulating the Payment thereof in such manner as that the Yearly Sum of Four hundred pounds appropriated out of the said Dutys to Services of this Nature be not exceeded in any one year from the time of appropriating thereof Taking care that in case of Excess the sum which the Petitioner shall want to be satisfyed on his said Bills by reason of such Excess be paid & placed to Account as part of the Four Hundred Pounds for the next succeeding Year . . . [3] Whitehall Treasury Chambers 31 July 1718

J. Aislabie

Geo. Baillie

Wm. Clayton

To Our Very Loving Freind
Sr Isaac Newton Knt
Master & Worker of his Mats Mint

450

(1) Mint/1, 8, p. 130.

(2) Pinckney's bill for the prosecution of currency offenders, covering the period from 29 July 1715 to 25 March 1718, amounted to £495.6s.10d, of which he had already been paid (in 1715 and 1717) £320 (ibid., pp. 122–9). Sandford approved his bill, while Newton and Bladen countersigned it as reasonable on 11 July 1718 (ibid., p. 129).

(3) The Warrant ends with the usual formula.

1294   MEMORANDA ON THE COPPER COINAGE
[JULY 1718]

From the holograph drafts in the Mint Papers[1]

Whereas a printed paper has been delivered at the door of the Honble House of Commons on behalf of the Petitioners for furnishing the Mint with manufactured Copper rejected some of wch were Mr Tho. Chambers, Mr. Geo. Clark, Mr John Essington, Mr Henry Robinson, Collonel Parker, Mr Rich. Jones, Mr Wood, Mr Holloway, & Mr Pye complaining [of the] coarseness of the copper money now coined these are to certify that in the reign of King Charles II the Copper was manufactured in Sweden excepting the stamping; that the Copper now coined beares the assay required & that this is the best assay wch can be had in the Mint: that Mr Chambers, Mr Clark, Mr Essington, Mr Robinson &c are partners & one of them insisted upon assaying his own Copper, another made a Proposal by wch they would have had 22¾[d] per pound weight for that which is now done for [18d].[2]

Two Letters written by Mr Richard Jones to Members of Parliament being published last Sessions at the door of the House of Commons, relating to ye coinage of copper money, & a third Letter to a member of Parliament being published on Saturday last in the same manner in wch the two formers Letters are cited as written by an author of note, whose advice the House of Commons should have followed, & complaint being made in this Letter yt the new copper coin is very defective, & the metal very base & inferior to the copper money coined [illegible] being less value by two pence in the pound weight, & that his Mats Officers have made a contract wch has hitherto been performed scandalously, all wch will be made appear if the Honble House of Commons shall please to appoint [a] Committee who shall send for & examin the Persons who deliv[er]ed in Proposals to ye Treasury some of wch are Mr Tho Chambers, Mr Geo Clark, Mr John Essington, Mr Henry Robinson, Collonel Parker Mr Richard Jones above mentioned Mr Wood, Mr Holloway & Mr Pye, several of whom proposed that the farthings & Halfe pence coined in

King Charles time should be the standard of the metal & that in six months
time there has been but three tons of copper delivered into the Mint tho the
people are in great want of Copper money: [3] in answer to all this it is represented that the Copper money coined in the reign of King Charles was of
Swedish Copper manufactured into blanks in Swedeland where the copper &
workmanship is cheaper then in England, that no finer Copper was proposed
by any of the persons above mentioned then that wch will spread red hot under
the great hammers of the battering mills till it be as thin as the half pence &
farthings, & that such copper is accounted as fine as King Charles' money,
that the copper now brought into the Mint is assayed by battering it both red
hot & cold till it be as thin as paper, & no better assay can be had in the Mint.
that Copper wch beares this assay is not inferior to the Copper wch beares
the assay of the battering Mill or not above a farthing in the pound weight
inferior to it, that [when the Officers of the Mint gave the Proposers a meeting
to heare them upon their Proposals Mr Wood who] [4] the working of Copper
into barrs of due size & fineness for making of money is a manufacture not
yet practised in England & therefore not yet ripe for a contract that Mr
Chambers, Mr Essington & Mr Robinson & (I think) Mr Clark are partners
& some of them proposed to do it by the battering mills first at $22\frac{3}{4}d$ & afterwards at $19d$ per Lwt, Mr Holloway proposed to do it by the battering mill
at a rate amounting to about $19d$ per Lwt & Mr Jones at $18\frac{1}{2}d$. Those that now
supply the Mint with copper do it at $18d$ per Lwt. Mr Pye in the name of Mr
Neale Marchant & Company proposed to deliver copper in barrs wch would
stand the assay at $17\frac{1}{2}d$, but could not under take the flatting & drawing thru
the mill to a proper size, & nealing & scouring at that rate. Mr Wood proposed to deliver Plates of fine rolled Copper fit for the Mint at $17\frac{1}{2}d$ per Lwt
but being absent in the country his son appeared for him at the Mint & went
away again without speaking with the officers of the Mint about his fathers
Proposal. [5] Mr Parker is partner with Mr Wood & these two have not since
offered to be imployed without entring into a contract. The persons now
imploying do not act upon a contract. They are only upon tryal, & if they do
not make good their Proposals, they may be laid aside at any time. To make
the Embosments & Puncheons & take off some Dyes from them was a work
of four months or above. Since that time they have begun to bring Copper
into the Mint & have made trials of several methods of manufacturing it, &
are still making further trials, & if it had not been for the importunity of the
people, no money would have been delivered till these trials had been over &
the best method fixed upon. [6] And now, if any of the Petitioners or any
worker in Copper will for a trial furnish the Mint with a Ton or two of Copper
monthly well sized & scoured & wch shall bear the assay now used in the Mint

452

at 18*d* per Lwt or under, he shall meet with encouragement suitable to the goodness of the workmanship & the lowness of the price. [7] But it is not reasonable that the persons whose Proposals have been rejected should be witnesses in their own cause.

that the Proposers here mentioned cannot be witnesses in their own cause but recourse must be had to the assays of the copper—that within these five years copper is risen 3*d* in the pound weight & the manufacture is now dearer in England then 40 years ago in Swedeland & a pound wch in the reign of King Charles was cut into 20*d* is now cut into no more then 23*d* [8]

*   *   *   *   *

To the Rt Honble the Lords Comm[ission]ers of his Mats
Treasury the Memorial of Sr Isaac Newton
most humbly sheweth

Cours copper when made read hot will not spread under the hammer but flyes in pieces. And in refining it, so soon as it begins to spread under the hammer when made red hot, it begins to be called fine copper, & then it is worth about 13*d* per lwt, & when it is fine enough to beare the great hammers of the battering mills, it is reputed fine copper free from all mixtures & is then valued at about 14½*d* per lwt. And of about this degree of fineness was the copper money coined in the reign of King Charles II, [And the complaint made the last winter by Mr Jones & Mr Essington & their partners was only [that] [9] the copper money then coined was not so fine by two pence in the pound weight] [10] it being of Swedish copper manufactured into blanks in Sweden. And of about the same degree of fineness is all such copper as being made red hot will then spread very thin under the hand hammer upon an anvill without cracking And by his Ma[jes]ties Warrant the money ought to be of this degree of fineness. [To beate it as thin as a leafe under an hand hammer is a Trial as great as to beat it as thin as a shilling under the great hammer of the battering mills.] [10] And this essay by the [hand hammer is established by his Mats Warrant to be the Rule by wch all the copper received into the Mint & all the money coined out of it was to be assayed, & has been assayed & by which the Copper Pix is to be tried.] [11] Copper may be refined to a much higher degree so as to be worth 2*s* or 2*s* 6*d* per lwt or above. But such copper is of little use except for making wire or leafe copper, & there is no certain assay or rule yet known by wch the fineness of it may be ascertained & the highest degree to wch it may be refined is not yet known. There have been blanks made of much finer copper then the money by people not

453

imployed by me & stamped in the Mint without my knowledge & then polished to give them a more beautifull gloss & shewn about to deceive people & bring the money into discredit, & brockage [12] hath also been picked out of the money & shewn about as a parcel of money received from the Mint, & there may have been blanks of coarse copper carried to the Mint & there stamped without my knowledge, or otherwise counterfeited, & mixed with the brockage. The goodnes of copper cannot be known by the looks alone. It must be assayed. By the colour & grain of the copper when it is newly broken some judgment may be made by a skilful person: but the surest trial is by the malleability of the copper hot or cold. And therefore this was made the assay of standard copper by his Mats. Warrant, & by this assay the Pix of what has been coyned is to be tried.

NOTES

(1) II, fos. 452 and 349r. The drafts are very rough, with much blotting, deletion and interlineation, as well as repetition. We have not attempted to improve them. Their date is uncertain, but it seems reasonable to suppose that they were written in the spring or summer of 1718.

(2) The figure is obliterated by a blot.

(3) If the 'last Sessions' of Parliament were those of 1717, Newton must be writing in 1718; and six months from the autumn of 1717, when the copper coinage began, indicates the spring of 1718, or thereabouts.

(4) The words between the brackets (in the original) are meant to be deleted.

(5) This seems to put the date of writing posterior to Letter 1286.

(6) A curious excuse coming from Newton, who had done his best to delay the 'trials' with copper, on the grounds that none was needed.

(7) Compare Number 1279, whence Newton obviously drew these phrases.

(8) These detached phrases appear, as here, at the end of the main draft.

(9) Editorial addition.

(10) Compare the preceding draft. Newton has deleted this sentence between (his own) brackets.

(11) Again, the words between Newton's brackets have been deleted. It appears that there was no Trial of the copper Pyx before 1722, when the money was found fully satisfactory in weight and quality (Mint Papers, II, fo. 317).

(12) Broken pieces, waste (Joseph Wright, *English Dialect Dictionary*).

1295 NEWTON TO DES MAIZEAUX
[*c.* AUGUST 1718]

From a draft in the University Library, Cambridge [1]

Sr

I have viewed the printed papers you left in my hands. The Remarks are only upon the Letter of Mr Leibnitz to Mr. l'Abbé Conti dated 9 Apr. 1716

454

[N.S.].[2] The letters to the Comtesse of Kilmansegger, & the Postscript of a Letter to Compt Bothmar are without an Answer, & I do not see that they need any.[3] None of his Letters were writ to me, & I had not answered any of them had not Mr l'Abbé Conti pressed me to write an Answer to the Postscript of a Letter to him that both might be shewed to the King.[4] But when I understood that Mr Leibnitz sent all the Letters open to Paris,[5] & his Answer came hither from thence, I declined writing any more Letters, & only wrote the Remarks [6] upon that Answer to satisfy my friends here that what he writ was easy to have been answered if it had come hither directly. The Commercium Epistolicum, notwit[h]standing any thing which has hitherto been said against it, remains in full force; & while that remains unshaken there is no need of writing any further about these matters.[7]

Mr Leibniz objected against it that in printing the ancient Letters & papers many things had been omitted wch made for him & against me:[8] but in offering twice to prove this by instances, he failed as often.

He objected also that the Committee of the R. Society strained things against him by fals interpretations but he never proved this in any one instance. He said indeed that they had maliciously interpreted the words of the Acta Eruditorum for January 1705, by saying: *sensus verborum est quod Newtonus fluxiones pro differentijs Leibnitianis substituit*: but he has misrepresented the place himself. For the Editors of the Acta in this Paragraph call Mr Leibnitz the INVENTOR, & thence deduce this conclusion. *Pro differentijs* IGITUR *Leibnitianis Newtonus adhibet semperque* [pro ijsdem] *adhibuit fluxiones——ijsque tum in suis Principijs Naturæ Mathematicis, tum in alijs postea editis* [pro differentijs] illis] *eleganter est usus* QUEMADMODUM ET *Honoratus Fabrius in sua Synopsi Geometrica motuum progressus Cavallerianæ methodo substituit.* This is the interpretation wch the Committee put upon the place, & the words *igitur* and *quemadmodum et* enforce it.[9]

It has been represented that in the Commercium Epistolicum Mr Leibnitz is complained of as having published my Method as his own; whereas he found the Method by himself. But second inventors have no right. Whether Mr Leibnitz found the Method by himself or not is not the Question. The Committee of the Royal Society did not enter into this Question, but on the contrary said:[10] *We take the proper Question to be, not who invented this or that Method but who was the first inventor of the method. And we beleive that those who have reputed Mr Leibnitz the first Inventor, knew little or nothing of his correspondence with Mr Collins & Mr Oldenburg long before, nor of Mr Newton's having the Method above 15 years before Mr Leibnitz began to publish it in the Acta eruditorum of Leipsic.* Here the Committee of the R. Society treat Mr Leibnitz as second Inventor and complain of him only for publishing the Method as his own without making any mention of the correspondence wch he had with me & my friends about it long before he published it, & by that concealment claiming the Method as first Inventor.

By that correspondence he had notice in June or July 1676 that I had a very general method of Analysis wch extended to the abstruser sorts of Problems in Geometry & that the Method of Tangents of Gregory & Slusius was but a branch or rather a Corollary thereof[11] & that it proceeded without any troublesome calculation & without sticking at surds & that I had it & had interwoven it with the Method of infinite series before I wrote my Letter of 10 Decem. 1672 & being tyred with these speculations had absteined from them five years when I wrote my Letter of 13 June 1676, & that the Art of Analysis by these methods became so general as to reach almost all sorts of Problems except some numeral ones like those of Diophantus.

But all this was not yet sufficient to make him understand the differential method. For in his Answer dated 13 [sic] Aug. 1676,[12] he wrote back. *Quod dicere videmini plerasque difficultates (exceptis Problematibus Diophantæis) ad series infinitas reduci; id mihi non videtur. Sunt enim multa usque adeo mira et implexa ut neque ab æquationibus pendeant neque ex quadraturis. Qualia sunt (ex multis alijs) Problemata methodi tangentium inversæ.* And in the same Letter he placed the perfection of Analysis in Analytical Tables of Tangents & the combinatory art: saying of the one, *Nihil est quod norim in tota Analysi momenti majoris*; & of the other, *Ea vero nihil differt ab Analysi illa suprema, ad cujus intima, quantum judicare possum, Cartesius non pervenit. Est enim ad eam constituendam opus Alphabeto Cogitationum humanarum.*

Mr James Bernoulli in the Acta Eruditorum for December 1691 & the Marquess de l'Hospital in the Introduction to his Analysis represented that the Differential Method was an improvement of Dr Barrow's method of tangents, & the Marquess represented further that the improvement lay in shewing how to proceed without stopping at surds. But the Marquess did not then know that Mr Leibnitz had notice of this improvement from me before he found it, & the Committee complained of him justly for concealing this & all the foregoing notices.

In October 1676 Mr Leibnitz came to London & there met with Dr Barrows Lectures & with my Letter to Mr Oldenburg dated 24 Oct. 1676, in wch the above mentioned notices were repeated & he was also told that the Tract in wch I interwove the two Methods was written five years before, that is, in the year 1671. And that before the plague wch raged in the year 1665 forced me from the University, I had found the method of converging series (including reductions by the binomial Rule and by division & extraction of roots both simple & affected,) so as to be able to deduce the areas of all figures

& ye lengths of all curve lines from their Abscissas & Ordinates & on the contrary to deduce the Abscissas & Ordinates & any other right lines from the areas & lengths of ye Curves. He was told also that when Mercators Logarithmotechnia came abroad, Dr Barrow sent Mr Collins a Compendium of the Method of these series, meaning the *Analysis per æquationes numero terminorum infinitas*. And at the same time he consulted Mr Collins to see what he could meet with in his hands of mine & Gregories Letters & Papers, & there saw a great part of our correspondence. And in his way from London to Hannover he was meditating how to improve the Method of Tangents of Slusius. For in a Letter to Mr Oldenburgh from Amsterdam dated $\frac{18}{28}$ Novem. 1676 he wrote:[13] Methodus Tangentium a Slusio publicata nondum rei fastigium tenet. Potest aliquid amplius præstari in eo genere, quod maximi foret usus ad omnis generis Problemata——Nimirum posset brevis quædam calculari circa Tangentes Tabula &c. He had not therefore yet found the right improvement, but at length he found it proprio Marte. For the next year in a Letter to Mr Oldenburgh dated 21 June 1677 he wrote:[14] Clarissimi Slusij methodum tangentium nondum esse absolutam celeberrimo Newtono assentior. Et jam a multo tempore rem tangentium generalius tractavi, scilicet per differentias Ordinatarum. And then he set down Dr Barrows method of Tangents with a new notation & shewed how it might be improved so as to give the method of Tangents of Slusius, & to proceed without stopping at surds; & then added: Arbitror quæ celare voluit Newtonus de Tangentibus ducendis, ab his non abludere. Quod addit, ex hoc eodem fundamento. Quadraturas quoque reddi faciliores, me in sententia hac confirmat.[15]

NOTES

(1) A large number of widely differing drafts, apparently all to Des Maizeaux in response to a request for comments on the proofs of volume II of the *Recueil*, appears in U.L.C. Add. 3968(27) and 3968(28) and elsewhere. These are difficult to date; it is not clear whether they are all drafts for the same letter, nor indeed whether a letter was in fact ever sent.

We date the drafts *c.* August 1718, because, according to Newton's own account (see his draft letter to Varignon, (Letter 1354, vol. VII), probably composed about the beginning of 1721), he received proof of the *Recueil* in July or August 1718, together with a letter asking for suggestions and improvements. Some proof-sheets of the first part of volume II (the section containing the correspondence with Leibniz and Conti, some additional letters, and Newton's 'Observations'—but lacking the final sheet of the latter) are now in U.L.C. Adv. d. 39.2, and are heavily corrected by Des Maizeaux, with a few minor annotations by Newton, some of which are in fact adopted in the *Recueil*. Hence Newton must at some time have been in contact with Des Maizeaux concerning these proof-sheets. Newton also made suggestions for re-ordering the letters (see Letter 1295 a below), which were, at least in part, adopted by Des Maizeaux. But of the greater part of Newton's draft suggestions and amendments—which include several sheets of 'Addenda & Corrigenda' (Add. 3968(29), fos. 428–35) and also an extensive 'Supplement to the Remarks' (Add. 3968(31), fos. 453–9) there is no trace in the *Recueil*, so we assume they were never transmitted to Des Maizeaux. However, it does seem that Newton in fact saw a second, complete proof of the *Recueil* before it was finally published; the page numbers in the 'Addenda & Corrigenda,' and in Letter 1295 *b* refer to the pagination as finally printed, not to the pagination of the proof-sheets mentioned above. So it appears that Newton may have planned, and possibly sent, a second letter to Des Maizeaux at a slightly later date. (See also Letter 1295 *a*.)

The most finished draft of a letter to Des Maizeaux seems to be Add. 3968(27), fos. 393–5, which we print here. None of the subject matter in this or the rougher drafts is essentially new, but we quote below excerpts from some of the variants, because they give some insight into Newton's attitude to the calculus dispute, and the extent of his involvement in it. They also show that he was by no means above concealing information when it suited his ends. If, indeed, all the drafts are of one intended letter, then Newton seems to have experimented with the form of his reply and to have changed his mind several times about what aspects of the dispute were worthy of inclusion. For example, the beginning of the draft on fo. 401r is very unlike the opening of the text we print:

Sr

  You know that when Mr l'Abbé Conti had received a Letter from Mr Leibnitz with a large Postscript against me full of accusations forreign to the Question, & the Postscript was shewed to the King, & I was pressed for an answer to be also shewed to his Majesty, & the same was afterwards sent to Mr Leibnitz: he sent it with his Answer to Paris declining to make good his charge & pretending that I was the Aggressor, & saying that he sent those Letters to Paris that he might have neutral & intelligent witnesses of what passed between us. I looked upon this as an indirect practise & forbore writing an Answer in the form of a Letter to be sent to him, & only wrote some Observations upon his Letter to satisfy my friends here that it was easy to have answered him had I thought fit to let him go on with his politicks. As soon as I heard that he was dead I caused the Letters & Observations to be printed least they should at any time come abroad imperfectly in France. You are now upon a designe of reprinting them with some other Letters written at the same time, whose Originals have been left in your hands for that purpose by Mr l'Abbé Conti for making that Controversy complete & I see no necessity of adding any thing more to what has been said, especially now Mr Leibnitz is dead.

(Further paragraphs from this draft are printed in Cohen, *Introduction*, pp. 295–6. Compare also the opening passages of the draft we print in note (7) below.)

The draft we print as our main text is largely concerned with discussion of the 1676 correspondence with Oldenburg, and with Leibniz's visits to London; other drafts mention both earlier issues—for example Newton's use of 'dot' notation (see note (13) below)—and more recent ones. We single out three excerpts as being of particular interest:

  (i) In a lengthy draft at Add. 3968, fo. 405, Newton mentions his use of fluxions in writing the *Principia*.

  By the inverse method of fluxions I found in the year 1677 the Demonstration of Keplers Astronomical Proposition vizt that the Planets move in Ellipses, wch is the Eleventh Proposition of the Book of Principles; & in the year 1683 at the importunity of Dr Halley I resumed the consideration thereof & added some more Propositions about the motions of the heavenly bodies wch were by him communicated to the R. Society & entred in their Letter-Book the winter following, & upon their request that those things might be published,

I wrote the Book of Principles in the years 1684, 1685, 1686 & in writing it, made much use of the method of fluxions direct & inverse, but did not set down the calculations in the Book it self because the book was written by the method of Composition, as all Geometry ought to be. [And ever since I wrote that Book I have been forgetting the Methods by which I wrote it.]

(In fact, it was in 1684, not 1683, that Halley visited Newton at Cambridge and encouraged him to work on the law of gravitation; see vol. II, p. 411.)

(ii) A draft at Add. 3968(27), fo. 385 includes criticism of Johann Bernoulli's 'Epistola pro Eminente Mathematico' (see Letter 1196):

There was also at the same time a Panegiric upon Mr J. Bernoulli & a Satyr upon Dr Keill written by an anonymous author & published in the Acta eruditorum for July 1716. The author has fathered it upon Mr John Bernoulli himself by calling a *formula* of his *meam formulam* pag. 34 lin. 27 & thereby has made Mr J Bernoulli call himself Eminentem Mathematicum, excelsum ingenium & virum ad abstrusa et abdita detegenda natum pag. 269 l. 13 & pag 298. l 32 & p. 301 lin 29. But its possible the author may have copied the words meam formulam out of one of Mr J. Bernoullis Letters. Mr Nicolas Bernoulli was in England in Autumn A.C. 1712 & . . .

(iii) In a draft at Add. 3968(27), fo. 398, Newton makes a comment on Wolf's 'Eloge' of Leibniz, (compare Letter 1305, note (8) (vol. VII)):

The Author of the Elogium of Mr Leibnitz published in the Acta Eruditorum for July 1717 said Commercio Epistolico Anglorum aliud quoddam suum, idemque amplius, [D. Leibnitius] opponere decreverat, et paucis ante obitum diebus Cl. Wolfio significavit se Anglos famam ipsius lacessentes reipsa refutaturum quamprimum enim a laboribus historicis vacaturus sit, daturum se aliquid in Analysi prorsus inexpectatum et cum inventis quæ hactenus in publicum prostant, sive Newtoni, sive aliorum nihil quicquam affine habens.

(2) Letter 1197.

(3) In another draft (Add. 3968(27), fo. 391) Newton dealt at length with the letters of Leibniz to Baroness Kilmansegge and to Count Bothmer (see Letter 1203), both of which were to be printed in Des Maizeaux's *Recueil*, and both of which contained lengthy accounts of Leibniz's view of the calculus dispute.

(4) See Letter 1187, note (1), p. 288.

(5) Leibniz sent his letters via Nicolas Rémond in Paris; see Letter 1170*a*, note (1), p. 255.

(6) That is, Newton's 'Observations' on Letter 1197; see Number 1211.

(7) Newton was more forceful in earlier drafts of this paragraph and also more open about his printing of the 'Observations', or 'Remarks', in the Appendix to Raphson, *History of Fluxions*. For example, the draft at Add. 3968(27), fo. 387 reads:

Sr

I have perused the printed sheets wch you left in my hands & beg leave to take notice that when the Postscript of Mr Leibnitz's first Letter to Mr l'Abbé Conti was shewed to me I did not think my self concerned to meddle with it, till at length I was pressed by Mr l'Abbe Conti to answer it that the Postscript wth my Answer might be shewed to the King. And when Mr Leibnitz sent it with his Answer to Mr Remond at Paris, & his Answer was sent open from Paris to Mr l'Abbé Conti I refused to write any thing more to be sent to him. For I perceived that as he formerly appealed from the Commercium Epistolicum & the judgment of the Committee of the R. Society upon it to the judgm[en]t of Mr John

Bernoulli, so he was now upon a designe of appealing to his friends at Paris his army of disciples & his interest at Court. But yet I wrote a Paper of Remarks upon his Letter & shewed it to some friends in private to let them see how easy it was to have answered his Letter had I thought fit to let him go on with his Politiques. And when I heard that he was dead I printed the Letters with these Remarks least they should be published imperfectly abroad. And since you are reprinting them with some other Letters written at the same time, & in some of those Letters he tells his own story at large; tho I will not write an Answer to those letters now he is dead; yet since in his Letter of 4 March 1711 when I knew not what had been printed against me six years before in the Acta Eruditorum nor was any way concerned in this controversy, he appealed to me against Dr Keill & would contend with no body but me nor let me rest till I set pen to paper: I think I may be allowed to tell the story my self, & leave it to be compared with his Narrations.

(8) In another draft (Add. 3968(27), fo. 390) Newton suggested that Leibniz tried to sidetrack the issue by appealing to Bernoulli, and by arguments with Clarke:

If Mr Leibnitz could have made a good objection against the Commercium Epistolicum, he might have done it in a short Letter without writing another book as big. But this book being matter of fact & unanswerable he treated it with approbrious language & avoided answering it by several excuses, & then endeavoured to lay it aside by appealing to the judgment of his friend Mr Bernoulli & by writing to his friends at Court, & by running the dispute into a squabble about a Vacuum, & Atoms, & universal gravity, & occult qualities, & Miracles, & the Sensorium of God, & the perfection of the world, & the nature of time & space, & the solving of Problemes, & the Question whether he did not find the Differential Method *proprio marte*: all which are nothing to the purpose. Mr James Gregory after a years study found the method of converging series *proprio marte*; but did not claim it because he had notice from England that there was such a general method before he searched for it, & by that notice knew that he was not the first inventor. *Gregorius autem*, said Mr Collins, *Newtonum primum ejus inventorem anticipare haud integram ducit*. The proper question is: Who was the first Inventor? Let it be proved that Mr Leibniz had the Method before he had any notice of it from England, & then let it be further proved that he had it before the date of my Letter of Decem. 10th 1672, & before the year 1671 in wch I wrote a tract upon it & before July 1699 &c; & by these steps, if they can be made the Commercium Epistolicum will at length begin to be shaken. For the proof lies upon the friends of Mr Leib. [For 1699 *read* 1669.]

(9) See Letter 830, note (2) (vol. v, pp. 115–16), where this passage is translated.

(10) *Commercium Epistolicum* (1st edition, London 1712), p. 122 and the Introduction to vol. v, p. xxvi.

(11) In Add. 3968(27), fo. 389, an earlier draft, this passage is rendered in a way which puts less emphasis on the originality of Newton's method: '. . . & that it was an improvement of the Method of Tangents of Gregory & Slusius, & that the method of Tangents of Gregory was an improvement of the Method of Tangents of Barrow & that I had it when I wrote my Letter of 10 Decem 1692 & that I had interwoven this Method with the method of infinite series before I wrote that Letter . . .'

(12) The correct date is 27 August 1676 N.S. The passages Newton quotes are to be found in vol. II, pp. 64 and 63 respectively (translations on pp. 71 and 70).

(13) See vol. II, p. 199 and (translation) p. 202.

(14) See vol. II, p. 213 and (translation) p. 219.

(15) In a draft at Add. 3968(27), fos. 389v–390v, Newton continues at greater length about his early use of fluxions:

It has been said that in the old Letters & Papers published in the Commercium Epistolicum there are no prickt letters. And indeed I seldome used prickt letters when I considered only first fluxions, as in the Introduction to the Book of Quadratures: but when I considered also second third & fourth fluxions, as in the body of the book, I distinguished them by the number of pricks. In the year 1692, at the request of Dr. Wallis I sent to him the first Proposition of the Book of quadratures with its solution & examples in first & second fluxions copied almost verbatim from the book, I sent him also the method of extracting fluents out of Equations involving fluxions, wch to the best of my memory was composed in the year 1671: & the Doctor printed them both the same year in the second Volume of his works which came abroad the next year A.C. 1693. And thence it may be understood that the Book of Quadratures was then in manuscript. In my Letter of 24 Octob. 1676 I set down the first Proposition of this Book verbatim in an Ænigma, & said that it was the foundation of the method there concealed, & that it gave me general Theorems for squaring of figures by series which sometimes break of & become finite, & how it gave me such series is explained in the first six Propositions of this Book, & I know no other method of finding them. And therefore I had the Method at that time so far as it is contained in those six Propositions. In the same Letter I copied also many Ordinates of Curves from a Table in the end of the tenth Proposition & upon the 7th 8th 9th & 10th Propositions I wrote to Mr Collins my Letter of 8 Novem. 1676 printed by Mr Jones. And from all this it may be understood that the book was then in manuscript. And as the notation used in this Book is the oldest, so it is the shortest & most expedite, but was not known to the Marquess de l'Hospital when he recommended the differential notation. If it be asked why I did not publish this book sooner, it was for the same reason that I did not publish the Theory of colours sooner, & I gave the reason in my Letter of 24 Octob. 1676

The first Proposition of the Book of Quadratures is certainly the foundation of the method of fluxions. That Proposition was comprehended verbatim in the Ænigma by wch in my Letter of Octob. 24. 1676 I concealed the foundation of the Method there spoken of, & therefore the Method there spoken of was the Method of fluxions. In that Letter I said that I had written a Tract on this Method & the Method of series togather five years before, but did not finish it, nor meddle any more wth these things till the year 1676, being tired with them. And in my Letter of June 13, 1676 I wrote to the same purpose. And this is the Method which I described in my Letter of Decem 10th 1672. In my Analysis per æquationes numero terminorum infinitas, I said of the Method described in that Tract: *Denique ad Analyticam merito pertinere censeatur cujus beneficio Curvarum areæ et longitudines &c (id modo fiet) exacte et Geometrice determinentur. Sed ista narrandi non est locus.* This relates to Quadratures by Series wch in some cases break off & become finite; as you may understand also by the Letter of Mr Collins to Mr Strode July 26 1672. And therefore before Dr Barrow sent that Tract of Analysis to Mr Collins, that is, before July 1699 [1669], I had the method of fluxions so far at the least as it is conteined in the first six Propositions of the Book of Quadratur[e]s.

The Committee of the R.S. said that I had this Method above 15 years before Mr Leibnitz began to publish it. And Mr Collins in a Letter to Mr Bertet dated Feb. 21. $16\frac{70}{71}$ said that about four years before that time I found a general Analysis for squaring all curvilinear spaces & doing what ever depends upon Quadratures. And in a Letter to Mr Strode dated 26 July 1672 he said that by the Analysis per æquationes numero terminorum in-

infinitas & other papers communicated before to Dr Barrow it appeared that I had the
Method & applyed it generally some years before the Doctor sent that Analysis to Mr
Collins, that is some years before July 1669. And in a Letter of Aug. 11 1676 to Mr David
Gregory the brother of Mr James Gr. Mr Collins wrote that a few months after the Logarithmotechnia of Mr Mercator came abroade, [wch was in Sept. 1688,] a copy of it being
sent to Dr Barrow at Cambridge he wrote back that this doctrine of infinite series was
invented by me about two years before the Publication of Mr Mercators Logarithmotechnia
& generally applied to all Curves & then sent to him my Manuscript. And I see no reason
why the testimony of Dr Barrow grounded upon what I had communicated to him from
time to time before the Logarithmotechnia came abroad, should be questioned in this
matter. Dr Wallis also who flourished in those days & was inquisitive & skilfull, & received
copies of my Letters from Mr Oldenburgh in ye year 1676; published the same thing in the
Preface to the first Volume of his works without being then contradicted, saying that in my
Letters of June 13 & Octob 24, 1676 I explained to Mr Leibnitz the Method invented by
me ten years before that time or above. And the testimony of these three ancient knowing
& credible witnesses may suffice to ex[c]use me for saying in the Introduction to the Book of
Quadratures that I found the Method by degrees in the years 1665 & 1666. I was then in the
prime of my age for invention & most intent upon & [*sic*] mathematicks & philosophy &
found out in those two years the methods of series & fluxions & the Theory of colours &
began [*draft ends here*].

Newton has cancelled the last sentence, extended in another draft into a longer passage of
reminiscence that has very often been quoted; see Cohen, *Introduction*, pp. 291–2.

1295*a* NEWTON TO DES MAIZEAUX
[*c.* AUGUST 1718]
From a draft in the University Library, Cambridge[1]

I have perused the printed sheets wch you left in my hands & beleive that
the Letter of Mr Leibnitz to Mr Remond wch is printed in the eigth place
should have been printed in the second,[2] the contents thereof relating to the
Postscript of the Letter of Mr Leibnitz to Mr l'Abbé Conti wch is printed
in the first place. Also the Letter of Mr Leibnitz to Mr l'Abbé Conti wch was
writ in answer to mine & is dated 9 April 1716 should have come before the
Letter of Mr Leibnitz to Madam la Comp[t]ess de Kilmansegger dated
18 April 1716.[3]

None of the letters being written to me I did not think it [*sic*] my self concerned to answer any of them . . .

NOTES

(1) Add. 3968(27) fo. 385: Newton's comments are clearly based on the early proof-
sheets sent to him by Des Maizeaux (see Letter 1295, note (1)), where he has renumbered the
letters in the margin. As in Letter 1295 *b*, the suggestions Newton makes here were not adopted,

but it is clear that Des Maizeaux did receive the information. In Add. 3968(36), fo. 508 is a list by Des Maizeaux entitled 'Table des Pieces selon l'ordre quelles doivent etre lûes, & qu'elles auroient dû etre imprimees' in which Newton's re-ordering is partially adopted. In the final printed version of the first edition there are also certain changes, but again Newton's re-ordering is not wholly adopted.

Another draft (Add. 3968(27), fo. 386v) indicates that at least at one stage Newton intended to send Des Maizeaux some prefatorial comments to the *Recueil*, or perhaps to have the letters reprinted himself. The draft is headed 'To the Reader' and contains the following passage, where Newton makes clear his authorship of the Appendix to Raphson's *History of Fluxions*, and also implies that he would like to reprint Bernoulli's 'Epistola pro Eminente Mathematico.'

There were some other Letters writ by Mr Leibnitz, Mr l'Abbe Conti, & Mr Remond, & by Mr l'Abbé Conti left in the hands of Mr de M[aizeaux] to be published together with what I had caused to be published before. And a few days since a copy of the same was brought to me printed off except the last sheet. But they being put together in wrong order I have caused them to be reprinted in due order of time, together with a Paper published about the same time in the Acta Lipsiensia by a nameless author who called a solution of Mr John Bernoulli solutionem *meam*.

(2) This was not done, but in the final printed version Des Maizeaux replaced the eighth letter with a different letter from Leibniz to Nicolas Rémond, dated 9 April 1716 N.S., and placed the letter in question here at p. 112, with a footnote to say it should have been printed in the second place. In the second edition of the *Recueil* the letter does appear in the second place.

(3) This was not done.

1295*b* NEWTON TO DES MAIZEAUX
[LATE 1718]

From a draft in the University Library, Cambridge[1]

Sr

I have run my eye over the printed Letters wch you have put into my hands, & since you have the Originalls I think it is right to let them come abroad. I have observed some faults which may be put into ye errata. Pag. 8 l. 13 dele c'est a dire, commune ou superficielle. Pag. 16. l. 5. write A Londres Feb. 26, $17\frac{15}{16}$ st. vet. Pag. 17. l. 13 write en Juillet 1714. P. 19. l. 20 desquelles P. 46. l. 11 en 1677. P. 75 lin 15 write Juin 1713 P. 78 l. 9 write 24.26. . . . .

NOTE

(1) Add. 3968(27), fo. 383. The errata Newton gives below do not appear in the printed edition; however the copy presented to the Royal Society by Des Maizeaux has them inserted in the latter's hand. Hence Newton must have communicated them to him at some time. The page numbers Newton gives refer to the volumes finally printed, not to the first incomplete page proofs sent to him by Des Maizeaux; see Letter 1295, note (1), p. 457.

INDEX

Bold figures refer to Letter numbers; italic figures refer to page numbers of major biographical notes. Letters and other documents printed in this volume are listed in capitals immediately under the writers name.

ABINGDON, Earl of
  WARRANT signed by: 3 August 1714, **1097**, 168–9
  and the coinage: 168–9
Académie Royale des Sciences
  books received from Newton: 59–60, 145–6
  and the calculus dispute: 314, 332, 333 n. 3
  and Flamsteed's *Historia Cœlestis*: 294 n. 1, 315
  Leibniz's election: 325
  see also under Mémoires de l'Académie Royale des Sciences
Act of Union: *see* Union, Act of
Acta Eruditorum
  and the calculus dispute: 89
  references to papers by
    Jakob Bernoulli; December 1691: 456; May 1697: 255 n. 6, 320 n. 7
    Johann [I] Bernouli; May 1697: 320 n. 7, 322; October 1698; 290, 323 n. 1; February and March 1713: 2, 4, 6 n. 10, 44–6, 64, 67–8, 73, 96 n. 2, 108, 113, 123, 145, 153–4, 189 n. 2, 202, 204, 275; July 1714: 359, 360, 362 n. 2; July 1716: *see* Johann [I] Bernoulli 'Epistola pro Eminente Mathematico'; June 1718: 304 n. 1
    Nikolaus [II] Bernoulli; May 1700: 321 n. 8, 322; May 1716: 320, 321 n. 10; June 1718: 356 n. 2, 438 n. 6, 448 n. 5
    Fatio; May 1700: 321 n. 8, 322
    Leibniz; October 1684: 94 n. 45, 287, 289 n. 18, 351 n. 38; January 1689: 351 n. 41; February 1689: *see* Leibniz 'Tentamen'; April 1691: 287; May 1700: 286, 291, 309, 347, 351 n. 40; October 1706: *see* Leibniz 'Tentamen', revised version; February 1712: 138 n. 5, 218 n. 1
    review of Newton's *Principia*: 298 n. 3
Admiralty, the, and the longitude: 266, 271–3
aether, and the Leibnizian vortex: 115, 116, 120 n. 2, 121 n. 5

air pump: 8, 9, 171
AISLABIE, JOHN: *443 n. 2*
  LETTERS
    to the Mint: 22 July 1718, **1292**, 450
    to Newton: 31 July 1718, **1293**, 450–1
ALESME, ANDRÉ D' *see* D'ALESME
ALLARDES or ALLARDICE, GEORGE: 263
ALLONVILLE, J-E D': *see* LOUVILLE, Chevalier de
Alva silver mine: *see* silver mine at Alva
ANISSON, JEAN: *43 n. 2*, 41–2, 96, 123
ANNE, Queen of England
  coins in her reign: 210
  coronation medals: 177, 179, 181
  death of: 168
  and Flamsteed's *Historia Cœlestis*: 268
anonymous
  LETTER to the Bishop of Worcester: 7 November 1713, **1021**, 34–6
  LETTERS from Newton to anonymous: [*c.* end of 1714], **1124**, 197–8; [22 March 1715], **1137**, 211–2; [end of 1715], **1176a**, 264–5
anonymous 'leading mathematician': *see* BERNOULLI, JOHANN [I]
ANSTIS, JOHN: 199, 244
APOLLONIUS and the calculus: 7, 8, 136, 250
APPLEBY, JOHN: 401 n. 3, 424
ARBUTHNOT, JOHN: xxxvii
  LETTER to Bolton: 14 January 1716, **1177**, 267–8
  WARRANT addressed to: 30 November 1715, **1171**, 255–6
Archbishop Tenison's Chapel: 381
ARCHIMEDES and the calculus dispute: xxxiii, 348
ARESKINE, ROBERT: *see* ERSKINE, ROBERT
ARGYLL, second duke of
  WARRANT signed by: 3 August 1714, **1097**, 168–9
ARISTOTLE: 261, 262 n. 5
ARNOLD, JOHN: xxxvi, *27 n. 1*, 228–9, 243–4, 254 n. 6, 304 n. 2

MONMORT, PIERRE RÉMOND DE: *216 n. 5*, 297
  LETTERS
    to Newton: 14 February 1717, **1234**,
      380; 16 March 1718, **1280**, 435–40
    to Taylor: 20 March 1716, **1194**, 299–300
    from Johann [I] Bernoulli: 28 March
      1717, **1237**, 383–4
    from Leibniz: 29 March 1716, **1198**, 314
  and Catherine Barton: 332, 333 n. 6, 380
  and Nikolaus [I] Bernoulli: 340 n. 2
  Critique sur la philosophie de M. Newton: 440,
    441 n. 8
  and the dispute between Taylor and
    Johann [I] Bernoulli: 362 n. 2
  'Dissertation' on philosophy: 441 n. 8
  intermediary in the calculus dispute: 313
    n. 1, 314, 332, 337, 339, 359, 360,
    362 n. 2, 435–40, 443, 459 n. 7
  and Leibniz's orthogonals problem: 356
    n. 2, 384 n. 1
  and Nicolas Rémond, confusion with:
    254 n. 2
  and Taylor's problem: 337, 339
MONTAGU, EDWARD WORTLEY: *see* WORTLEY
MONTAGUE, CHARLES: *see* HALIFAX, first Earl
  of
MONTGOMERY, JOHN: 226, 263–5
Moon
  appearance during solar eclipse: 224–5
  and the determination of longitude: xxiii,
    161, 194, 211
  lunar theory
    and Fatio: 393 n. 5
    and Flamsteed: xxxvii, 33
MOORE, SIR JONAS (the elder)
  and Flamsteed's instruments: 33, 69, 71
    nn. 4 & 5
MOORE, SIR JONAS (the younger): 69, 71 n. 6
MORTIMER, Earl: *see* OXFORD, Earl of
MOSCATELLI, DORICIGLIO: 421 nn. 1 & 3
MOSES: 34
motion in a resisting medium: 26–7, 44, 45,
    108, 289 n. 21
  Johann [I] Bernoulli's work on: 152, 153,
    202, 204
  see also Leibniz: on vortex theory
MOUTON, GABRIEL: 88, 329 n. 4
MUET, LE: *see* LE MUET

NEAL, Mr: 401 n. 1, 405, 452

NEILE, Sir PAUL: 10 n. 10
NEILE, WILLIAM: 8–9, 10 n. 10
Neuer Büchersaal der Gelehrten Welt, article on
  the calculus dispute: 133, 135 n. 2
New England: 46, 47 n. 3
NEWPORT, THOMAS: *301 n. 3*, 412 n. 2
  WARRANTS signed by: 26 March 1716,
    1195, 300–1; 11 September 1717,
    1260, 411–12
  and the copper coinage: 411–12, 434
  and Flamsteed's *Historia Cœlestis:* xxxvii
  and the Mint engravers: 300–1
NEWTON, Sir ISAAC
  ACCOUNT FOR THE *Historia Cœlestis:* 13
    January 1716, **1177a**, 269
  GOLD AND SILVER MONIES COINED, 1702–
    17: **1267a**, 433–4
  for LETTERS to Newton: *see* AISLABIE,
    BAILLIE, BALLE, HANNAH BARTON,
    BENTLEY, N. [I] BERNOULLI, BIGNON,
    BURCHETT, CARLISLE, CHAMBERLAYNE,
    CLAYTON, CONTI, CORKER, COTES,
    CUNNINGHAM, 'D. S.', D'ALESME,
    DERHAM, EDGCUMBE, FATIO, FLAM-
    STEED, FONTENELLE, HALDANE,
    HAYNES, KEILL, LEEUWENHOEK,
    LEVESQUE DE POUILLY, LOWNDES,
    'M. M.', MENSHIKOV, MONMORT, Sir
    JOHN NEWTON, WILLIAM NEWTON,
    NORTHEY, ONSLOW, OXFORD, 'P. B.',
    PARRY, POPPLE, KATHERINE RASTALL,
    ST QUINTIN, 'SGRAVESANDE, SHREWS-
    BURY, SMITHSON, SOUTHWELL, C.
    STANHOPE, Lord STANHOPE, STRINGER,
    TAYLOR, TAYLOUR, THORPE, TILSON,
    TREASURY, VARIGNON, H. WAL-
    POLE, Sir ROBERT WALPOLE, WARREN,
    WOOD, WORTLEY, Sir CHRISTOPHER
    WREN
  for LETTERS from Newton: *see* anonymous,
    BIGNON, BOLTON, BRANDSHAGEN,
    BURCHETT, CHAMBERLAYNE, CONTI,
    Coronation Committee, COTES,
    D'AUMONT, DES MAIZEAUX, DITTON,
    DRUMMOND, GRIGSBY, HALDANE,
    HALIFAX, JOHNSON, KEILL, MENS-
    HIKOV, Sir JOHN NEWTON, OXFORD,
    POPPLE, 'SGRAVESANDE, SHREWSBURY,
    SLOANE, Lord STANHOPE, THORPE,
    TOWNSHEND, TREASURY, TRON

WOLF, CHRISTIAN, LETTERS (*cont.*)
 1715, **1153**, 234–5; 20 September
 1715, **1159**, 239–40; 4 March 1716,
 1192, 297–8
 from Johann [I] Bernoulli: 28 March
 1716, **1196**, 301–4; 30 August 1717,
 1258, 409–10
 from Leibniz: 22 March 1715, **1136**,
 211; 7 May 1715, **1142**, 222–3;
 [Autumn 1715], **1160**, 240
 and Johann [I] Bernoulli's 'Epistola pro
 Eminente Mathematico': xxxvii, 257
 n. 4, 302–3, 334–5, 359, 360, 409–10
 on the calculus dispute: 179–80, 206–7,
 216–18, 234–5, 239–40, 297–8, 334–5,
 419–20
 Keill, criticism of: 234–5, 239–40, 297–8,
 334–5
 and Leibniz
 biography of Leibniz planned: 377
 and the *Charta Volans*: 20 n. 1, 32 n. 4,
 114, 131, 133, 216–18, 313 n. 4
 his 'Éloge' of Leibniz: 459 n. (ii)
 and the *Historia et Origo*: 205 n. 2
 the 'Remarques': 20 n. 1
 and the 'Tentamen': 141 n. 3
 and the *Recensio*: 297–8, 419–20
WOOD, WILLIAM
 LETTERS to Newton: 7 May 1718, **1286**,
 445–6; 12 May 1718, **1287**, 446–7

and the copper coinage: xxiv, 401 n. 3, 412,
 424, 434–5, 449, 451–2
WOODWARD, JOHN: 102, 260
woollen manufacturers of Taunton: *see* under
 Taunton
Worcester, Bishop of: 35 n. 1, 280
 LETTER from anonymous writer: 7 November 1713, **1021**, 34–6
 and Newton's chronological writings: 34–6
WORTLEY, EDWARD: *238 n. 5*
 LETTERS to Newton: 12 September 1715,
 1157, 237–8; 21 September 1715,
 1161, 241–2
 and coronation medals: 241–2
 and the prosecution of counterfeiters:
 237–8
WOTTON, Mr: 103, 105
wrecking: 380 n. 2
WREN, Sir CHRISTOPHER
 LETTERS
 to Bolton: 14 January 1716, **1177**,
 267–8
 to Newton: 30 November 1714, **1120**,
 93
 from Bolton: 30 November 1715, **1171**,
 255–6
 Leibniz's opinion of: 253
 on the longitude: 192
WREN, CHRISTOPHER, Junior: 192 n. 1